# Advanced Control Systems:

## Theory and Applications

# River Publishers Series in Automation, Control and Robotics

---



Indexing: all books published in this series are submitted to the Web of Science Book Citation Index (BkCI), to SCOPUS, to CrossRef and to Google Scholar for evaluation and indexing

The "River Publishers Series in Automation, Control and Robotics" is a series of comprehensive academic and professional books which focus on the theory and applications of automation, control and robotics. The series focuses on topics ranging from the theory and use of control systems, automation engineering, robotics and intelligent machines.

Books published in the series include research monographs, edited volumes, handbooks and textbooks. The books provide professionals, researchers, educators, and advanced students in the field with an invaluable insight into the latest research and developments.

Topics covered in the series include, but are by no means restricted to the following:

- Robots and Intelligent Machines
- Robotics
- Control Systems
- Control Theory
- Automation Engineering

# Advanced Control Systems:
## Theory and Applications

Editors

Yuriy P. Kondratenko

Vsevolod M. Kuntsevich

Arkadii A. Chikrii

Vyacheslav F. Gubarev

**Published 2021 by River Publishers**
River Publishers
Alsbjergvej 10, 9260 Gistrup, Denmark
www.riverpublishers.com

**Distributed exclusively by Routledge**
4 Park Square, Milton Park, Abingdon, Oxon OX14 4RN
605 Third Avenue, New York, NY 10158

First published in paperback 2024

*Advanced Control Systems: Theory and Applications* / by Yuriy P. Kondratenko, Vsevolod M. Kuntsevich, Arkadii A. Chikrii, Vyacheslav F. Gubarev.



*Routledge is an imprint of the Taylor & Francis Group, an informa business*

Publisher's Note
The publisher has gone to great lengths to ensure the quality of this reprint but points out that some imperfections in the original copies may be apparent.

While every effort is made to provide dependable information, the publisher, authors, and editors cannot be held responsible for any errors or omissions.

ISBN: 978-87-7022-341-6 (hbk)
ISBN: 978-87-7004-308-3 (pbk)
ISBN: 978-1-003-33701-0 (ebk)

DOI: 10.1201/9781003337010

# Contents

# Preface

The monograph consists of extended versions of the selected papers presented and discussed at *XXV International Conference on Automatic Control* "Automatics 2018" (September 18–19, 2018, Lviv, Ukraine) which is the main Ukrainian Control Conference organized by Ukrainian Association on Automatic Control (National member organization of IFAC – International Federation on Automatic Control) and Lviv National University "Lvivska Politechnica."

More than 100 papers were discussed at the conference in the main directions: mathematical problems of control, optimization, and game theory; control and identification under uncertainty; automated control of technical, technological, and biotechnical objects; controlling the aerospace craft, marine vessels and other moving objects; intelligent control and information processing; mechatronics and robotics; information measuring technologies in automation; automation and IT training of personnel; the Internet of things and the latest technologies.

The subject of the book is to present systematized research, description, and analysis of new theoretical results and applications of advanced control theory and various types of advanced control systems into different fields of human activity.

In term of the structure, the 14 chapters of the book are grouped into two sections: (1) Advances in Theoretical Research on Automatic Control and (2) Advances in Control Systems Applications.

The ***first part***, "Advances in Theoretical Research on Automatic Control," includes seven contributions.

The chapter "On Descriptor Control Impulsive Delay Systems that Arise in Lumped-Distributed Circuits," by Chikrii A., Rutkas A., and Vlasenko L., deals with descriptor control impulsive delay systems. The properties of a class of descriptor control systems are discussed. The results are illustrated on an example of a descriptor system that describes transient states in a radio technical filter with lossless transmissions.

A.A. Chentsov, A.G. Chentsov, and A.N. Sesekin, in "An Extremal Routing Problem with Constraints and Complicated Cost Functions," consider one routing problem with precedence conditions and complicated cost functions. The authors focus on engineering problem connected with dismantling of finite system of radiation sources.

In "Principle of Time Stretching for Motion Control in Condition of Conflict," G.Ts. Chikrii discusses the dynamic games of pursuit, described by a system of general form, encompassing a wide range of the functional-differential systems. It is shown that, in the case of time delay of current information on the process state to the pursuer, this game is equivalent to the game with complete information with the changed dynamics and the terminal set. On the basis of this result, the time stretching principle is developed which allows to expand a class of games to which the direct methods of pursuit can be applied. The detailed analysis of the model example of soft meeting of two controlled mathematical pendulums is provided to support the obtained result.

The chapter "Bio-Inspired Algorithms for Optimization of Fuzzy Control Systems: Comparative Analysis," by Oleksiy Kozlov and Yuriy Kondratenko, is devoted to the research and comparative analysis of the bio-inspired algorithms of synthesis and optimization of fuzzy control systems, in particular, ant colony optimization and genetic algorithms adapted for automatic rule base synthesis with the determination of optimal consequents for the Mamdani-type fuzzy systems. The studies and comparative analysis are conducted on a specific example, namely, at developing various rule base configurations of the fuzzy control system for the multipurpose mobile robot able to move on inclined and vertical ferromagnetic surfaces.

Inverse dynamic models approach, considered in the chapter "Inverse Model Approach to Disturbance Rejection Problem," by L. Lyubchyk, allows one to study the disturbance rejection and output tracking problem in linear multivariable control systems. The properties of closed-loop control system with inverse model-based controller have been investigated for the purpose of attainable disturbance rejection accuracy assessment. Nonsingularity conditions of controlled plant structure are established, allowing eliminating disturbance estimates from decoupling compensator equation. Peculiarities of applying inverse model approach for discrete-time systems are also considered and procedures for structural-parametric synthesis of discrete disturbance observers and compensators are suggested.

Kiforenko B.N. and Kiforenko S.I., in the chapter "Invariant Relations in the Theory of Optimally Controlled Systems," introduce a new concept of price-target invariance of dynamical systems optimal control. Several types of optimization problems are discussed for which it is possible to obtain relations between control functions and criterion for assessing the quality of control. These relationships appear to violate the Leibniz sufficient reason principle. The relevance of determining the nature of these relationships is due to the need to solve the problem of the possibility of their practical use, which greatly simplifies the structure of the facility's control system.

The chapter "Robust Adaptive Controls for a Class of Nonsquare Memory-Less Systems," by L. Zhiteckii and K. Solovchuk, deals with the discrete-time robust adaptive control for a class of uncertain multivariable memory-less (static) plants with arbitrary unmeasurable bounded disturbances. The cases of nonsquare systems, where the number of the control inputs is less than the number of the outputs, are studied. It is assumed that the plant parameters defining the elements of its gain matrix are unknown. Simulation results are presented to support the theoretic study.

The ***second part***, "Advances in Control Systems Applications," also includes seven contributions.

Vyacheslav Gubarev, Victor Romanenko, and Yurii Miliavskyi, in the chapter "Advanced Identification of Impulse Processes in Cognitive Maps," consider the identification problem of a complex system using data obtained experimentally. Impulse processes in cognitive maps are very important and widespread class of such systems. Regularization procedure was introduced in standard subspace methods which allows finding an approximate solution consistent with errors in available data. Application of the developed methods is demonstrated on the example.

In "Strategy for Simulation Complex Hierarchical Systems Based on the Methodologies of Foresight and Cognitive Modeling," Pankratova N.D., Gorelova G.V., and Pankratov V.A. consider the strategy of complex hierarchical systems simulation based on the foresight and cognitive modeling methodologies. Authors propose a two-step procedure: (1) at the first stage, to apply the foresight methodology with the aim of creating alternatives of scenarios and (2) to use the obtained results at the second stage as initial data for cognitive modeling for constructing scenarios of the desired future and ways of their implementation. Scenarios of the education system future are modeled taking into account the coronavirus pandemic and the unstable economy of the modern world.

The chapter "Special Cases in Determining the Spacecraft Position and Attitude Using Computer Vision System," by Gubarev V., Salnikov N., Melnychuk S., Shevchenko V., and Maksymyuk L., investigates the accuracy of the relative pose estimation using monocular computer vision systems. The orientation and position of the target solid are determined by a set of control points located on its surface, which are recognized on the captured digital image. Their measured positions contain a sampling error, the propagation of which leads to inaccurate solution. The magnitude of the final error depends on the spatial location of the used control points and therefore varies with the movement and rotation of the target. In such an application, as measuring the relative pose of the target spacecraft during the approach and docking in space, it is necessary to control the magnitude of the solution errors and ensure that the specified accuracy of the solution is obtained.

Dmitriy V. Lebedev, in chapter "On Determining the Spacecraft Orientation by Information from a System of Stellar Sensors," focuses on increasing requirements on the accuracy of information on the spacecraft orientation parameters when solving problems both onboard the spacecraft and on the ground processing information coming from orbit that stimulate the search for such algorithmic solutions for processing information from onboard orientation sources that would satisfy these requirements. The problem of minimizing the influence of residual uncertainty in the angular position of stellar sensors in the coordinate system associated with the spacecraft on the accuracy of the information systems under consideration is discussed. Accuracy characteristics of satellite orientation, which are realized by using the proposed algorithm, are studied. Simulation confirms the effectiveness of the proposed algorithmic solutions.

The chapter "Control Synthesis of Rotational and Spatial Spacecraft Motion at Approaching Stage of docking," by Volosov V., Melnychuk S., Salnikov N., and Shevchenko V., is devoted to the control of rotational and translational motion for shock-free docking with a noncooperative rotating spacecraft. It is assumed that the relative position and orientation of target spacecraft are measured by a computer vision system installed on an active spacecraft. The angular motion parameters of the target spacecraft, including ratios of the body inertia moments are estimated by the proposed ellipsoidal estimation algorithm. The spacecraft orbital motion parameters are assumed to be known. The control problem is solved by using Lyapunov method separately for kinematic and dynamic equations of the relative spacecraft motion.

In "Intelligent Algorithms for the Automation of Complex Biotechnical Objects," V. Lysenko, N. Zaiets, A. Dudnik, T. Lendiel, and K. Nakonechna present the contributions to the problem of energy-efficient control of biotechnological objects. The modern intelligent algorithms for processing control object information and applying the results are considered. Appropriate control strategies allow maximizing the production profit of the poultry houses and greenhouses. Proposed structure of intelligent control system allows to formulate control strategies for greenhouse complexes, which significantly reduce energy costs in the vegetable production.

Maksymov M.V., Brunetkin O.I., Beglov K.V., Alyokhina S.V., and Butenko O.V., in the chapter "Automatic Control for the Slow Pyrolysis of Organic Materials with Variable Composition," consider the chemical and energy equipment which use raw materials of a certain composition and requires fine-tuning. A slight change in the composition of the same type of substance leads to failure or even a stoppage of technological processes. The proposed method of organic substances processing allows optimally transforming various organic substances into the gaseous state. An imitation mathematical model describes the dynamic properties of the pyrolysis unit. Modeling of automatic control system has shown that at this stage, a set of single-circuit regulators is enough. Shutting off any of the regulators leads to the deterioration of regulation quality, but the system as a whole continues to function steadily.

The chapters selected for this book provide an overview of (1) some problems in the area of advanced control systems design, modeling, engineering, and implementation, and (2) the approaches and techniques that relevant research groups within this area are employing to try to solve them.

We would like to express our deep appreciation to all authors for their contributions as well as to reviewers for their timely and interesting comments and suggestions. We certainly look forward to working with all contributors again in nearby future.

July 12, 2020

*Editors:*
Yuriy P. Kondratenko
Vsevolod M. Kuntsevich
Arkadii A. Chikrii
Vyacheslav F. Gubarev

# List of Contributors

**S.V. Alyokhina,** *A. Podgorny Institute of Mechanical Engineering Problems of the National Academy of Sciences of Ukraine, Kharkiv, Ukraine. V.N. Karazin Kharkiv National University, Ukraine. E-mail: svitlana.alyokhina@gmail.com*

**K.V. Beglov,** *Odesa National Polytechnic University, Shevchenko ave. 1, Odessa, Ukraine, 65044; E-mail: beglov.kv@opu.ua*

**O.I. Brunetkin,** *Odessa National Polytechnic University, Shevchenko ave. 1, Odessa, Ukraine, 65044; E-mail:a.i.brunetkin@gmail.com*

**O.V. Butenko,** *Odessa National Polytechnic University, Shevchenko ave. 1, Odessa, Ukraine, 65044*

**A.A. Chentsov,** *N.N. Krasovskii Institute of Mathematics and Mechanics, Ural Branch of Russian Academy of Sciences, 16 S. Kovalevskaya, Yekaterinburg, 620108, Russia, Ural Federal University, 19 Mira st., Yekaterinburg, 620002, Russia; E-mail: chentsov.a@binsys.ru*

**A.G. Chentsov,** *N.N. Krasovskii Institute of Mathematics and Mechanics, Ural Branch of Russian Academy of Sciences, 16 S. Kovalevskaya, Yekaterinburg, 620108, Russia, Ural Federal University, 19 Mira st., Yekaterinburg, 620002, Russia; E-mail: chentsov@imm.uran.ru*

**A. Chikrii,** *Glushkov Institute of Cybernetics, Kiev, Ukraine; E-mail: g.chikrii@gmail.com*

**G.Ts. Chikrii,** *Glushkov Institute of Cybernetics, Kiev, Ukraine; E-mail: g.chikrii@gmail.com*

**A. Dudnyk,** *Heroiv Oborony Str.15, building 11, of.311, Kyiv, National University of Life and Environmental Sciences of Ukraine, Ukraine, 03041*

**G.V. Gorelova,** *Engineering and Technology Academy of the Southern Federal University, Russia*

**V. Gubarev,** *Space Research Institute, National Academy of Sciences of Ukraine and State Space Agency of Ukraine, Glushkov Av. 40, 4/1, Kyiv, 03680, Ukraine; E-mail: v.f.gubarev@gmail.com*

**B. Kiforenko,** *S.P. Timoshenko Institute of mechanics of NASU, P. Nesterov str., 3, Kiev, 03057, Ukraine; E-mail: bkifor@ukr.net*

**S. Kiforenko,** *International Research and Training Center for Information Technologies and Systems of the National Academy of Sciences of Ukraine and Ministry of Education and Science of Ukraine, 40 Glushkov ave.,Kiev, 03680 GSP, Ukraine; E-mail: skifor@ukr.net*

**Yuriy Kondratenko,** *Petro Mohyla Black Sea National University, 10 68th Desantnykiv st., Mykolayiv, 54003, Ukraine; E-mail: y_kondrat2002@yahoo.com, yuriy.kondratenko@chmnu.edu.ua*

**Oleksiy Kozlov,** *Petro Mohyla Black Sea National University, 10 68th Desantnykiv st., Mykolayiv, 54003, Ukraine; E-mail: kozlov_ov@ukr.net*

**Dmitriy V. Lebedev,** *International Research and Training Center for Information Technologies and Systems, Kiev, Ukraine; E-mail: ldv1491@gmail.com*

**T. Lendiel,** *Heroiv Oborony Str.15, building 11, of.311, Kyiv, National University of Life and Environmental Sciences of Ukraine, Ukraine, 03041*

**V. Lysenko,** *Heroiv Oborony Str.15, building 11, of.311, Kyiv, National University of Life and Environmental Sciences of Ukraine, Ukraine, 03041; E-mail: lysenko@nubip.edu.ua*

**L. Lyubchyk,** *National Technical University "Kharkiv Polytechnic Institute", Kirpitcheva str., 2, Kharkiv, 61002, Ukraine; E-mail: Leonid.Liubchyk@khpi.edu.ua*

**M.V. Maksymov,** *Odessa National Polytechnic University, Shevchenko ave. 1, Odessa, Ukraine, 65044; E-mail:prof.maksimov@gmail.com*

**L. Maksymyuk,** *Space Research Institute, National Academy of Sciences of Ukraine and State Space Agency of Ukraine, Glushkov Av. 40, 4/1, Kyiv, 03680, Ukraine; E-mail: hatahatky@gmail.com*

**S. Melnychuk,** *Space Research Institute, National Academy of Sciences of Ukraine and State Space Agency of Ukraine, Glushkov Av. 40, 4/1, Kyiv, 03680, Ukraine; E-mail: melnychuk89s@gmail.com*

**Yurii Miliavskyi,** *"Institute for Applied System Analysis" of National Technical University of Ukraine "Igor Sikorsky Kyiv Polytechnic Institute", 37a Peremohy av., Kyiv, 03056, Ukraine; E-mail: yuriy.milyavsky@gmail.com*

**K. Nakonechna,** *Heroiv Oborony Str.15, building 11, of.311, Kyiv, National University of Life and Environmental Sciences of Ukraine, Ukraine, 03041*

**V.A. Pankratov,** *Institute for Applied System Analysis, Igor Sikorsky Kyiv Polytechnic Institute, Ukraine*

**N.D. Pankratova,** *Institute for Applied System Analysis, Igor Sikorsky Kyiv Polytechnic Institute, Ukraine; E-mail: natalidmp@gmail.com*

**Victor Romanenko,** *"Institute for Applied System Analysis" of National Technical University of Ukraine "Igor Sikorsky Kyiv Polytechnic Institute", 37a Peremohy av., Kyiv, 03056, Ukraine*

**A. Rutkas,** *Kharkov National University of Radio Electronics, Kharkov, Ukraine*

**N. Salnikov,** *Space Research Institute, National Academy of Sciences of Ukraine and State Space Agency of Ukraine, Glushkov Av. 40, 4/1, Kyiv, 03680, Ukraine; E-mail: salnikov.nikolai@gmail.com*

**A.N. Sesekin,** *N.N. Krasovskii Institute of Mathematics and Mechanics, Ural Branch of Russian Academy of Sciences, 16 S. Kovalevskaya, Yekaterinburg, 620108, Russia, Ural Federal University, 19 Mira st., Yekaterinburg, 620002, Russia; E-mail: sesekin@list.ru.*

**V. Shevchenko,** *Space Research Institute, National Academy of Sciences of Ukraine and State Space Agency of Ukraine, Glushkov Av. 40, 4/1, Kyiv, 03680, Ukraine; E-mail: vovan_16@ukr.net, vovan@gmail.com*

**K. Solovchuk,** *Poltava Scientific Research Forensic Center of the MIA of Ukraine, Rybalskiy lane, 8, Poltava, 36011, Ukraine; E-mail: solovchuk_ok@ukr.net*

**L. Vlasenko,** *Kharkov National University of Radio Electronics, Kharkov, Ukraine*

**V. Volosov,** *Space Research Institute, National Academy of Sciences of Ukraine and State Space Agency of Ukraine, Glushkov Av. 40, 4/1, Kyiv, 03680, Ukraine; E-mail: wwolosov@gmail.com*

**N. Zaiets,** *Heroiv Oborony Str.15, building 11, of.311, Kyiv, National University of Life and Environmental Sciences of Ukraine, Ukraine, 03041*

**L. Zhiteckii,** *International Research and Training Center for Information Technologies and Systems of the National Academy of Science of Ukraine and Ministry of Education and Sciences of Ukraine, Acad. Glushkova av., 40, Kiev, 03187, Ukraine; E-mail: leonid_zhiteckii@i.ua*

# List of Figures

# List of Tables

# List of Abbreviations

| | |
|---|---|
| $\alpha_i(Q)$ | Singular value of a matrix $Q$ with number $i$ |
| $\det Q$ | Determinant of a $Q$ |
| $\ell_\infty$ | Space of all bounded scalar sequences |
| $\emptyset$ | Empty set |
| $\lambda_i(Q)$ | Eigenvalue of a matrix $Q$ with number $i$ |
| $\mathbf{R}^{l \times s}$ | Set of $l \times s$ matrices with real-valued elements |
| $\mathbf{R}^l$ | Set of $l$-dimensional vectors whose components are real numbers |
| $\text{sign } a$ | Function taking the value 1 if $a > 0$, 0 if $a = 0$ and $-1$ if $a < 0$ |
| $\|\|w\|\|_2$ | Euclidean norm of a vector $w$ |
| $0_s$ | $s$-tuple of zeros ($s$-dimensional zero vector) |
| $I_s$ | Identity $s \times s$ matrix |
| $Q(q^{(ij)})$ | Matrix $Q$ with elements $q^{(ij)}$ |
| $Q^+$ | Pseudoinverse matrix caused by $Q$ |
| A | Avner Friedman |
| ACO | Ant colony optimization |
| ACS | Automatic control systems |
| ASC | Active spacecraft |
| BDE | Bundesverband der Deutschen Entsorgungs |
| BF | Body frame |
| BN | Boris Nikolayevich Pshenichny |
| BSS | Basic stellar sensor |
| CES | Complex education system |
| CHS | Complex hierarchical systems |
| CMLS | Cognitive modeling large system |
| CP | Center of projection |
| CRF | Camera reference frame |
| CS | Coordinate system |
| CVS | Computer vision system |
| DC | Disturbance compensator |

| | |
|---|---|
| DDC | Disturbance decoupling compensator |
| DDP | Disturbance decoupling problem |
| DDPM | Disturbances decoupling problem by measurement feedback |
| DDPS | Disturbance decoupling problem with stability |
| DO | Disturbance observer |
| DP | Dynamic programming |
| EGF | Earth's gravitational field |
| EU | European Union |
| FACS | Fuzzy automatic control system |
| FC | Fuzzy controller |
| FFC | Feedforward compensator |
| GA | Genetic algorithm |
| GTSP | Generalized traveling salesman problem |
| i.i.d. | Independently identically distributed |
| IMC | Internal model control |
| InvMC | Inverse model control |
| IRF | Inertial reference frame |
| IVSC | Integral variable structure control |
| L | Leonard Berkovitz |
| LMI | Linear matrix inequalities |
| LOC | Local optimal control |
| LOF | Local orbital frame |
| LS | Lev Semyonovich Pontryagin |
| LT | Linguistic terms |
| MF | Membership functions |
| MIMO | Multi-input multi-output system |
| MISO | Multi-input single-output system |
| MR | Mobile robot |
| MRP | Modified Rodrigues parameters |
| NN | Nikolaj Nikolayevich Krasovskii |
| O | Otomar Hayek |
| OP | Ordered pair |
| PnP | Perspective-n-points |
| PSC | Passive spacecraft |
| QR | Quaternion of rotation |
| R Isaacs | Rufus Isaacs |
| RB | Rule base |
| RHP | Rodrigues–Hamilton parameters |

| | |
|---|---|
| SC | Spacecraft |
| SIMO | Single-input multi-output system |
| SME | Sliding mode equivalence |
| SS, SS1, SS2 | Stellar sensors |
| SSA | Sequential search algorithm |
| TSP | Traveling salesman problem |
| TSP-PC | Traveling salesman problem with precedence conditions |
| UIO | Unknown input observer |
| Upper script | "T" Transposition operator |
| US | United States (of America's) |
| VSDO | Variable structure disturbance observer |
| VSFC | Variable structure feedforward compensator |
| VSS | Variable structure system |
| WRF | World reference frame |

# Part I

# Advances in Theoretical Research on Automatic Control

# 1

# On Descriptor Control Impulsive Delay Systems that Arise in Lumped-Distributed Circuits

**A. Chikrii[1], A. Rutkas[2] and L. Vlasenko[2]**

[1]Glushkov Institute of Cybernetics, Kiev, Ukraine
[2]Kharkov National University of Radio Electronics, Kharkov, Ukraine
Corresponding author: A. Chikrii <g.chikrii@gmail.com>

### Abstract

Results on properties of a class of descriptor control systems are obtained. The system state is described by a nonlinear impulsive delay equation not solved with respect to the derivative of the state. The matrix coefficient in the time derivative is allowed to be noninvertible. We indicate conditions to solve this equation. The conditions are formulated in terms of the resolvent of characteristic matrix pencil. The results are illustrated on an example of a descriptor system that describes transient states in a radio technical filter with lossless transmissions.

**Keywords**. Descriptor control system, impulsive delay differential equation, characteristic pencil, resolvent, circuit with transmission lines.

## 1.1 Introduction

The theory of descriptor control systems is an important part of the general theory of control systems; for example, see the bibliography in the books [1–3]. The subject of this paper is to investigate a descriptor control system

Acknowledgment
This work was partially supported by the National Research Foundation of Ukraine. Grant No. 2020.02/0121.

whose states are described by a nonlinear impulsive delay equation not solved with respect to the derivative of the state. Such equations arise in the circuit theory with transmission lines. As in [4], the delay effect is due to long line segments. By structure and location of lumped elements, generally, the equation is not solved with respect to the derivative (differential algebraic equation) [5]. Also we suppose that the outputs of lines are subject to pulse perturbations. Thus, we deal with a descriptor control system with delays and pulse variations. Here, we develop the theory of such systems and apply it to investigate circuits with lossless transmission lines under pulse perturbations.

We consider the semilinear delay equation

$$\frac{d}{dt}[A_0y(t)] + \sum_{j=0}^{N} B_jy(t-\omega_j) = f(t,y(t)) + Ku(t), \;\; a.a. \; t \geq t_0, \tag{1.1}$$

with the initial conditions

$$y(t) = g(t), \;\; a.a. \; t_0 - \omega \leq t \leq t_0, \;\; (A_0y)(t_0+0) = y_0, \tag{1.2}$$

and the pulse actions

$$\begin{aligned} &\Delta_k[A_0y(t)] = \Im_k(A_0y)(t_k-0), \;\; k = 1,2,..., \\ &\Delta_k[A_0y(t)] \equiv (A_0y)(t_k+0) - (A_0y)(t_k-0). \end{aligned} \tag{1.3}$$

Here $A_0, B_j$ are complex matrices of dimensions $n \times n$, $K$ is a complex matrix of dimension $n \times m$, the control $\mathrm{u(t)}$ is an $\mathrm{m}$-dimensional vector-function with values in $\mathbf{C}^m$, $f(t,v) : [t_0,\infty) \times \mathbf{C}^n \to \mathbf{C}^n$, $g(t) : [t_0 - \omega, t_0] \to \mathbf{C}^n$, $\Im_k : \Omega_k \to \mathbf{C}^n$, $\Omega_k \subset \mathbf{C}^n$, and $\Omega_k \supset A_0\mathbf{C}^n$. Let delays be ordered $0 = \omega_0 < \omega_1 < \cdots < \omega_N = \omega$, times $\{t_k\}_{k=0}^{\infty}$ be enumerated $t_0 < t_1 < t_2 < \cdots$ and $t_k \to \infty$ as $k \to \infty$. The specific feature of Equation (1.1) is that the matrix $A_0$ may be noninvertible. In the case of a noninvertible $A_0$, the equation is degenerate. A similar situation happens for the neutral functional differential equation with nonatomic difference operator (see Definition in [4] and theoretical foundations on such equations in [6–8]).

We use the following notation: $\mathrm{Ker}$ is a kernel of matrix; $\mathrm{Lin}\{...\}$ is a linear span of vectors; $Y_1 \dot{+} Y_2$ is a direct sum of two subspaces; $E$ is a unit matrix; $\mathrm{y}^{\mathrm{tr}}$ is a transposed vector; $L_1(a,b;\mathbf{C}^n)$ is the space of $\mathbf{C}^n$-valued functions, which are Lebesgue integrable on $[a,b]$; $W_1^1(a,b;\mathbf{C}^n)$ is the space of $\mathbf{C}^n$-valued functions, which belong to $L_1(a,b;\mathbf{C}^n)$ together with their distributional derivatives of first order;$L_{1,\mathrm{loc}}(a,\infty;\mathbf{C}^n)$ is the space of functions, which belong to $L_1(a,b;\mathbf{C}^n)$ for all $b > a$.

The paper is organized as follows. In Section 1.2, we describe a circuit with lossless transmission lines and show the importance of considering

Equation (1.1) with degenerate matrix $A_0$. In Section 1.3, we introduce some basic assumptions and obtain an existence and uniqueness result for problems (1.1)–(1.3). In Section 1.4, we use the results of Section 1.3 to describe transient states in the circuit from Section 1.2.

## 1.2 Example of Descriptor Control System

Let us consider the lumped-distributed circuit presented in Figure 1.1. At the inputs of the transmission lines, there are voltage sources $E_1, E_2$; at the outputs, there are an inductor $L$ and a resistor $r$; a nonlinear function $\psi(z) : \mathbf{C} \to \mathbf{C}$ gives the voltage $\psi(I_\psi)$ in the indicated box. The problem is described by the well-known telegraph equations

$$L_j \frac{\partial I_j(x_j,t)}{\partial t} = -\frac{\partial U_j(x_j,t)}{\partial x_j}, \quad C_j \frac{\partial U_j(x_j,t)}{\partial t} = -\frac{\partial I_j(x_j,t)}{\partial x_j}, \qquad (1.4)$$
$$0 \le x_j \le 1, \quad t \ge t_0, \quad j = 1, 2,$$

where $L_j, C_j$ are positive constants. The boundary conditions for partial differential equation (1.4) are defined by the following way.

Write down the Kirchhoff equations for currents and voltages:

$$U_1(0,t) = E_1(t), \;\; U_2(0,t) = E_2(t), \qquad (1.5)$$

$$I_r(t) = I_\psi(t), \;\; U_L(t) = U_r(t) + U_\psi(t),$$
$$U_1(1,t) = U_2(1,t) = U_L(t), \;\; I_L(t) = I_1(1,t) + I_2(1,t) - I_r(t). \qquad (1.6)$$

Taking into account Equation (1.6) and the Ohm laws

$$U_r(t) = rI_r(t), \;\; U_\psi(t) = \psi(I_\psi(t)),$$

the oscillation equation for inertial element

$$U_L(t) = L\frac{d}{dt}[I_L(t)]$$

becomes

$$U_1(1,t) = U_2(1,t) = rI_r(t) + \psi(I_r) = L\frac{d}{dt}[I_1(1,t) + I_2(1,t) - I_r(t)]. \qquad (1.7)$$

Thus, relations (1.5) and (1.7) are the boundary conditions for partial differential equations (1.4).

In a way analogous to that used in [4, Introductions], problems (1.4), (1.5), and (1.7) are transformed to the delay differential equation (1.1).

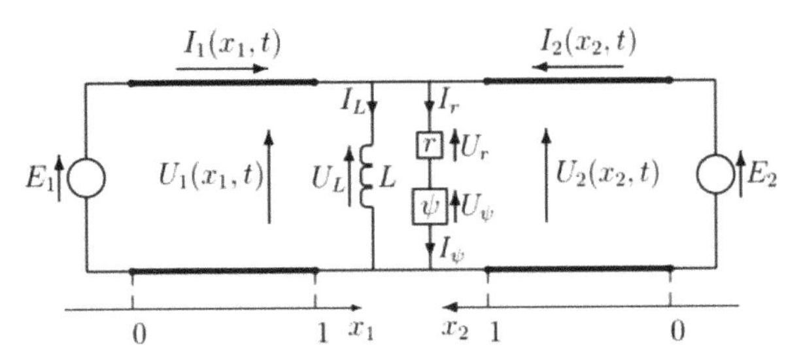


**Figure 1.1** The two-conductor transmission line.

For this purpose, the values of the functions $I_j(x_j,\cdot), U_j(x_j,\cdot)$ at the point $x_j = 0$ are expressed by their values at the point $x_j = 1$ and substituted in the boundary conditions. Then the boundary conditions have the form of Equation (1.1).

Really, the general solution of Equation (1.4) has the form

$$\begin{aligned} U_j(x_j,t) &= \Phi_j(x_j - a_j t) + \Psi_j(x_j + a_j t), \\ I_j(x_j,t) &= \frac{1}{\rho_j}[\Phi_j(x_j - a_j t) - \Psi_j(x_j + a_j t)], \ \ j = 1,2, \end{aligned} \tag{1.8}$$

where $\Phi_j, \Psi_j$ are smooth functions, and the values

$$\rho_j = \sqrt{\frac{L_j}{C_j}}, \quad a_j = \frac{1}{\sqrt{L_j C_j}}, \quad j = 1,2, \tag{1.9}$$

are the wave impedances and the wave velocities. The functions $\Phi_j, \Psi_j$ are expressed by $I_j(1,\cdot), U_j(1,\cdot)$:

$$\begin{aligned} \Phi_j(x_j - a_j t) &= \tfrac{1}{2}(U_j + \rho_j I_j)(1, t - (x_j - 1)/a_j), \\ \Psi_j(x_j + a_j t) &= \tfrac{1}{2}(U_j - \rho_j I_j)(1, t + (x_j - 1)/a_j). \end{aligned}$$

These expressions are substituted in the general solution (1.8):

$$\begin{aligned} U_j(x_j,t) =& \frac{1}{2}[(U_j + \rho_j I_j)(1, t - (x_j - 1)/a_j) \\ &+ (U_j - \rho_j I_j)(1, t + (x_j - 1)/a_j)], \\ I_j(x_j,t) =& \frac{1}{2\rho_j}[(U_j + \rho_j I_j)(1, t - (x_j - 1)/a_j) \\ &- (U_j - \rho_j I_j)(1, t + (x_j - 1)/a_j)]. \end{aligned} \tag{1.10}$$

In Equation (1.10), we set $x_j = 0$ and change $t$ for $t - 1/a_j$. We obtain the relations for the voltages

$$U_j(0, t - 1/a_j) = \frac{1}{2}[(U_j + \rho_j I_j)(1, t) + (U_j - \rho_j I_j)(1, t - 2/a_j)]$$

and for the currents

$$I_j(0, t-1/a_j) = \frac{1}{2\rho_j}[(U_j + \rho_j I_j)(1, t) - (U_j - \rho_j I_j)(1, t-2/a_j)]. \quad (1.11)$$

Substitute $U_j(0, t - 1/a_j)$in Equation (1.5) where $t$ is replaced by $t - 1/a_j$. Relation (1.5) take the form

$$\begin{aligned} &U_j(1, t) + \rho_j I_j(1, t) + U_j(1, t - \omega_j) - \rho_j I_j(1, t - \omega_j) \\ &= 2E_j(t - \omega_j/2), \quad j = 1, 2, \end{aligned} \quad (1.12)$$

where

$$\omega_j = 2/a_j = 2\sqrt{L_j C_j}, \quad j = 1, 2. \quad (1.13)$$

Define the state $y(t)$, the control $u(t)$, and the nonlinear function $f(t, v)$:

$$y(t) = \begin{pmatrix} U_1(1, t) \\ I_1(1, t) \\ I_2(1, t) \\ I_r(t) \end{pmatrix}, \quad u(t) = \begin{pmatrix} E_1(t - \omega_1/2) \\ E_2(t - \omega_2/2) \end{pmatrix},$$
$$f(t, v) = \begin{pmatrix} 0 \\ 0 \\ 0 \\ \psi(v_4) \end{pmatrix}. \quad (1.14)$$

Introduce into consideration the matrices

$$A_0 = \begin{pmatrix} 0 & 0 & 0 & 0 \\ 0 & 0 & 0 & 0 \\ 0 & L & L & -L \\ 0 & 0 & 0 & 0 \end{pmatrix}, \quad B_0 = \begin{pmatrix} 1 & \rho_1 & 0 & 0 \\ 1 & 0 & \rho_2 & 0 \\ -1 & 0 & 0 & 0 \\ 1 & 0 & 0 & -r \end{pmatrix},$$
$$B_1 = \begin{pmatrix} 1 & -\rho_1 & 0 & 0 \\ 0 & 0 & 0 & 0 \\ 0 & 0 & 0 & 0 \\ 0 & 0 & 0 & 0 \end{pmatrix}, \quad B_2 = \begin{pmatrix} 0 & 0 & 0 & 0 \\ 1 & 0 & -\rho_2 & 0 \\ 0 & 0 & 0 & 0 \\ 0 & 0 & 0 & 0 \end{pmatrix}, \quad K = \begin{pmatrix} 2 & 0 \\ 0 & 2 \\ 0 & 0 \\ 0 & 0 \end{pmatrix}. \quad (1.15)$$

Now, we can rewrite the boundary conditions (1.5) and (1.12) in the form of Equation (1.1).

We suppose that in the circuit, there is a pulse generator that produces momentary disturbances at discrete times. These disturbances stimulate jumps of currents and voltages. The pulse generator is not represented in Figure 1.1 since it influences only at discrete times. We assume that the pulse variations of the vector $y(t)$ in Equation (1.14) are realized by the rule (1.3).

## 1.3 Restrictions, Definitions, and States of System

The characteristic pencil significantly affects the dynamic of descriptor system [9]. In the case of the delay equation (1.1), we consider the characteristic matrix pencil $\lambda A_0 + B_0$. We suppose that the characteristic pencil is regular, namely $\det(\lambda A_0 + B_0) \not\equiv 0$, and the resolvent

$$R(\lambda) = (\lambda A_0 + B_0)^{-1} = \{l_{kj}(\lambda)\}_{k,j=1}^{n} \tag{1.16}$$

satisfies the estimate

$$|l_{kj}(\lambda)| \leq C_1, \ \ |\lambda| \geq C_2 \tag{1.17}$$

with positive constants $C_1, \ C_2$. In this case, the space $\mathbf{C}^n$ is decomposed into the direct sum [9]:

$$\mathbf{C}^n = Y_1 \dot{+} Y_2, \ \ Y_1 = A_0 \mathbf{C}^n, \ \ Y_2 = B_0 \mathrm{Ker} A_0.$$

Let $Q_1$ be the projective matrix on $Y_2$ parallel to $Y_1$. Restriction (1.17) for resolvent (1.16) is valid for the circuit example under the consideration. The constructions of the projective matrices $Q_1, Q_2$ can be found in [9]:

$$Q_1 = \frac{1}{2\pi i} \oint\limits_{|\lambda|=C_2} A_0(\lambda A_0 + B_0)^{-1} d\lambda, \qquad Q_2 = E - Q_1.$$

As in [5], we introduce the auxiliary matrix

$$G = A_0 + Q_2 B_0.$$

The following properties hold:

$$GY_1 \subset Y_1, \ \ GY_2 \subset Y_2, \ \ \exists\, G^{-1}, \ \ A_0 G^{-1} = Q_1, \ \ B_0 G^{-1} Q_2 = Q_2. \tag{1.18}$$

Concerning the functions $f(t, v)$, $u(t)$ in Equation (1.1) and $g(t)$ in Equation (1.2), we suppose that $f(t, v)$ as a function of $t$ for each $v \in \mathbf{C}^n$ is an element of $L_{1,\mathrm{loc}}(t_0, \infty; \mathbf{C}^n)$, $u(t) \in L_{1,\mathrm{loc}}(t_0, \infty; \mathbf{C}^m)$, and $g(t) \in$

$L_1(t_0 - \omega, t_0; \mathbf{C}^n)$. A function $y(t) \in L_{1,\text{loc}}(t_0 - \omega, \infty; \mathbf{C}^n)$ such that $A_0 y(t) \in W_1^1(t_k, t_{k+1}; \mathbf{C}^n)$ for $k = 0, 1, \ldots$ is a solution of problems (1.1)–(1.3) on $[t_0 - \omega, \infty)$ if $y(t)$ satisfies initial condition (1.2) for almost all $t \in [t_0 - \omega, t_0]$, $A_0 y(t)$ satisfies Equation (1.2) at $t = t_0 + 0$, $y(t)$ satisfies Equation (1.1) for almost all $t \in [t_0, \infty)$, and $A_0 y(t)$ satisfies Equation (1.3).

First, we investigate problems (1.1) and (1.2) on a finite segment $[t_0 - \omega, t_0 + \tau_0]$ $(\tau_0 > 0)$ without pulse actions. A function $y(t) \in L_1(t_0 - \omega, t_0 + \tau_0; \mathbf{C}^n)$ is a solution of problems (1.1) and (1.2) on $[t_0 - \omega, t_0 + \tau_0]$ if $A_0 y(t) \in W_1^1(t_0, t_0 + \tau_0; \mathbf{C}^n)$, $y(t)$ satisfies the first initial condition in Equation (1.2) for almost all $t \in [t_0 - \omega, t_0]$, $A_0 y(t)$ satisfies the second condition in Equation (1.2) at $t = t_0 + 0$, and $y(t)$ satisfies Equation (1.1) for almost all $t \in [t_0, t_0 + \tau_0]$.

**Theorem 1.1.** *Let the following assumptions for problems (1.1) and (1.2) be valid.*

*Restriction (1.17) for resolvent (1.16) holds.*

*The control $u(t) : [t_0, t_0 + \tau_0] \to \mathbf{C}^m$ belongs to the class $u(t) \in L_1(t_0, t_0 + \tau_0; \mathbf{C}^m)$.*

*The function $f(t, v) : [t_0, t_0 + \tau_0] \times \mathbf{C}^n \to \mathbf{C}^n$ as a function of $t$ belongs to $L_1(t_0, t_0 + \tau_0; \mathbf{C}^n)$ for each $v \in \mathbf{C}^n$ and satisfies the Lipschitz condition*

$$||f(t, v) - f(t, w)|| \le M||v - w||, \quad \forall v, w \in \mathbf{C}^n, \quad a.a.\ t \in [t_0, t_0 + \tau_0], \tag{1.19}$$

*with a constant $M$ independent of $t$ and such that*

$$M||G^{-1}Q_2|| < 1. \tag{1.20}$$

*In Equation (1.2), the initial vector $y_0$ satisfies the restriction*

$$Q_2 y_0 = 0 \tag{1.21}$$

*and the initial function $g(t) : [t_0 - \omega, t_0] \to \mathbf{C}^n$ belongs to the classg$(t) \in L_1(t_0 - \omega, t_0; \mathbf{C}^n)$.*

*Then, there exists a unique solution $y(t)$ of problems (1.1) and (1.2) on $[t_0 - \omega, t_0 + \tau_0]$. The solvability of problems (1.1) and (1.2) on $[t_0 - \omega, t_0 + \tau_0]$ is equivalent to the solvability of the equation*

$$\begin{gathered} y(t) = \Phi_1(y)(t) \equiv G^{-1}\{e^{-B_0 G^{-1}(t-t_0)} y_0 + \\ + \int_{t_0}^{t} e^{-B_0 G^{-1}(t-\tau)} Q_1[f(\tau, y(\tau)) + Ku(\tau) - \sum_{j=1}^{N} B_j y(\tau - \omega_j)]d\tau + \\ + Q_2[f(t, y(t) + Ku(t) - \sum_{j=1}^{N} B_j y(t - \omega_j)]\}, \quad a.a.\ t \in [t_0, t_0 + \tau], \end{gathered} \tag{1.22}$$

*with initial data* (1.2).

**Remark 1.1.** *Since $Y_1 = A_0\mathbf{C}^n$, condition (1.21) on the initial vector $y_0$ in Equation (1.2) is necessary.*

**Proof of Theorem 1.1.** By applying the projectors $Q_1, Q_2$ to Equation (1.1) and by using properties (1.18), Equation (1.1) splits into the equations

$$\frac{d}{dt}[A_0y(t)] + B_0G^{-1}[A_0y(t)] = Q_1[f(t, y(t)) + Ku(t) - \sum_{j=1}^{N} B_jy(t-\omega_j)], \tag{1.23}$$

$$Q_2B_0y(t) = Q_2[f(t, y(t)) + Ku(t) - \sum_{j=1}^{N} B_jy(t-\omega_j)]. \tag{1.24}$$

By the method of variation of constants for Equation (1.23) with unknown function $A_0y(t)$ and initial vector $y_0$, we obtain its equivalent form

$$A_0y(t) = e^{-B_0G^{-1}(t-t_0)}y_0 + \int_{t_0}^{t} e^{-B_0G^{-1}(t-\tau)}Q_1[f(\tau, y(\tau)) + Ku(\tau) - \sum_{j=1}^{N} B_jy(\tau-\omega_j)]d\tau. \tag{1.25}$$

It follows from Equations (1.24) and (1.25) and property (1.18) that problems (1.1) and (1.2) on $[t_0-\omega, t_0+\tau_0]$ is equivalent to problems (1.22) and (1.2).

Let $\bar{\omega} = \min\{\tau_0, \omega_1\}$. In the space $\Lambda_1 = L_1(t_0-\omega, t_0+\tau_1; \mathbf{C}^n)$, where $\tau_1 \in (0, \bar{\omega}]$ is determined below, we consider the mapping $\Phi$, defined on functions $v(t)$ from the closed subset $\Lambda_0 = \{v \in \Lambda_1 : v(t) = g(t),\ a.a.\ t \in [t_0-\omega, t_0]\}$:

$$\Phi(y)(t) = \Phi_1(y)(t),\ a.a.\ t \in [t_0, t_0+\tau_1],$$

$$\Phi(y)(t) = g(t),\ a.a.\ t \in [t_0-\omega, t_0],$$

where $\Phi_1$ is determined by Equation (1.22). It is clear that $\Phi$ maps $\Lambda_0$ into itself. With the help of Equation (1.19), we estimate the norm

$$||\Phi(v) - \Phi(w)||_{\Lambda_1} \leq [M||G^{-1}|| \cdot ||Q_1|| + \int_0^{\tau_1} e^{||B_0G^{-1}||\tau}d\tau + M||G^{-1}Q_2||]||v-w||_{\Lambda_1}.$$

By Equation (1.20), we can choose a number $\tau_1 \in (0, \bar{\omega}]$ such that the mapping $\Phi$ is contractive:

$$M||G^{-1}|| \cdot ||Q_1|| + \int_0^{\tau_1} e^{||B_0 G^{-1}||\tau} d\tau < 1 - M||G^{-1}Q_2||.$$

By virtue of the contractive mapping theorem, $\Phi$ has a unique fixed point $y(t) \in \Lambda_0$, which is a solution of problems (1.22) and (1.2) on $[t_0 - \omega, t_0 + \tau_1]$ and hence a solution of problems (1.1) and (1.2) on $[t_0 - \omega, t_0 + \tau_1]$. If $\tau_1 < \tau_0$, then we uniquely extend this solution to $[t_0 + \tau_1, t_0 + \min\{2\tau_1, \tau_0\}]$. To this end, we apply the arguments presented above to Equation (1.1) on $[t_0 + \tau_1, t_0 + \min\{2\tau_1, \tau_0\}]$ with the initial function $y(t)$ on $[t_0 + \tau_1 - \omega, t_0 + \tau_1]$, which is just found, and the initial condition $(A_0 y)(t_0 + \tau_1 + 0) = (A_0 y)(t_0 + \tau_1 - 0)$. It is clear that, in a finite number of steps, we uniquely extend the solution to the whole segment $[t_0 - \omega, t_0 + \tau_0]$.

The theorem is proved.

We now establish conditions for the one-valued solvability of impulsive delay problems (1.1)–(1.3).

**Theorem 1.2.** *Let the following assumptions for problems* (1.1)–(1.3) *be valid.*

*Restriction (1.17) for resolvent (1.16) holds.*

*The control $u(t) : [t_0, \infty) \to \mathbf{C}^m$ belongs to the class $u(t) \in L_{1,loc}(t_0, \infty; \mathbf{C}^m)$.*

*The function $f(t, v) : [t_0, \infty) \times \mathbf{C}^n \to \mathbf{C}^n$ as a function of t belongs to $L_{1,\mathrm{loc}}(t_0, \infty; \mathbf{C}^n)$ for each $v \in \mathbf{C}^n$ and for $k = 1, 2, ...$ satisfies the Lipschitz conditions*

$$||f(t, v) - f(t, w)|| \leq M_k||v - w||, \ \forall v, w \in \mathbf{C}^n, \ a.a. \ t \in [t_{k-1}, t_k], \tag{1.26}$$

*with constants $M_k$ independent of t and such that*

$$M_k||G^{-1}Q_2|| < 1, \ \ k = 1, 2, .... \tag{1.27}$$

*Relation (1.21) is true.*

*The initial function $g(t) : [t_0 - \omega, t_0] \to \mathbf{C}^n$ in Equation* (1.2) *belongs to the class $g(t) \in L_1(t_0 - \omega, t_0; \mathbf{C}^n)$.*

*The pulse actions $\Im_k : \Omega_k \to \mathbf{C}^n$ in Equation (1.3) satisfy the restrictions*

$$Q_2 \Im_k(v) = 0, \ \ v \in \Omega_k, \ \ k = 1, 2, .... \tag{1.28}$$

*Then, there exists a unique solution $y(t)$ of problems (1.1)–(1.3) on $[t_0 - \omega, \infty)$. This solution satisfies the equation*

$$\begin{gathered} y(t) = G^{-1}\{e^{-B_0G^{-1}(t-t_0)}y_0 + \\ + \int_{t_0}^{t} e^{-B_0G^{-1}(t-\tau)}Q_1[f(\tau, y(\tau)) + Ku(\tau) - \sum_{j=1}^{N} B_j y(\tau - \omega_j)]d\tau + \\ + Q_2[f(t, y(t) + Ku(t) - \sum_{j=1}^{N} B_j y(t - \omega_j)] + \\ + \sum_{t_0<t_k<t} e^{-B_0G^{-1}(t-t_k)}\Im_k((A_0 y)(t_k - 0))\}, \quad a.a. \ t \geq t_0. \end{gathered} \tag{1.29}$$

**Proof.** Let $\hat{y}_k(t)$ be a solution of delay equation (1.1) on $[t_{k-1} - w, t_k]$ with the initial data

$$\begin{gathered} (A_0\hat{y}_k)(t_{k-1} + 0) = y_{k-1}, \ \hat{y}_k(t) = \hat{y}_{k-1}(t), \ a.e. \ t \in [t_{k-1} - \omega, t_{k-1}], \\ y_k = (A_0\hat{y}_k)(t_k - 0) + \Im_k((A_0\hat{y}_k)(t_k - 0)), \ k = 1, 2, ..., \\ \hat{y}_0(t) = g(t), \ a.a. \ t \in [t_0 - \omega, t_0]. \end{gathered} \tag{1.30}$$

We build a solution $y(t)$ of impulsive problems (1.1)–(1.3) successively on the segments $[t_{k-1}, t_k]$, $k = 1, 2, ...$, with the help of the solutions $\hat{y}_k(t)$.

By Theorem 1.1, there exists a unique solution $\hat{y}_1(t)$ of problems (1.1) and (1.2) on $[t_0 - w, t_1]$. Obviously, $Q_2A_0\hat{y}_1(t) = 0$. Taking into account (1.28), we get $Q_2y_1 = 0$. We again apply Theorem 1.1 to prove the existence and uniqueness of a solution $\hat{y}_2(t)$ of Equation (1.1) on $[t_1 - w, t_2]$ with initial data (1.30) when $k = 2$. Further, we find successively for $k = 3, ...$ a solution $\hat{y}_k(t)$ of Equation (1.1) on $[t_{k-1} - w, t_k]$ with initial data (1.30). The function

$$y(t) = g(t), \ a.e. \ t \in [t_0 - \omega, t_0],$$

$$y(t) = \hat{y}_k(t), \ a.a. \ t \in [t_{k-1}, t_k], \ k = 1, 2, ...$$

is a unique solution of problems (1.1)–(1.3) on $[t_0 - w, \infty)$.

To complete the proof, make sure that Equation (1.29) is true. In view of Theorem 1.1, the solution $\hat{y}_k(t)$ satisfies the equation

$$\begin{gathered} \hat{y}_k(t) = G^{-1}\{e^{-B_0G^{-1}(t-t_{k-1})}y_{k-1} + \\ + \int_{t_{k-1}}^{t} e^{-B_0G^{-1}(t-\tau)}Q_1[f(\tau, \hat{y}_k(\tau)) + Ku(\tau) - \sum_{j=1}^{N} B_j\hat{y}_k(\tau - \omega_j)]d\tau + \\ + Q_2[f(t, \hat{y}_k(t) + Ku(t) - \sum_{j=1}^{N} B_j\hat{y}_k(t - \omega_j)]\}, \ a.a. \ t \in [t_{k-1}, t_k]. \end{gathered} \tag{1.31}$$

By Equation (1.31) for $k = 1$, relation (1.29) is valid for almost all $t \in [t_0, t_1]$. Then, we find $A_0y(t_1 - 0) = A_0y_1(t_1 - 0)$ and substitute this expression in

$y_1$ (1.30). Now, by means of Equation (1.31) for $k = 2$, we establish that the solution $y(t)$ satisfies Equation (1.29) for almost all $t \in [t_1, t_2]$. Repeating similar arguments for the segments $[t_2, t_3]$,$[t_3, t_4]$,..., we obtain the required result.

The theorem is proved.

**Remark 1.2.** *By Remark 1.1 and the representation of the initial vector $y_k$ in form (1.30), restrictions (1.28) on the pulse actions $\Im_k(v)$ are necessary for $v = A_0 y(t_k - 0)$ to solve impulsive problems (1.1)–(1.3) with a noninvertible matrix $A_0$ when $Q_2 \neq 0$.*

## 1.4 A Nonlinear Circuit with Transmission Lines in the Presence of Pulse Perturbations

Return to the circuit example described in Section 2. We have the following result.

**Proposition 1.1.** *Let the following assumptions for the radio technical filter in Figure 1.1 be valid.*

*The functions of input voltage sources $E_j(t) : [t_0 - \sqrt{L_j C_j}, \infty) \to \mathbf{C}$, $j = 1, 2$, satisfy the restrictions*

$$E_j(t) \in L_{1,\mathrm{loc}}(t_0 - \sqrt{L_j C_j}, \infty). \tag{1.32}$$

*The function $\psi(z) : \mathbf{C} \to \mathbf{C}$ satisfies the Lipschitz condition*

$$|\psi(z_1) - \psi(z_2)| \leq M|z_1 - z_2|, \ \ \forall\, z_1, z_2 \in \mathbf{C} \tag{1.33}$$

*with a sufficiently small constant $M$ independent of $z$.*

*The voltage $U_1(1,t)$ and the currents $I_1(1,t), I_2(1,t), I_r(t)$ are given on $[t_0 - \omega, t_0]$:*

$$\begin{gathered} U_1(1,t) = g_1(t) \in L_1(t_0 - \omega, t_0), \ \ I_1(1,t) = g_2(t) \in L_1(t_0 - \omega, t_0), \\ I_2(1,t) = g_3(t) \in L_1(t_0 - \omega, t_0), \ \ I_r(t) = g_4(t) \in L_1(t_0 - \omega, t_0), \\ \omega = 2 \max_{j=1,2}\{\sqrt{L_j C_j}\}. \end{gathered} \tag{1.34}$$

*The current $I_L(t)$ is given at $t_0$:*

$$I_L(t_0 + 0) = I_0. \tag{1.35}$$

*At times* $t_1 < t_2 < \cdots$ *(*$t_k \to \infty$ *as* $k \to \infty$*), the current* $I_L(t)$ *is subject to the pulse actions*

$$I_L(t_k + 0) - I_L(t_k - 0) = d_k(I_L)(t_k - 0), \;\; k = 1, 2, ..., \\ d_k(z) : \mathbf{C} \to \mathbf{C}. \tag{1.36}$$

*Then, the voltage* $U_1(1,t)$ *and the currents* $I_1(1,t), I_2(1,t), I_r(t)$ *can be uniquely determined on* $[t_0 - \omega, \infty)$ *such that* $U_1(1,t), I_1(1,t), I_2(1,t), I_r(t) \in L_{1,\mathrm{loc}}(t_0 - \omega, \infty)$ *and the relation holds*

$$I_L(t) = I_1(1,t) + I_2(1,t) - I_r(t) \in W_1^1(t_{k-1}, t_k), \;\; k = 1, 2, .... \tag{1.37}$$

**Proof.** To prove Proposition 1.1, we apply Theorem 1.2. Recall that the vector function $y(t) = (U_1(1,t), I_1(1,t), I_2(1,t), I_r(t))^{tr}$ satisfies the delay equation (1.1) in the notations (1.13)–(1.15) where$A_0, B_0, B_1, B_2$ are $4 \times 4$ matrices, $K$ is a $4 \times 2$ matrix, the values of the control $\mathrm{u(t)}$ belong to $\mathbf{C}^2$, $f(t,v) : [t_0, \infty) \times \mathbf{C}^4 \to \mathbf{C}^4$, and $\omega_1, \omega_2$ are delays with $\omega = \max\{\omega_1, \omega_2\}$ defined in Equation (1.34). Introduce the initial vector function $g(t) : [t_0 - \omega, t_0] \to \mathbf{C}^4$ and the initial vector $y_0 \in \mathbf{C}^4$ in initial conditions (1.2), the maps $\Im_k(v) : \mathbf{C}^4 \to \mathbf{C}^4$ in pulse actions (1.3):

$$g(t) = \begin{pmatrix} g_1(t) \\ g_2(t) \\ g_3(t) \\ g_4(t) \end{pmatrix}, \; y_0 = \begin{pmatrix} 0 \\ 0 \\ LI_0 \\ 0 \end{pmatrix}, \; \Im_k(v) = \begin{pmatrix} 0 \\ 0 \\ Ld_k\left(\frac{v_3}{L}\right) \\ 0 \end{pmatrix}. \tag{1.38}$$

Taking into account the Kirchhoff law $I_L(t) = I_1(1,t) + I_2(1,t) - I_r(t)$ in Equation (1.6) and relations (1.34)–(1.36), we obtain that the vector function $y(t)$ is a solution of problems (1.1)–(1.3).

Let us verify that the conditions of Theorem 1.2 hold for problems (1.1)–(1.3) in the notations (1.13)–(1.15) and (1.38). The characteristic matrix pencil

$$\lambda A_0 + B_0 = \begin{pmatrix} 1 & \rho_1 & 0 & 0 \\ 1 & 0 & \rho_2 & 0 \\ -1 & \lambda L & \lambda L & -\lambda L \\ 1 & 0 & 0 & -r \end{pmatrix}$$

is regular. Really,

$$\det(\lambda A_0 + B_0) = \lambda L(\rho_1\rho_2 + r\rho_1 + r\rho_2) + r\rho_1\rho_2 \not\equiv 0,$$

where the wave impedances $\rho_1, \rho_2$ are defined in Equation (1.9). Introducing the notation

$$q(\lambda) = \frac{r}{\lambda L} + \frac{r}{\rho_1} + \frac{r}{\rho_2} + 1,$$

by Equation (1.16), we find the resolvent

$$(\lambda A_0 + B_0)^{-1} = \frac{1}{q(\lambda)} \begin{pmatrix} \frac{r}{\rho_1} & \frac{r}{\rho_2} & -\frac{r}{\lambda L} & 1 \\ \frac{q(\lambda)\rho_1 - r}{\rho_1^2} & -\frac{r}{\rho_1\rho_2} & \frac{r}{\lambda L \rho_1} & -\frac{1}{\rho_1} \\ -\frac{r}{\rho_1\rho_2} & \frac{q(\lambda)\rho_2 - r}{\rho_2^2} & \frac{r}{\lambda L \rho_2} & -\frac{1}{\rho_2} \\ \frac{1}{\rho_1} & \frac{1}{\rho_2} & -\frac{1}{\lambda L} & \frac{1 - q(\lambda)}{r} \end{pmatrix}.$$

Estimate (1.17) holds. We find the subspaces

$$\mathrm{Ker} A_0 = \mathrm{Lin}\left\{ \begin{pmatrix} 1 \\ 0 \\ 0 \\ 0 \end{pmatrix}, \begin{pmatrix} 0 \\ 1 \\ 0 \\ 1 \end{pmatrix}, \begin{pmatrix} 0 \\ 0 \\ 1 \\ 1 \end{pmatrix} \right\},$$

$$Y_1 = A_0 \mathbf{C}^4 = \mathbf{Lin}\left\{ \begin{pmatrix} 0 \\ 0 \\ 1 \\ 0 \end{pmatrix} \right\},$$

$$Y_2 = B_0 \mathbf{Ker} A_0 = \mathbf{Lin}\left\{ \begin{pmatrix} 1 \\ 1 \\ -1 \\ 1 \end{pmatrix}, \begin{pmatrix} \rho_1 \\ 0 \\ 0 \\ -r \end{pmatrix}, \begin{pmatrix} 0 \\ \rho_2 \\ 0 \\ -r \end{pmatrix} \right\}.$$

According to the direct decomposition $\mathbf{C}^4 = Y_1 \dot{+} Y_2$, we have the corresponding mutually complementary projective matrices $Q_1, Q_2$ on $Y_1, Y_2$:

$$Q_1 = \begin{pmatrix} 0 & 0 & 0 & 0 \\ 0 & 0 & 0 & 0 \\ \frac{r\rho_2}{\alpha} & \frac{r\rho_1}{\alpha} & 1 & \frac{\rho_1\rho_2}{\alpha} \\ 0 & 0 & 0 & 0 \end{pmatrix}, \quad Q_2 = \begin{pmatrix} 1 & 0 & 0 & 0 \\ 0 & 1 & 0 & 0 \\ -\frac{r\rho_2}{\alpha} & -\frac{r\rho_1}{\alpha} & 0 & -\frac{\rho_1\rho_2}{\alpha} \\ 0 & 0 & 0 & 1 \end{pmatrix},$$

$$\alpha = \rho_1\rho_2 + r\rho_1 + r\rho_2.$$

Now, we can find the matrix

$$G = A_0 + Q_2 B_0 = \begin{pmatrix} 1 & \rho_1 & 0 & 0 \\ 1 & 0 & \rho_2 & 0 \\ -1 & \beta & \beta & -\beta \\ 1 & 0 & 0 & -r \end{pmatrix}, \quad \beta = L - \frac{r\rho_1\rho_2}{\alpha},$$

and its inverse matrix

$$G^{-1} = \frac{1}{\alpha L} \begin{pmatrix} \beta r \rho_2 & \beta r \rho_1 & -r\rho_1\rho_2 & \beta\rho_1\rho_2 \\ \beta(r+\rho_2)+r\rho_2 & -\beta r & r\rho_2 & -\beta\rho_2 \\ -\beta r & \beta(r+\rho_1)+r\rho_1 & r\rho_1 & -\beta\rho_1 \\ \beta\rho_2 & \beta\rho_1 & -\rho_1\rho_2 & \gamma \end{pmatrix},$$

$$\gamma = -\beta(\rho_1+\rho_2) - \rho_1\rho_2.$$

By Equation (1.32), we have $u(t) \in L_1(t_0, \infty; \mathbf{C}^2)$. Here, $f(t, v) : [t_0, \infty) \times \mathbf{C}^4 \to \mathbf{C}^4$ in Equation (1.14) is a constant of $t$. On account of Lipschitz condition (1.33), the function $f(t, v)$ satisfies Lipschitz conditions (1.26) with the constants $M_k = M$. We suppose that the constant $M$ is sufficiently small such that $M||G^{-1}Q_2|| < 1$. Then, restrictions (1.27) of Theorem 1.2 are fulfilled. The initial vector $y_0$ in Equation (1.38) meets requirement (1.21). By Equation (1.34), we obtain that $g(t)$ in Equation (1.38) satisfies $g(t) \in L_1(t_0 - \omega, t_0; \mathbf{C}^4)$. The pulse action$\Im_k(v) : \mathbf{C}^4 \to \mathbf{C}^4$ in Equation (1.38) satisfies restrictions (1.28).

Thus, the conditions of Theorem 1.2 for problems (1.1)–(1.3) in notations (1.13)–(1.15) and (1.38) are established. By Theorem 1.2, we uniquely determine functions $U_1(1,t), I_1(1,t), I_2(1,t), I_r(t) \in L_{1,\text{loc}}(t_0-\omega, \infty)$ such that relations (1.37) are valid.

The proposition is proved.

## 1.5 Conclusion

We derive results for a special class of control systems, namely, descriptor systems. Mention here the books [10], where results on different aspects of control systems theory and its applications were collected. In particular, the conflict-controlled systems are considered in the papers [11–13] and references therein. Problems considered here are related to studies on control systems with delay.

Up till now, problems (1.1)–(1.3) were studied in finite-dimensional spaces, when $A_0, B_j, K$ are matrices. Many physical and technical processes are described by partial differential equations. In the general case, partial differential equations are not solvable with respect to the leading time derivative – equations of non-Kovalevskaya or Sobolev type, for example [14]. In abstract form, these equations are implicit operator differential equations with differential operators in spaces of functions. For these reasons, problems (1.1)–(1.3) can be considered with operator coefficients $A_0, B_j, K$ in abstract Banach or Hilbert spaces. The results of Section 3 are extended to this case and this extension can be applied to investigate partial differential equations with delay and impulse actions.

In the light of classical mathematical system theory [15, 16], every dynamical control system obeys an equation of state and an equation of output. This is the case for the game dynamic problems [17–20].

Regarding the example of control radio engineering system from Section 1.2, we give differential algebraic equation (1.1) for the state $y(t)$ (1.14). Here, we also use notation (1.15). Let us take note of the fact that Equation (1.11), where $j = 1, 2$, presents the relations for the output

$$\eta(t) = \begin{pmatrix} I_1(0, t - \omega_1/2) \\ I_2(0, t - \omega_2/2) \end{pmatrix}.$$

Really, introduce the notation

$$M_0 = M_0^+ = M_1^+ = M_2^+ = \begin{pmatrix} 0 & 0 & 0 & 0 \\ 0 & 0 & 0 & 0 \end{pmatrix}, \quad M_0^- = \frac{1}{2}\begin{pmatrix} \frac{1}{\rho_1} & 1 & 0 & 0 \\ \frac{1}{\rho_2} & 0 & 1 & 0 \end{pmatrix},$$

$$M_1^- = \frac{1}{2}\begin{pmatrix} -\dfrac{1}{\rho_1} & 1 & 0 & 0 \\ 0 & 0 & 0 & 0 \end{pmatrix}, \quad M_2^- = \frac{1}{2}\begin{pmatrix} 0 & 0 & 0 & 0 \\ -\dfrac{1}{\rho_2} & 0 & 1 & 0 \end{pmatrix}, \quad N = 2.$$

Then, Equation (1.11) for $j = 1, 2$ gives the equation for the output in the form

$$\eta(t) = \frac{d}{dt}[M_0 y(t)] + \sum_{j=0}^{N} \left[M_2^- y(t - \omega_j) + M_2^+ y(t + \omega_j)\right] + Ku(t).$$

## References


[1] R.F. Curtain, H. Zwart, An Introduction to Infinite-Dimensional Linear Systems Theory, Springer-Verlag, New York, 1995.

[2] B.M. Chen, Z. Lin, Y. Shamash, Linear systems theory: a structural decomposition approach, Birkhauser, Boston, Basel, Berlin, 2004.

[3] G.R. Duan, Analysis and Design of Descriptor Linear Systems, Springer, New York, Dordrecht, Heidelberg, London, 2010, p. 45-99.

[4] J.K. Hale, S.M. Verduyn Lunel, Introduction to Functional Differential Equations, Springer, New York, 1993.

[5] L. Vlasenko, 'Implicit linear time-dependent differential-difference equations and applications', Mathematical Methods in the Applied

Sciences, 2000, vol. 23, issue 10, pp. 937-948, https://doi.org/10.1002/1099-1476(20000710)23:10<937::AID-MMA144>3.0.CO;2-B

[6] F. Kappel, K.P. Zhang, 'On neutral functional differential equations with nonatomic difference operator', J. Math. Anal. and Appl., 1986, vol. 113, pp. 311-343.

[7] E.M. Fernàndez-Berdaguer, 'Solutions for a class of neutral differential equations with nonatomic D-operator', Applied Mathematics and Optimization, 1986, vol. 14, N 1, pp. 95-105.

[8] L.A. Vlasenko, A.G. Rutkas, 'On a class of impulsive functional-differential equations with nonatomic difference operator', Mathematical Notes, 2014, vol. 95, issue 1-2, pp. 32-42, https://doi.org/10.1134/S0001434614010040

[9] A.G. Rutkas, 'Spectral methods for studying degenerate differential-operator equations. I', Journal of Mathematical Sciences, 2007, vol. 144, issue 4, pp. 4246-4263. http://doi.org/10.1007/s10958-007-0267-2

[10] V.M. Kuntsevich, V.F. Gubarev, Y.P. Kondratenko, D.V. Lebedev, V.P. Lysenko (Eds), Control Systems: Theory and Applications. Series in Automation, Control and Robotics, River Publishers, Gistrup, Delft, 2018. https://www.riverpublishers.com/book_details.php?book_id=668

[11] G.T. Chikrii, 'Using the effect of information delay in differential games of pursuit', Cybernetics and Systems Analysis, 2007, vol. 43, issue 2, pp. 244-245, https://doi.org/10.1007/s10559-007-0042-x

[12] G.Ts. Chikrii, 'On One problem of approach for damped oscillations', Journal of Automation and Information Sciences, 2009, vol. 4, issue 10, pp. 1-9, https://doi.org/10.1615/JAutomatinfScien.v41.i10.10

[13] G.T. Chikrii, 'Principle of time stretching in evolutionary games of approach', Journal of Automation and Information Sciences, 2016, vol. 48, N 5, pp. 12-26. http://doi.org/10.1615/JAutomatinfScien.v48.i5.20

[14] A. Rutkas, L. Vlasenko, 'On a differential game in a non-damped distributed system', Mathematical Methods in the Applied Sciences, 2019, vol. 42, issue 18, pp. 6155-6164. http://doi.org/10.1002/mma.5712

[15] R.E. Kalman, P.L. Falb, M.A. Arbib, Topics in Mathematical System Theory, Mc Graw-Hill Book Company, New York, San Francisco, St. Louis, Toronto, London, Sydney, 1969.

[16] Philip Hartman, Ordinary Differential Equations, John Wiley& Sons, New York London Sydney, 1964.

[17] O. Hayek, Pursuit Games, Academic Press, New York, 1975.
[18] R. Isaacs, Differential Games, John Wiley & Sons, New York, 1965.
[19] A.N. Krasovskii, N.N. Krasovskii, Control under Lack of Information, Birkhauser, 1995.
[20] A. Friedman, Differential Games, Wiley- Interscience, New York, 1971.

# 2

# An Extremal Routing Problem with Constraints and Complicated Cost Functions

**A.A. Chentsov, A.G. Chentsov, and A.N. Sesekin**

N.N. Krasovskii Institute of Mathematics and Mechanics, Ural Branch of Russian Academy of Sciences, 16 S. Kovalevskaya, Yekaterinburg, 620108, Russia, Ural Federal University, 19 Mira st., Yekaterinburg, 620002, Russia.
E-mail: chentsov.a@binsys.ru, chentsov@imm.uran.ru, sesekin@list.ru.

## Abstract

One routing problem with precedence conditions and complicated cost functions is considered. The natural application can be connected with the engineering problem of dismantling of radiation sources. We must choose starting point, route (index permutation), and concrete trajectory of process. In addition, our index permutation defines the sequence of task. The concrete trajectory must be coordinated with this permutation. In addition, different constraints arise. In particular, the choice of the above-mentioned permutation must satisfy to precedence conditions. For introduction of these conditions, the corresponding system of ordered pairs is specified. These ordered pairs are called address. In our mathematical setting, additive criterion is used. This criterion is formed with employment of cost functions with the task list dependence. In the large, the investigated problem can be considered as a control problem with discrete time for that admissible solutions have the hierarchical structure. In this article, we focus on engineering problem connected with dismantling of finite system of radiation sources; for this problem, the above-mentioned task list dependence has the following nature. Namely, in every time, the corresponding executor is affected to those and only those sources that were not dismantled at this time. For solving this applied problem, the widely understood dynamic programming is used. On this foundation, optimal algorithm for PC is constructed. The computing experiment was realized.

**Keywords**. Precedence conditions, route optimization, Bellman function, trace.

## 2.1 Introduction

In this article, the routing problem oriented to engineering applications connected with nuclear energy industry is investigated. Some situations arising under accidents on nuclear power plants are considered. We consider executor operations under sequential dismantling of radiation sources. As a result, the very complicated routing problem arises. In this problem, constraints and complicated cost functions are presented. Among all constraints, precedence conditions are selected (these conditions can be used for decreasing the computing complexity). The complication of cost functions is connected with sequence of dismantling tasks. General questions concerning the problems of the work organization under increased radiation are considered in [1–3]. In addition, natural elements of a routing arise. Indeed, it is natural to consider the question connected with the decrease of the total radiation dose obtained by executor. This dose depends on the ordering of operation. Therefore, the extreme routing problem arises. This problem is complicated by constraints and the task list dependence in cost functions.

There are many options for setting routing problems generated by specific problems that arise in various fields of activity (see, for example, [4–6] and the bibliography thereto).

Of course, the well-known traveling salesman problem (TSP) can be considered as a prototype of our routing problem. Therefore, we recall some investigations devoted to TSP; see [7–12]. In this connection, we separately note TSP-PC; this is TSP with precedence conditions. We note some investigations connected with TSP-PC; see [13, 14].

Moreover, we note investigations connected with generalized TSP or GTSP; see [15, 16]. However, our study is essentially different from TSP and problems of type TSP. Namely, we consider the problem with multivariate movements between visited objects. This feature motivates the use of the model of megacities (finite nonempty sets). Each visiting to a megacity is accompanied by the implementation of some work called internal. These works depend on point of arrival to megacity and point of departure. Of course, the choice of permutation not yet defines the process current. The additional choice of a trajectory is required. So, a nonempty bundle of trajectories is connected with every permutation, called a route.

Therefore, we have some discrete analog of controlled differential inclusion. This interpretation is in good agreement with the hierarchical structure of used solutions. We choose a route (permutation of indexes) and a trajectory from bundle corresponding to the route. So, we operate with bundles of trajectories. Moreover, we choose a starting point for our process. Finally, we use the dynamic programming (DP) variant that is a serious development of the Bellman procedure [7] which is widely used in control theory. In discrete optimization, usually the Held–Karp procedure [8] is used (the Bellman procedure is retrograde and the Held–Karp procedure is direct). We note that our DP variant ascending to the Bellman scheme of [7] has the property of universality with respect to starting point. This singularity permits to realize the procedure of the starting point choice.

So, in our investigation, the complex including starting point, route (permutation of indexes), and trajectory is optimized.

## 2.2 General Notions and Designations

In the following, a very complicated extreme routing problem is considered. For correct mathematical setting, the very developed formalization is required. In this connection, we recall several set-theoretical notions and designations that will be used in the following.

So, we use standard set-theoretical symbolism (quantifiers, propositional connectives, and so on). For all objects $\alpha$ and $\beta$, by $\{\alpha;\beta\}$, we denote the unique set containing $\alpha$ and $\beta$, and not containing any other elements. If $\gamma$ is an object, then $\{\gamma\} \triangleq \{\gamma;\gamma\}$ is singleton containing $\gamma$. Any set is an object. Therefore, for all objects $u$ and $v$, we suppose $(u, v) \triangleq \{\{u\};\{u;v\}\}$ for ordered pair (OP) with the first element $u$ and the second element $v$; in this definition, we follow [17]. If $h$ is an OP, then by $\mathrm{pr}_1(h)$ and $\mathrm{pr}_2(h)$, we denote the first and the second elements of $h$ uniquely defined by condition $h= (\mathrm{pr}_1(h), \mathrm{pr}_2(h))$. If $x$, $y$, and $z$ are objects, then $(x, y, z) \triangleq ((x, y),z)$; see [18, Ch. I] (of course, $x$ is the first element of triplet $(x, y, z)$, $y$ is the second element, and $z$ is the third element of this triplet).

If $H$ is a set, then by $\mathcal{P}(H)$ (by $\mathcal{P}'(H)$), we denote the family of all (all nonempty) subsets of $H$; $\mathrm{Fin}(H)$ is the family of all nonempty finite subsets of $H$. Of course, $\mathcal{P}'(H) =\mathcal{P}(H)\setminus\{\emptyset\}$ and, under finite set $H$,

$\mathrm{Fin}(H) = \mathcal{P}'(H)$. If $A$ and $B$ are nonempty sets, then $B^A$ (see [17]) is the set of all mappings from $A$ into $B$. In addition, the expressions $g \in B^A$ and $g$: $A \to B$ are equivalent; if $\mathbf{g} \in B^A$ and $a \in A$, then $\mathbf{g}(a) \in B$ is the value of $\mathbf{g}$ at the point $a$. If, moreover, $\varphi \in B^A$ and $C \in \mathcal{P}(A)$, then $\varphi^1(C) \triangleq \{\varphi(x) : x \in C\} \in \mathcal{P}(B)$ (see [17]) is the image of the set $C$ under operation $\varphi$; $\varphi^1(C) \neq \emptyset$ under $C \neq \emptyset$.

For every nonempty sets $A$, $B$, and $C$, $A \times B \times C \triangleq (A \times B) \times C$ (see [18, Ch. I]) and, under $h \in D^{A \times B \times C}$, where $D$ is a nonempty set, $x \in A \times B$, and $y \in C$, the value $h(x, y) \in D$ is defined and denoted also as $h(x_1, x_2, y)$, where $x_1 \triangleq \mathrm{pr}_1(x) \in A$ and $x_2 \triangleq \mathrm{pr}_2(x) \in B$. Suppose that $\mathbb{R}$ is a real line, $\mathbb{N} \triangleq \{1;2;...\}$, $\mathbb{N}_0 \triangleq \{0\} \cup \mathbb{N} \in \mathcal{P}'(\mathbb{R})$, and $\mathbb{R}_+ \triangleq \{\xi \in \mathbb{R} |\ 0 \leq \xi\}$.

If $K$ is a nonempty finite set, then $|K| \in \mathbb{N}$ is the cardinality of $K$ by definition; we suppose that $|\emptyset| \triangleq 0$. We suppose that

$$\overline{p, q} \triangleq \{k \in \mathbb{N}_0 |\ (p \leq k)\&(k \leq q)\} \quad \forall p \in \mathbb{N}_0 \quad \forall q \in \mathbb{N}_0.$$

In addition, $\overline{1,n} = \{k \in \mathbb{N} |\ k \leq n\} \in \mathcal{P}'(\mathbb{R}_+)$ under $n \in \mathbb{N}$ and $\overline{1,0} = \emptyset$. For every nonempty finite set $K$, by $(\mathrm{bi})[K]$, we denote the set of all bijections [19, §3] from $\overline{1, |K|}$ onto K. Then, under $n \in \mathbb{N}$, $(\mathrm{bi})[\overline{1,n}]$ is the set of all permutations [19, §3] of the index set $\overline{1,n}$.

If $S$ is a nonempty set, then $\mathcal{R}_+[S]$ is the set of all real-valued non-negative functions defined on $S$; so $\mathcal{R}_+[S] \triangleq (\mathbb{R}_+)^S$.

## 2.3 General Routing Problem and Its Specific Variant

In the following, we fix a nonempty set $X$, a set $X^0 \in \mathrm{Fin}(X)$ (so, $X^0$ is a nonempty finite subset of $X$; elements of $X^0$ play the role of starting points), a number $N \in \mathbb{N}$ for which $N \geq 2$, and nonempty (finite) subsets of $X$

$$M_1 \in \mathrm{Fin}(X), ..., M_N \in \mathrm{Fin}(X). \tag{2.1}$$

We consider the sets (2.1) as megacities. Moreover, we fix relations

$$\mathbb{M}_1 \in \mathcal{P}'(M_1 \times M_1), ..., \mathbb{M}_N \in \mathcal{P}'(M_N \times M_N). \tag{2.2}$$

So, under $j \in \overline{1,N}$, in the form of $\mathbb{M}_j$, we have a nonempty subset of $M_j \times M_j$. We consider megacities (2.1) as objects of a visiting. Under visiting to $M_j$, executor implements some works called interior. The concrete variant of this work is defined by OP $z = (x, y) \in \mathbb{M}_j$, where $x$ is input point and $y$ is departure point. So, $\mathbb{M}_j$ determines the possible options for performing interior work under visiting to $M_j$.

The sequence of the megacity visiting is defined by index permutation called route. Then $\mathbb{P}=(\mathrm{bi})[\overline{1,N}]$ is the set of all permutations of the index set $\overline{1,N}$. Of course, elements of $\mathbb{P}$ and only they (in the next account) are routes. We consider the following processes:

$$x^0 \to (x_1^{(1)} \in M_{\alpha(1)} \rightsquigarrow x_1^{(2)} \in M_{\alpha(1)}) \\ \to \cdots \to (x_N^{(1)} \in M_{\alpha(N)} \rightsquigarrow x_N^{(2)} \in M_{\alpha(N)}), \tag{2.3}$$

where $x^0 \in X^0$ and $\alpha \in \mathbb{P}$. We must choose

$$(x^0, \alpha, x_1^{(1)}, x_1^{(2)}, ..., x_N^{(1)}, x_N^{(2)})$$

for minimization of some criterion. This choice must satisfy some constraints. In addition, $x^0$ is an element of $X^0,\ \ X^0 \subset X$. Moreover,

$$z_1 = (x_1^{(1)}, x_1^{(2)}) \in \mathbb{M}_{\alpha(1)}, ..., z_N = (x_N^{(1)}, x_N^{(2)}) \in \mathbb{M}_{\alpha(N)}. \tag{2.4}$$

In fact, with the employment of Equations (2.3) and (2.4), we obtain the next variant of our process

$$\left(x^0,\ x^0\right) \ \to\ z_1\ \in\ \mathbb{M}_{\alpha(1)} \to \cdots \to\ z_N \in \mathbb{M}_{\alpha(N)}.$$

Finally, the choice of $\alpha \in \mathbb{P}$ can be constrained by precedence conditions. For introduction of these conditions, we use a set

$$\mathbf{K} \in \mathcal{P}(\overline{1,N} \times \overline{1,N})$$

of OP from elements of $\overline{1,N}$ (the case $\mathbf{K}=\emptyset$ is not excluded). Elements of $\mathbf{K}$ are called address pairs. For $z \in \mathbf{K}$, the (first) element $\mathrm{pr}_1(z) \in \overline{1,N}$ is called a sender and $\mathrm{pr}_2(z) \in \overline{1,N}$ is called a receiver (of load, information, and so on). The precedence conditions consist in the following: $\alpha \in \mathbb{P}$ is admissible (by precedence conditions) if for every $z=(i,j) \in \mathbf{K}$, a visiting to $M_i = M_{\mathrm{pr}_1(z)}$ must precede a visiting to $M_j$. We suppose that

$$\forall \mathbf{K}_0 \in \mathcal{P}'(\mathbf{K})\ \exists z_0 \in \mathbf{K}_0\colon\ \mathrm{pr}_1(z_0) \neq \mathrm{pr}_2(z) \quad \forall z \in \mathbf{K}_0. \tag{2.5}$$

The discussion about this (not restrictive) condition is reduced in [22]. If $\alpha \in \mathbb{P}$, then $\alpha^{-1} \in \mathbb{P}$ is the permutation inverse to $\alpha$. Then (under condition (2.5)),

$$\mathbf{A} \triangleq \{\alpha \in \mathbb{P}|\ \forall z \in \mathbf{K}\ \ \forall t_1 \in \overline{1,N}\ \ \forall t_2 \in \overline{1,N}\ (z=(\alpha(t_1),\alpha(t_2))) \Rightarrow \\ \Rightarrow (t_1<t_2)\}=\{\alpha \in \mathbb{P}|\ \alpha^{-1}(\mathrm{pr}_1(z)) < \alpha^{-1}(\mathrm{pr}_2(z))\ \forall z \in \mathbf{K}\} \in \mathcal{P}'(\mathbb{P}). \tag{2.6}$$

So, $\mathbf{A} \neq \emptyset$ and $\mathbf{A} \subset \mathbb{P}$. But, as it is indicated in Equation (2.3), the choice of $\alpha \in \mathbf{A}$ does not define the process current: the notion of trajectory is required. For this, previously, we introduced the sets

$$\mathfrak{M}_j \triangleq \{\mathrm{pr}_1(z) : z \in \mathbb{M}_j\} \in \mathcal{P}'(M_j)$$

(the set of all input points) and

$$\mathbf{M}_j \triangleq \{\mathrm{pr}_2(z) : z \in \mathbb{M}_j\} \in \mathcal{P}'(M_j) \tag{2.7}$$

(the set of all departure points), where $j \in \overline{1,N}$. In terms of Equation (2.7), we introduce the sets

$$(\mathbb{X} \triangleq X^0 \cup (\bigcup_{i=1}^{N} \mathfrak{M}_i)) \text{ and } (\mathbf{X} \triangleq X^0 \cup (\bigcup_{i=1}^{N} \mathbf{M}_i)). \tag{2.8}$$

We consider the set $\mathbb{X} \times \mathbf{X}$ as phase space of our processes (2.3) and (2.4). Moreover, we introduce the set $\mathfrak{Z}$ of all collections

$$(z_i)_{i \in \overline{0,N}} \colon \overline{0,N} \to \mathbb{X} \times \mathbf{X}.$$

Then, under $x \in X^0$ and $\alpha \in \mathbb{P}$, we suppose that

$$\mathcal{Z}_\alpha[x] \triangleq \{(z_t)_{t \in \overline{0,N}} \in \mathfrak{Z} |\ (z_0 = (x,x)) \& (z_t \in \mathbb{M}_{\alpha(t)}\ \forall t \in \overline{1,N})\}; \tag{2.9}$$

of course, $\mathcal{Z}_\alpha[x] \in \mathrm{Fin}(\mathfrak{Z})$. Elements of Equation (2.9) are trajectories coordinated with route $\alpha$ and starting from point $x$. Then, for $x \in X^0$, in the form of

$$\widetilde{\mathbf{D}}[x] \triangleq \{(\alpha, \mathbf{z}) \in \mathbf{A} \times \mathfrak{Z} |\ \mathbf{z} \in \mathcal{Z}_\alpha[x]\} \in \mathrm{Fin}(\mathbf{A} \times \mathfrak{Z}), \tag{2.10}$$

we obtain the set of all admissible solutions with the starting point $x$. Moreover, we consider the routing problem for which $x \in X^0$ may vary for optimization of a criterion. For a considered routing problem,

$$\mathbf{D} \triangleq \{(\alpha, \mathbf{z}, x) \in \mathbf{A} \times \mathfrak{Z} \times X^0 |\ (\alpha, \mathbf{z}) \in \widetilde{\mathbf{D}}[x]\} \in \mathrm{Fin}(\mathbf{A} \times \mathfrak{Z} \times X^0) \tag{2.11}$$

is the set of all admissible solutions for our more general problem. Such admissible solutions are triplets including route, trajectory, and starting point.

Now, we introduce the corresponding additive criterion. For this, under $\mathfrak{N} \triangleq \mathcal{P}'(\overline{1,N})$, we suppose that (in general setting)

$$\begin{gathered} \mathbf{c} \in \mathcal{R}_+ [\mathbf{X} \times \mathbb{X} \times \mathfrak{N}], c_1 \in \mathcal{R}_+ [\mathbb{X} \times \mathbf{X} \times \mathfrak{N}], ..., \\ c_N \in \mathcal{R}_+ [\mathbb{X} \times \mathbf{X} \times \mathfrak{N}], f \in \mathcal{R}_+ [\mathbf{X}]. \end{gathered} \tag{2.12}$$

In connection with Equation (2.12), we note that elements of $\mathfrak{N}$ (nonempty subsets of $\overline{1,N}$) play the role of a list of tasks not completed at the current time. Each such list indicates a collection of sources that have not been dismantled at this moment; it is these sources that provide the real radiation exposure on the performer. We use the function $\mathbf{c}$ for estimation of exterior movements, the functions $c_j,\ \ j \in \overline{1,N}$, for estimation of interior works, and $f$ for estimation of terminal state (the point $x_N^{(2)}$ in Equation (2.3)). If $x \in X^0$, $\alpha \in \mathbb{P}$, and $(z_i)_{i\in\overline{0,N}} \in \mathcal{Z}_\alpha[x]$, then

$$\mathfrak{C}_\alpha[(z_i)_{i\in\overline{0,N}}] \triangleq \sum_{t=1}^{N} [\mathbf{c}(\mathrm{pr}_2(z_{t-1}), \mathrm{pr}_1(z_t), \alpha^1(\overline{t,N}))+ \\ +c_{\alpha(t)}(z_t, \alpha^1(\overline{t,N}))] + f(\mathrm{pr}_2(z_N)); \tag{2.13}$$

of course, for us, in Equation (2.13), a variant $\alpha \in \mathbf{A}$ is essential. Then, the general problem can be written in the form

$$\mathfrak{C}_\alpha[\mathbf{z}] \to \min, \quad (\alpha, \mathbf{z}, x) \in \mathbf{D}; \tag{2.14}$$

the value $V \in \mathbb{R}_+$ of Equation (2.14) is the smallest of numbers $\mathfrak{C}_\alpha[\mathbf{z}],\ (\alpha, \mathbf{z}, x) \in \mathbf{D}$. Then, in the form of the set

$$\mathbf{SOL} \triangleq \{(\alpha^0, \mathbf{z}^0, x^0) \in \mathbf{D} |\ \mathfrak{C}_{\alpha^0}[\mathbf{z}^0] = V\} \in \mathrm{Fin}(\mathbf{D}), \tag{2.15}$$

we obtain a (nonempty) set of all optimal solutions of the problem (2.14). If $x \in X^0$, then we consider the next $x$-problem

$$\mathfrak{C}_\alpha[\mathbf{z}] \to \min, \quad (\alpha, \mathbf{z}) \in \tilde{\mathbf{D}}[x]; \tag{2.16}$$

for Equation (2.16), the value $\tilde{V}[x] \in \mathbb{R}_+$ of this problem is defined as the smallest of numbers $\mathfrak{C}_\alpha[\mathbf{z}],\ (\alpha, \mathbf{z}) \in \tilde{\mathbf{D}}[x]$, and

$$(\mathrm{sol})[x] \triangleq \{(\alpha^0, \mathbf{z}^0) \in \tilde{\mathbf{D}}[x] |\ \mathfrak{C}_{\alpha^0}[\mathbf{z}^0] = \tilde{V}[x]\} \in \mathrm{Fin}(\tilde{\mathbf{D}}[x]). \tag{2.17}$$

So, we select routing problems with fixed starting point. Of course, in our case, the equality

$$V = \min_{x\in X^0} \tilde{V}[x] \tag{2.18}$$

is realized. We note that for a point $x^0 \in X^0$ with $\tilde{V}(x^0) = V$ and $(\alpha^0, \mathbf{z}^0) \in (\mathrm{sol})[x^0]$, we obtain that

$$(\alpha^0, \mathbf{z}^0, x^0) \in \mathbf{SOL}. \tag{2.19}$$

Our goal consists in the determination of the value (2.18) and a solution from the set (2.15); we use Equation (2.19) also. For solving problem (2.14), we use DP.

Now, we note some singularities of costs functions (2.12) for the considered applied problem, connected with dismantling of the radioactive elements. These cost functions are defined here (see Equation (2.2)) on wider sets than required. Actually, further, the values of the function $\mathbf{c}$ will be needed to estimate the movements between megacities, as well as the movements from $x \in X^0$ to megacities. The values of the functions $c_j$ are essential under conditions when their arguments are chosen from the sets $\mathbb{M}_j \times \mathfrak{N}$. The values $f(x)$ are significant when $x \in \mathbf{M}_j$ for $j \in \overline{1,N}$. For all other cases, the values of $\mathbf{c}, c_1, \ldots, c_N, f$ can be given arbitrarily and, in particular, are proposed to be zero. In this part, we use constructions of [20] and [21]. So, in this special variant of our setting, the function $\mathbf{c}$ is defined by [20, (6.1), (6.3), (6.20)]. We recall these constructions of [20] very briefly. Namely, for $x \in \mathbf{X}$, $y \in \mathbb{X}$, and $K \in \mathfrak{N}$, we suppose that

$$\mathbf{c}(x, y, K) = \sum_{k \in K} \mathbf{c}(x, y, \{k\}); \tag{2.20}$$

we sum the values of cost functions for separate sources. If $k \in K$, then $\mathbf{c}(x, y, \{k\})$ is defined by integration of nonlinear function along the trajectory of the executor permutation from $x$ to $y$. The corresponding nonlinear function is defined as dependence inversely proportional to square of the Euclidean distance between executor and the source with index $k$. This integration procedure is reduced to table formulas; see [20, (6.20)]. So, we obtain the radiation dose obtained by executor under (exterior) permutation from $x$ to $y$ (see Equation (2.20)).

As already noted, for $j \in \overline{1,N}$, the values of $c_j(\overline{x}, \overline{y}, \overline{K})$, where $\overline{x} \in \mathbb{X}$, $\overline{y} \in \mathbf{X}$, and $\overline{K} \in \mathfrak{N}$, are essential for $\overline{x} \in \mathfrak{M}_j$, $\overline{y} \in \mathbf{M}_j$, and $j \in \overline{K}$. Now, we discuss only this case. Namely, we sum doses obtained by executor under permutations and under work by dismantling of source with index $j$. In addition, the permutation dose is realized as the sum of two numbers. The first number corresponds to the dose obtained by executor under permutation from the arrival point $\overline{x}$ to source. Under this permutation, we take into account the radioactive influence of all sources with numbers from the set $K$ including source with number $j$ oneself. The second number, used under calculation of the permutation dose, is defined for trajectory from the source to point of departure $\overline{y}$ (we keep in mind the departure from $M_j$). In this case, the corresponding dose is defined without the source with index $j$. So, for this calculation, we take into account the set $K\setminus\{j\}$ of sources. For determination of the concrete individual doses for separate sources, we use the integration procedures similar to those discussed under construction of the function $\mathbf{c}$ (we keep in mind the integration of nonlinear functions).

Finally, in $c_j(\overline{x}, \overline{y}, \overline{K})$, the component estimating the immediate work including dismantling is used. We keep in mind the work that is not connected with the motion of executor. This component is the concrete dose obtained by executor under the dismantling realization. This dose depends from the time interval $\triangle t_j$ corresponding to the time required for the dismantling realization for source with index $j$. Moreover, this dose depends from intensity of all sources which are not dismantled on the given time moment. All intensities operate during $\triangle t_j$. Of course, the action of source with index $j$ is taken into account since executor is situated near to this source. As a result, $c_j(\overline{x}, \overline{y}, \overline{K})$ is realized as the sum of all the above-mentioned components.

Finally, the function $f$ defines residual radiation dose estimating the executor motion after the ending of all works. In particular, the variant $f(x) \equiv 0$ is possible.We suppose that $f$ is independent on starting point $x \in X^0$. This is possible when executor is transposed to a given fixed point after the ending of all works. However, such a case is also possible when there are sources that are not included in the initial list for one reason or another, that their disposal would lead to an unacceptable dose of radiation, and therefore the real plan was limited in this case only to the sources mentioned above. Nevertheless, at the stage of removing the contractor (s) from the operating mode, the mentioned remaining sources affect this contractor (s).

So, we obtain the very difficult extremal problem with constraints and complicated cost functions. For solving this problem, we use nonstandard DP procedure. Therefore, for mathematical setting and the consequent solution construction, the serious formalization is required. For this, the sufficiently detailed summary of mathematical constructions will be reduced (partially, it was made in Section 2.1 and at the beginning of this section). In the following section, we introduce required notions for the used variant of DP.

## 2.4 Dynamic Programming, 1

We recall that $\mathfrak{N}$ is the family of all nonempty subsets of $\overline{1,N}$; of course, $\mathfrak{N} \neq \emptyset$. If $K \in \mathfrak{N}$, then we suppose that

$$\Xi[K] \triangleq \{z \in \mathbf{K} |\ (\mathrm{pr}_1(z) \in K)\&(\mathrm{pr}_2(z) \in K)\};$$

so, $\Xi[K] \subset \mathbf{K}$ and, for $(i,j) \in \mathbf{K}$

$$((i,j) \in \Xi[K]) \Leftrightarrow ((i \in K)\&(j \in K)).$$

Now, we introduce $\mathbf{I} \in \mathfrak{N}^{\mathfrak{N}}$; so, $\mathbf{I}$: $\mathfrak{N} \to \mathfrak{N}$ is defined by the rule

$$\mathbf{I}(\tilde{K}) \triangleq \tilde{K} \setminus \{\mathrm{pr}_2(z) : z \in \Xi[\tilde{K}]\} \quad \forall \tilde{K} \in \mathfrak{N}. \tag{2.21}$$

So, under $K \in \mathfrak{N}$, the set $\mathbf{I}(K)$ consists of all indexes $i_0 \in K$ for which $i_0 \neq j$ under every $(i,j) \in \Xi[K]$. Of course, the case $K=\overline{1,N}$ is possible. Then $\mathbf{I}(\overline{1,N}) \neq \emptyset$ and $\mathbf{I}(\overline{1,N}) \subset \overline{1,N}$. In addition, $\Xi[\overline{1,N}] = \mathbf{K}$ and

$$\mathbf{I}(\overline{1,N}) = \overline{1,N} \setminus \{\mathrm{pr}_2(z) : \; z \in \mathbf{K}\}.$$

We recall that, for $\alpha \in (\mathrm{bi})[K]$ and $m \in \overline{1,|K|}$, in the form of $\alpha^1(\overline{m,|K|})$, we have the set of all indexes $\alpha(j), \; j \in \overline{m,|K|}$; of course, the set $\mathbf{I}(\alpha^1(\overline{m,|K|}))$ is defined by the rule (2.21), where the stipulation $\tilde{K}=\alpha^1(\overline{m,|K|}) = \{\alpha(j) : \; j \in \overline{m,|K|}\}$ is used. Then, we can introduce admissible (by deletion in sense (2.21)) routes for visiting megacities $M_k, \; k \in K$. Namely, if $K \in \mathfrak{N}$, then

$$(\mathbf{I}-\mathrm{bi})[K] \triangleq \{\alpha \in (\mathrm{bi})[K] | \; \alpha(m) \in \mathbf{I}(\alpha^1(\overline{m,|K|})) \; \forall m \in \overline{1,|K|}\} \tag{2.22}$$

is the set of all admissible routes for visiting $M_k, \; k \in K$. Then [22, part 2]

$$\begin{aligned}\mathbf{A} = (\mathbf{I}-\mathrm{bi})[\overline{1,N}] &= \{\alpha \in (\mathrm{bi})[\overline{1,N}] | \; \alpha(m) \in \mathbf{I}(\alpha^1(\overline{m,N})) \; \forall m \in \overline{1,N}\} = \\ &= \{\alpha \in \mathbb{P} | \; (\alpha(1) \in \mathbf{I}(\overline{1,N}))\&(\alpha(m) \in \mathbf{I}(\alpha^1(\overline{m,N})) \; \forall m \in \overline{2,N})\} = \\ &= \{\alpha \in \mathbb{P} | \; (\alpha(1) \in \mathbf{I}(\overline{1,N}))\&(\alpha(m) \in \mathbf{I}(\alpha^1(\overline{1,N} \setminus \overline{1,m-1})) \; \forall m \in \overline{2,N})\} = \\ &= \{\alpha \in \mathbb{P} | \; (\alpha(1) \in \mathbf{I}(\overline{1,N}))\&(\alpha(m) \in \mathbf{I}(\overline{1,N} \setminus \alpha^1(\overline{1,m-1})) \; \forall m \in \overline{2,N})\}.\end{aligned} \tag{2.23}$$

So, for complete routes, the precedence admissibility and the deletion admissibility are equivalent. As a result, we have the same supply (2.23) of valid routes. Let $\widetilde{\mathbb{X}} \triangleq \mathbb{X} \cup \mathbf{X}$. Then, under $K \in \mathfrak{N}$, by $\widetilde{\mathbb{Z}}_K$, we denote the set of all collections

$$(z_i)_{i \in \overline{0,|K|}} \colon \; \overline{0,|K|} \to \widetilde{\mathbb{X}} \times \widetilde{\mathbb{X}}.$$

If $x \in \mathbf{X}$, $K \in \mathfrak{N}$ and $\alpha \in (\mathrm{bi})[K]$, then

$$\begin{aligned} Z[x;K;\alpha] \triangleq \{(z_i)_{i \in \overline{0,|K|}} \in \widetilde{\mathbb{Z}}_K | \; (z_0 = (x,x)) \text{ and} \\ (z_t \in \mathbb{M}_{\alpha(t)} \; \forall t \in \overline{1,|K|})\} \in \mathrm{Fin}(\widetilde{\mathbb{Z}}_K). \end{aligned} \tag{2.24}$$

Of course, the set (2.24) is defined under $K = \overline{1,N}$. In addition, $\mathbb{P} = (\mathrm{bi})[\overline{1,N}]$ and $\mathfrak{Z} \subset \widetilde{\mathbb{Z}}_{\overline{1,N}}$; therefore, $Z[x;\overline{1,N};\alpha]$ is defined for $x \in X^0$ and $\alpha \in \mathbb{P}$ (of course, $X^0 \subset \mathbf{X}$). And what is more, under $x \in X^0$ and $\alpha \in \mathbb{P}$

$$\mathcal{Z}_\alpha[x] = Z[x;\overline{1,N};\alpha]. \tag{2.25}$$

We use Equations (2.23) and (2.25) in a totality. Now, we consider the natural definition of the Bellman function. For this, we suppose that

$$v(x,\emptyset) \triangleq f(x) \quad \forall x \in \mathbf{X}. \tag{2.26}$$

For definition $v(x,K)$ under $x \in \mathbf{X}$ and $K \in \mathfrak{N}$, we introduce the corresponding value of a partial routing problem. We begin with definition of a criterion. Namely, for $x \in \mathbf{X}$, $K \in \mathfrak{N}$, $\alpha \in (\mathbf{I} - \mathrm{bi})[K]$, and $(z_i)_{i \in \overline{0,|K|}} \in Z[x;K;\alpha]$

$$\begin{aligned}\mathfrak{C}_\alpha[(z_i)_{i\in\overline{0,|K|}}|K] \triangleq \textstyle\sum_{t=1}^{|K|} [\mathbf{c}(\mathrm{pr}_2(z_{t-1}), \mathrm{pr}_1(z_t), \alpha^1(\overline{t,|K|}))+\\ +c_{\alpha(t)}(z_t, \alpha^1(\overline{t,|K|}))] + f(\mathrm{pr}_2(z_{|K|})).\end{aligned} \tag{2.27}$$

If $x \in \mathbf{X}$ and $K \in \mathfrak{N}$, then

$$\begin{aligned}D(x,K) \triangleq \{(\alpha,\mathbf{z}) \in (\mathbf{I} - \mathrm{bi})[K] \times \widetilde{\mathbb{Z}}_K |\, \mathbf{z}\\ \in Z[x;K;\alpha]\} \in \mathrm{Fin}((\mathbf{I} - \mathrm{bi})[K] \times \widetilde{\mathbb{Z}}_K).\end{aligned} \tag{2.28}$$

Then, under $x \in \mathbf{X}$ and $K \in \mathfrak{N}$, the next problem

$$\mathfrak{C}_\alpha[(z_i)_{i\in\overline{0,|K|}}|K] \to \min, \quad (\alpha, (z_i)_{i\in\overline{0,|K|}}) \in D(x,K) \tag{2.29}$$

is defined (see Equations (2.27) and (2.28)). We associate with $x \in \mathbf{X}$ (see Equation (2.8)) and $K \in \mathfrak{N}$ the value

$$v(x,K) \triangleq \min_{(\alpha,\mathbf{z})\in D(x,K)} \mathfrak{C}_\alpha[\mathbf{z}|K] \in \mathbb{R}_+ \tag{2.30}$$

of the problem (2.29). Since $\mathcal{P}(\overline{1,N}) = \mathfrak{N} \cup \{\emptyset\}$, by Equations (2.26) and (2.30), the Bellman function

$$v \in \mathcal{R}_+[\mathbf{X} \times \mathcal{P}(\overline{1,N})] \tag{2.31}$$

is defined. We note the obvious connection (2.30) and the function

$$\tilde{V}[\cdot] \triangleq (\tilde{V}[x])_{x\in X^0} \in \mathcal{R}_+[X^0]. \tag{2.32}$$

Namely, by Equations (2.10), (2.22), (2.25), and (2.28), we have the equality

$$\tilde{D}[x] = D(x, \overline{1,N}) \quad \forall x \in X^0. \tag{2.33}$$

Moreover, with employment of Equations (2.22) and (2.25), we obtain (see Equations (2.13) and (2.27)) that, under $x \in X^0$, $\alpha \in \mathbf{A}$, and $\mathbf{z} \in \mathcal{Z}_\alpha[x]$

$$\mathfrak{C}_\alpha[\mathbf{z}] = \mathfrak{C}_\alpha[\mathbf{z}|\overline{1,N}] \tag{2.34}$$

(we use the obvious equality $|\overline{1,N}| = N$). Then, using Equations (2.30), (2.33), and (2.34), we obtain that

$$\tilde{V}[x] = v(x, \overline{1,N}) \quad \forall x \in X^0. \tag{2.35}$$

So, in fact, $\tilde{V}[\cdot]$ (2.32) is a construction of $v$ (2.31) to $X^0$; more precisely, $\tilde{V}$ is the construction of $v(\cdot, \overline{1,N})$ to $X^0$. We note that under $K \in \mathfrak{N}$, $j \in \mathbf{I}(K)$ and $z \in \mathbb{M}_j$, and we obtain $\mathrm{pr}_2(z) \in \mathbf{M}_j$ and, in particular (see Equation (2.8)), $\mathrm{pr}_2(z) \in \mathbf{X}$; as a result, $v(\mathrm{pr}_2(z), K\backslash\{j\})$ is defined.

**Proposition 1.** If $x \in \mathbf{X}$ and $K \in \mathfrak{N}$, then

$$v(x, K) = \min_{j \in \mathbf{I}(K)} \min_{z \in \mathbb{M}_j} [\mathbf{c}(x, \mathrm{pr}_1(z), K) + c_j(z, K) + v(\mathrm{pr}_2(z), K\backslash\{j\})]. \tag{2.36}$$

**Proof.** By $\omega$, we denote the expression on the right side of Equation (2.36). We omit the simplest variant of our argument corresponding to the case $|K| = 1$ (see [25, p. 66]). So, we suppose that $n \triangleq |K| \geq 2$ . Then $n-1 \in \overline{1, N-1}$. Using Equation (2.30), we choose $(\alpha^0, (z_t^0)_{t \in \overline{0,n}}) \in D(x, K)$ for which

$$\mathfrak{C}_{\alpha^0}[(z_i^0)_{i \in \overline{0,|K|}}|K] = v\,(\mathrm{x}, \mathrm{K})\,. \tag{2.37}$$

Then $\alpha^0 \in (\mathbf{I}-\mathrm{bi})\,[K]$ and

$$(z_t^0)_{t \in \overline{0,n}} \in Z\left[x; \mathrm{K}; \alpha^0\right].$$

By Equation (2.22), we obtain that $\alpha^0 \in (\mathrm{bi})\,[K]$ and, in particular,

$$\alpha^0 \,:\, \overline{1,n} \,\rightarrow K.$$

Then $\alpha^0\,(1) \in K$, $\boldsymbol{K} \triangleq K\backslash\left\{\alpha^0\,(1)\right\} \in \mathfrak{N}$, and $|\boldsymbol{K}| = n-1$. Of course, $\alpha^0\,(1) \in \mathbf{I}(K)$ and $z_1^0 \in \mathbb{M}_{\alpha^0(1)}$. Therefore, by definition of $\omega$

$$\omega \leq \mathbf{c}(x, \mathrm{pr}_1(z_1^0), K) + c_{\alpha^0(1)}(z_1^0, K) + v(\mathrm{pr}_2(z_1^0), \boldsymbol{K}). \tag{2.38}$$

It is easy to verify that

$$\alpha_0 \triangleq (\alpha^0(i+1))_{i\in\overline{1,n-1}} \in (\mathbf{I} - \text{bi})\,[\boldsymbol{K}]\,. \tag{2.39}$$

So, $\alpha_0\,(t) = \alpha^0(t+1)$ and $z^0_{t+1} \in \mathbb{M}_{\alpha_0(t)}$ under $t \in \overline{1, n-1}$ (we use Equation (2.24); indeed, $z^0_j \in \mathbb{M}_{\alpha^0(j)}$ for $j \in \overline{1,n}$). Now, we introduce the procession

$$(z^{\#}_t)_{t\in\overline{0,n-1}} : \overline{0, n-1} \to \widetilde{\mathbb{X}} \times \widetilde{\mathbb{X}}$$

by the following rule

$$\left(z^{\#}_0 \triangleq \left(\text{pr}_2\left(z^0_1\right), \text{pr}_2\left(z^0_1\right)\right)\right) \text{ and } \left(z^{\#}_t \triangleq z^0_{t+1} \;\; \forall t\in\overline{1,n-1}\right). \tag{2.40}$$

Then, it is easily to verify that by Equations (2.24) and (2.40)

$$(z^{\#}_t)_{t\in\overline{0,n-1}} \in Z[\text{pr}_2\left(z^0_1\right); \boldsymbol{K}; \alpha_0] \tag{2.41}$$

(we use the equality $|\boldsymbol{K}| = n-1$). Of course,

$$\left(\alpha_0, \left(z^{\#}_t\right)_{t\in\overline{0,n-1}}\right) \in D\left(\text{pr}_2\left(z^0_1\right), \boldsymbol{K}\right). \tag{2.42}$$

As a corollary, by Equations (2.30) and (2.42), we obtain the inequality

$$v(\text{pr}_2\left(z^0_1\right), \boldsymbol{K}) \leq \mathfrak{C}_{\alpha_0}[(z^{\#}_t)_{i\in\overline{0,\text{n}-1}}|\boldsymbol{K}].$$

Therefore, by Equation (2.27), the following inequality is realized:

$$v\left(\text{pr}_2\left(z^0_1\right), \boldsymbol{K}\right) \leq \textstyle\sum_{t=1}^{n-1} [\mathbf{c}(\text{pr}_2(z^{\#}_{t-1}), \text{pr}_1(z^{\#}_t), \alpha^1_0(\overline{t, n-1})) +$$
$$+ c_{\alpha_0(t)}(z^{\#}_t, \alpha^1_0(\overline{t, n-1}))] + f(\text{pr}_2(z^{\#}_{n-1})).$$

With the employment of Equations (2.39) and (2.40), we obtain that

$$v\left(\text{pr}_2\left(z^0_1\right), \boldsymbol{K}\right) \leq \sum_{t=1}^{n-1} [\mathbf{c}(\text{pr}_2(z^0_t), \text{pr}_1(z^0_{t+1}), (\alpha^0)^1(\overline{t+1, n})) +$$
$$+ c_{\alpha^0(t+1)}(z^0_{t+1}, (\alpha^0)^1(\overline{t+1, n}))] + f(\text{pr}_2(z^0_n)) =$$
$$= \sum_{t=1}^{n-1} \mathbf{c}(\text{pr}_2(z^0_t), \text{pr}_1(z^0_{t+1}), (\alpha^0)^1(\overline{t+1, n})) +$$

$$+\sum_{t=2}^{n} c_{\alpha^0(t)}(z_t^0, (\alpha^0)^1(\overline{t,n})) + f(\mathrm{pr}_2(z_n^0)) =$$

$$=\sum_{t=2}^{n} \mathbf{c}(\mathrm{pr}_2(z_{t-1}^0), \mathrm{pr}_1(z_t^0), (\alpha^0)^1(\overline{t,n}))+$$

$$+\sum_{t=2}^{n} c_{\alpha^0(t)}(z_t^0, (\alpha^0)^1(\overline{t,n})) + f(\mathrm{pr}_2(z_n^0)). \qquad (2.43)$$

From Equations (2.38) and (2.43), the following representation is realized:

$$\begin{aligned}\omega \leq\ & \mathbf{c}(\mathrm{x}, \mathrm{pr}_1(z_1^0), K)) + c_{\alpha^0(1)}(z_1^0, K) \\ & + \textstyle\sum_{t=2}^{n} [\mathbf{c}(\mathrm{pr}_2(z_{t-1}^0), \mathrm{pr}_1(z_t^0), (\alpha^0)^1(\overline{t,n}))+ \\ & + c_{\alpha^0(t)}(z_t^0, (\alpha^0)^1(\overline{t,n}))] + f(\mathrm{pr}_2(z_n^0)).\end{aligned} \qquad (2.44)$$

Recall that by Equation (2.34) and choice of $(z_t^0)_{t\in\overline{0,n}}$, we obtain that $z_0^0 = (x, x)$; moreover, by bijectivity of $\alpha^0$, we have the equality $K = (\alpha^0)^1(\overline{1,n})$. As a corollary, from Equation (2.44), the next inequality follows:

$$\omega \leq \mathbf{c}(\mathrm{pr}_2(z_0^0), \mathrm{pr}_1(z_1^0), (\alpha^0)^1(\overline{1,n})) + c_{\alpha^0(1)}(z_1^0, (\alpha^0)^1(\overline{1,n}))+$$

$$+\sum_{t=2}^{n} \left[\mathbf{c}\left(\mathrm{pr}_2\left(z_{t-1}^0\right), \mathrm{pr}_1\left(z_t^0\right), (\alpha^0)^1\left(\overline{t,n}\right)\right) + c_{\alpha^0(t)}\left(z_t^0, (\alpha^0)^1\left(\overline{t,n}\right)\right)\right]$$

$$+f\left(\mathrm{pr}_2\left(z_n^0\right)\right) =$$

$$=\sum_{t=1}^{n} [\mathbf{c}(\mathrm{pr}_2(z_{t-1}^0), \mathrm{pr}_1(z_t^0), (\alpha^0)^1(\overline{t,n})) + c_{\alpha^0(t)}(z_t^0, (\alpha^0)^1(\overline{t,n}))]+$$

$$+ f\left(\mathrm{pr}_2\left(z_n^0\right)\right). \qquad (2.45)$$

Using Equations (2.27) and (2.45), we obtain that

$$\omega \leq \mathfrak{C}_{\alpha^0}\left[\left(z_{\mathrm{t}}^0\right)_{i\in\overline{0,\mathrm{n}}} \,\middle|\, K\right], \qquad (2.46)$$

where $n = |K|$. From Equations (2.37) and (2.46), the next inequality follows:

$$\omega \leq v\,(x, K)\,. \qquad (2.47)$$

With employment of definition of $\omega$, we choose $q \in \boldsymbol{I}(K)$ and $\boldsymbol{u} \in \mathbb{M}_q$ such that

$$\omega = \mathbf{c}\,(x, \mathrm{pr}_1\,(\boldsymbol{u})\,, K) + c_q\,(\boldsymbol{u}, K) + v\,(\mathrm{pr}_2\,(\boldsymbol{u})\,, Q)\,, \qquad (2.48)$$

where $Q \triangleq K\backslash\{q\}$. Then, $\mathrm{pr}_2(\boldsymbol{u}) \in \mathbf{M}_q$. In particular, $\mathrm{pr}_2(\boldsymbol{u}) \in \mathbf{X}$. Moreover, $|Q| = n-1$. With the employment of Equation (2.28)

$$D(\mathrm{pr}_2(\boldsymbol{u}), Q) = \left\{(\alpha, \boldsymbol{z}) \in (\mathbf{I} - \mathrm{bi})[Q] \times \widetilde{\mathbb{Z}}_Q \;\middle|\; \boldsymbol{z} \in Z[\mathrm{pr}_2(\boldsymbol{u}); Q; \alpha]\right\} \in$$

$$\in \mathrm{Fin}\left((\mathbf{I} - \mathrm{bi})[Q] \times \widetilde{\mathbb{Z}}_Q\right),$$

where $\widetilde{\mathbb{Z}}_Q$ is the set of all mapping

$$(z_i)_{i\in\overline{0,n-1}} : \overline{0,n-1} \to \widetilde{\mathbb{X}} \times \widetilde{\mathbb{X}}$$

(we use the equality $|Q| = n-1$). By Equation (2.30),

$$v(\mathrm{pr}_2(\boldsymbol{u}), Q) = \min_{(\alpha,\mathbf{z})\in D(\mathrm{pr}_2(\boldsymbol{u}),Q)} \mathfrak{C}_\alpha[\mathbf{z} \mid Q] \in \mathbb{R}_+. \qquad (2.49)$$

Using Equation (2.49), we choose OP

$$(\beta_0, (w_i)_{i\in\overline{0,n-1}}) \in D(\mathrm{pr}_2(\boldsymbol{u}), Q)$$

for which

$$v(\mathrm{pr}_2(\boldsymbol{u}), Q) = \mathfrak{C}_{\beta_0}\left[(w_i)_{i\in\overline{0,n-1}} \;\middle|\; Q\right] \in \mathbb{R}_+. \qquad (2.50)$$

Then, $\beta_0 \in (\mathbf{I}-\mathrm{bi})[Q]$ and $(w_i)_{i\in\overline{0,n-1}} \in Z[\mathrm{pr}_2(\boldsymbol{u}); Q; \beta_0]$. In this case, $(w_i)_{i\in\overline{0,n-1}} \in \widetilde{\mathbb{Z}}_Q$ and therefore

$$(w_i)_{i\in\overline{0,n-1}} : \overline{0, n-1} \to \widetilde{\mathbb{X}} \times \widetilde{\mathbb{X}}.$$

Moreover, by Equation (2.24), we obtain that

$$(w_0 = (\mathrm{pr}_2(\boldsymbol{u}), \mathrm{pr}_2(\boldsymbol{u}))) \text{ and } (w_t \in \mathbb{M}_{\beta_0(t)} \;\; \forall t \in \overline{1, n-1}). \qquad (2.51)$$

We recall that $\beta_0 \in (\mathrm{bi})[Q]$ and, in particular,

$$\beta_0 : \overline{1, n-1} \to Q;$$

moreover, for $\beta_0$, the following property takes place:

$$\beta_0(m) \in \mathbf{I}\left((\beta_0)^1\left(\overline{m, n-1}\right)\right) \;\; \forall m \in \overline{1, n-1}. \qquad (2.52)$$

We introduce $\beta^0 : \overline{1,n} \to K$ by the following rule:

$$(\beta^0(1) \triangleq q) \text{ and } \left(\beta^0(t) \triangleq \beta_0(t-1) \;\; \forall t \in \overline{2,n}\right). \qquad (2.53)$$

Then, as it is easy to verify (see Equation (2.22)),

$$\beta^0 \in (\mathbf{I} - \text{bi})\,[K]\,. \tag{2.54}$$

Moreover, we introduce an auxiliary procession

$$(\tilde{w}_t)_{t\in\overline{1,n}} \,:\, \overline{1,n} \to \widetilde{\mathbb{X}} \times \widetilde{\mathbb{X}}$$

by the following rule:

$$\left(\tilde{w}_1 \triangleq \boldsymbol{u}\right) \text{ and } \left(\tilde{w}_t \triangleq w_{t-1} \;\; \forall t \in \overline{2,n}\right). \tag{2.55}$$

By the choice of $\boldsymbol{u}$, we have the property $\boldsymbol{u} \in \mathbb{M}_{\beta^0(1)}$. In addition, from Equations (2.51) and (2.55), we have under $t \in \overline{2,n}$ that $t-1 \in \overline{1,n-1}$ and

$$\tilde{w}_t = w_{t-1} \,\in\, \mathbb{M}_{\beta_0(t-1)}$$

and according to Equation (2.53), $\tilde{w}_t \in \mathbb{M}_{\beta^0(t)}$. As a result, we get the property:

$$\tilde{w}_t \in \mathbb{M}_{\beta^0(t)} \;\; \forall\, t\in\overline{1,n}. \tag{2.56}$$

Recall that $\widetilde{\mathbb{Z}}_K$ is the set of all mappings from $\overline{0,n}$ to $\widetilde{\mathbb{X}} \times \widetilde{\mathbb{X}}$ (we take into account that $n = |K|$). Moreover, according to Equation (2.56), we have, in particular, that $\tilde{w}_t \in \widetilde{\mathbb{X}} \times \widetilde{\mathbb{X}}$ for $t \in \overline{1,n}$. Besides, $(x,x) \in \widetilde{\mathbb{X}} \times \widetilde{\mathbb{X}}$. With this in mind, we assume that

$$(\hat{w}_t)_{t\in\overline{0,n}} \,:\, \overline{0,n} \to \widetilde{\mathbb{X}} \times \widetilde{\mathbb{X}}$$

defined by the following rules:

$$\left(\hat{w}_0 \triangleq (x,x)\right) \text{ and } \left(\hat{w}_t \triangleq \tilde{w}_t \;\; \forall t \in \overline{1,n}\right). \tag{2.57}$$

It is clear that $(\hat{w}_t)_{t\in\overline{0,n}} \in \widetilde{\mathbb{Z}}_K$. In view of Equation (2.24) and (2.56), we obtain

$$(\hat{w}_t)_{t\in\overline{0,n}} \in Z\left[x; K; \beta^0\right]. \tag{2.58}$$

From Equations (2.28), (2.54), and (2.58), it follows that

$$(\beta^0, (\hat{w}_t)_{t\in\overline{0,n}}) \in D(x,K). \tag{2.59}$$

Then, by virtue of Equations (2.30) and (2.59), we have the inequality

$$v\,(x,K) \leq \mathfrak{C}_{\beta^0}\left[(\hat{w}_t)_{t\in\overline{0,n}}\,\middle|\, K\right]. \tag{2.60}$$

This means (see Equation (2.27)) that we have the estimate

$$v(x,K) \leq \sum_{t=1}^{n} \left[ \mathbf{c}\left(\mathrm{pr}_2(\hat{w}_{t-1}), \mathrm{pr}_1(\hat{w}_t), (\beta^0)^1(\overline{t,n})\right) \right.$$
$$\left. + c_{\beta^0(t)}\left(\hat{w}_t, (\beta^0)^1(\overline{t,n})\right)\right] +$$
$$+ f(\mathrm{pr}_2(\hat{w}_n)) = \mathbf{c}(\mathrm{pr}_2(\hat{w}_0), \mathrm{pr}_1(\hat{w}_1), (\beta^0)^1(\overline{1,n}))$$
$$+ c_{\beta^0(1)}(\hat{w}_1, (\beta^0)^1(\overline{1,n})) +$$
$$+ \sum_{t=2}^{n} \left[ \mathbf{c}\left(\mathrm{pr}_2(\hat{w}_{t-1}), \mathrm{pr}_1(\hat{w}_t), (\beta^0)^1(\overline{t,n})\right) \right.$$
$$\left. + c_{\beta^0(t)}\left(\hat{w}_t, (\beta^0)^1(\overline{t,n})\right)\right] +$$
$$+ f(\mathrm{pr}_2(\hat{w}_n)) = \mathbf{c}(x, \mathrm{pr}_1(\boldsymbol{u}), K) + c_q(\boldsymbol{u}, K) +$$
$$+ \sum_{t=2}^{n} [\mathbf{c}(\mathrm{pr}_2(\hat{w}_{t-1}), \mathrm{pr}_1(\hat{w}_t), (\beta^0)^1(\overline{t,n}))$$
$$+ c_{\beta^0(t)}(\hat{w}_t, (\beta^0)^1(\overline{t,n}))] + f(\mathrm{pr}_2(\hat{w}_n)), \tag{2.61}$$

where equalities $\hat{w}_0 = (x,x)$, $\hat{w}_1 = \boldsymbol{u}$, and $(\beta^0)^1(\overline{1,n}) = K$ are taken into account (see Equations (2.54), (2.55), and (2.57)). Moreover, according to Equation (2.53), for $t \in \overline{2,n}$

$$(\beta^0)^1(\overline{t,n}) = \{\beta^0(1) : 1 \in \overline{t,n}\} = \{\beta_0(l-1) : l \in \overline{t,n}\} =$$
$$= \{\beta_0(k) : k \in \overline{t-1, n-1}\} = (\beta_0)^1(\overline{t-1,n-1}). \tag{2.62}$$

Further, for $t \in \overline{2,n}$, by virtue of Equations (2.55) and (2.57), we have the equality

$$\hat{w}_t = w_{t-1}.$$

At the same time, $\hat{w}_1 = \tilde{w}_1 = \boldsymbol{u}$ and therefore for $t = 2$

$$\mathrm{pr}_2(\hat{w}_{t-1}) = \mathrm{pr}_2(\hat{w}_1) = \mathrm{pr}_2(\boldsymbol{u}) = \mathrm{pr}_2(w_0)$$

(see Equation (2.51)); as a result, we obtain for $t = 2$

$$\mathbf{c}\left(\mathrm{pr}_2(\hat{w}_{t-1}), \mathrm{pr}_1(\hat{w}_t), (\beta^0)^1(\overline{t,n})\right) =$$
$$= \mathbf{c}\left(\mathrm{pr}_2(w_{t-2}), \mathrm{pr}_1(w_{t-1}), (\beta_0)^1(\overline{t-1,n-1})\right). \tag{2.63}$$

If $t \in \overline{2,n}$ and $t \neq 2$, then $t-1 \in \overline{2,n-1}$ and $\hat{w}_{t-1} = \tilde{w}_{t-1} = w_{t-2}$; in addition, $\hat{w}_t = \tilde{w}_t = w_{t-1}$. As a result

$$\mathbf{c}\left(\mathrm{pr}_2\left(\hat{w}_{t-1}\right), \mathrm{pr}_1\left(\hat{w}_t\right), \left(\beta^0\right)^1\left(\overline{t,n}\right)\right) =$$

$$= \mathbf{c}\left(\mathrm{pr}_2\left(w_{t-2}\right), \mathrm{pr}_1\left(w_{t-1}\right), \left(\beta_0\right)^1\left(\overline{t-1,n-1}\right)\right), \tag{2.64}$$

$$c_{\beta^0(t)}\left(\hat{w}_t, \left(\beta^0\right)^1\left(\overline{t,n}\right)\right) = c_{\beta_0(t-1)}\left(w_{t-1}, \left(\beta_0\right)^1\left(\overline{t-1,n-1}\right)\right). \tag{2.65}$$

Then, by Equations (2.61) and (2.63)–(2.65), we obtain that

$$v(x,K) \leq \mathbf{c}(x, \mathrm{pr}_1(\boldsymbol{u}), K) + c_q(\boldsymbol{u}, K) +$$

$$+ \sum_{t=2}^{n} [\mathbf{c}\left(\mathrm{pr}_2\left(w_{t-2}\right), \mathrm{pr}_1\left(w_{t-1}\right), \left(\beta_0\right)^1\left(\overline{t-1,n-1}\right)\right) +$$

$$+ c_{\beta_0(t-1)}\left(w_{t-1}, \left(\beta_0\right)^1\left(\overline{t-1,n-1}\right)\right)] + f\left(\mathrm{pr}_2\left(w_{n-1}\right)\right) =$$

$$= \mathbf{c}\left(x, \mathrm{pr}_1\left(\boldsymbol{u}\right), K\right) + c_q\left(\boldsymbol{u}, K\right)$$

$$+ \sum_{\tau=1}^{n-1} [\mathbf{c}\left(\mathrm{pr}_2\left(w_{\tau-1}\right), \mathrm{pr}_1\left(w_\tau\right), \left(\beta_0\right)^1\left(\overline{\tau,n-1}\right)\right) +$$

$$+ c_{\beta_0(\tau)}\left(w_\tau, \left(\beta_0\right)^1\left(\overline{\tau,n-1}\right)\right)] + f\left(\mathrm{pr}_2\left(w_{n-1}\right)\right). \tag{2.66}$$

However, according to Equations (2.27) and (2.49), the equality

$$\mathfrak{C}_{\beta_0}[\left(w_i\right)_{i\in\overline{0,n-1}} \,\Big|\, Q] = \sum_{t=1}^{n-1} [\mathbf{c}\left(\mathrm{pr}_2\left(w_{t-1}\right), \mathrm{pr}_1\left(w_t\right), \left(\beta_0\right)^1\left(\overline{t,n-1}\right)\right) +$$

$$+ c_{\beta_0(t)}\left(w_t, \left(\beta_0\right)^1\left(\overline{t,n-1}\right)\right)] + f\left(\mathrm{pr}_2\left(w_{n-1}\right)\right)$$

holds, and therefore from (2.66) follows the inequality

$$v\left(x,K\right) \leq \mathbf{c}\left(x, \mathrm{pr}_1\left(\boldsymbol{u}\right), K\right) + c_q\left(\boldsymbol{u}, K\right) + \mathfrak{C}_{\beta_0}[(w_i)_{i\in\overline{0,n-1}} \mid Q].$$

In view of Equation (2.50), we obtain the following estimate:

$$v\left(x,K\right) \leq \mathbf{c}\left(x, \mathrm{pr}_1\left(\boldsymbol{u}\right), K\right) + c_q\left(\boldsymbol{u}, K\right) + v(\mathrm{pr}_2\left(\boldsymbol{u}\right), Q). \tag{2.67}$$

Then, it follows from Equations (2.48) and (2.67) that

$$v(x, K) \leq \omega.$$

The last inequality, taking into account Equation (2.47), means the validity of the required equality

$$v(x, K) = \omega.$$

So, Equation (2.36) is established (under $|K| \geq 2$).

From Proposition 1, the next equality follows: under $x \in X^0$,

$$\tilde{V}[x] = \min_{j \in \mathbf{I}(\overline{1,N})} \min_{z \in \mathbb{M}_j} \left[ \mathbf{c}\left(x, \mathrm{pr}_1(z), \overline{1,N}\right) \right.$$
$$\left. + c_j\left(z, \overline{1,N}\right) + v\left(\mathrm{pr}_2(z), \overline{1,N} \setminus \{j\}\right)\right]. \qquad (2.68)$$

Of course, the construction of all function $v$ is connected with serious difficulties of the computing nature. But, rational employment of the precedence conditions supply some new possibilities in this direction (we keep in mind the lowering of the computing complexity). These possibilities are considered in the next section.

## 2.5 Dynamic Programming, 2

In our section, we consider the approach of [20–23] connected with construction and employment of the Bellman function layers. For this, at first, we introduce the set

$$\mathcal{G} \triangleq \{K \in \mathfrak{N} | \ \forall z \in \mathbf{K} \ \ (\mathrm{pr}_1(z) \in K) \Rightarrow (\mathrm{pr}_2(z) \in K)\}. \qquad (2.69)$$

Moreover, let $\mathcal{G}_s \triangleq \{K \in \mathcal{G} | \ \ s = |K|\} \quad \forall s \in \overline{1,N}$. Then, the family

$$\{\mathcal{G}_1; \ldots; \mathcal{G}_N\} \qquad (2.70)$$

is a partition of $\mathcal{G}$ (2.69). In addition, $\mathcal{G}_N = \{\overline{1,N}\}$ and

$$\mathcal{G}_1 = \{\{t\}: \ t \in \overline{1,N} \setminus \mathbf{K}_1\},$$

where $\mathbf{K}_1 = \{\mathrm{pr}_1(z) : \ z \in \mathbf{K}\}$. Moreover, by [24, (4.6)],

$$\mathcal{G}_{s-1} = \{K \setminus \{t\}: K \in \mathcal{G}_s, \ \ t \in \mathbf{I}(K)\} \ \forall s \in \overline{2,N}. \qquad (2.71)$$

So, by Equation (2.71), the recurrent procedure is defined: $\mathcal{G}_N$ is known and every transformation for families (2.70) is realized by Equation (2.71).

**The layers of the position space.** We consider $(x, K)$, where $x \in X$ and $K \in \mathfrak{N}$, as a position. In the position space, we construct the sets $D_0, D_1, ..., D_N$. In addition, $D_0$ and $D_N$ are defined very simply. Namely, $D_0 \triangleq \{(x, \emptyset) : x \in \widetilde{\mathcal{M}}\}$ under

$$\widetilde{\mathcal{M}} \triangleq \bigcup_{i \in \overline{1,N} \setminus \mathbf{K}_1} \mathbf{M}_i.$$

Moreover, $D_N \triangleq \{(x, \overline{1,N}) : x \in X^0\}$. For the construction of intermediate layers, the special procedure is used. Namely, under $s \in \overline{1,N-1}$, we introduce sequentially for $K \in \mathcal{G}_s$ that

$$J_s(K) \triangleq \{j \in \overline{1,N} \setminus K | \ \{j\} \cup K \in \mathcal{G}_{s+1}\},$$

$$\mathcal{M}_s[K] \triangleq \bigcup_{j \in J_s(K)} \mathbf{M}_j;$$

$$\mathbb{D}_s[K] \triangleq \{(x, K) : x \in \mathcal{M}_s[K]\},$$

and we consider the last set as a cell of the position space. Next, we suppose that under $s \in \overline{1,N-1}$

$$D_s \triangleq \bigcup_{K \in \mathcal{G}_s} \mathbb{D}_s[K]. \tag{2.72}$$

So, $D_i$ has a cell-like structure ($i = 1, \ldots, N$). As a result, we obtain

$$D_0 \neq \emptyset, \ D_1 \neq \emptyset, ..., D_N \neq \emptyset, \tag{2.73}$$

where $D_j \subset \mathbf{X} \times \mathcal{P}(\overline{1,N}) \quad \forall j \in \overline{0, N}$. We note that (see [23, Section 4])

$$(\mathrm{pr}_2(z), K \setminus \{j\}) \in D_{s-1} \ \forall s \in \overline{1,N} \ \ \forall (x, K) \in D_s \ \forall j \in \mathbf{I}(K) \ \ \forall z \in \mathbb{M}_j. \tag{2.74}$$

**The layers of Bellman function.** Now, we use Equations (2.31) and (2.72). So, for $l \in \overline{0, N}$, we introduce $v_l \in \mathcal{R}_+[D_l]$ by the following rule:

$$v_l(x, K) \triangleq v(x, K) \ \ \forall (x, K) \in D_l. \tag{2.75}$$

Then, $v_0 \in \mathcal{R}_+[D_0]$ is defined by the values $v(x, \emptyset), \ x \in \widetilde{\mathcal{M}}$, where $\widetilde{\mathcal{M}} \subset \mathbf{X}$. Then, by Equation (2.26)

$$v_0(x, \emptyset) = f(x) \ \ \forall x \in \widetilde{\mathcal{M}}. \tag{2.76}$$

With employment of Equation (2.74), we obtain the following property: if $s \in \overline{1,N}$, $(x, K) \in D_s$, $j \in \mathbf{I}(K)$, and $z \in \mathbb{M}_j$, then the value $v_{s-1}(\mathrm{pr}_2(z), K \setminus \{j\}) \in \mathbb{R}_+$ is defined correctly. In addition, by Proposition 1, we have the next property: if $s \in \overline{1,N}$, then transformation $v_{s-1} \to v_s$ is defined by the rule

$$\begin{aligned} v_s(x, K) = \min_{j \in \mathbf{I}(K)} \min_{z \in \mathbb{M}_j} [&\mathbf{c}(x, \mathrm{pr}_1(z), K) + c_j(z, K) \\ &+ v_{s-1}(\mathrm{pr}_2(z), K \setminus \{j\})] \qquad \forall (x, K) \in D_s. \end{aligned} \tag{2.77}$$

So, we obtain the following recurrent procedure:

$$v_0 \to v_1 \to \cdots \to v_N, \tag{2.78}$$

where $v_0$ is defined by Equation (2.76) and, for $s \in \overline{1,N}$, the transformation $v_{s-1}$ to $v_s$ is realized by Equation (2.77).

We note that by Equations (2.45) and (2.75),

$$\tilde{V}[x] = v_N(x, \overline{1,N}) \quad \forall x \in X^0 \tag{2.79}$$

(we use the above-mentioned representation of $D_N$). From Equations (2.77)–(2.79), the important property of our DP procedure follows: the procedure (2.77)–(2.79) is universal with respect to $x \in X^0$. Namely, dependence on the initial state from $X^0$ arises only at the last stage of Equation (2.78). This dependence is defined by Equation (2.79).

Now, we note two possible variants of employment for procedure (2.78).

1) Algorithm for determination of the global extremum. In this part, we consider an analog of the scheme of [24–26]. Namely, we use the variant of Equation (2.78) with the layers overwriting. For this variant, in the computer memory, at every stage of the procedure, only one of the functions from Equation (2.78) is situated. This permits to economize the memory resources. Namely, the function $v_0$ is defined by Equation (2.76). In addition, in fact, $v_0$ is required "part" of the terminal function $f$. In connection with employment (2.77), we note the following possibility. If $s \in \overline{1,N-1}$ and the function $v_{s-1}$ is already constructed, then we determine $v_s$ by Equation (2.77). Then, $v_{s-1}$ is replaced by $v_s$; the function $v_{s-1}$ is destroyed. These processes of realization of the Bellman function layers continue until exhaustion of the index set $\overline{1,N}$. As a result, we obtain the function $v_N \in \mathcal{R}_+[D_N]$. Therefore, we obtain the values

$$\tilde{V}[x] = v_N(x, \overline{1,N}),$$

where $x \in X^0$ (we recall that $(x, \overline{1,N}) \in D_N$ under $x \in X^0$ and the values $v_N(x, \overline{1,N}) \in \mathbb{R}_+$ are already defined). Now, for the determination of $V$, we use Equation (2.18). The obtained value $V$ can be used for justified prediction of the process quality and for possible heuristics testing.

2) Algorithm for construction of the optimal solution. In this part, we consider the procedure of construction of some solution from the set (2.15). Namely, we suppose that all functions (2.78) are already constructed. In addition, the procedure of finding of the functions (2.78) is constructed on the basis of Equation (2.77). In this part, we suppose that all functions (2.78) are saved in the computer memory. So, $v_0, v_1, ..., v_N$ are known. Then, by Equation (2.79), we have all values $\tilde{V}[x], \quad x \in X^0$. Then, we solve the problem

$$\tilde{V}[x] \to \min, \quad x \in X^0. \tag{2.80}$$

So, we find $x^0 \in X^0$ for which $\tilde{V}[x^0] = V$ (see Equation (2.18)). Now, we consider the problem (2.16) under $x = x^0$. In addition, we construct an element of $(\mathrm{sol})[x^0]$ (see Equation (2.17)) by the DP procedure.

Namely, suppose that $\mathbf{z}^{(0)} \triangleq (x^0, x^0)$. Then, by Equations (2.77) and (2.79)

$$\begin{aligned} V = \tilde{V}[x^0] = v_N(x^0, \overline{1,N}) = \min_{j \in \mathbf{I}(\overline{1,N})} \min_{z \in \mathbb{M}_j} [\mathbf{c}(x^0, \mathrm{pr}_1(z), \overline{1,N}) + \\ + c_j(z, \overline{1,N}) + v_{N-1}(\mathrm{pr}_2(z), \overline{1,N} \setminus \{j\})]. \end{aligned} \tag{2.81}$$

In connection with Equation (2.81), we solve the problem

$$\begin{aligned} \mathbf{c}(x^0, \mathrm{pr}_1(z), \overline{1,N}) + c_j(z, \overline{1,N}) + v_{N-1}(\mathrm{pr}_2(z), \overline{1,N} \setminus \{j\}) \to \min, \\ j \in \mathbf{I}(\overline{1,N}), \quad z \in \mathbb{M}_j. \end{aligned} \tag{2.82}$$

We find $\eta_1 \in \mathbf{I}(\overline{1,N})$ and $\mathbf{z}^{(1)} \in \mathbb{M}_{\eta_1}$ as a solution for the problem (2.82). So

$$\mathbf{c}(x^0, \mathrm{pr}_1(\mathbf{z}^{(1)}), \overline{1,N}) + c_{\eta_1}(\mathbf{z}^{(1)}, \overline{1,N}) + v_{N-1}(\mathrm{pr}_2(\mathbf{z}^{(1)}), \overline{1,N} \setminus \{\eta_1\}) = V \tag{2.83}$$

(we use Equation (2.81)). In addition, by Equation (2.74),

$$(\mathrm{pr}_2(\mathbf{z}^{(1)}), \overline{1,N} \setminus \{\eta_1\}) \in D_{N-1} \tag{2.84}$$

(in Equation (2.84), we take into account that $(x^0, \overline{1,N}) \in D_N$). Therefore, by Equations (2.77) and (2.84),

$$\begin{aligned} & v_{N-1}(\mathrm{pr}_2(\mathbf{z}^{(1)}), \overline{1,N} \setminus \{\eta_1\}) = \\ = & \min_{j \in \mathbf{I}(\overline{1,N} \setminus \{\eta_1\})} \min_{z \in \mathbb{M}_j} [\mathbf{c}(\mathrm{pr}_2(\mathbf{z}^{(1)}), \mathrm{pr}_1(z), \overline{1,N} \setminus \{\eta_1\}) + \\ & + c_j(z, \overline{1,N} \setminus \{\eta_1\}) + v_{N-2}(\mathrm{pr}_2(z), \overline{1,N} \setminus \{\eta_1; j\})]. \end{aligned} \tag{2.85}$$

In connection with Equation (2.85), we consider the following problem:

$$\mathbf{c}(\mathrm{pr}_2(\mathbf{z}^{(1)}), \mathrm{pr}_1(z), \overline{1,N} \setminus \{\eta_1\}) + c_j(z, \overline{1,N} \setminus \{\eta_1\}) + \\ + v_{N-2}(\mathrm{pr}_2(z), \overline{1,N} \setminus \{\eta_1; j\}) \to \min, \quad j \in \mathbf{I}(\overline{1,N} \setminus \{\eta_1\}), \ z \in \mathbb{M}_j. \tag{2.86}$$

We find $\eta_2 \in \mathbf{I}(\overline{1,N} \setminus \{\eta_1\})$ and $\mathbf{z}^{(2)} \in \mathbb{M}_{\eta_2}$ as a solution for the problem (2.86):

$$v_{N-1}(\mathrm{pr}_2(\mathbf{z}^{(1)}), \overline{1,N} \setminus \{\eta_1\}) = \mathbf{c}(\mathrm{pr}_2(\mathbf{z}^{(1)}), \mathrm{pr}_1(\mathbf{z}^{(2)}), \overline{1,N} \setminus \{\eta_1\}) + \\ + c_{\eta_2}(\mathbf{z}^{(2)}, \overline{1,N} \setminus \{\eta_1\}) + v_{N-2}(\mathrm{pr}_2(\mathbf{z}^{(2)}), \overline{1,N} \setminus \{\eta_1; \eta_2\}). \tag{2.87}$$

Of course, the next inclusion takes place:

$$(\mathrm{pr}_2(\mathbf{z}^{(2)}), \overline{1,N} \setminus \{\eta_1; \eta_2\}) \in D_{N-2} \tag{2.88}$$

(in connection with Equation (2.88), we recall Equations (2.74) and (2.54)). From Equations (2.83) and (2.87), we obtain that

$$V = \mathbf{c}(x^0, \mathrm{pr}_1(\mathbf{z}^{(1)}), \overline{1,N}) + \mathbf{c}(\mathrm{pr}_2(\mathbf{z}^{(1)}), \mathrm{pr}_1(\mathbf{z}^{(2)}), \overline{1,N} \setminus \{\eta_1\}) + \\ + c_{\eta_1}(\mathbf{z}^{(1)}, \overline{1,N}) + c_{\eta_2}(\mathbf{z}^{(2)}, \overline{1,N} \setminus \{\eta_1\}) + v_{N-2}(\mathrm{pr}_2(\mathbf{z}^{(2)}), \overline{1,N} \setminus \{\eta_1; \eta_2\}). \tag{2.89}$$

**Remark 2.1.** If $N = 2$, then by Equation (2.89), the pair $((\eta_i)_{i \in \overline{1,2}}, (\mathbf{z}^{(i)})_{i \in \overline{0,2}})$ is an element of $(\mathrm{sol})[x^0]$ (the corresponding proof is very simple).

Returning to general case of $N$, $N \geq 2$, we note that the procedures similar to Equations (2.82) and (2.86) must continue until exhaustion of the index set $\overline{1,N}$. By these procedures, an admissible solution $(\eta, (\mathbf{z}^{(j)})_{j \in \overline{0,N}}) \in \widetilde{\mathbf{D}}[x^0]$,

$$\eta = (\eta_j)_{j \in \overline{1,N}} \in \mathbf{A},$$

with $\mathfrak{C}_\eta[(\mathbf{z}^{(j)})_{j \in \overline{0,N}}] = \tilde{V}[x^0]$ will be constructed. Of course, by Equation (2.17),

$$(\eta, (\mathbf{z}^{(j)})_{j \in \overline{0,N}}) \in (\mathrm{sol})[x^0]$$

and by the choice of $x^0$ $\mathfrak{C}_{\eta/}[(\mathbf{z}^{(j)})_{j \in \overline{0,N}}] = V$. Therefore, we obtain that

$$(\eta, (\mathbf{z}^{(j)})_{j \in \overline{0,N}}, x^0) \in \mathbf{SOL}.$$

So, optimal solution of the problem (2.14) is constructed.

## 2.6 Computational Experiment

This section is dedicated to calculation and computer realization of the theoretical constructions, which were considered in previous sections of current article. Now, we consider simplified interpretation of the practical problem about optimization of the sequence of dismantling acts of the radiation sources, for example, with respect to a decontamination of nuclear power station after accident. We will use the model in which each radiation source is situated in some room or building with the set of doors that are used for input or output purposes; the sets of inputs and outputs are associated with the above-mentioned megacities. In other words, sets of inputs/outputs correspond to radiation sources that must be dismantled. The decontamination machine is started from the basic (starting) point and moves to input of the first (with respect to sequence of visiting) megacity; after that, it moves to the point, which is situated near radiation source and which we want to turn off and implement work associated with dismantling of this radiation source (this process takes some time). The distance from point, which is used for dismantling work, up to radiation source corresponds to the ability of manipulator to turn off this radiation source. After that, the machine moves to the exit (output point) for the corresponding megacity. From this point, the dismantling device moves to the input point of the next (in the order of visiting) megacity and the process of turning off the corresponding radiation source and leaving the room (building) after that is realized. As described above, the process of visiting the megacities and dismantling the radiation sources are repeated as long as the set of nondismantled radiation sources is not exhausted. After that, it is supposed that process of dismantling is finished. In our model, we use an additive criterion of cost aggregation. In this problem, the main measure of costs is a dose of radiation. Momentary radiation dose is inversely proportional of the square of distance up to the active radiation source; we must summarize the doses of radiation from each active (nondismantled) radiation source. The above-mentioned process of moving and dismantling works is estimated by summary radiation dose. We integrate radiation doses from each active radiation source at each moment of time during the moving and dismantling procedure along the trajectory of the decontamination machine. We will consider the following two types of movements.

I) External displacements: displacements from the base point (parking point of the dismantling machine) to the entrance of the first (in the order of visiting) megacity or displacement from the exit point of a megacity after dismantling the corresponding radiation source to the entrance of the next (in the order of visiting) megacity.

II) Internal movements: move from the entrance of the megacity to the disassembly point next to the corresponding radiation source (we consider a uniform grid on a circle centered at the radiation source; each point of this grid corresponds to each entry point of the megacity) and indicate the exit point of the megacity after disabling the disassembly radiation source. Recall that the dismantling process that occurs at the dismantling point takes some time; this time, the durations, and radii of circles around the sources are parameters of our problem.

We want to minimize the summary dose of radiation (in other words, harm to health or equipment) by selection of a route from the set (2.6) and a trace from the set (2.9) of visiting megacities and by choosing the basic point of motion from the set $X^0$.

We define some parameters of the above-mentioned problem.

1) Let $\Gamma_s, \quad s \in \overline{1,N}$, be an intensity of radiation of source with index $s$.
2) In addition, let $v_i, \ i \in \overline{1,N}$, be the speed of the motion in the megacity $i$ and $v_{i \to j}, \quad i \in \overline{0,N}, \ j \in \overline{1,N}$, be the speed of motion from basic point to first megacity (if $i=0$) in the order of motion or speed of motion from megacity $i$ to megacity $j$ (we consider only motion with constant speed).
3) Also, let $t$ be a current time and $T_{i,j}, \ i \in \overline{0,N}, j \in \overline{1,N}$, be the duration of moving from initial point or megacity $i$ to megacity $j$ (if $i=j$, then we consider internal moving in megacity $i$).
4) In addition, let $r$ be a current distance (for measuring a distance, we use Euclidean metric) and let indexes at $r$ define indexes of objects, between which we measure distance; $\rho_s(t)$ is a distance from radiation source $s, \ s \in \overline{1,N}$, up to the position of dismantling the machine or executor at time moment $t$ (we consider only constant speeds and keep in mind only cases I) or II).
5) Let $\Delta t_i, \ i \in \overline{1,N}$, be time durations which need to decontaminate the machine or executor for complete dismantling of the radiation source with index $i$ and $\overline{r}_i, \ i \in \overline{1,N}$, be radius of circle of a dismantling position around radiation source (see II).

We consider a model in which two principles of calculation of momentary dose of radiation are used:

a) Radiation dose from the dismantling radiation source with index $s$ in the distance $r$ to this source when the executor or machine moves from the input point of megacities to the dismantling point at the corresponding radiation source. In this case, at each time moment, radiation impact from this source is equal to $2\Gamma_s/r^2$.
b) Radiation dose from the radiation source with index $s$ in the distance $r$ to this source when moving is not related to moving to dismantling point that corresponds source with index $s$. The value of radiation dose at each moment of time in this situation is equal to $\Gamma_s/r^2$. This principle of dose calculation takes place at each moment of time, when executor stands up at the dismantling point near the dismantling radiation source with index $s$: the value of radiation dose is equal to $\Gamma_s/\overline{r}_s^2$, where $\overline{r}_s$ is explained below.

In addition, the radiation dose, which the executor or dismantling machine gets during the dismantling process when it is standing at the radiation source with index $s$ in the distance $\overline{r}_s$ to this source during time $\triangle t_s$ (this time is needed for realization of dismantling process), is equal to $\triangle t_s\Gamma_s/\overline{r}_s^2$.

The theoretical constructions that were considered in the previous parts of this article are realized in the form of program for a personal computer (we use the above-mentioned algorithmic approach with respect to the practical problem of decreasing the radiation dose during dismantling jobs on the radiation hazardous objects). This program works under Windows 64-bit operation system (version Windows 7 or older) in multithread mode (calculation part is executed in a thread separate from user interface). It is possible to represent (in planar case) results of calculation in the graphical form and save it to the file of graphical type bmp; the scale of areas of the picture can be increased. Source data and results of calculation are saved in the text file with defined structure. We use this program for calculation experiment, which is performed on a personal computer with CPU Intel Core i7, size of RAM 64 GB, and installed operational system Windows 7 Ultimate SP1.

We consider one example of solving a testing problem which is used for the demonstration of the above-mentioned program working. Let 31 finite planar sets be given ($N = 31$), which are uniform grids on rectangles and circles, and let set $X^0$ consist of 5 elements:

$$X^0=\{(-20,90);(0,0);(90,-115);(-90,-100);(-90,90);(30,-30)\}.$$

Suppose that set **K** consists of 34 elements. Let speeds of moving between megacities (case I) is more than 4 times the speeds of moving inside megacities (case II). Let values of intensities of radiation of sources be in an interval between 1.3 and 5.5; distances for dismantling of radiation sources are more than 1.0 and less than 1.6.

We get the following results: optimal value of radiation dose is 182.38, selected basic point is (−90, −100), and the duration of calculation is 11 hour 45 minutes 7 seconds. A graphical view of route and trace can be seen in Figure 2.1.

From Figure 2.2, a fragment of the executor trajectory can be seen: we can see two radiation sources A and B, and the executor moves to the input point A' of megacity corresponding to source A (this is external moving) and after that it moves to the dismantling point A* near radiation source A (this is internal moving) and stands in this position for some time, which is needed for the realization of the dismantling process. After finishing the dismantling process, the executor moves to the output point of the above-mentioned megacity A" (this is internal moving too) and moves to the next megacity with the corresponding radiation source B (it moves to the input point B'; this is external moving). After that, it realizes actions that are analogous to the above-mentioned operations for the radiation source A (moves to the dismantling point B*, stands at this point, and moves to the output point B").

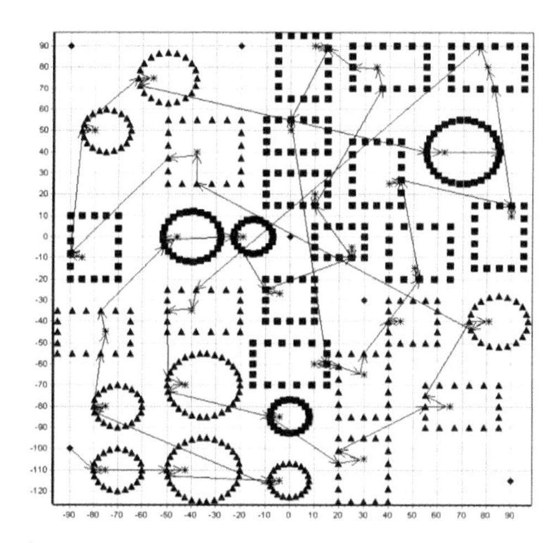


**Figure 2.1** Route and trace of moving the dismantling machine.

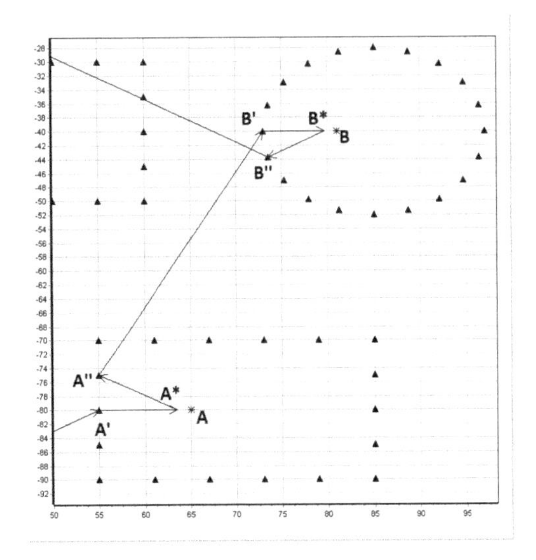


**Figure 2.2** Fragment of route and trace of moving of dismantling machine.

## 2.7 Conclusion

In the given investigation, the very complicated setting with elements of routing is considered. This setting is motivated by an actual engineering problem connected with nuclear power. Taking into account the features of the engineering problem, our statement leads to the emergence of cost functions depending on the list of tasks. In addition, in problems of this kind, precedence conditions naturally arise, which imposes restrictions on the choice of the method of numbering the sources to be visited for dismantling. Each source has a near zone with high radiation, the boundary of which is sampled. As a result of this, a megacity (nonempty finite set) is formed. It is also possible that some sources are in association with an input–output system, the totality of which (in this special case) is a megacity. In all cases, work near the source requires a separate description, which is achieved by introducing the cost functions of internal works. The terminal component of the criterion in the simplest case (but not always) can vanish. However, we limited ourselves to consideration the case when the mentioned terminal component is independent of the starting point. Under the mentioned conditions, the paper indicates the procedure of a universal (with respect to the choice of a starting point) DP, delivering a global extremum of the problem under it consideration.

We note that the considered constructions have many other applications. Now, we restrict oneself to indicate only one important direction of engineering applications connected with sheet cutting (see [27–35]). In these problems, precedence conditions are connected (in particular) with necessity of earlier cutting of internal detail contours in comparison with cutting of external contour. The task list dependence (for cost functions) can arise by reasons penalty under violation of different constraints of dynamic nature. The proposed method can be used as a solution for such types of problems (in this connection, see [27–35]).

## 2.8 Acknowledgment

The research was supported by Russian Foundation for Basic Research under project no. 20-08-00873 and by Act 211 Government of the Russian Federation under contract 02.A03.21.0006.

## References

[1] O.L. Tashlykov. Dose costs of personnel in the nuclear power industry. Analysis. By reducing. Optimization. LAP Lambert. [in Russian], 2011.

[2] O.L. Tashlykov. Organization and technology of nuclear energetics. Ekaterinburg: GOU VPO UGTU-UPI, [in Russian], 2005.

[3] V.V. Korobkin, A.N. Sesekin, O.L. Tashlykov, A.G. Chentsov. Methods of routing and their appendix in problems of increase of efficiency and safety of operation of nuclear power plants. Under general editorship of I.A. Kaliaev. Moscow: Novye Tekhnologii, [in Russian], 2012.

[4] B. Werners, Y. Kondratenko, Alternative Fuzzy Approaches for Efficiently Solving the Capacitated Vehicle Routing Problem in Conditions of Uncertain Demands. In: Complex Systems: Solutions and Challenges in Economics, Management and Engineering, Christian Berger-Vachon et al. (Eds.), Studies in Systems, Decision and Control, Vol. 125, Berlin, Heidelberg: Springer International Publishing, pp. 521-543, 2018.

[5] M. Gendrean, G. Ghiani, E. Guerriero. Time-dependent routing problem: A review. Computers & Operations Research, vol. 64. pp. 189-197, 2015.

[6] G. Laporte. The vehicle routing problem: an overview of exact and approximate algorithms. Eur. J. Oper. Res. **59**(3), pp. 345–358, 1992.

[7] R. Bellman. Dynamic programming treatment of the travelling salesman problem, J. ACM, vol. 9, issue 1, pp. 61–63, 1962.
[8] M. Held, R.M. Karp. A dynamic programming approach to sequencing problems, Journal of the Society for Industrial and Applied Mathematics, vol. 10, issue 1, pp. 196–210, 1962.
[9] L.D.C. Little, K.G. Murty, D.W. Sweeney, C. Karel. An algorithm for the traveling salesman problem, Operations Research, vol. 11, issue 6, pp. 972–989, 1963.
[10] G. Gutin, A.P. Punnen. The traveling salesman problem and its variations, Boston: Springer US, 2007.
[11] W.J. Cook. In pursuit of the traveling salesman. Mathematics at the limits of computation, Princeton, New Jersey: Princeton University Press, 248 p., 2012.
[12] E.Kh. Gimadi, M.Yu. Khachai. Extremal problems on sets of permutations, Yekaterinburg: UMC UPI, 220 p. [in Russian], 2016.
[13] Z.H. Ahmed, S.N. Narahari Pandit. The Travelling Salesman Problem with Precedence Constraints OPSEARCH, Vol. 38, Ño 3, P. 299, 2001.
[14] L. Bianco, S. Mingozzi, S. Ricciardelli, M. Spadoni. Exact and Heuristic Procedures for the Travelling Salesman Problem with Precedence Constraints, Based on Dynamic Programming. Information Systems and Operational Research, 32, pp. 19–32, 1994.
[15] F. Lokin. Procedures for travelling salesman problems with additional constraints. European Journal of Operational Research. Vol.3, no. 2. pp. 135–141, 1979.
[16] K. Castelino, R. D'Souza, P.K. Wright. Toolpath optimization for minimizaing airtime during machining. Journal of Manufacturing Systems. Vol. 22, pp. 173-180, 2003.
[17] K. Kuratowski, A. Mostowski. Set theory, Amsterdam: North-Holland, xi+417 p., 1967.
[18] J. Dieudonne. Foundations of modern analysis, New York: Academic Press, 361 p., 1960
[19] T. Cormen, C. Leizerson, R. Rivest. Introduction to algorithms, Cambridge: MIT Press, 1990.
[20] A.G. Chentsov, A.A. Chentsov. A model variant of the problem about radiation sources utilization (iterations based on optimization insertions). Izvestiya Instituta Matematiki i Informatiki Udmurtskogo Gosudarstvennogo Universiteta. Vol.50. pp. 83–109. [in Russian], 2017.

[21] A.G. Chentsov, A.M. Grigoryev. Optimizing multi-insertions in routing problems with constraints, Vestn. Udmurtsk. un-ta. Matem. Meh. Komp'jut. nauki, Vol. 28. N. 4. pp. 513–530, [in Russian], 2018.
[22] A.G. Chentsov. Extremal problems of routing and distribution of tasks: questions of theory, Moscow-Izhevsk: RHD, 238 p., [in Russian], 2008.
[23] A.G. Chentsov, P.A. Chentsov. Routing under constraints: problem of visit to megalopolises. Automation and Remote Control. N 11. pp. 1957—1974, 2016.
[24] A.G. Chentsov, A.A. Chentsov. The routing problem, complicated by the dependence of cost functions and "current" restrictions on the job list. Modelirovanie i analiz informazionnyh system. Vol. 23. N 2, pp. 211-227, [in Russian], 2016.
[25] A. A. Chentsov, A. G. Chentsov. The iterations method in generalized courier problem with singularity in the definition of cost functions, Vestn. Udmurtsk. Univ. Mat. Mekh. Komp. Nauki. N 3. pp. 88–113, [in Russian], 2013.
[26] E.L. Lawler. Efficient implementation of dynamic programming algorithms for sequencing problems: Tech. Rep.: BW 106/79: Stichting Mathematisch Centrum. pp. 1–16, 1979.
[27] E. Bohez, S. S. Makhanov, K. Sonthipermpoon. Adaptive Nonlinear Tool Path Optimization for Five-axis Machining." International Journal of Production Research 38 (17): 4329–4343, 2000.
[28] R. Dewil, P. Vansteenwegen, D. Cattrysse. Construction Heuristics for Generating Tool Paths for Laser Cutters. International Journal of Production Research 52 (20): 5965–5984, 2014.
[29] R. Dewil, P. Vansteenwegen, D. Cattrysse. An Improvement Heuristic Framework for the Laser Cutting Tool Path Problem. International Journal of Production Research 53 (6): 1761–1776, 2015.
[30] R. Dewil, P. Vansteenwegen, D. Cattrysse. A review of cutting path algorithms for cutters. International Journal of Advanced Manufactoring Technology 87: 1865–1884 2016.
[31] A.A. Petunin, Modelling of Tool Path for the CNC Sheet Cutting Machines. AIP Conference Proceedings 1690: 060002(1)–060002(7), 2015.
[32] A.A. Petunin, C. Stylios. 2016. Optimization Models of Tool Path Problem for CNC Sheet Metal Cutting Machines. IFAC - PaperOnLine 49 (12): 23–28, 2016.
[33] S.Q.J. Xie, J. Gan, G. G.Wang. Optimal Process Planning for Compound Laser Cutting and Punch using Genetic Algorithms. International

Journal of Mechatronics and Manufacturing Systems 2 (1–2): 20–38, 2009.

[34] W.B. Yang, Y. W. Zhao, J. Jing, W. W. Liang. An Effective Algorithm for Tool-path Airtime Optimization during Leather Cutting. Advanced Materials Research 102–104: 373–377, 2010.

[35] M.-K. Lee, K.-B. Kwon. Cutting path Optimization in CNC Cutting Processes using a Two-step Genetic Algorithm. International Journal of Production Research 44 (24): 5307–5326, 2006.

# 3

# Principle of Time Stretching for Motion Control in Condition of Conflict

**G.Ts. Chikrii**

Glushkov Institute of Cybernetics, Kiev, Ukraine
E-mail: g.chikrii@gmail.com

## Abstract

This paper deals with the dynamic games of pursuit, described by a system of general form, encompassing a wide range of the functional-differential systems. It is shown that in the case of time delay of current information on the process state to the pursuer, this game is equivalent to the game with complete information with the changed dynamics and the terminal set. On the basis of this equivalence, the principle of time stretching is developed to analyze the dynamic games of pursuit, for which classic Pontryagin's condition, lying at the heart of all direct pursuit methods, does not hold. It is based on transition to the auxiliary game with special kind information delay, constructed with the help of the so-called function of time stretching, and further transition to the equivalent game. Investigation is performed in the frames of the First Direct method, providing bringing of the trajectory of conflict-controlled process to the cylindrical terminal set at a finite moment of time. In so doing, construction of the pursuer's control is accomplished on the basis of the Filippov–Castaing theorem on measurable choice that ensures realization of the process of pursuit in the class of stroboscopic strategies by Hajek. We establish sufficient conditions for termination of the games, for which Pontryagin's condition does not hold, and specify them for the cases differential and integro-differential dynamics. Also, detailed analysis of the model examples of soft meeting of two controlled mathematical pendulums and the integro-differential game is provided to support the obtained result.

**Keywords**: dynamic game, time-variable information delay, Pontryagin's condition, Aumann's integral, principle of time stretching, Minkowski difference, integro-differential game, soft meeting.

## 3.1 Introduction

In the theory of dynamic games, a number of efficient methods are created to make a decision under conditions of conflict and uncertainty. They are originated in fundamental works of R Isaacs [1], LS Pontryagin [2], NN Krasovskii [3], L Berkovitz [4], A Friedman [5], O Hajek [6], BN Pshenitchny [7], and their disciples and based on various mathematical ideas respective to the availability of information to opposing sides in the course of the game [8–10]. There exists a wide range of mechanical, economical, and biological processes which can be described by dynamic systems of various kinds, in particular, by the ordinary differential, difference, difference-differential, integral, integro-differential, partial differential, and fractional equations, as well as by impulse systems, depending on the process nature [11–14]. Any disturbance, counteraction, or inaccuracy readily leads to game situation. The deciding factor in the study of dynamic games is availability of information on the current state of the process, its prehistory, or various kinds of counter-part's discrimination, which results in the problems of pursuit-evasion by position or in the class of stroboscopic, quasi-, or $\varepsilon$-strategies.

In real systems, information often arrives with delay in time. It is shown that the dynamic game of pursuit with variable information delay is equivalent to certain perfect-information games with the changed dynamics and terminal set. This fundamental result was first proved for the linear differential games with constant delay of information and later on for the case of variable information delay [15]. This opened up possibilities for the application of classic methods to analyze the games with delay of information [15, 16].

Pontryagin's condition [2] lies at the heart of the First Direct method, developed for solving the linear differential pursuit games. It reflects an advantage of the pursuer over the evader in control resources in terms of the game parameters. However, there is a number of cases in which this condition does not hold, e.g. the problems of soft meeting (simultaneous coincidence of geometric coordinates and velocities of objects), pursuit problems for oscillatory processes or different inertia systems, etc. (see [17–19]).

Analysis of Pontryagin's condition, performed by Nikolskij [17], significantly advanced its understanding and was a contributory factor to this condition modification [20], prescribing construction of the current pursuer control on the basis of the evader one in the past.

Establishment of close relation of the modified condition with the transition from the original game with perfect information to an auxiliary game with delayed information [21] gave impetus to the development of efficient approach to solving complicated games of pursuit, namely those for which Pontryagin's condition does not hold [21, 22]. This approach bears the name of the principle of time stretching.

This paper analyzes the dynamic games of pursuit, described by a system of general form that encompasses a wide range of the functional-differential systems [23]. The gist of the time-stretching principle consists in artificial worsening of the availability of information on the current evader control to the pursuer. It is assumed that information about behavior of the evader reaches the pursuer with the time delay. In fact, the transition is made from the original game with complete information to the game with the same dynamics and the terminal set, yet with special kind of information delay. This delay is a function of time, decreasing as the game trajectory approaches the terminal set and vanishing as it hits the target. The central idea of the time-stretching principle consists in the introduction of certain functions, called the time-stretching function, in which terms the time delay is expressed in explicit form. The obtained game with delayed information is analyzed on the basis of its equivalence to the perfect-information game with the changed dynamics. An important point is that Pontryagin's condition for the latter game involves the time-stretching function.

The time-stretching principle proved its efficiency in solving the problems of soft meeting in various cases of second-order dynamics, for which formula the time-stretching function is deduced in explicit form, in their number, in the case of different kinds of dynamics of the pursuer and the evader [24, 25]. Simple conditions on the game parameters ensuring feasibility of the pursuit termination are deduced. In the paper, the time-stretching principle is specified for the integro-differential games. To this end, we derive solution of the integro-differential system in the Cauchy form. In so doing, the method of successive approximations is used to solve the Volterra integro-differential equation of second order. To illustrate the suggested technique, an example of integro-differential pursuit game is examined in detail.

Also, a detailed analysis of the model example of soft meeting of two controlled mathematical pendulums and integro-differential game is provided to support the obtained result.

## 3.2 Equivalence of the Pursuit Game with Delay of Information to the Game with Complete Information

Although analysis of the pursuit game begins with the presentation of a system of equations, describing the conflict-controlled process, subsequent study of the game, as a rule, deals only with the formula of the system solution. In the case of differential game, it is the Cauchy formula.

We present a trajectory of the conflict-controlled process in the form

$$z(t) = g(t) - \int_{t_0}^{t} \left(f_1(t, \theta, u(\theta)) - f_2(t, \theta, v(\theta))\right) d\theta, t \in [t_0, +\infty). \quad (3.1)$$

Here $z(t) \in R^n$, where $R^n$ is the real $n$-dimensional Euclidean space and $g : [t_0, +\infty) \to R^n$ is the continuous vector function. Controls $u$ and $v$ are picked by the players at each instant of time from the compacts $U$ and $V$ in a way that their realizations in time be Lebesgue measurable functions. Functions $f_1(t, \theta, u)$ and $f_2(t, \theta, v)$, $f_1 : \Delta(t_0) \times U \to R^n$, $f_2 : \Delta(t_0) \times V \to R^n$, where $\Delta(t_0) = \{(t, \theta) : 0 \leq t_0 < \theta \leq t \leq +\infty\}$, are assumed to be Lebesgue measurable both in $t$ and $\theta$, and continuous in $u$ and $v$, respectively; $U \in K(R^n)$, $V \in K(R^n)$, whereby $K(R^n)$ is denoted as the set of all non-empty compacts from $R^n$.

Besides, a terminal set $M_*$ having a cylindrical form is given:

$$M_* = M_0 + M. \quad (3.2)$$

Here $M_0$ is a linear subspace of $R^n$ and $M$ – a convex compact from the orthogonal complement to $M_0$ in $R^n$, i.e. $M \in coK(L)$. By $coK(L)$, the set of all convex sets from $K(R^n)$ is meant.

Let us denote by $\Omega_U$ and $\Omega_V$ the sets of all measurable functions taking their values in the compacts $U$ and $V$, respectively. In the sequel, they are referred to as the sets of admissible controls of the pursuer and the evader, respectively.

We will analyze the game, standing on the pursuer side. The goal of the pursuer is at a finite moment of time to bring a trajectory of the system (3.1) to the terminal set $M_*$ under arbitrary admissible control of the evader. By the moment of the game termination, the first moment of time $t$ when $z(t) \in M_*$ is meant. Such a dynamic game is called the game of pursuit.

Let us define by $\pi$ the operator of orthogonal projection from $R^n$ onto $L$, $\pi : R^n \to L$. Then bringing the system trajectory to the terminal set is equivalent to the inclusion of $\pi z(t) \in M$. It is supposed that $g(t_0) \notin M_*$

and the players know the parameters of conflict-controlled process (3.1), namely vector functions$g(t)$, functions $f_1(t,\theta,u)$ and $f_2(t,\theta,v)$, the control domains $U$, $V$, and the terminal set $M^*$.

To begin with, we demonstrate the impact of information delay on the dynamic game of pursuit, that is, the equivalence of the game with variable information delay to certain complete-information game with the changed object dynamics and the terminal set, and then, on its basis, outline the time-stretching principle.

Let us suppose that current information on the game state becomes available to the pursuer with the delay in time $\tau(t)$. The function $\tau :[t_0+\tau_0,+\infty) \to R$, $\tau(t_0+\tau_0) = \tau_0$, is assumed to be piecewise-continuous; besides, it can have no more than countable number of the first-order discontinuities and is absolutely continuous on the intervals of its continuity. Also we assume that $\dot{\tau}(t) < 1$ at the points, where the derivative $\dot{\tau}(t)$ exists. It is easy to verify that the last condition provides for access of fresh information to the pursuer in the course of the game.

The game starts at the moment $t_0$, but information on the evader control becomes available to the pursuer only beginning from the moment$t_0+\tau_0$,$\tau_0 > 0$. In the course of the game, i.e. at each current instant of time$t$, $t \geq t_0+\tau_0$, the pursuer has access to information on the evader control at the moment$t-\tau(t)$. Denote by $u^t(\cdot)$ the realization of the pursuer control on the half-interval$[t-\tau(t),t)$,

$$u^t(s) = \{u(s) : s \in [t-\tau(t),t)\}.$$

We name the pair $(g(t), u^t(\cdot))$ by position of the game at the moment $t$. Let us suppose that on the initial half-interval $[t_0, t_0+\tau_0)$, the pursuer applies some admissible control $u^{t_0+\tau_0}(\cdot)$, $u^{t_0+\tau_0}(\cdot) = \{u(s) : s \in [t_0, t_0+\tau_0)\}$, $u^{t_0+\tau_0}(\cdot) \in \Omega_U$. Then the pair $(g(t_0), u^{t_0+\tau_0}(\cdot))$ is the initial position of the delayed-information game.

From the pursuer's point of view, the attainability set of process (3.1) at the moment $t$ under the fixed controls of the pursuer $u(\cdot)$ on $[t_0,\ t]$ and $v(\cdot)$ on $[t_0,\ t-\tau(t)]$ can be presented in the form

$$Z(t) = \left\{ z : z = g(t) - \int_{t_0}^{t} f_1(t,\theta,u(\theta))\,d\theta + \right.$$

$$\left. + \int_{t_0}^{t-\tau(t)} f_2(t,\theta,v(\theta))\,d\theta - \int_{t-\tau(t)}^{t} f_2(t,\theta,v_1(\theta))\,d\theta,\ v_1(\theta) \in \Omega_V \right\}.$$

We say that the above-described game with the variable information delay terminates at the moment $t$ if the inclusion $Z(t) \subset M_*$ is fulfilled.

In the sequel, the notion of Aumann integral of set-valued mapping is used [26].

**Definition 3.1** *Let $F(t)$ be a measurable mapping, $F: [t_0, T] \to P(R^n)$, where $P(R^n)$ is a set of all closed subsets of the space $R^n$. The union of integrals, taken over all its measurable selections $f(t)$, $f(t) \in F(t)$, namely,*

$$\bigcup_{f(\cdot)\in F(\cdot)} \int_{t_0}^{T} f(t)\, dt,$$

*is called the Aumann integral of set-valued mapping $F(t)$.*

We will use the following notation for the Aumann integral:

$$V(\tau(t)) = \left\{ x : x = \int_{t-\tau(t)}^{t} f_2(t, \theta, v(\theta))\, d\theta,\ v(\cdot) \in \Omega_V \right\}.$$

Let us introduce an auxiliary variable

$$\tilde{z}(t) = g(t) - \int_{t_0}^{t_0+\tau_0} f_1\left(t, \theta, u^{t_0+\tau_0}(\theta)\right) d\theta - \int_{t_0+\tau_0}^{t} f_1(t, \theta, u(\theta))\, d\theta$$
$$+ \int_{t_0}^{t-\tau(t)} f_2(t, \theta, v(\theta))\, d\theta.$$

We substitute $\theta = \theta_1 - \tau(\theta_1)$into the integral in the last term of the above formula and make a notation

$$\tilde{g}(t) = g(t) - \int_{t_0}^{t_0+\tau_0} f_1\left(t, \theta, u^{t_0+\tau_0}(\theta)\right) d\theta.$$

Then, taking into account that $\tau(t_0+\tau_0)=\tau_0$, $\dot{\tau}(t)<1$, we obtain

$$\begin{aligned}\tilde{z}(t) =& \tilde{g}(t) - \int_{t_0+\tau_0}^{t} f_1(t,\theta,u(\theta))\,d\theta \\ &+ \int_{t_0+\tau_0}^{t} (1-\dot{\tau}(\theta))\, f_2(t,\theta-\tau(\theta),v(\theta-\tau(\theta)))\,d\theta, \qquad (3.3)\\ &t\in[t_0+\tau_0,t],\ \tilde{z}(t_0+\tau_0)=\tilde{g}(t_0+\tau_0).\end{aligned}$$

It is evident that

$$Z(t)=\tilde{z}(t)+V(\tau(t)).$$

In the sequel, we employ the operation of geometric subtraction by Minkowski [5].

**Definition 3.2.** *Let $X$ and $Y$ be non-empty sets from $R^n$. The geometric difference of sets is defined by the formula*

$$X \underline{*} Y=\{z: z+Y\subset X\}=\bigcap_{y\in Y}(X-y),\ X\in R^n, Y\in R^n.$$

Let us introduce into consideration the following set-valued mappings:

$$M(\tau(t))=M\underline{*}\,V(\tau(t)),\ M_*(t)=M_0+M(\tau(t)). \qquad (3.4)$$

**Condition 3.1.** *The set-valued mapping $M(\tau(t))$ has non-empty images for all $t\geq t_0+\tau_0$.*

It is clear that, as soon as $Z(t)\subset M_*$, $\tilde{z}(t)\in M_*(t)$ and vice versa. Let us consider the perfect-information pursuit game with $\tilde{z}(t)$ as a current state, evolving on $[t_0+\tau_0,+\infty]$, and the terminal set $M_*(t)$. We see that, by virtue of Condition 3.1, the terminal set $M_*(t)$ has a solid variable set component $M(\tau(t))$.

The above reasoning results in the following theorem.

**Theorem 3.1.** *Let in the game (3.1)–(3.2) with variable information delay $\tau(t)$ Condition 3.1 hold. Then this game can be terminated at the moment $T$, $T\geq t_0+\tau_0$, starting from the initial position $(g(t_0), u^{t_0+\tau_0}(\cdot))$, if and only if the game (3.3)–(3.4) with complete information can be terminated at the same time $T$.*

## 3.3 Principle of Time Stretching in Dynamic Games of Pursuit

Here we outline an approach to solving games of pursuit, for which Pontryagin's condition does not hold, in the frames of the First Direct method [2], and derive sufficient conditions ensuring the game termination at some finite time, which, generally speaking, are not optimal. In such case, they say about guaranteed result. It is achieved by using counter-controls by Krasovskii [3], prescribed by stroboscopic strategies by Hajek [6]. We name this approach by the time-stretching principle.

The First Direct method, created to study the linear differential games of pursuit, is based on the Pontryagin's condition that reflects an advantage of the pursuer over the evader expressed in terms of the game parameters. The linear differential game is described by the system of linear differential equations

$$\dot{z}(t) = Az(t) - u(t) + v(t), \tag{3.5}$$

where$z \in R^n$,$z(0) = z_0$, and$A$ is a quadratic matrix of order $n$. It presents a particular case of the conflict-controlled process (3.1) with

$$g(t) = e^{tA} z_0, f_1(t, \theta, u) = e^{(t-\theta)A} u, f_2(t, \theta, v) = e^{(t-\theta)A} v,$$

whereby$e^{tA}$is denoted as the exponent of the matrix$tA$.

**Condition 3.2 (Pontryagin's).** *The set-valued mapping*

$$W(t) = \pi e^{tA} U \underline{*} \pi e^{tA} V \tag{3.6}$$

*has non-empty images at all*$t \in [0, +\infty)$.

Note that in the case of linear differential game, the state variable $\tilde{z}(t)$ of the equivalent game with complete information, evolving on the half-axis$[\tau_0, +\infty)$, satisfies the differential equation

$$\dot{\tilde{z}}(t) = A\,\tilde{z}(t) - u(t) + (1 - \dot{\tau}(t))\, e^{\tau(t)\,A} v(t - \tau(t)),$$

$$\tilde{z}(\tau_0) = e^{\tau_0 A} z_0 - \int_0^{\tau_0} e^{(\tau_0 - \theta)A} u(\theta)\, d\theta.$$

The terminal set of the equivalent game appears as the set$M_*(t)$ (3.4) with

$$V(\tau(t)) = \left\{ x : x = \int_0^{\tau(t)} e^{\theta\,A} v(\theta)\, d\theta,\ v(\cdot) \in \Omega_V \right\}.$$

The set-valued mapping $W(t)$ (3.6) is applied in the First Direct method in the following way. It is shown that if at some time$t_1$ the initial state of the game $z_0$ satisfies the inclusion

$$\pi\ e^{t_1 A} z_0 \in \int\limits_0^{t_1} W(\theta)\, d\theta,$$

then the game of pursuit can be terminated at the moment $t_1$ under arbitrary admissible controls of the evader.

Below, we present generalization of Pontryagin's condition to the case of dynamics of general form (3.1).

**Condition 3.3.** The set-valued mapping

$$W(t,\theta) = \pi\ f_1(t,\theta,U) \underset{-}{*} \pi\ f_2(t,\theta,V)$$

has non-empty images for all $t_0 \le \theta \le t < +\infty$.

In the case where Condition 3.3 does not hold, we propose its modification, constructed with the help of the function of time stretching.

**Definition 3.3.** *By the function of time stretching is named a non-negative, monotonically increasing function of time $I(t)$, $t \in [0,+\infty)$, $I(0) = 0$, $I(t) > t$, $t > 0$, which can have at most countable number of discontinuities and all discontinuities are of the first order, absolutely continuous on the intervals of its continuity, and such that $\sup\limits_{t \in [0,+\infty)\setminus\Delta} \dot{I}(t) < +\infty$, where $\Delta$ is the set of $I(t)$ discontinuity and non-differentiability points.*

**Condition 3.4.** *There exists time-stretching function$I(t)$ such that the set-valued mapping*

$$W_1(t,\theta) = \pi\ f_1(t_0 + I(t), t_0 + I(t) - \theta, U) \underset{-}{*} \dot{I}(\theta)\ \pi\ f_2 \\ (t_0 + I(t), t_0 + I(t) - I(\theta), V)$$

*has non-empty images for all $0 \le t_0 < +\infty$,$0 \le \theta \le t < +\infty$.*

The following theorem provides sufficient conditions for the game termination, starting from a given initial state.

**Theorem 3.2.** *Let in the perfect-information game (3.1)–(3.2) Condition 3.4 hold and let for the given initial state $g(t_0)$ there exist a finite moment* *of time* $t_1$*:*

$$t_1 = \left\{ \min t : t \geq 0, \pi \left( g\left(t_0 + I\left(t\right)\right) - \int\limits_{t_0}^{t_0+I(t)-t} f_1\left(t_0 + I\left(t\right), \theta, U\right) d\theta \right) \right.$$
$$\left. \bigcap \left( M + \int\limits_0^t W_1\left(t,\theta\right) d\theta \right) \neq \emptyset \right\} \tag{3.7}$$

*Then a trajectory of process (3.1) can be brought by the pursuer to the terminal set* $M_*$ *(3.2) at the moment of time*$t_0 + I\left(t_1\right)$*, under any admissible controls of the evader.*

*Proof.* Let us set $\tau_0 = I\left(t_1\right) - t_1$ and divide the interval of time $\left[t_0,\, t_0 + I\left(t_1\right)\right]$ into two parts – the initial half-interval $\left[t_0,\, t_0 + \tau_0\right)$ and the closed interval$\left[t_0 + \tau_0, t_0 + I\left(t_1\right)\right]$. In view of the assumptions concerning non-emptiness of the intersection in the definition of time $t_1$ (3.7) and non-emptiness of the images of the set-valued mapping $W_1\left(t,\theta\right)$ (Condition 3.3), there exist an admissible control $u^{t_0+\tau_0}\left(\theta\right)$, $\theta \in \left[t_0, t_0 + \tau_0\right)$, point $m$, $m \in M$, and measurable selection $w_1\left(t_1,\theta\right)$, $w_1\left(t_1,\theta\right) \in W_1\left(t_1,\theta\right)$, $\theta \in \left[0, t_1\right)$ such that the following equation is fulfilled:

$$\pi \left( g\left(t_0 + I\left(t_1\right)\right) - \int\limits_{t_0}^{t_0+I(t_1)-t_1} f_1\left(t_0 + I\left(t_1\right), \theta,\, u^{t_0+\tau_0}\left(\theta\right)\right) d\theta \right)$$
$$= m + \int\limits_0^{t_1} w_1\left(t_1,\theta\right) d\theta. \tag{3.8}$$

We set the pursuer control on the initial time interval$\left[t_0,\, t_0 + \tau_0\right)$, equal to$u^{t_0+\tau_0}\left(\cdot\right)$. The trajectory of the conflict-controlled process (3.1) on the interval $\left[t_0 + \tau_0, t_0 + I\left(t_1\right)\right]$ can be presented in the form

$$\begin{aligned} z\left(t_0 + \tau_0 + t\right) =& g\left(t_0 + \tau_0 + t\right) \\ & - \int\limits_0^t \left(f_1\left(t_0 + \tau_0 + t, t_0 + \tau_0 + \theta, u\left(t_0 + \tau_0 + \theta\right)\right)\right. \\ & \left. - f_2\left(t_0 + \tau_0 + t, t_0 + \tau_0 + \theta, v\left(t_0 + \tau_0 + \theta\right)\right)\right) d\theta. \end{aligned}$$

Let us build current control of the pursuer at each instant of time$t_0 + \tau_0 + \theta, \theta \in \left[0, t_1\right]$, on the basis of the evader control at the instant$t_0 + I\left(t_1\right) - I\left(t_1 - \theta\right)$.

It is easy to see that

$$t_0 + I(t_1) - I(t_1 - \theta) = t_0 + \tau_0 + \theta - (I(t_1 - \theta) - (t_1 - \theta)).$$

One can observe that control of the pursuer at current instant$t_0 + \tau_0 + \theta$,$\theta \in [0, t_1]$, is constructed on the basis of the evader control on the time $I(t_1 - \theta) - (t_1 - \theta)$earlier. Thus, on the interval $[t_0 + \tau_0, t_0 + I(t_1)]$, a transition is made from the original game with complete information to the auxiliary game with the same dynamics and the same terminal set but with variable delay of information

$$\tau(t_0 + \tau_0 + \theta) = I(t_1 - \theta) - (t_1 - \theta), \theta \in [0, t_1]. \tag{3.9}$$

By Theorem 3.1, this game with information delay is equivalent to the perfect-information game having the dynamics

$$\begin{aligned}
&\tilde{z}(t_0 + \tau_0 + t) = \tilde{g}(t_0 + \tau_0 + t) - \\
&\int_0^t (f_1(t_0 + \tau_0 + t, t_0 + \tau_0 + \theta, u(t_0 + \tau_0 + \theta)) + \\
&\dot{I}(t_1 - \theta) f_2\big(t_0 + \tau_0 + t, t_0 + I(t_1) - \\
&I(t_1 - \theta), v(t_0 + I(t_1) - I(t_1 - \theta))\big)\, d\theta,
\end{aligned}$$

$$\tilde{g}(t_0 + \tau_0 + t) = g(t_0 + \tau_0 + t) - \int_{t_0}^{t_0+\tau_0} f_1\left(t_0 + \tau_0 + t, \theta, u^{t_0+\tau_0}(\theta)\right) d\theta. \tag{3.10}$$

The terminal set of the equivalent game has the form (3.4), in which

$$M_*(t_0 + \tau_0 + t) = M_0 + M(\tau(t_0 + \tau_0 + t)),$$

$$M(\tau(t_0 + \tau_0 + t)) = M_{\underline{*}}\, V(\tau(t_0 + \tau_0 + t)) = M_{\underline{*}}\, V(I(t_1 - t) - (t_1 - t)).$$

In view of the formula for the time delay (3.9),$\tau(t_0 + I(t_1)) = \tau(t_0 + \tau_0 + t_1) = 0$; therefore, $M_*(t_0 + I(t_1)) = M_*$. Thus, at the moment of the game termination $t_0 + I(t_1)$ the terminal set of the equivalent game is the set $M_*$.

Let us prescribe control of the pursuer on the interval $[t_0 + \tau_0, t_0 + I(t_1)]$ to be equal to a measurable solution of the equation:

$$\begin{aligned}
&\pi f_1(t_0 + I(t_1),\, t_0 + \tau_0 + \theta,\, u(t_0 + \tau_0 + \theta)) \\
&\quad = \dot{I}(t_1 - \theta)\, \pi f_2\big(t_0 + I(t_1), t_0 + I(t_1) - I(t_1 - \theta), v(t_0 + I(t_1) \\
&\qquad - I(t_1 - \theta))\big) + w(t_1, \theta).
\end{aligned}$$

It exists by virtue of the Filippov–Castaing theorem on a measurable choice [27]. Using the above formula, from the formula (3.10), one can easily deduce the relation

$$\pi\tilde{z}\left(t_0+\tau_0+t_1\right)=\pi\tilde{g}\left(t_0+\tau_0+t_1\right)-\int\limits_0^{t_1} w\left(t_1,\theta\right)d\theta .$$

With account of formula (3.8), it yields that$\pi\,\tilde{z}\left(t_0+I\left(t_1\right)\right)=m$, where$m\in M$. Therefore, at the moment $t_0+I\left(t_1\right)$, the trajectory of the game (3.10) with complete information as well as the trajectory of its equivalent – the game (3.1) with variable information delay (3.9) – hits the terminal set$M_*$. It points to the termination of original game (3.1) with complete information at the moment of time$t_0+I\left(t_1\right)$. It should be noted that in the course of the game, the pursuer constructs its current control on the basis of the evader's control in the past. The theorem is proved.

In the case of the linear differential game, Condition 3.4 and Theorem 3.2 take the forms of Condition 3.5 and Theorem 3.3, respectively.

**Condition 3.5** [20]. *There exists a time-stretching function*$I\left(t\right)$ *such that*

$$W_1\left(t\right)=\pi\ e^{tA}U \underset{-}{*} \dot{I}\left(t\right)\pi\ e^{I(t)A}V\neq\emptyset, 0\leq t<+\infty .$$

**Theorem 3.3.** *Let the linear differential game (3.1)–(3.2) satisfy Condition 3.5 and suppose that for the given initial state* $z_0$, *there exists a finite instant of time*$t_1$,

$$t_1=\min\left\{t\geq 0:\pi\left(e^{I(t)A}z_0-\int\limits_0^{I(t)-t}e^{(I(t)-\theta)A}U d\theta\right)\bigcap\right.$$
$$\left.\left(M+\int\limits_0^{t}W_1\left(\theta\right)d\theta\right)\neq\emptyset\right\}. \tag{3.11}$$

*Then the game can be terminated by the pursuer at the time instant* $I\left(t_1\right)$ *under arbitrary admissible controls of the evader.*

To this end, on the half-interval $\left[0,\tau_0\right)$, control of the pursuer is set equal to $u^{\tau_0}\left(\theta\right)$ and on the interval $\left[\tau_0,\tau_0+t_1\right]$, it is built in the form of a measurable solution of the equation

$$\pi\, e^{(t_1-\theta)A}\, u\,(\tau_0+\theta) =$$

$$= \dot{I}\,(t_1-\theta)\;\pi\, e^{I(t_1-\theta)A} v\,(I\,(t_1) - I\,(t_1-\theta)) + w_1\,(t_1-\theta)\,, \theta \in [0, t_1]\,.$$

$w_1\,(\theta)\,,\, w_1\,(\theta) \in W_1\,(\theta)\,, \theta \in [0, t_1]$ are determined by the formula (3.11).

## 3.4 Integro-Differential Game of Pursuit

Let us consider two controlled systems evolving in $R^n$accordingly to the integro-differential equations, respectively:

$$\dot{x}\,(t) = A_1 x\,(t) + \lambda \int_0^t K\,(t,s)\, x\,(s)\; ds + f_1\,(u)\,,\;\; u \in U,\;\; x\,(0) = x_0, \tag{3.12}$$

$$\dot{y}\,(t) = A_2 y\,(t) + \mu \int_0^t L\,(t,s)\, y\,(s)\; ds + f_2\,(v)\,,\;\; v \in V, y\,(0) = y_0. \tag{3.13}$$

Here, $A_1$and $A_2$ are constant matrices, $K\,(t,s)$ and $L\,(t,s)$ are matrix functions whose elements are continuous on the set$\Delta = \{(t,s) : 0 \le s < t < +\infty\}$, $f_1\,(u) : U \to R^n$ and $f_2\,(v) : V \to R^n$, are continuous vector functions, $u$ and $v$ are control parameters of the pursuer and the evader, respectively, and $\lambda$ and $\mu$ are real numbers.

The goal of the pursuer is in the shortest time to achieve meeting with the evader, i.e.$x\,(t) = y\,(t)$,$t < +\infty$, and the evader tries to escape or maximally postpone the meeting.

We assume that the pursuer constructs its control on the basis of information about initial states of the players and current control of the evader, i.e. employs counter-controls by Krasovskii [3].

Solutions to Equations (3.12) and (3.13), under the initial state$x(0) = x_0$, $y(0) = y_0$ and chosen controls $u\,(\theta)\,,\, v\,(\theta)\,,\, \theta \in [0, t]$, can be presented in the following forms, respectively:

$$x\,(t) = e^{A_1 t} x_0 + \int_0^t e^{A_1(t-\theta)} f_1\,(u\,(\theta))\, d\theta$$

$$+\lambda\int_0^t e^{A_1(t-\theta)}\left(\int_0^\theta K(\theta,s)\,x(s)\,ds\right)ds,$$

$$y(t)=e^{A_2t}x_0+\int_0^t e^{A_2(t-\theta)}f_2(\nu(\theta))\,d\theta$$

$$+\mu\int_0^t e^{A_2(t-\theta)}\left(\int_0^\theta L(\theta,s)\,y(s)\,ds\right)ds.$$

Let us interchange, by virtue of the Dirichlet rule [28], the order of integration in the second terms of the both expressions. Then we obtain

$$\int_0^t e^{A_1(t-\theta)}\left(\int_0^\theta K(\theta,s)\,x(s)\,ds\right)d\theta=\int_0^t\left(\int_s^t e^{A_1(t-\theta)}K(\theta,s)\,d\theta\right)x(s)\,ds,$$

$$\int_0^t e^{A_2(t-\theta)}\left(\int_0^\theta L(\theta,s)\,y(s)\,ds\right)d\theta=\int_0^t\left(\int_s^t e^{A_2(t-\theta)}L(\theta,s)\,d\theta\right)y(s)\,ds.$$

These equations are the linear integral Volterra equations of second order [29]

$$x(t)=\lambda\int_0^t \widehat{K}(t,s)\,x(s)\,ds+g_1(t), \tag{3.14}$$

$$y(t)=\mu\int_0^t \widehat{L}(t,s)\,y(s)\,ds+g_2(t). \tag{3.15}$$

Here,

$$g_1(t)=e^{A_1t}x_0+\int_0^t e^{A_1(t-s)}f_1(u(\theta))\,d\theta, \tag{3.16}$$

$$g_2(t)=e^{A_2t}y_0+\int_0^t e^{A_2(t-s)}f_2(v(\theta))\,d\theta, \tag{3.17}$$

$$\hat{K}(t,s)=\int_0^t e^{A_1(t-s)}K(\theta,s)\,ds,$$

$$\overset{\wedge}{L}(t,s) = \int_0^t e^{A_2(t-s)} L(\theta, s)\, ds.$$

Using the method of successive approximations [29], one can find solutions to Equations (3.14) and (3.15):

$$x(t) = \lambda \int_0^t \overset{\wedge}{R_1}(t,s;\lambda)\, g_1(s)\, ds + g_1(t), \tag{3.18}$$

$$y(t) = \mu \int_0^t \overset{\wedge}{R_2}(t,s;\mu)\, g_2(s)\, ds + g_2(t). \tag{3.19}$$

$\overset{\wedge}{R_1}(t,s;\lambda)$ and $\overset{\wedge}{R_2}(t,s;\mu)$ are the resolvents of the integral equations (3.14) and (3.15) and are defined by the Neumann rows

$$\overset{\wedge}{R_1}(t,s;\lambda) = \sum_{i=1}^{\infty} \lambda^{n-1} \overset{\wedge}{K_i}(t,s), \tag{3.20}$$

$$\overset{\wedge}{R_2}(t,s;\lambda) = \sum_{i=1}^{\infty} \mu^{n-1} \overset{\wedge}{L_i}(t,s). \tag{3.21}$$

These rows are absolutely converging and the iterated kernels $\overset{\wedge}{K_i}(t,s)$, $\overset{\wedge}{L_i}(t,s)$, $i = 1,\ 2, ...$, are given by the recursive formulas

$$\overset{\wedge}{K_1}(t,s) = K(t,s),\ \overset{\wedge}{K_i}(t,s) = \int_s^t \overset{\wedge}{K_1}(t,\varsigma)\, \overset{\wedge}{K_{i-1}}(\varsigma, s)\, d\varsigma,$$

$$\overset{\wedge}{L_1}(t,s) = \hat{L}(\theta,s),\ \overset{\wedge}{L_i}(t,s) = \int_0^t \overset{\wedge}{L_1}(t,\varsigma)\, \overset{\wedge}{L_{i-1}}(\varsigma,s)\, d\varsigma, i = 2, ....$$

Let us substitute formulas (3.16) and (3.17) for $g_1(t)$ and $g_2(t)$into the expressions (3.18) and (3.19). Then we have

$$x(t) = e^{A_1 t} x_0 + \int_0^t e^{A_1(t-\theta)} f_1(u(\theta))\, d\theta + \lambda \int_0^t \overset{\wedge}{R_1}(t,s;\lambda)\, e^{A_1 s} ds \cdot x_0 +$$

$$+\lambda\int_0^t \overset{\wedge}{R}_1(t,s;\lambda)\left(\int_0^s e^{A_1(s-\theta)} f_1(u(\theta))\,d\theta\right)ds,$$

$$y(t)=e^{A_2 t}y_0+\int_0^t e^{A_2(t-\theta)} f_2(\nu(\theta))\,d\theta+\mu\int_0^t \overset{\wedge}{R}_2(t,s;\mu)\,e^{A_2 s}ds\cdot y_0+$$

$$+\mu\int_0^t \overset{\wedge}{R}_2(t,s;\mu)\left(\int_0^s e^{A_2(s-\theta)} f_2(\nu(\theta))\,d\theta\right)ds.$$

Upon application of the Dirichlet formula in the last terms of the above expressions, we obtain

$$\begin{aligned} x(t)&=\left(e^{A_1 t}+\lambda\int_0^t \overset{\wedge}{R}_1(t,s;\lambda)\,e^{A_1 s}ds\right)x_0+\\ &+\int_0^t\left(e^{A_1(t-\theta)}+\lambda\int_\theta^t \overset{\wedge}{R}_1(t,s;\lambda)\,e^{A_1(s-\theta)}ds\right)f_1(u(\theta))\,d\theta, \end{aligned} \tag{3.22}$$

$$\begin{aligned} y(t)&=\left(e^{A_2 t}+\lambda\int_0^t \overset{\wedge}{R}_2(t,s;\mu)\,e^{A_2 s}ds\right)y_0+\\ &+\int_0^t\left(e^{A_2(t-\theta)}+\mu\int_\theta^t \overset{\wedge}{R}_2(t,s;\mu)\,e^{A_2(s-\theta)}ds\right)f_2(u(\theta))\,d\theta. \end{aligned} \tag{3.23}$$

With the use of notations,

$$\Phi_1(t,\theta)=e^{A_1(t-\theta)}+\lambda\int_\theta^t \overset{\wedge}{R}_1(t,s;\lambda)\,e^{A_1(s-\theta)}ds, \tag{3.24}$$

$$\Phi_2(t,\theta)=e^{A_2(t-\theta)}+\mu\int_\theta^t \overset{\wedge}{R}_2(t,s;\mu)\,e^{A_2(s-\theta)}ds, \tag{3.25}$$

the solutions to Equations (3.12) and (3.13) can be presented in explicit form:

$$x(t)=\Phi_1(t,0)\,x_0+\int_0^t \Phi_1(t,\theta)\;f_1(u(\theta))\,d\theta, \tag{3.26}$$

$$y(t) = \Phi_2(t,0)\, y_0 + \int_0^t \Phi_2(t,\theta)\; f_2(\nu(\theta))\, d\theta. \qquad (3.27)$$

Let us pass from the game under study (3.12) and (3.13) to the equivalent game with the origin as the terminal set and the state vector $z = y - x$, evolving in$R^n$according to Equation (3.1) in which

$$g(t) = \Phi_2(t,0) - \Phi_1(t,0)\,, f_1(t,\theta,u(\theta)) = \Phi_1(t,\theta)\, u(\theta)\,,$$

$$f_2(t,\theta,v(\theta)) = \Phi_2(t,\theta)\, v(\theta)\,.z_0 = y_0 - x_0.$$

Then Conditions 3.3 and 3.4 and Theorem 3.2 reduce to Conditions 3.6 and 3.7 and Theorem 3.4, respectively.

**Condition 3.6.** *The set-valued mapping*

$$\tilde{W}(t,\theta) = \Phi_1(t,\theta)\; f_1(U) \underline{*} \Phi_2(t,\theta)\, f_2(V)$$

*has non-empty images for all* $0 \le \theta \le t < +\infty$.

**Condition 3.7.** *There exists a function of time stretching*$I(t)$ *such that the set-valued mapping*

$$\tilde{W}_1(t,\theta) = \Phi_1(I(t),\theta)\; f_1(U) \underline{*} \dot{I}(t-\theta)\; \Phi_2(I(t), I(t) - I(t-\theta))\, f_2(V)$$

*has non-empty images for all* $0 \le \theta \le t < +\infty$.

**Theorem 3.4.** *Let for the pursuit games* (3.12) and (3.13) Condition 3.7 *hold and let for the given initial states*$x_0$ *and* $y_0$ *there exist a finite moment*$t_1$,

$$t_1 = t_1(x_0, y_0) =$$

$$\left\{ \min t : t \ge 0, \left( \Phi_2(I(t),0)\, y_0 - \Phi_1(I(t),0)\, x_0 - \int_0^{I(t)-t} \Phi_1(I(t),\theta)\; f_1(U)\, d\theta \right) \right.$$

$$\left. \cap \int_0^t \tilde{W}_1(t,\theta)\; d\theta \ne \emptyset \right\}. \qquad (3.28)$$

*Then the pursuer can terminate the game at the moment of time* $I(t_1)$ *for arbitrary admissible control of the evader.*

It should be noted that in the course of pursuit, beginning from the moment$\tau_0 = I(t_1) - t_1$, the pursuer constructs its current control on the basis of the evader's control at the moment on the time $I(t_1-\theta)-(t_1-\theta)$earlier, defined by the formula

$$\Phi_1\left(I\left(t_1\right), \tau_0+\theta\right) f_1\left(u\left(\tau_0+\theta\right)\right)=$$
$$=\dot{I}\left(t_1-\theta\right) \Phi_2\left(I\left(t_1\right), I\left(t_1\right)-I\left(t_1-\theta\right)\right) f_2\left(v\left(I\left(t_1\right)\right.\right.$$
$$\left.\left.-I\left(t_1-\theta\right)\right)\right)+w\left(t_1, t_1-\theta\right), \theta \in\left[0, t_1\right].$$

## 3.5 Illustrative Example of the Integro-Differential Game of Pursuit

Let us consider the game, in which motions of the pursuer and the evader are described by the integro-differential equations, respectively:

$$\dot{x}(t)=\lambda \int_0^t x(s)\, ds+u,\ u \in U,\ x(0)=x_0, \|u\| \leq \rho \tag{3.29}$$

$$\dot{y}(t)=\mu \int_0^t y(s)\, ds+v,\ v \in V, y(0)=y_0, \|v\| \leq \sigma. \tag{3.30}$$

It is a particular case of the games (3.12) and (3.13). We see that in the game in study$K_1(t,s)$, $K_2(t,s)$ appears as the unit matrices and$A_1$, $A_2$ as zero matrices. Using the formulas for iterated kernels, one can easily evaluate

$$\overset{\wedge}{K}_1(t,s)=(t-s)\, E, \overset{\wedge}{L}_1(t,s)=(t-s)\, E.$$

Let us denote

$$\omega_1=\sqrt{-\lambda}, \omega_2=\sqrt{-\mu}.$$

In view of formulas (3.20) and (3.21), the resolvents of Equations (3.29) and (3.30) have the following forms, respectively:

$$\overset{\wedge}{R}_1(t,s)=\frac{1}{\omega_1} \sin \omega_1(t-s), \overset{\wedge}{R}_2(t,s)=\frac{1}{\omega_2} \sin \omega_2(t-s), \text{ if } \lambda<0,\ \mu<0,$$

$$\overset{\wedge}{R}_1(t,s)=\frac{1}{\omega_1} sh \omega_1(t-s),\ \overset{\wedge}{R}_2(t,s)=\frac{1}{\omega_2} sh \omega_2(t-s), \text{ if } \lambda>0,\ \mu>0.$$

Then, by formulas (3.24) and (3.25),

$$\Phi_1(t, \theta)=\begin{cases}\cos \omega_1(t-s) \cdot E & if\ \lambda<0 \\ ch \omega_1(t-s) \cdot E & if\ \lambda>0,\end{cases}$$

$$\Phi_2(t,\theta) = \begin{cases} \cos\omega_2(t-s)\cdot E & if\ \mu < 0 \\ ch\omega_2(t-s)\cdot E & if\ \mu > 0. \end{cases}$$

Let us analyze Conditions 3.6 and 3.7 for various combinations of $\lambda$ and $\mu$ signs:

1) $\lambda < 0$, $\mu > 0$. Condition 3.6 reduces to the form

$$\rho\left|\cos\omega_1 t\right| - \sigma\left|ch\omega_2 t\right| \geq 0.$$

This inequality is not fulfilled for all $t \geq 0$; therefore, Condition 3.6 does not hold. The principle of the time stretching is not applicable because

$$ch\omega_2 t = \frac{e^{\omega_2 t} + e^{-\omega_2 t}}{2} \to +\infty, \text{ as } t \to +\infty.$$

2) $\lambda > 0$, $\mu < 0$. Condition 3.6 has the form

$$\rho\left|ch\omega_1 t\right| - \sigma\left|\cos\omega_2 t\right| \geq 0 \forall t \geq 0.$$

It holds if $\rho \geq 2\sigma$.

3) $\lambda > 0$, $\mu > 0$. Condition 3.6 has the form

$$\rho\left|ch\omega_1 t\right| - \sigma\left|ch\omega_2 t\right| \geq 0 \forall t \geq 0.$$

It holds if $\rho \geq \sigma$ and $\omega_1 > \omega_2$.

4) $\lambda < 0$, $\mu = 0$. Condition 3.6 takes the form

$$\rho\left|\cos\omega_1 t\right| - \sigma \geq 0 \forall t \geq 0.$$

It does not hold and the time stretching is not applicable.

5) $\lambda = 0$, $\mu < 0$. Condition 3.6 reduces to the form

$$\rho - \sigma\left|\cos\omega_2 t\right| \geq 0 \forall t \geq 0.$$

It holds if $\rho \geq \sigma$.

6) $\lambda = 0$, $\mu > 0$. Condition 3.6 has the form of the inequality $\rho - \sigma\left|ch\omega_2 t\right| \geq 0$ and does not hold for all $t \geq 0$. The principle of time stretching is not applicable since $ch\omega_2 t \to +\infty$ as $t \to +\infty$.

7) $\lambda > 0$, $\mu = 0$. Condition 3.6 takes the form $\rho\left|ch\omega_2 t\right| - \sigma \geq 0 \forall t \geq 0$ and is fulfilled if $\rho \geq 2\sigma$.

The case $\lambda < 0$, $\mu < 0$ presents special interest and is analyzed in detail. Condition 3.6 has the form

$$\rho\left|\cos\omega_1 t\right| \cdot S \underline{*}\, \sigma\left|\cos\omega_2 t\right| \cdot S \neq \emptyset \forall t \geq 0.$$

Here$S$,$S \in R^n$, is the ball of unit radius centered at the origin. Evidently, this condition does not hold. Let us apply the time-stretching principle. In this case, Condition 3.7 (with the time stretching) looks as follows:

$$\tilde{W}_1(t) = \rho \left|\cos \omega_1 t\right| S \underline{*}\, \sigma \dot{I}(t) \left|\cos \omega_2 I(t)\right| \cdot S \neq \emptyset \forall t \geq 0.$$

It is equivalent to the inequality

$$\rho \left|\cos \omega_1 t\right| - \sigma \dot{I}(t) \left|\cos \omega_2 I(t)\right| \geq 0 \forall t \geq 0. \quad (3.31)$$

Let us assume that

$$\omega_1 > \omega_2 \quad (3.32)$$

and set

$$I(t) = \frac{k\pi}{2\omega_2} - \frac{\omega_1}{\omega_2} t, t \in \left[\frac{(k-1)\pi}{2\omega_2}, \frac{k\pi}{\omega_2}\right), k = 1, 2, \ldots.$$

Then the relation (3.31) reduces to the inequality:

$$\left(\rho - \sigma \frac{\omega_1}{\omega_2}\right) \left|\cos \omega_1 t\right| \geq 0. \quad (3.33)$$

Let us impose, in addition to the condition (3.32), the following constraint on the game parameters:

$$\frac{\rho}{\omega_1} \geq \frac{\sigma}{\omega_2}. \quad (3.34)$$

It is easy to see that, under the conditions (3.32) and (3.34), the inequality (3.35) holds for all $t \geq 0$. This means the fulfillment of Condition 3.7.

Now we will show that assumption (3.28) of Theorem 3.4 is satisfied for arbitrary initial states$x_0$ and $y_0$ of the objects in study. To this end, we set control of the pursuer equal to zero on the initial half-interval of time, i.e. $u^0(\theta) = 0$,$\theta \in [0,\ I(t_1) - t_1)$. Then, in view of Theorem 3.4, it remains to show that there exists a finite instant of time$t_1$,$0 < t_1 < +\infty$, at which the following inclusion is true:

$$\pi e^{I(t)A} z_0 \in \int_0^t \tilde{W}_1(\theta)\, d\theta.$$

In the example under study, this inclusion reduces to the form

$$\cos \omega_2 I(t)\, y_0 - \cos \omega_1 I(t)\, x_0 \in \int_0^t \left(\rho \left|\cos \omega_1 t\right| - \sigma \dot{I}(t) \left|\cos \omega_2 I(t)\right|\right) d\theta \cdot S. \quad (3.35)$$

Substituting $I(t)$ into the inclusion (3.35), we convert it into the relationship

$$\left\| \cos\omega_1 t \cdot y_0 - \cos\frac{{\omega_1}^2}{\omega_2} t \cdot x_0 \right\| \leq \left(\rho - \sigma\frac{\omega_1}{\omega_2}\right) \int_0^t |\cos\omega_1\theta|\, d\theta. \quad (3.36)$$

One can see that the left-hand part of the above inequality is less than or equal to $\|x_0\| + \|y_0\|$, while the right-hand part is greater than or equal to $\frac{1}{\omega_1}\left(\rho - \sigma\frac{\omega_1}{\omega_2}\right)\left[\frac{t}{\pi/\omega_1}\right]$, whereby symbol $[\cdot]$ denotes the integer part of a number. Therefore, there exists a moment of time $t_1$ at which inequality (3.36) is satisfied and, therefore, inclusion (3.35) holds. Thus, by virtue of Theorem 3.4, under conditions (3.32) and (3.34) on the game parameters, the pursuer can approach the evader at a finite moment of time, for arbitrary initial states of the players.

## 3.6 Soft Meeting of Mathematical Pendulums

Let us investigate the problem of soft meeting of two conflict-controlled objects described by the systems, respectively

$$\ddot{x} + a^2 x = \rho u, \; x \in R^n, \quad (3.37)$$

$$\ddot{y} + b^2 y = \sigma v, \; y \in R^n. \quad (3.38)$$

Here, $x$ and $y$ are geometric coordinates of the players, $u$ and $v$ are their controls, $\|u\| \leq 1$, $\|v\| \leq 1$, $a$ and $b$ are the proper angular oscillation frequencies of the systems (3.37) and (3.38),$\rho$ and$\sigma$ are force coefficients, and $a, b, \rho, \sigma$are positive numbers, $a > b$. The initial states and velocities are given:

$$x(0) = x_0, \; \dot{x}(0) = \dot{x}_0, \; y(0) = y_0, \; \dot{y}(0) = \dot{y}_0. \quad (3.39)$$

It should be emphasized that in the case of dynamics (3.37) and (3.38), even the problem of meeting in geometric coordinates presents a challenge [18, 19]. Pontryagin's condition in this case has the form

$$W(t) = \bigcap_{\|v\|\leq 1} \bigcup_{\|u\|\leq 1} \left(\frac{\rho}{a}|\sin at|\, u - \frac{\sigma}{b}|\sin bt|\, v\right) \neq \emptyset \quad \forall t \geq 0.$$

To fulfill this relationship, rather burdensome conditions should be met [18], namely, $b = (2k - 1)\ a$, where $k$ is a natural number, and

$$\rho \frac{b}{a} \left| \frac{tgat}{tgbt} \right| \geq \sigma, \quad t \in \left[ 0, \frac{\pi}{2a} \right].$$

Otherwise, Pontryagin's condition can be satisfied only periodically in time.

Here, we study significantly a more complicated problem, namely, the problem of simultaneous coincidence of the object geometric coordinates and velocities at some finite moment of time (the problem of soft meeting).

Let us introduce new variables

$$z_1 = x, \quad z_2 = \dot{x}, \quad z_3 = y, \quad z_4 = \dot{y},$$

and pass from systems (3.37) and (3.38) of second order to a system of first order of the form (3.5):

$$\begin{aligned} \dot{z}_1 &= z_2, \\ \dot{z}_2 &= -a^2 z_1 + \rho\, u, \\ \dot{z}_3 &= z_4, \\ \dot{z}_4 &= -b^2 z_3 + \sigma\, v. \end{aligned}, \tag{3.40}$$

The initial conditions (3.39) transform into the following one:

$$z_1\,(0) = x_0, \; z_2\,(0) = \dot{x}_0, \; z_3\,(0) = y_0, \; z_4\,(0) = \dot{y}_0.$$

One can see that the terminal set$M$, operator $\pi$, and control sets take the forms:

$$M = \{\, (z_1, z_2, z_3, z_4)\,,\; z_1, z_2, z_3, z_4 \in R^n \,:\; z_1 = z_3,\; z_2 = z_4 \}\,, \tag{3.41}$$

$$\pi = \begin{pmatrix} E & 0 & -E & 0 \\ 0 & E & 0 & -E \end{pmatrix}, U = \begin{pmatrix} O & \rho S & O & O \end{pmatrix},$$

$$U = \begin{pmatrix} O & O & O & \sigma S \end{pmatrix}.$$

Then the soft meeting of objects (3.37) and (3.38), at the moment of time, is equivalent to the bringing of the trajectory of system (3.40) to terminal set (3.41), i.e. $\pi\, z(t) = 0$.

The fundamental matrix of the united system is (see [31])

$$e^{At} = \begin{pmatrix} \cos at\, \cdot E & \frac{1}{a} \sin at \cdot E & O & O \\ -a \sin at \cdot E & \cos at \cdot E & O & O \\ O & O & \cos bt \cdot E & \frac{1}{b} \sin bt \cdot E \\ O & O & -b \sin bt \cdot E & \cos bt \cdot E \end{pmatrix},$$

where $E$ and $O$ are unit and zero square matrices of order $n$, respectively.

Condition 3.2 for this game has the form

$$W(t) = \bigcap_{\|v\|\le 1} \bigcup_{\|u\|\le 1} \begin{pmatrix} \frac{\rho}{a}|\sin at|\,u - \frac{\sigma}{b}|\sin bt|\,v \\ \rho|\cos at|\,u - \sigma|\cos bt|\,v \end{pmatrix} \ne \emptyset \quad \forall t \ge 0.$$

One can easily see that it does not hold. To solve the problem of soft meeting in hand, let us take advantage of Condition 3.5:

$$W_1(t) = \bigcap_{\|v\|\le 1} \bigcup_{u\le 1} \begin{pmatrix} \frac{\rho}{a}|\sin at|\,u - \frac{\sigma}{b}\dot{I}(t)\,|\sin bI(t)|\,v \\ \rho|\cos at|\,u - \sigma\dot{I}(t)\,|\cos bI(t)|\,v \end{pmatrix} \ne \emptyset \quad \forall t \ge 0. \tag{3.42}$$

This condition is met if a pair of $n$-dimensional vectors$d_1(t),\ d_2(t)$exist such that for each vector $v,\ v \in R^n,\ \|v\| \le 1$, one can find vector $u,\ u \in R^n,\ \|u\| \le 1$, for which the following equalities are simultaneously satisfied:

$$\begin{aligned} \tfrac{\sigma}{b}\dot{I}(t)\,|\sin bI(t)|\,v + d_1(t) &= \tfrac{\rho}{a}|\sin at|\,u, \\ \sigma\dot{I}(t)\,|\cos bI(t)|\,v + d_2(t) &= \rho|\cos at|\,u. \end{aligned} \tag{3.43}$$

In particular, these equalities should hold at$v \equiv 0$. Then we have

$$d_1(t) = \frac{\rho}{\alpha}\ |\sin at|\ u_0, d_2(t) = \rho\ |\cos at|\ u_0,$$

where $u_0$ is the control which corresponds to $v \equiv 0$. Upon substitution of$d_1(t)$and$d_2(t)$ into system (3.41), we obtain the system

$$\begin{aligned} \tfrac{\sigma}{b}\dot{I}(t)\,|\sin bI(t)|\,v + d_1(t) &= \tfrac{\rho}{a}|\sin at|\,(u - u_0), \\ \sigma\dot{I}(t)\,|\cos bI(t)|\,v + d_2(t) &= \rho|\cos at|\,(u - u_0), \end{aligned}$$

whence follows the equality for the function of time stretching:

$$\frac{1}{a}|\sin at|\ |\cos bI(t)| = \frac{1}{b}|\cos at|\ |\sin bI(t)|. \tag{3.44}$$

We will build function$I(t)$ step by step, beginning with the half-interval of time$\left[0, \frac{\pi}{2a}\right)$, where$\cos at = |\cos at|$and$|tgat| = tgat$. By definition, the time-stretching function $I(t)$ is strictly monotone and such that$I(0) = 0$, $I(t) \ge t$ for $t \ge 0$. Let us seek function $I(t)$ such that $\cos bI(t) > 0$, $t \in \left[0, \frac{\pi}{2a}\right)$. Dividing both sides of equality (6.6) by the positive product $|\cos at|\ |\cos bI(t)|$, we come to the relationship

$$\frac{1}{a}tgat = \frac{1}{b}\ |tgbI(t)|. \tag{3.45}$$

Function $tgat$ has discontinuities at $t = (2k-1)\frac{\pi}{2a}$, $k = 1, 2, \ldots$. It takes positive values on the open interval $\left(0, \frac{\pi}{2a}\right)$ and at $t = 0$ turns into zero. Concerning function $tgbI(t)$, there are two possibilities:

1) $tgbI(t) \geq 0$, $t \in \left[0, \frac{\pi}{2a}\right)$, then formula (3.45) transforms into equality $\frac{1}{a}tgat = \frac{1}{b}\,tgbI(t)$, whence follows the formula

$$I(t) = \frac{1}{b}arctg\left(\frac{b}{a}tgat\right). \tag{3.46}$$

By symbol $arctg$ is meant the principal value of argument. We observe that on $\left[0, \frac{\pi}{2a}\right)$ function $I(t)$ (3.46) has the derivative:

$$\dot{I}(t) = \frac{a^2}{a^2 - (a^2 - b^2)\sin^2 at}. \tag{3.47}$$

Since $\dot{I}(t) > 0$,$t \in \left(0, \frac{\pi}{2a}\right)$, and$I(0) = 0$, then $I(t)$ is not decreasing on $\left[0, \frac{\pi}{2a}\right)$.

2) $tgbI(t) < 0$; then from formula (3.45), it follows that $tgbI(t) = -\frac{b}{a}tgat$ and

$$I(t) = \frac{1}{b}arctg\left(-\frac{b}{a}tgat\right) = \frac{1}{b}\left[\pi - arctg\left(\frac{b}{a}tgat\right)\right].$$

We see that $I(0) = \frac{\pi}{b}$, which contradicts with the condition$I(0) = 0$. Therefore, this case is rejected.

Thus, the function of time stretching $I(t)$ of the form (3.46) is constructed on the half-interval$[0, \pi/2a)$ where it has a continuous derivative. Let us set$I\left(\frac{\pi}{2a}\right) = \frac{\pi}{2b}$. On the half-interval$(\pi/2a, \pi/a,]$, where$tgat < 0$, formula (3.44) transforms into the equality

$$|tgbI(t)| = -\frac{b}{a}tgat.$$

Analogously to the previous reasoning, we consider two cases:

1) $tgbI(t) \geq 0$; then$tgbI(t) = -\frac{b}{a}tgat$ and

$$I(t) = \frac{1}{b}arctg\left(-\frac{b}{a}tgat\right), \dot{I}(t) = -\frac{a^2}{a^2 - (a^2 - b^2)\sin^2 at}.$$

Since $a > b$, then $\dot{I}(t) < 0$, which contradicts the requirement for monotone non-decreasing of the time-stretching function$I(t)$.

2) $tgbI(t) < 0$; then$I(t)$ is defined by the formula $tgbI(t) = \frac{b}{a}tgat$, whence follows the validity of formula (3.46) for $I(t)$ on $\left[\frac{\pi}{2a}, \frac{\pi}{a}\right)$. On the interval$\left(\frac{\pi}{2a}, \frac{\pi}{a}\right)$the condition for monotone non-decreasing of function $I(t)$ and condition $I(t) > t$ are readily fulfilled.

Thus, on the half-interval of time $\left[0, \frac{\pi}{a}\right)$, we constructed the function of time stretching $I(t)$ of the form (3.46) that has continuous derivative. In an analogous manner, we analyze the relationship (3.45) on the next half-interval$\left[\frac{\pi}{a}, \frac{2\pi}{a}\right)$. To fulfill equality $I\left(\frac{\pi}{a}\right) = \frac{\pi}{b}$, providing continuous extension of function$I(t)$, we take advantage of the arctangent multi-valence and set

$$I(t) = \frac{\pi}{b} + \frac{1}{b}arctg\left(\frac{b}{a}tgat\right), t \in \left[\frac{\pi}{a}, \frac{2\pi}{a}\right).$$

Using analogous consideration, we come to the formula for the function of time stretching

$$I(t) = \frac{1}{b}\left[(k-1)\pi + arctg\left(\frac{b}{a}tgat\right)\right],$$
$$t \in [(k-1)\pi/a, k\pi/a), k = 1, 2, \ldots. \tag{3.48}$$

We see that function $I(t)$is continuous on the semi-axis$[0, +\infty)$. Note that the derivative $\dot{I}(t)$ on the time interval$\left[0, \frac{\pi}{2a}\right]$ is increasing from unit to $\frac{a^2}{b^2}$ and is decreasing from $\frac{a^2}{b^2}$ to unit on$\left[\frac{\pi}{2a}, \frac{\pi}{a}\right]$. It is the case for all pairs of the intervals.

$$\left[(k-1)\frac{\pi}{a}, (2k-1)\frac{\pi}{2a}\right], \left[(2k-1)\frac{\pi}{2a}, k\frac{\pi}{a}\right], k = 1, 2, \ldots.$$

Now we proceed to deducing sufficient conditions for the game termination. In the game under study, the set $W_1(t)$ (3.2) can be presented in the following form:

$$W_1(t) = \rho \bigcup_{s \in R^n,\ \|s\| \le 1} \begin{pmatrix} \frac{1}{a}|\sin at|\ s \\ |\cos at|\ s \end{pmatrix} \overset{*}{-} \sigma\dot{I}(t) \bigcup_{s \in R^n,\ \|s\| \le 1} \begin{pmatrix} \frac{1}{a}|\sin bI(t)|\ s \\ |\cos bI(t)|\ s \end{pmatrix}. \tag{3.49}$$

By virtue of formula (3.48), function$I(t)$ satisfies the equation

$$tgbI(t) = \frac{b}{a}tgat.$$

Upon its differentiation, we obtain

$$\cos^2 bI(t) = \dot{I}(t)\cos^2 at.$$

From the last two equalities, there follow the relationships

$$\begin{aligned} |\cos bI(t)| &= \sqrt{\dot I(t)}\,|\cos at|, \\ \tfrac{1}{b}|\sin bI(t)| &= \tfrac{1}{a}\sqrt{\dot I(t)}\,|\sin at|. \end{aligned} \tag{3.50}$$

One can observe that the following formula is true for the game under study:

$$\pi e^{At}U = \left(\dot I(t)\right)^{1/2} \pi e^{AI(t)}V.$$

Therefore, formula for $W_1(t)$ (3.49) takes the form

$$W_1(t) = \left(\rho - \sigma\left(\dot I(t)\right)^{3/2}\right)\bar S(t),$$

where

$$\bar S(t) = \bigcup_{s\in R^n,\ \|s\|\le 1} \begin{pmatrix} \frac{1}{a}|\sin at|\ s \\ |\cos at|\ s \end{pmatrix}.$$

Since $a > b$, in view of formula (3.47), we have

$$\sup_{t\in[0,+\infty)} \dot I(t) = \frac{a^2}{b^2}.$$

That is why the set $W_1(t)$contains the set

$$\left(\rho - \sigma\alpha^3/\beta^3\right)\bar S(t).$$

Hence, the following constraints on the game parameters:

$$a > b,\ \frac{\rho}{a^3} \ge \frac{\sigma}{b^3} \tag{3.51}$$

provide fulfillment of condition (3.42) for non-emptiness of the set $W_1(t)$ (Condition 3.5 for the game under study).

Now we prove existence of the moment$t_1$, defined by formula (3.11). Let us set control of the pursuer $u^0(\theta) \equiv 0,\ \theta \in [0, \tau_0)$. Then, $t_1$ is the first moment of time $t$, at which the following inclusion is fulfilled:

$$\begin{pmatrix} \cos(bI(t))E & \frac{1}{b}\sin(bI(t))E \\ -b\sin(bI(t))E & \cos(bI(t))E \end{pmatrix}\cdot\begin{pmatrix} y_0 \\ \dot y_0 \end{pmatrix} - \begin{pmatrix} \cos(aI(t))E & \frac{1}{a}\sin(aI(t))E \\ -\sin(aI(t))E & \cos(aI(t))E \end{pmatrix}\cdot\begin{pmatrix} x_0 \\ \dot x_0 \end{pmatrix} \in \int_0^t W_1(\theta)\,d\theta \tag{3.52}$$

The matrices, standing at the left-hand side of this inclusion and acting at the vectors of initial states$(x_0, \dot{x}_0)$,$(y_0, \dot{y}_0)$, appear as the operators of rotation. That is why, as time $t$ grows, the vector, standing at the left-hand side of the inclusion, does not leave certain ball$rS$,$rS \in R^{2n}$, of radius $r$, centered at the origin. From previous considerations, it follows that the set, standing at the right-hand side of the inclusion (3.52), contains the set

$$\overline{W_1}\left(t\right)=\left(\rho-\sigma\frac{a^3}{b^3}\right)S\left(t\right),$$

where

$$S\left(t\right)=\bigcup_{s\in R^n,\ \|s\|\leq 1}\begin{pmatrix}\frac{1}{a}\int\limits_0^t\left|\sin a\theta\right|d\theta\cdot s\\ \int\limits_0^t\left|\cos a\theta\right|d\theta\cdot s\end{pmatrix}.$$

The following estimates are true:

$$\int_0^t\frac{1}{a}\left|\sin a\theta\right|d\theta\geq\frac{1}{a^2}\left[\frac{t}{\pi/2a}\right],\int_0^t\left|\cos a\theta\right|d\theta\geq\frac{1}{a}\left[\frac{t}{\pi/2a}\right]$$

Hence, it follows that the set $\overline{W_1}\left(t\right)$ contains the ball of radius

$$\left(\rho-\frac{a^3}{b^3}\sigma\right)\frac{\sqrt{a^2+1}}{a^2}\left[\frac{t}{\pi/2a}\right],$$

centered at the origin of space$R^{2n}$. As $t\to+\infty$,it tends to a ball of infinite radius centered at the origin. Therefore, at some finite moment of time $t_1$,the set$\overline{W_1}\left(t\right)$ absorbs the ball $rS$ and the inclusion (3.50) holds true.

Hence, under condition (3.51) on the game parameters, the pursuer can achieve soft meeting with the evader at a finite instant of time, calculated at the very beginning of the game.

As far as the problem of coincidence of only geometric coordinates is concerned, significantly less stringent conditions are required, namely $a>b$, $\rho/a\geq\sigma/b$.

## 3.7 Conclusion

It is shown that the dynamic game of pursuit with separated control blocks of the players and variable delay of information is equivalent to certain perfect-information games. Based on this fact, an original approach is developed to study game dynamic problems of pursuit with complete information for which classic Pontryagin's condition, reflecting an advantage of the pursuer over the evader in control resources, does not hold. The time-stretching modification of this condition, proposed in the paper, forms a basis for the time-stretching principle, which makes it feasible to obtain sufficient conditions for bringing the game trajectory to the terminal set at a finite moment of time. It is applicable to a wide range of conflict-controlled functional-differential systems. In the paper, sufficient conditions for guaranteed capture are obtained in the case of integro-differential dynamics of objects. By way of illustration, examples of integro-differential game of pursuit and the problem of soft meeting of two mathematical pendulums are examined in detail. Simple relationships in the form of simple inequalities between dynamic parameters and control resources of the players are deduced which provide achievement of the game goal, under arbitrary initial states of the players.

The time-stretching principle offers promise as an efficient tool for probing complicated problems of conflict counteraction of moving objects.

## References


[1] R.F. Isaacs. Differential Games, New York-London-Sydney: Wiley Interscience, 479 p., 1965.

[2] L.S. Pontryagin. Selected Scientific Papers, 2, Moscow: Nauka, 576 p., 1988 (in Russian).

[3] N.N. Krasovskii. Game Problems on the Encounter of Motions, Moscow: Nauka, 420 p., 1970 (in Russian)

[4] L.D. Berkovitz. Differential games of generalized pursuit and evasion, SIAM, Control and Optimization, pp. 361-373, v.24, N53, 1986.

[5] O. Hayek. Pursuit Games, New York: Academic Press, 266 p., 1975.

[6] A. Friedman, Differential Games, New York: Wiley Interscience, 350 p., 1971.

[7] B.N. Pshenitchny. $\varepsilon$-strategies in Differential Games, Topics in Differential Games, New York, London, Amsterdam: North Holland Publ. Co., pp. 45-49, 1973.

[8] A.A. Chikrii. An analytic method in dynamic games, Proceedings of the Steklov Institute of Mathematics, pp. 69-85, v. 271, 2010.

[9] K.G. Dziubenko, A.A. Chikrii. An approach problem for a discrete system with random perturbations, Cybernetics and Systems Analysis, pp. 271-281, v.46, N2, 2010.

[10] K.G. Dzyubenko, A.A. Chikriy. On the game problem of searching moving objects for the model of semi-markovian type, Journal of Automation and Information Sciences, pp. 1-11, v. 38, N 9, 2006.
[11] G. Siouris. Missile Guidance and Control Systems. NewYork: Springer-Verlag, 666 p., 2004.
[12] L.A. Vlasenko. Existence and uniqueness theorems for an implicit delay differential equations, Differential Equations, pp. 689-694, v. 36, N5 2000.
[13] L.A. Vlasenko, A.G. Rutkas. Stochastic impulse control of parabolic systems of Soblev type, Differential Equations, p. 1498-1507, v.47, N 10, 2011.
[14] L.A. Vlasenko, A.G. Rutkas. Optimal control of a class of random distributed Sobolev type systems with aftereffect, Journal of Automation and Information Sciences, pp. 66-76, v. 45, N 9, 2013.
[15] G.Ts. Chikrii. On a problem of pursuit under variable information time lag on the availability of a state vector, Dokl. Akad. Nauk Ukrainy, pp. 855-857, v. 10, 1979 (in Russian).
[16] G.Ts. Chikrii, An approach to solution of linear differential games with variable information delay, Journal of Automation and Information Sciences, pp. 163-170, 27 (N 3&4), 1995.
[17] M.S. Nikolskij. Application of the first direct method in the linear differential games. Izvestia Akad. Nauk SSSR, pp. 51-56, v. 10, 1972 (in Russian).
[18] A.A. Chikrii. Conflict-Controlled Processes, Boston, London, Dordrecht: Springer Science & Business Media, 424 p., 2013.
[19] A.V. Mezentsev. On some class of differential games, Izvestia AN SSSR, Techn. kib., pp. 3-7, № 6, 1971 (in Russian).
[20] D. Zonnevend. On one method of pursuit, Doklady Akademii Nauk SSSR, pp. 1296-1299, v. 204, № 6, 1972 (in Russian).
[21] G.Ts. Chikrii. Using impact of information delay for solution of game problems of pursuit, Dopovidi Natsional'noi Akademii Nauk Ukrainy, pp. 107-111, N12, 1999.
[22] G.Ts. Chikrii. Using the effect of information delay in differential pursuit games, Cybernetics and Systems Analysis, pp. 233-245, v.43, N2, 2007.
[23] A. Chikrii. Control of moving objects in condition of conflict, in book "Control Systems: Theory and Applications", River Publishers, Denmark, pp. 17-42, 2018.

[24] G.Ts. Chikrii. On one problem of approach for damped oscillations, Journal of Automation and Information Sciences, pp. 1-9, v.41, N4, 2009.
[25] G.Ts. Chikrii. Principle of time stretching in evolutionary games of approach, Journal of Automation and Information Sciences, pp. 12-26, v. 48, N5, 2016.
[26] R.J. Aumann. Integrals of set-valued functions, J. Math. Anal. Appl., pp.1-12, v. 12, 1965.
[27] A.F. Filippov. Differential equations with discontinuous righthand sides, Dordrecht, Boston: Kluwer Publishers, 258 p., 1988.
[28] A.N. Kolmogorov, S.V. Fomin. Elements of Theory of Functions and Functional Analysis, Moscow: Nauka, 624 p., 1989 (in Russian).
[29] M.L. Krasnov, A.I. Kiseliov, G.I. Makarenko. Integral Equations, Moscow: Nauka, 192p., 1968 (in Russian).
[30] S.D. Eidelman, A.A. Chikrii, A.G. Rurenko. Linear integro-differential games of approach, Journal of Automation and Information Sciences, pp. 1-6, 31, N 1-3, 1999.
[31] N.V. Vasilenko. Theory of Oscillations, Kiev: Vyshcha Shkola, 430 p., 1992 (in Russian).

# 4

# Bio-Inspired Algorithms for Optimization of Fuzzy Control Systems: Comparative Analysis

**Oleksiy Kozlov, Yuriy Kondratenko**

Petro Mohyla Black Sea National University, 10 68th Desantnykiv st., Mykolayiv, 54003, Ukraine
E-mail: kozlov_ov@ukr.net, y_kondrat2002@yahoo.com, yuriy.kondratenko@chmnu.edu.ua

## Abstract

This paper is devoted to the research and comparative analysis of the bio-inspired algorithms of synthesis and optimization of fuzzy control systems, in particular, ant colony optimization and genetic algorithms adapted for automatic rule base synthesis with the determination of optimal consequents for the Mamdani-type fuzzy systems. The studies and comparative analysis are conducted on a specific example, namely, at developing various rule base configurations of the fuzzy control system for the multipurpose mobile robot able to move on inclined and vertical ferromagnetic surfaces. For the quantitative assessment of the effectiveness and comparative analysis of the investigated algorithms, the specialized criteria are introduced. The recommendations for using the presented bio-inspired algorithms for synthesis and optimization of fuzzy control systems of various types and dimension are formulated on the basis of obtained research results.

**Keywords**. Fuzzy control system, synthesis and optimization method, bio-inspired algorithms, fuzzy controller, rule base, multipurpose mobile robot.

## 4.1 Introduction

In recent years, intelligent automatic control systems have become widespread due to the rapid development of information technology and computer hardware [1–3]. These systems are designed and successfully used for automation of complex dynamic plants with non-stationary, non-linear, or uncertain parameters in order to increase the efficiency of their functioning [4–6]. A separate class of intelligent control systems includes fuzzy control systems that are based on the theory of fuzzy sets, fuzzy logic, and soft computing, proposed by Zadeh [7–9]. Initially, systems of this class were used for the control of non-stationary plants, for which there was considerable experience in manual control gained by their operators [10–12]. Such control systems were built primarily on the basis of fuzzy logical inference of the Mamdani type [13–15] and designed solely on the basis of formalization of operators' knowledge and expert estimates. In this case, the expert-operator carried out the synthesis of the structure and parameters of fuzzy automatic control systems (FACSs) in manual mode, namely, (a) selected: input and output variables; the number of linguistic terms (LTs) for each input and output variable; types of membership functions for each LT; types of aggregation, activation, accumulation, and defuzzification methods; (b) composed the production rule base (RB), consisting of antecedents and consequents, as well as (c) determined the parameters of LT membership functions for each input and output variable [16–18]. This approach in some cases had significant limitations since with the presence of a large number of adjustable parameters of FACSs, even sufficiently experienced operators could not always effectively carry out their design.

The next step in the development of the fuzzy control theory, which allowed to expand the field of application of FACSs as well as to increase the efficiency of their development process, was the use of systems based on the fuzzy inference of Takagi–Sugeno [19], the rules consequents of which are impulse-type membership functions and are sums of weighted instantaneous values of input variables. These systems began to be effectively used to automate complex non-linear and non-stationary objects, for which simulation models based on experimental data were previously developed [20–22]. In the process of synthesizing FACSs of this type, the steps of choosing the number of LTs of output variables, types and parameters of their membership functions, as well as compiling a base of linguistic rules were replaced by the step of finding a set of weighting coefficients in the rules consequents using optimization procedures based on mathematical programming methods. This made it possible to partially automate the design process of fuzzy systems and reduce the negative impact of the subjective factor on it. However, the remaining stages of the synthesis were carried out on the basis of expert knowledge in the same way as for FACSs of the Mamdani type.

A further improvement of systems based on Takagi–Sugeno fuzzy inference was their hybridization with neural networks of direct signal propagation, as a result of which adaptive neuro-fuzzy systems appeared, for example, the adaptive-network-based fuzzy inference system (ANFIS) [23] and others [24, 25]. FACSs based on such a concept preserved the interpretability of the rules of fuzzy systems "IF ..., THEN ..." in the control process and gained the ability of training as traditional neural networks using algorithms of the error back propagation [24, 26]. Thus, in the design process, the expert sets the structure of systems of this type, and all the parameters are determined automatically in the training process based on training samples.

Modern fuzzy control systems are developed mainly using progressive methods of automated design, which contain structural-parametric optimization procedures based on objective functions and can use expert knowledge to formulate preliminary hypotheses about the FACS structure and parameters [27, 28]. These systems can be built on the basis of various types of fuzzy inference (Mamdani, Takagi–Sugeno, etc.) and use their potential most effectively to automate complex dynamic objects of various classes. Currently, one of the main directions of research and development of advanced FACSs is the development, improvement, and testing of intelligent methods of their automated design and structural-parametric optimization [29–31]. In turn, the studies of bio-inspired intelligent methods of synthesis and optimization of fuzzy control systems are the most important and promising in this field [32–34].

This paper is devoted to the research and comparative analysis of the bio-inspired algorithms of synthesis and optimization of fuzzy control systems, in particular, algorithms of automatic RB synthesis with the determination of optimal consequents for the Mamdani-type FACSs. The structure of the paper is organized as follows. Section 4.2 presents the statement of the research problem, the purpose of this work, and the analysis of publications in the researched field. Section 4.3 sets out in detail several algorithms of the RB automatic synthesis developed by the authors, which are presented as sequential computational algorithms, as well as introduced criteria for the quantitative assessment of their effectiveness. Section 4.4 presents the results of the effectiveness study and comparative analysis of the proposed bio-inspired algorithms obtained on the specific example, in particular, when developing various RB configurations of the fuzzy control system for the multipurpose mobile robot (MR), which is able to move on inclined and vertical ferromagnetic surfaces. The conclusions and references list, in turn, are given at the end of this paper.

## 4.2 Related Works and Problem Statement

Many examples of development and successful applications of FACSs for control of different plants in various areas are presented in scientific and technical literature [35–37]. Also, quite many research papers are dedicated to synthesis and structural-parametric optimization of fuzzy systems using different approaches and methods [38, 39]. In particular, procedures and algorithms of FACS synthesis and structural optimization based on the optimal choice of the LT membership functions, RB interpolation and reduction, and methods of defuzzification are presented in [40, 41]. Special attention is also paid to the designing methods that include parametric optimization of the LT membership functions of the FACSs of Mamdani type, as well as finding of weight coefficients for the rule consequents of Takagi–Sugeno systems [42–44]. The obtained results of the conducted research show that progressive bio-inspired methods of multiagent and evolutionary optimization [45–47], which include algorithms, that model the interacting behavior of collective animals, insects, and microorganisms, as well as different genetic methods [48–50], are quite effective and promising for conducting the FACS synthesis and training [51–53]. The given bio-inspired algorithms relate to the algorithms of stochastic global optimization, which have a number of advantages in comparison with the classical optimization approaches: (a) give the opportunity to effectively optimize FACSs of large dimension; (b) allow detailed study of large, multimodal, and non-smooth search space, excluding looping at local minima; (c) do not impose any restrictions on the FACS objective functions [54–56].

Presented bio-inspired algorithms can be conditionally divided into two main groups: (a) multiagent algorithms, which include algorithms of ant colony optimization (ACO) [57], bee colony optimization [58], particle swarm optimization [59], bacterial foraging optimization [60], etc.; (b) evolutionary algorithms which relate genetic algorithms (GAs) [61], evolutionary strategies [62], as well as algorithms of differential evolution [63], artificial immune systems [64], biogeography-based optimization [65], and others. These algorithms can be effectively applied for the synthesis

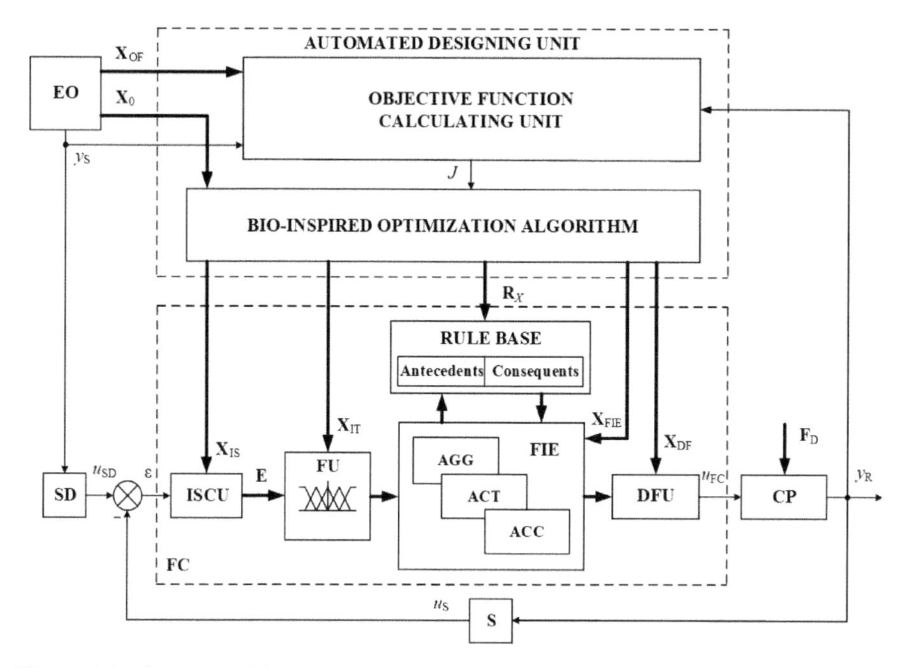


**Figure 4.1** Structure of the generalized FACS with automated designing unit including bio-inspired optimization algorithm.

and optimization of various fuzzy inference devices: controllers, identifiers, observers, signal filters, adaptive devices, etc. [66–68]. Next, consider the generalized single-input single-output (SISO) FACS with the fuzzy controller and automated designing unit, including bio-inspired optimization algorithm (Figure 4.1).

In turn, the following notations are used in Figure 4.1: EO is the expert-operator; SD is the setting device; ISCU is the input signals calculating unit; FU is the fuzzification unit; FIE is the fuzzy inference engine; AGG, ACT, and ACC are the aggregation, activation, and accumulation operations, respectively; DFU is the defuzzification unit; CP is the control plant; S is the sensor; $y_S$ and $y_R$ are the set and real values of the control variable $y$; $\mathbf{X}_{OF}$ is parameters vector of the objective function $J$; $\mathbf{X}_0$ is the preliminary value (preliminary hypothesis) of the vector $\mathbf{X}$ that determines the FC structure and parameters; $\varepsilon$ is the FACS control error; $\mathbf{X}_{IS}$ is the vector that defines the FC input signals vector $\mathbf{E}$, which is calculated on the basis of the error $\varepsilon$; $\mathbf{X}_{IT}$ is the vector that determines the number, types, and parameters of LTs of the FC input variables; $\mathbf{R}_X$ is the RB consequents vector; $\mathbf{X}_{FIE}$ is the vector that defines the operations of aggregation, activation, and accumulation of the fuzzy inference engine; $\mathbf{X}_{\mathrm{DF}}$ is the vector that determines the number, types, and parameters of LTs of the FC output variable $u_{\mathrm{FC}}$, as well as the defuzzification method; $\mathbf{F}_{\mathrm{D}}$ is the vector of disturbances that affect the control plant; $u_{\mathrm{SD}}$ and $u_{\mathrm{S}}$ are the output signals of the FACS setting device and the sensor, respectively.

In the presented fuzzy control system (Figure 4.1), the input signals calculating unit ISCU [69], based on the vector $\mathbf{X}_{\mathrm{IS}}$ forms the vector **E**, which can consist of various combinations of error $\varepsilon$ signals, its derivatives of various orders, integral, etc. For example, the vector **E** can be represented by the expression

$$E = \left\{ k_{\mathrm{P}}\varepsilon, k_{\mathrm{D}}\dot{\varepsilon}, k_{2\mathrm{D}}\ddot{\varepsilon}, k_{\mathrm{I}} \int \varepsilon dt \right\}, \tag{4.1}$$

where $k_{\mathrm{P}}$, $k_{\mathrm{D}}$, $k_{2\mathrm{D}}$, and $k_{\mathrm{I}}$ are the normalizing coefficients of the $\mathbf{X}_{\mathrm{IS}}$ vector that are used to convert the FC input signals into relative units from their maximum values.

The fuzzification unit FU defines the membership degrees of the numerical values of the input signals vector **E** to the corresponding FC input LTs [22], the number, types, and parameters of which are determined by the vector $\mathbf{X}_{\mathrm{IT}}$. In turn, the fuzzy inference engine FIE, based on the fuzzified signals and received data from the RB, sequentially performs aggregation, activation, and accumulation operations, the types of which are specified by the $\mathbf{X}_{\mathrm{FIE}}$ vector [25]. The FC RB is a set of linguistic rules consisting of certain antecedents and consequents. Moreover, for the Mamdani-type FC with an input vector (4.1), the RB rules are defined by the expression

$$\begin{gathered}\text{IF “}k_{\mathrm{P}}\varepsilon = A_1\text{” AND “}k_{\mathrm{D}}\dot{\varepsilon} = A_2\text{” AND “}k_{2\mathrm{D}}\ddot{\varepsilon} = A_3\text{”} \\ \text{AND “}k_{\mathrm{I}} \textstyle\int \varepsilon dt = A_4\text{” THEN “}u_{\mathrm{FC}} = A_5\text{”,}\end{gathered} \tag{4.2}$$

where $A_1$, $A_2$, $A_3$, $A_4$, and $A_5$ are corresponding LTs of the FC input and output signals.

The defuzzification unit DFU implements the conversion of a consolidated fuzzy inference into a crisp numerical FC output signal [15].

The problem of the fuzzy controller development of presented FACS (Figure 4.1) is solved with the help of parallel or sequential search of the optimal values of the $\mathbf{X}_{\mathrm{IS}}$, $\mathbf{X}_{\mathrm{IT}}$, $\mathbf{R}_{X}$, $\mathbf{X}_{\mathrm{FIE}}$, and $\mathbf{X}_{\mathrm{DF}}$ vectors by means of the automated designing unit, which includes the objective function calculation unit and a certain bio-inspired optimization algorithm [70–72]. Wherein, the expert-operator EO sets only the parameters vector of the objective function $\mathbf{X}_{\text{OF}}$ and the initial values (initial hypotheses) $\mathbf{X}_0$ vector of the FC structure and parameters.

At solving the given problem of structural-parametric synthesis of a Mamdani-type FACS, the task of automatic RB development with the determination of optimal consequents at insufficient initial information deserves a special attention [73–75].

The antecedents of the RB rules are various combinations of LTs of the FC input variables [76, 77]. In turn, the consequent $\text{LT}_r$ of each $r$th RB rule is selected from the set of all possible rules consequents $\{\text{LT}^1, \text{LT}^2, ..., \text{LT}^\nu\}$, which includes $v$ LTs of the FC output variable $u_{\text{FC}}$

$$\text{LT}_r \in \{\text{LT}^1, \text{LT}^2, ..., \text{LT}^\nu\}. \tag{4.3}$$

For example, when choosing the number of LTs of the FC output variable $v$ = 7, the following set is formed:

$$\text{LT}_r \in \{\text{LT}^1, \text{LT}^2, \text{LT}^3, \text{LT}^4, \text{LT}^5, \text{LT}^6, \text{LT}^7\} = \{\text{BN}, \text{N}, \text{SN}, \text{Z}, \text{SP}, \text{P}, \text{BP}\}, \tag{4.4}$$

where BN is big negative; N is negative; SN is small negative; Z is zero; SP is small positive; P is positive; BP is big positive.

The vector of consequents $\mathbf{R}_X$ can be formed in different ways; here, the task of the consequents optimizing is reduced to the task of finding the optimal vector of consequents $\mathbf{R}_{\text{opt}}$ from the set of all possible alternative variants, which provides optimal quality indicators of the fuzzy control system [73]. The consequents vector $\mathbf{R}_X$ for the $x$th alternative variant of the RB in general terms can be represented as follows [74]:

$$\begin{gathered} R_X = \{\text{LT}_{X1}, \text{LT}_{X2}, ..., \text{LT}_{Xr}, ..., \text{LT}_{Xs}\}, \\ \text{LT}_{Xr} \in \{\text{LT}^1, \text{LT}^2, ..., \text{LT}^\nu\}, X \in \{1, 2, ..., \nu^s\}, \end{gathered} \tag{4.5}$$

where $v^s$ is total number of all possible variants of the vector $\mathbf{R}_X$, which is defined as a number of the output variable LT $v$, raised to the power of the total number of the RB rules $s$.

Thus, the task of synthesis of the FACS RB consequents is reduced to determination of such a vector of the RB consequents $\mathbf{R}_X = \mathbf{R}_{\text{opt}}$, at which the value of the objective function $J$ of the FACS will be optimal ($J = J_{\text{opt}}$) [73].

This task is a complicated discrete optimization task of large dimension, the solution of which requires an effective bio-inspired method of global search that takes into account the peculiarities of the consequents formation of the RB rules in the conditions of information uncertainty and insufficient initial information [73, 74, 78]. By solving this task by exhaustive search of all possible vectors $\mathbf{R}_X$ $(X = 1, 2, ..., \nu^s)$, the value of the objective function $J$ will need to be calculated $v^s$ times, which even at a small dimension of the RB requires significant computational and time costs [74]. For example, for the FC with enough simple structure with seven rules ($s$ = 7) and five LTs of the output variable ($v$ = 5), the value of the objective function $J$ needs to be calculated $5^7$ = 78,125 times during the exhaustive search of the vector $\mathbf{R}_X$, which requires a lot of time and computing resources.

The main aim of this work is efficiency research and comparative analysis of the bio-inspired algorithms of automatic synthesis and optimization of RBs for the Mamdani-type FACSs.

## 4.3 Bio-inspired Algorithms of Synthesis and Optimization of Rule Bases for Fuzzy Control Systems

Analysis of existing bio-inspired intelligent algorithms of global optimization shows that ACO and GAs are most effective for solving complex discrete optimization problems of large dimension [74, 79–81]. Particularly, a number of published papers present examples of the successful application of various types and modifications of ACO and GA for solving different discrete optimization problems such as: traveling salesman problem [82, 83], scheduling problem [84–86], optimization of vehicle routes [87–89], graph coloring problem [90–92], etc. [93–95]. Thus, it is expedient to carry out research and comparative analysis of the ACO and GAs adapted to the specifics of the task of automatic synthesis and optimization of RBs for the Mamdani-type FACSs. Also, it is advisable to compare these bio-inspired algorithms with a fairly simple method of automatic RB synthesis based on sequential search [73] in this work. For a comparative analysis of the above algorithms, it is advisable to introduce and use specialized criteria to evaluate their effectiveness.

Next, consider the presented algorithms in detail.

### 4.3.1 ACO Algorithm for Synthesis and Optimization of Rule Bases for the Mamdani-Type FACS

Ant colony algorithms simulate the behavior of interacting ants as social insects that are members of one large colony at the joint solving the problem of finding the shortest path [57, 79, 82]. The models of ants are represented by interacting agents of one self-organizing system which move along the graph of solutions to find the optimum of the problem being solved [82, 87]. In this work, to adapt the ACO algorithms for solving the task of RB synthesis and optimization, it is necessary to represent the structure of the RB of FC in the form of a special graph, as well as the consequents of the RB rules $LT_r^j$ $(r \in \{1, ..., s\}; j \in \{1, ..., \nu\})$ in the form of graph destinations, on which the agent ants move [74].

This graph consists of layers, nodes, and edges. The path of each *z*th agent passes through certain nodes and edges of the graph on each separate *n*th iteration. The path length that a particular agent has traveled is an abstract quantity and is determined by the value of the objective function *J* of a fuzzy control system.

The structure of this graph corresponds to the structure of the RB of FC and is determined by the total number of fuzzy rules *s* and the number of their possible consequents *v* [74]. The RB graph consists of one zero layer, which has one node and is a starting point for all $z_{\max}$ population agents, and *s* main layers, each of which has a serial number *r* and corresponds to the certain *r*th RB rule ($r = 1, 2, \ldots, s$). In addition, each *r*th layer has *v* nodes, which corresponds to the number of possible consequents of RB rules and, therefore, to the number of LTs of the FC output variable $u_{\mathrm{FC}}$. In turn, each *rj*th node of the layer with the serial number *r* of this graph corresponds to *j*th possible consequent of the *r*th fuzzy rule of the formed RB of FC ($r = 1, 2, \ldots, s$; $j = 1, 2, \ldots, v$) [74].

After each *z*th agent passes through certain nodes of all *s* layers of this graph on each separate iteration with number *n*, the corresponding FC RB with a specific vector of consequents of the fuzzy rules $\mathbf{R}^z(n)$ ($z = 1, \ldots, z_{\max}$; $n = 1, \ldots, n_{\max}$) is formed.

The peculiarity of this RB graph is that its edges are installed only between the nodes of different layers (there are no edges between nodes of one layer) [74]. The transfer of the *z*th agent from any *rj*th node of the layer with the serial number *r* is possible only to any $(r + 1)i$th node of the next layer with the serial number $r + 1$ $(r \in \{1, ..., s\}; j, i \in \{1, ..., \nu\})$. Thus, the movement of each *z*th agent on each separate iteration with the number *n* starts from the zero layer and is carried out sequentially through all the *s* layers of the graph (from the first to the last) only in the direction of increasing the serial number of the layers *r*. The number of transitions of each *z*th agent over the nodes of the graph at one iteration is equal to the total number of layers of the graph and, accordingly, to the fuzzy rules *s*.

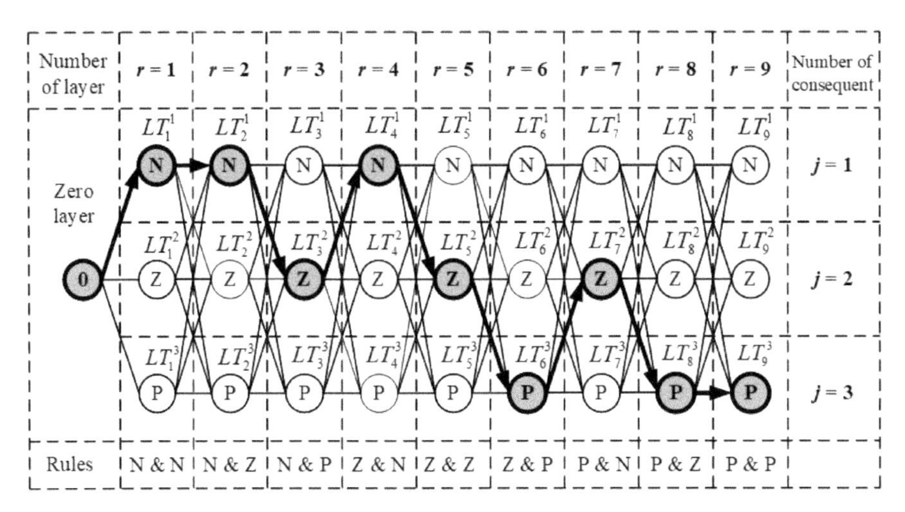


**Figure 4.2** Rule base graph for the FC with $s = 9$ and $v = 3$.

For example, Figure 4.2 shows the RB graph for the FC with two input variables, $\varepsilon$ and $\dot{\varepsilon}$, each of which is estimated by the corresponding three LT (N, Z, and P), and with the output variable $u_{\mathrm{FC}}$, which also has three LT (N, Z, and P). Thus, the number of RB rules and, respectively, layers of the graph is determined by the number of all possible combinations of LTs of input variables and equals 9 ($s = 9$). The number of nodes in each $r$th layer, in turn, is $v = 3$.

Antecedents of RB rules are formed by the following combinations of LTs of the input variables $\varepsilon$ and $\dot{\varepsilon}$: 1) N and (AND) N; 2) N and Z; 3) N and P; 4) Z and N; 5) Z and Z; 6) Z and P; 7) P and N; 8) P and Z; 9) P and P. In turn, the vector of consequents of the fuzzy rules $\mathbf{R}^z(n)$ formed as a result of the passage of the graph by the $z$th agent at iteration with number $n$ (bold in Figure 4.2) is described by the expression

$$R^z(n) = \{\mathrm{N}, \mathrm{N}, \mathrm{Z}, \mathrm{N}, \mathrm{Z}, \mathrm{P}, \mathrm{Z}, \mathrm{P}, \mathrm{P}\}. \tag{4.6}$$

The probability of the transition of the $z$th agent from $rj$th node of the layer with the serial number $r$ to $(r+1)i$th node of the next layer with the serial number $r + 1$ ($r \in \{1, ..., s\}$ ; $j, i \in \{1, ..., \nu\}$) at iteration $n$ ($n = 1, \ldots, n_{\max}$) is determined by the expression [57]

$$P^z_{rj,(r+1)i}(n) = \frac{\left[\tau_{rj,(r+1)i}(n)\right]^{\alpha} \cdot \left[\eta_{rj,(r+1)i}\right]^{\beta}}{\sum\limits_{rj,(r+1)i=rj,(r+1)1}^{rj,(r+1)v} \left[\tau_{rj,(r+1)i}(n)\right]^{\alpha} \cdot \left[\eta_{rj,(r+1)i}\right]^{\beta}}, \tag{4.7}$$

where $\tau_{rj,(r+1)i}(n)$ is the intensity of the pheromone on the edge between the nodes *rj* and (*r*+1)*i* at the iteration *n*; $\eta_{rj,(r+1)i}$ is an inverse of the relative distance $D_{rj,(r+1)i}$ between the nodes *rj* and (*r*+1)*i*; α is an adjustable parameter that sets the relative importance of the pheromone trace on the edge when selecting the next node; β is an adjustable parameter that gives relative importance of the distance $D_{rj,(r+1)i}$ between the nodes *rj* and (*r*+1)*i*.

In turn,

$$\eta_{rj,(r+1)i} = \frac{1}{D_{rj,(r+1)i}}; \tag{4.8}$$

$$D_{rj,(r+1)i} = \left| j_r - i_{(r+1)} \right| + 1, \tag{4.9}$$

where $j_r$ is a serial number of the *j*th node in the layer *r*; $i_{(r+1)}$ is a serial number of the *i*th node in the layer *r* + 1.

The relative distance $D_{01,\ 1i}$ between a single node of the zero layer and any *i*th node of the first layer (*i* = 1, . . ., *v*) equals one [74].

After each *z*th agent passes through the certain *rj*th nodes of all *s* layers of the RB graph on each iteration with the number *n*, the path length of the *z*th agent for the generated vector of the consequents of fuzzy rules $\mathbf{R}^z(n)$ is calculated, which is determined by the value of the FC objective function $J^z(n)$.

The amount of pheromone left on each edge *rj*,(*r*+1)*i* of the graph by the *z*th agent at the iteration *n* is determined on the basis of the value of objective function $J^z(n)$ [79] in the following way:

$$\begin{cases} \Delta\tau^z_{rj,(r+1)i}(n) = \frac{Q}{J^i(n)}, & \text{at } rj,(r+1)i \in H^z(n); \\ \Delta\tau^z_{rj,(r+1)i}(n) = 0, & \text{at } rj,(r+1)i \notin H^z(n), \end{cases} \tag{4.10}$$

where *Q* is an adjustable parameter that refers to the amount of pheromone, left in the way by an agent; $H^z(n)$ is a set of edges (path) that the *z*th agent has passed at the iteration *n*. It is advisable to choose the parameter *Q* of one order with the optimal value of the FC objective function $J_{\text{opt}}$. The value of pheromone $\Delta\tau^z_{rj,(r+1)i}(n)$ left by the *z*th agent on the edge *rj*,(*r*+1)*i* at the iteration *n* depends inversely proportional on the value of objective function $J^z(n)$: small value of objective function is characterized by a high concentration of pheromone, large – by low.

After calculation of the pheromone values $\Delta\tau^z_{rj,(r+1)i}(n)$ left by each *z*th agent on each edge *rj*,(*r*+1)*i* of the graph, based on the dependence (4.10), the obtained results are used to increase the pheromone on the edges of the graph

[57] according to the expression

$$\tau_{rj,(r+1)i}\left(n\right) = \tau_{rj,(r+1)i}\left(n-1\right) + \rho \cdot \sum_{z=1}^{z_{\max}} \Delta\tau^{z}_{rj,(r+1)i}\left(n\right), \tag{4.11}$$

where $\rho$ is a ratio of the amount of pheromone that agents leave on the way (it is advisable to choose in the range from 0 to 1).

To use the elite strategy of ant colony algorithms in this method, the so-called elite agents are introduced, which allows to increase significantly the rate of convergence of the method [57, 74]. In addition, at each iteration, an additional increase of pheromone is applied to the edges, which are included in the best path at this iteration with the smallest value of objective function $J_{\min}(n)$. The additional amount of pheromone left on each edge, included in the best path at iteration $n$, is determined based on the value of its objective function $J_{\min}(n)$ and the number of elite agents in population $e$ as follows:

$$\Delta\tau_{e}\left(n\right) = \frac{e \cdot Q}{J_{\min}\left(n\right)}. \tag{4.12}$$

At the end of each iteration $n$ for each edge $rj,(r+1)i$ of the graph, the operation of evaporation of pheromone [74] in accordance with the expression (4.13) is applied

$$\tau_{rj,(r+1)iF}\left(n\right) = \tau_{rj,(r+1)i}\left(n\right) \cdot \left(1-\rho\right), \tag{4.13}$$

where $\tau rj,(r+1)iF(n)$ is a final value of the amount of pheromone on the edge $rj,(r+1)i$ of the graph at the end of the iteration $n$ after applying the evaporation operation based on the expression (4.13). When moving from the iteration $n$ to the iteration $n + 1$ ($n = 1, \ldots, n_{\max}$) on each edge $rj,(r+1)i$ $\left(r \in \{1, ..., s\}; j, i \in \{1, ..., \nu\}\right)$ of the graph, the pheromone is updated according to the expression

$$\tau_{rj,(r+1)i}\left(n+1\right) = \tau_{rj,(r+1)iF}\left(n\right). \tag{4.14}$$

Thus, ACO algorithm adapted to the specifics of the task of automatic synthesis and optimization of RBs for the Mamdani-type FACSs consists of the following successive steps [74].

**Step 1:** The ACO algorithm initialization. At this stage, the construction of the RB graph of the fuzzy control system is implemented on the basis of a preliminary choice of the number of LTs for each input variable and LTs $v$ for the FC output variable $u_{\mathrm{FC}}$. The number of the main layers of the graph is equal to the total number of RB rules $s$, the number of the nodes in each $r$th layer is chosen equal to the number of all possible consequents $v$ of the rules $LT^j$ ($j = 1, \ldots, v$) and accordingly to the number of all LT of the output variable $u_{FC}$. Also, in this step, the FC objective function $J$ is selected as well as its optimal value $J_{opt}$. Moreover, an agent population is created and the main parameters of the algorithm are established at this stage: the number of agents in the population $z_{max}$, adjustable parameters $\alpha$, $\beta$, $Q$, and $\rho$, maximum number of iterations $n_{max}$, and number of elite agents in the population $e$. In addition, at this step, a small positive initial value of the pheromone is established before the agents begin to move.

**Step 2:** Movement of agents along the nodes of the FC RB graph. The movement of each $z$th agent of the population, created in Step 1, starts from the node of the zero layer and is carried out sequentially through all $s$ layers of the graph (from the first to the last) only in the direction of increasing the serial number of the layers $r$. The transition of the $z$th agent from any $rj$th node of the layer with the serial number $r$ is possible only to any $(r + 1)i$th node of the next layer with the serial number $r + 1$. The number of transitions of each $z$th agent by nodes of a graph at one iteration is equal to the total number of the layers of the graph $s$. If $z$th agent is in the node $rj$ and $r < s$, then to determine the following edge of the path $rj,(r + 1)i$, the probability $P^z_{rj,(r+1)i}(n)$ of the transition of the given agent into the node $(r + 1)i$ is calculated according to the expression (4.7). In turn, the relative distance $D_{rj,(r+1)i}$ between nodes $rj$ and $(r+1)i$, as well as its inverse $\eta_{rj,(r+1)i}$, are calculated based on dependencies (4.9) and (4.8), respectively. Step 2 continues until every $z$th agent of population passes all $s$ layers of the graph.

**Step 3:** Calculation of the values of the FC objective function. At this stage for each vector of consequents of the fuzzy rules $\mathbf{R}^z(n)$, formed by each $z$th agent of population ($z = 1, \ldots, z_{max}$) during its movement in Step 2, the value of the FC objective function $J_z(n)$, selected in Step 1, is calculated.

**Step 4:** Algorithm completion check. At this stage, one carries out the choice of the best vector of consequents of the RB fuzzy rules $\mathbf{R}_{best}(n)$, formed during the movement of agents in Step 2, by the smallest value of the FC objective function $J_{min}(n)$ from the values calculated in Step 3. For the selected best vector of consequents of the RB fuzzy rules $\mathbf{R}_{best}(n)$, the algorithm completion is checked, which can be determined by the optimal value of the FC objective function ($J_{min}(n) \leq J_{opt}$) or by limiting the maximum number of the iterations $n_{max}$. Completion can also be considered achieved if the value of the objective function $J$ has not decreased for a certain number of iterations. If this checking is positive, then go to Step 8. Otherwise, go to Step 5.

**Step 5:** Increase of the pheromone on the edges of the FC RB graph. At this stage, on the basis of the values of objective function $J^z(n)$, calculated in Step 3, the amount of pheromone left on each edge $rj,(r+1)i$ of the graph by each $z$th agent ($z = 1, \ldots, z_{\max}$) while moving in Step 2 is calculated based on Equation (4.10). The calculated values of the pheromone amount $\Delta\tau^z_{rj,(r+1)i}$ left by each $z$th agent on each edge $rj,(r+1)i$ of the graph while moving are used to increase the pheromone on the edges of the graph according to expression (4.11). Also, according to the elite strategy, an additional increase of pheromone is applied at this step to the edges, which are included in the best path at this iteration with the lowest value of objective function $J_{\min}(n)$. This pheromone value is calculated based on Equation (4.12).

**Step 6:** Evaporation of the pheromone on the edges of the RB graph. For each edge $rj$, $(r+1)i$ of the graph, the pheromone evaporation operation according to expression (4.13) is applied and its final value $\tau_{rj,(r+1)iF}$ on each edge $rj,(r+1)i$ of the graph is calculated.

**Step 7:** Transition to the next iteration of the algorithm. Return of all $z_{\max}$ agents of population to the node of the zero layer of the graph as well as updating the pheromone on each edge $rj,(r+1)i$ of the graph according to expression (4.14) is carried out. Move on to Step 2.

**Step 8:** Algorithm completion.

### 4.3.2 Genetic Algorithm for Synthesis and Optimization of Rule Bases for the Mamdani-Type FACS

GAs relate to evolutionary computation algorithms that solve optimization problems using mechanisms of natural evolution, such as selection, crossover, mutation, and inheritance. These algorithms operate with populations consisting of a finite set of individuals at solving the optimization problem [45, 67, 80]. The individuals included in the population are represented by the chromosomes with sets of task parameters encoded in them, i.e. solutions, which can also be called points in the search space [81, 89]. In turn, each chromosome consists of a number of genes, each of which is the specific parameter of the optimization problem. To evaluate the chromosome fitness and correspondingly the goodness of the optimization task solution, the fitness function is used [83].

In this work, to adapt the GA for solving the problem of automatic synthesis of the RB for the fuzzy control systems, the RB consequents vector $\mathbf{R}_X$ is represented as a chromosome, each gene of which is the specific consequent $\mathrm{LT}_r$ ($r$ = 1, 2, ..., $s$), that is selected from the set of all LTs $v$ of the fuzzy control system output variable. The fitness function calculation is implemented on the basis of the FACS objective function $J$. Since GAs are function of maximization algorithms, then to minimize the value of the objective function $J$ of the fuzzy control system, the fitness function $f$ for the $z$th chromosome at the iteration $n$ should be calculated as follows:

$$f^z(n) = \frac{1}{J^z(n)}. \tag{4.15}$$

Thus, GA adapted to the specifics of the task of automatic synthesis and optimization of RBs for the Mamdani-type FACSs consists of the following successive steps.

**Step 1:** The GA initialization. At this stage, the structure of the RB of the FC is constructed, and parameters and genetic operator types of the GA are selected. Also, structure, parameters, and optimal value of the objective function of the fuzzy control system, as well as the optimization completion criterion, are selected. Here, the structure of the FC RB is constructed based on a preliminary selection of the number of LTs for each input variable [15]. The chromosome size is determined by the rules total number $s$ and the chromosome genes are the rules consequents $\mathrm{LT}_r$, which are selected from the set of all possible terms $v$ of the FC output variable $u_{\mathrm{FC}}$. In turn, for the selection of genetic operator, it is advisable to use the proportional selection, for the crossover – the one-point crossover operator and for the mutation operator – the simple mutation, for which a gene is randomly selected in the mutating chromosome, and then randomly replaced. Also, at this stage, the probabilities of crossover $P_{\mathrm{C}}$ and mutation $P_{\mathrm{M}}$, as well as the size of initial population $z_{\max}$, are specified. The fitness function value $f$ is calculated based on the objective function value $J$ according to Equation (4.15).

As for the optimization completion criterion, achievement of the following conditions can be used [45, 80].

*Condition 1*: Achievement of the optimal value of the fitness function $f_{\mathrm{opt}}$ of the FACS. If the fitness function value $f$ of some chromosomes reaches $f_{\mathrm{opt}}$ in the process of GA functioning, then the criterion is considered achieved.

*Condition 2*: Exceeding the maximum allowed number of iterations $n_{\max}$. In this case, the maximum number of iterations should be set previously, before starting of the genetic search.

*Condition 3*: Degeneration of the population. In this case, previously, before starting the GA operating, the number of iterations $n_p$ and the threshold of the coefficient of improvement of the values of the fitness functions of the best chromosome $\rho_p$ should be set. Starting from the iteration $n_p$, at each subsequent iteration ($n_p$+1), the improvement coefficient ρ is calculated and compared with the threshold value $\rho_p$. In turn, the coefficient of improvement ρ is calculated as

$$\rho = \frac{f_{\text{best}(n)} - f_{\text{best}(n-1)}}{f_{\text{best}(n-1)}}, \tag{4.16}$$

where $f_{\text{best}(n)}$ is the best value of the fitness function $f$ at the iteration $n$; $f_{\text{best}(n-1)}$ is the best value of the fitness function $f$ at the iteration $n$–1. If the value ρ is less than $\rho_p$, then the completion criterion is considered reached.

**Step 2:** Encoding of the RB consequents vector into the chromosome. At this stage, the RB consequents vector $\mathbf{R}_X$ of the FC is encoded into the chromosome that consists of genes, each of which is the specific consequent $\text{LT}_r^j$ $(r = \{1, ..., s\}; j \in \{1, ..., \nu\})$. In this case, real encoding is used [67].

**Step 3:** Creation of the initial population of chromosomes. The initial population of $z_{\max}$ chromosomes is created containing information on $z_{\max}$ different variants (finite set of admissible solutions to a problem) of the RB consequents vector $\mathbf{R}_X$ for the fuzzy control system at this stage. In turn, the initial population $P_0$ is described by the expression

$$P_0 = \{R_1, R_2, ..., R_z, ..., R_{z_{\max}}\}, \tag{4.17}$$

where $\mathbf{R}_z$ is the population's $z$th chromosome. It is advisable to randomly set the values of the chromosomes genes in this case.

**Step 4:** Evaluation of the chromosomes of the current population. The chromosome evaluation is carried out by the following steps.

*Step 4.1:* Decoding of chromosomes. At this stage, decoding of each chromosome of the population into the RB consequents vector $\mathbf{R}_X$ is implemented.

*Step 4.2:* Building the RB of the fuzzy control system. Building RBs with the consequent vectors that meet the estimated chromosomes is implemented at this stage.

*Step 4.3:* Calculation of the objective (fitness) function of the fuzzy control system. At this stage, simulation of the FACS with developed RB is carried out and the fitness function $f_z(n)$ value is calculated for each $z$th chromosome of the population ($z$ = 1, .., $z_{\max}$). Wherein, the simulation of the FACS is carried out in all possible operating modes (at different input signals and disturbances) and the fitness function $f$ is calculated on the basis of the objective function $J_z(n)$ according to Equation (4.15).

**Step 5:** Checking of the optimization completion criteria. The search can expire when one of the conditions of the completion criterion, given in Step 1, occurs. If *Condition 1* is satisfied, the transition to Step 10 is implemented. If *Conditions 2* or *3* are satisfied, the transition to Step 1 is carried out.

**Step 6:** Selection of chromosomes for crossover. At this stage, the population chromosomes are selected for crossover to generate new solutions on the basis of the values of the fitness function $f_z(n)$. In this case, it is advisable to use the proportional selection [81], which is implemented by the following steps.

*Step 6.1:* Determination of the average value of the fitness function of the population. Determination of the average value of the fitness function $f_{\mathrm{M}}(n)$ of the population as the average of the arithmetic values of the fitness functions of all individuals is implemented according to the following equation [67]:

$$f_{\mathrm{M}}(n) = \frac{1}{z_{\max}} \sum_{z=1}^{z_{\max}} f_z(n) . \tag{4.18}$$

*Step 6.2:* Calculation of the selection ratio. At this stage, the value of the selection ratio $P_S^z$ for each $z$th individual is calculated as follows [52]:

$$P_S^z(n) = \frac{f_z(n)}{f_{\mathrm{M}}(n)} . \tag{4.19}$$

*Step 6.3:* Forming an array of chromosomes for crossover. The array of individuals admitted to crossover is formed at this step depending on the $P_S^z$ value (if $P_S^z > 1$, then, the individual is considered well adapted and allowed to crossover).

**Step 7:** Crossover of the chromosomes. At this stage, the crossover operator for the chromosomes selected in the previous step is applied. In this case, random selection of parent couples is applied [83].

*Step 7.1*: Numbering of the chromosomes. The numbering of all the representatives of the current population is implemented in an arbitrary manner at this stage.

*Step 7.2:* Choosing the first parent. At this stage, the number on the interval [0; 1] is randomly chosen for each $z$th chromosome, starting from the first one. The first parent in the pair will be the first chromosome, for which the given random number is not less than probability of crossover $P_C$ (selected at Step 1).

*Step 7.3*: Choosing the second parent. The browsing of the population is continued, starting from the decision following after the first parent (selected in previous step) until the randomly chosen number again is not less than $P_C$. The chromosome for which such a condition is fulfilled will be the second parent. Step 7.3 is continued until the required number of pairs of parents is selected.

*Step 7.4*: Crossover of the selected parent chromosomes. At this stage, the one-point cross-operator is applied for each parent pair [94]. In turn, the crossing point in the crossover operation is chosen randomly.

**Step 8:** Mutation of chromosomes. The mutation operator is applied for the chromosomes, selected at Step 6 according to the probability of mutation $P_M$ (selected at Step 1).

*Step 8.1:* Copying the parent chromosomes into a chromosome-heir. At this stage, the parent chromosomes, admitted to mutations, are copied into the chromosome-heirs.

*Step 8.2:* Choosing the mutant gene. The mutant gene of each chromosome (the consequent of the $r$th rule), admitted to mutation, is chosen randomly [114].

*Step 8.3*: Choosing the new gene value. Choosing a new gene value that is not equal to the current one is implemented from the set of all possible LTs $v$ of the FC output variable for each mutating gene.

**Step 9:** Forming a new generation. A new generation of elite chromosomes and chromosome-heirs obtained by applying crossover and mutation is formed at this stage. Then the transition to Step 4 is carried out.

**Step 10:** Algorithm completion.

### 4.3.3 Algorithm of Automatic Rule Base Synthesis for the Mamdani-Type FACS Based on Sequential Search

The presented algorithm provides an iterative optimization of the RB consequents vector $\mathbf{R}_X$ by means of a sequential search of the consequents of each rule of the FC RB. The given sequential search algorithm (SSA) of automatic RB synthesis consists of the following steps [73].

**Step 1:** Algorithm initialization. The preliminary synthesis of the FC RB is carried out on the basis of the previously selected LTs of the input variables, and the set of possible consequents for each rule $\{\mathrm{LT}^1, \mathrm{LT}^2, ..., \mathrm{LT}^\nu\}$ is determined based on the previously selected LTs of the output variable $u_{\mathrm{FC}}$. The initial vector of the RB consequents $\mathbf{R}_X$, in turn, is determined randomly and is set in the FC. Furthermore, the maximum number of iterations at the implementation of the algorithm $n_{\max}$ is set at this stage, and the objective function $J$ of the FACS is selected.

**Step 2:** Transition to the first rule of the FC RB. The transition to the first rule of the RB is carried out at this stage to initiate the iterative procedures for finding the optimal vector of the RB consequents $\mathbf{R}_{\mathrm{opt}}$.

**Step 3:** Checking the checklist. All vectors of the RB consequents $\mathbf{R}_X$, for which the objective function $J$ has been already calculated during the implementation of the algorithm, as well as the corresponding values of the objective function $J$, are entered in the checklist. If the current vector of the RB consequents $\mathbf{R}_X$ is already placed in the checklist, then the transition is carried out to Step 6, and in the opposite case, the transition is performed in Step 4.

**Step 4:** Calculation of the value of the objective function for the FC with the current vector of the RB consequents. The calculation of the value of the objective function $J$ for the FC with the current vector of the RB consequents $\mathbf{R}_X$ as well as recording of this information in the checklist is carried out at this stage.

**Step 5:** Checking the completion of the previous optimization process of the current rule. The optimization calculations for the current $r$th rule are considered complete if the values of the objective function $J$ for each $j$th consequent from the set of all $v$ possible consequents for this rule has been calculated. If the checking has given a positive result, then the transition to Step 7 is carried out. In the opposite case, the transition to Step 6 is performed.

**Step 6:** Setting of the next consequent $j + 1$ from the set of all $v$ possible consequents in the current rule. After that, the transition to Step 3 is carried out.

**Step 7:** Choosing the consequent for which the value of the objective function is the smallest. The choice of the consequent, for which the value of the objective function $J$ is the smallest among those obtained at the optimization calculations for the current $r$th rule, and its fixation in this rule is carried out at this stage.

**Step 8:** Checking the completion of the RB optimization. The RB optimization is considered complete if the optimal value of the objective function is achieved ($J = J_{\mathrm{opt}}$) or if the maximum number of iterations $n_{\max}$, previously set at Step 1, is carried out at Step 8. If the checking, carried out at this stage, has given a positive result, then the transition to Step 10 is performed. In the opposite case, the transition to Step 9 is carried out. In the case, when all of the $s$ rules of this FC RB have been optimized at the previous steps of the algorithm, and the checking at Step 8 does not give a positive result, then the transition to Step 2 is carried out, and iterative procedures for finding the optimal vector of the RB consequents $\mathbf{R}_X$ continue again, starting from the first rule of the RB.

**Step 9:** Transition to the next rule of the FC RB. After that, the transition to Step 3 is carried out.

**Step 10:** Algorithm completion.

At the implementation of the given algorithm, the consistent iterative procedures for optimizing the rules of the RB from the first to the $s$th can be carried out $l$ rounds (times) before finding the optimal vector of the RB consequents $\mathbf{R}_{\mathrm{opt}}$, at which $J \leq J_{\mathrm{opt}}$, or performing the maximum number of iterations $n_{\max}$, that is set at Step 1. The checklist and its checking at Step 3 is used to avoid recurring calculations of the objective function $J$ for the FACS with the same vector of the RB consequents $\mathbf{R}_X$ [73]. This allows to get rid of unnecessary iterations, the number of which is equal to $s - 1$ for one round. Analyzing peculiarities of the above algorithms, it can be concluded that their computational and time costs substantially depend on the complexity of the synthesized RBs. Here, the calculation of the objective function values requires significantly greater computational costs than the other calculations used in the algorithms (selection, crossover, and mutation of the chromosomes in GA as well as movement of agents, increase and evaporation of the pheromone in ACO, etc.). This is due to the fact that for the calculation of the objective function value on each iteration, it is necessary to conduct simulation of the FACS functioning with the control plant in all possible operating modes (under the action of various inputs and disturbing influences) to efficiently synthesize consequents of all the used RB rules. The complexity of the used computer models of the control plants also has a significant effect in this case. Thus, it is advisable to use the parameter $\upsilon_{Jopt}$ for generalized assessment of the computational and time costs of each algorithm, which is the total number of calculations of the objective function $J$ required to achieve its optimal value $J_{opt}$. In turn, for the SSA, $\upsilon_{Jopt}$ is equal to the total number of iterations $n_{Jopt}$ required to achieve the optimal value of the objective function ($\upsilon_{Jopt} = n_{Jopt}$). For ACO and GA, it is defined as the product of the number of agents or individuals $z_{max}$ by the number of iterations $n_{Jopt}$

$$\upsilon_{Jopt} = z_{max} n_{Jopt}. \tag{4.20}$$

As for the complexity of the synthesized RBs, it is expedient to evaluate it on the basis of the total number of the RB rules $s$ and the number of their all possible consequents $\nu$. In turn, for the SSA, the maximum number of necessary iterations $n_{max}$ directly depends on the number of rules $s$ and consequents $\nu$ as follows:

$$n_{max} = l\left[s\nu - (s-1)\right]. \tag{4.21}$$

For ACO and GA, the number of iterations $n_{Jopt}$ required to achieve the optimal value of the objective function also implicitly depends on the complexity of the synthesized RBs. Therefore, for the generalized quantitative assessment of the RB complexity, it is advisable to introduce a universal indicator $C_{RB}$, which directly depends on the number of RB rules $s$, as well as on the number of their possible consequents $\nu$. In turn, this indicator can be calculated as follows:

$$C_{RB} = s\nu. \tag{4.22}$$

The comparative analysis of the different algorithms can be conducted on the basis of the dependence of computational costs on the complexity of synthesized RBs

$$\upsilon_{Jopt} = f\left(C_{RB}\right). \tag{4.23}$$

Thus, an effective algorithm of the RB synthesis should have a small value of parameter $\upsilon_{Jopt}$ at a sufficiently large value of indicator $C_{RB}$.

For the efficiency assessment of the above algorithms for the RB synthesis, it is also expedient to introduce and use a relative performance indicator that is calculated on the basis of the equation

$$K_{\mathrm{RPI}} = \frac{C_{\mathrm{RB}}}{v_{J\mathrm{opt}}}. \tag{4.24}$$

Therefore, the greater the coefficient $K_{\mathrm{RPI}}$, the more effective the researched algorithm is.

Also, the dependence of the given indicator on the complexity of synthesized RBs can be used for the algorithm comparison at different structures of the RB

$$K_{\mathrm{RPI}} = f\left(C_{\mathrm{RB}}\right). \tag{4.25}$$

Efficiency research and comparative analysis of the above-given bio-inspired algorithms is performed in this work on a specific example, in particular, when developing various RB configurations of the fuzzy control system for the multipurpose MR that is able to move on inclined and vertical ferromagnetic surfaces [96–98].

## 4.4 Development of the Rule Base of the Fuzzy Control System for the Multipurpose Mobile Robot

The considered types of multipurpose MRs, which are able to move on inclined and vertical ferromagnetic surfaces, can be successfully used in shipbuilding, ship repair, gas and oil refining industry, agriculture, etc. for implementation of specific labor-intensive technological operations without human intervention in automatic mode [96, 99]. In particular, such robots help to replace human labor in hazardous health and life-threatening conditions, as well as significantly improve the quality and speed of execution of operations such as: welding, painting, rust removal, cleaning, and others. Furthermore, to perform the above-given technological operations, the multipurpose MRs have to move along pre-set trajectories at a certain speed and operate under uncertainty of the working surface caused by its specific features: structural damage, presence of other obstacles, etc. [96–98]. Thus, automatic control of the main motion variables (position, speed, angle, etc.) under the action of coordinate and parametrical disturbances is one of the most important and complicated automation tasks of MRs of such types.

In this work, the caterpillar MR and the fuzzy control system of its linear speed are considered as an example, for which an RB needs to be developed by means proposed in the previous section algorithms. In turn, the speed control system of the multipurpose caterpillar MR based on the fuzzy controller has the same structural organization as the generalized FACS shown in Figure 4.1. The linear speed of MR $V_{\mathrm{MR}}$ acts as a controlled coordinate *y* in this system.

The designing of the RB for the speed controller is carried out for the caterpillar MR with the following main parameters: MR's total mass of 150 kg, length and width of the hull are 1 and 0.8 m, driving wheel radius of 0.15 m, pre-set linear speed of movement of 0.2 m/s, two drive motors 2PB132MH, and gear ratio of 105 [69, 96]. The mathematical model of the MR, which is used in the calculations, consists of the following main equations [69]:

$$\begin{aligned} M_{\mathrm{EM}a} = {} & \frac{D_{\mathrm{M}}}{\eta_{\mathrm{M}}}\frac{d\omega_{\mathrm{M}}}{dt} + \frac{1}{k_{\mathrm{R}}\eta_{\mathrm{MR}}}\left[D_{\Sigma\mathrm{W}}\frac{d\omega_{\mathrm{W}}}{dt} + R_{\mathrm{W}}\left(\zeta\frac{G\cos\gamma+F}{2} + \right.\right. \\ & + G\sin\gamma\left(\frac{\cos\varphi_{\mathrm{MR}}}{2} + b\frac{x_O}{B}\sin\varphi_{\mathrm{MR}} + b\zeta\frac{h_{\mathrm{C}}}{B}\sin\varphi_{\mathrm{MR}}\right) - \\ & - b\frac{\mu_{\mathrm{T}}L(G\cos\gamma+F)}{4B}\left(1+\frac{4x_O^2}{L^2}\right) + \\ & + F_{\mathrm{TO}}\left(\frac{1}{2}\cos\beta_{\mathrm{F}} - \frac{(x_1-x_O)}{B}\sin\beta_{\mathrm{F}}\right) + \left(\frac{G\cos\gamma+F}{2g} + bG\sin\gamma\sin\varphi_{\mathrm{MR}}\frac{h_{\mathrm{C}}}{\mathrm{g}B}\right)\frac{dV_{\mathrm{C}a}}{dt} + \\ & \left.\left. + \left(m_{\mathrm{MR}} + \frac{F}{g}\right)\frac{L^2+B^2}{12B}\frac{d\omega_{\mathrm{MR}}}{dt}\right)\right]; \end{aligned} \tag{4.26}$$

$$F = \frac{F_{\mathrm{TO}} + G\left(\zeta\cos\gamma + \sin\gamma - \xi\cos\gamma\right) + m_{\mathrm{MR}}c\lambda}{\xi-\zeta}; \tag{4.27}$$

$$V_{\mathrm{C1}} = \omega_{\mathrm{MR}}\left(R_{\mathrm{T}} - 0.5B\right) = \omega_{\mathrm{W1}}R_{\mathrm{W}}; \tag{4.28}$$

$$V_{\mathrm{C2}} = \omega_{\mathrm{MR}}\left(R_{\mathrm{T}} + 0.5B\right) = \omega_{\mathrm{W2}}R_{\mathrm{W}}; \tag{4.29}$$

$$V_{\mathrm{MR}} = \left(V_{\mathrm{C2}} + V_{\mathrm{C1}}\right)/2; \tag{4.30}$$

$$\omega_{\mathrm{MR}} = V_{\mathrm{MR}}/R_{\mathrm{T}} = \left(V_{\mathrm{C2}} - V_{\mathrm{C1}}\right)/B; \tag{4.31}$$

$$R_{\mathrm{T}} = \frac{0.5(V_{\mathrm{C2}} + V_{\mathrm{C1}})}{(V_{\mathrm{C2}} - V_{\mathrm{C1}})}, \tag{4.32}$$

where $F_{\mathrm{TO}}$ and $F$ are the values of the specific forces of technological operation and the clamping magnets; $G$ is the total weight of the robot and technological equipment; $\gamma$ is the surface inclination angle; $\zeta$ is the coefficient of rolling friction; $\xi$ is the coefficient of adhesion; $m_{\mathrm{MR}}$ is the robot mass; $\lambda$ is a combined mass ratio, $\lambda = 1.15 + 0.001k_{\mathrm{R}}^2$; $k_{\mathrm{R}}$ is the gear ratio; $c$ is the robot acceleration; $V_{\mathrm{C1}}$, $V_{\mathrm{C2}}$ are the current linear velocities of lagging and running caterpillars; $R_{\mathrm{T}}$ is the robot turning radius; $V_{\mathrm{MR}}$, $\varphi_{\mathrm{MR}}$ are the current linear speed and course of robot; $\omega_{\mathrm{MR}}$ is the rotation speed of the MR, $\omega_{\mathrm{MR}}$

**Table 4.1** Structure configurations of the MR's speed fuzzy controller.

| **Structure number** | **Number of inputs** | **Vector E** | **Number of inputs LT** | **Number of output LT** $v$ | **Number of rules** $s$ |
|---|---|---|---|---|---|
| 1 | 2 | $\{k_P\varepsilon, k_D\dot{\varepsilon}\}$ | $5 \times 5$ | 5 | 25 |
| 2 | 2 | $\{k_P\varepsilon, k_D\dot{\varepsilon}\}$ | $5 \times 5$ | 7 | 25 |
| 3 | 2 | $\{k_P\varepsilon, k_D\dot{\varepsilon}\}$ | $7 \times 7$ | 7 | 49 |
| 4 | 3 | $\left\{k_P\varepsilon, k_D\dot{\varepsilon}, k_I \int \varepsilon dt\right\}$ | $5 \times 5 \times 3$ | 7 | 75 |
| 5 | 3 | $\left\{k_P\varepsilon, k_D\dot{\varepsilon}, k_I \int \varepsilon dt\right\}$ | $5 \times 5 \times 5$ | 7 | 125 |
| 6 | 3 | $\left\{k_P\varepsilon, k_D\dot{\varepsilon}, k_I \int \varepsilon dt\right\}$ | $7 \times 5 \times 5$ | 9 | 175 |

$= d\varphi_{MR}/dt$; $B$ is the distance between the centers of caterpillars; $\omega_{W1}$, $\omega_{W2}$ are angular velocities of lagging and running wheels; $R_W$ is the radius of the driving wheel; $a$ and $b$ are the coefficients that take into account direction of rotation of the robot, $a = 1$, $b = 1$ – for a lagging caterpillar; $a = 2$, $b = -1$ – for a running caterpillar; $M_{EMa}$ ($a = 1, 2$) are the electromagnetic torques of drive motors; $\eta_M$, $\eta_{MR}$ are the efficiency coefficients of the motor and MR; $D_M$ is the moment of inertia of the motor anchor; $D_{\Sigma W}$ is the total moment of inertia of two wheels and caterpillar; $L$, $h_C$ are the length of robot and height of its center of gravity; $x_O$, $x_1$ are the distances from the transverse axis of the robot to the turning centers of the caterpillars and to the point of fixing the technological equipment; $\beta_F$ is the angle of deviation of the force $F_{TO}$ from the longitudinal axis of the robot; $\mu_T$ is the cornering resistance, which depends on the turning radius.

In this work, to conduct the effectiveness study and comparative analysis of the presented bio-inspired algorithms, a synthesis of the RBs for the various structure configurations of the MR's speed fuzzy controller is carried out. In particular, the studies are carried out by the above three algorithms (ACO, GA, and SSA) for six different structures, the configurations of which are summarized in Table 4.1.

In turn, the total number of the RB rules $s$ for the fuzzy controller is determined by the number of all possible combinations of LT of the input signals. Each $r$th rule for each RB is represented by the expression (4.2). The normalizing coefficients that are used to convert the FC input signals into relative units from their maximum values have the following values: $k_P = 5$; $k_D = 0.33$; $k_I = 60$.

Moreover, for each input and output variable of the speed FC when using three, five, seven, and nine LTs, the following sets of LT of the triangular type are used:

$$\{\mathrm{N}, \mathrm{Z}, \mathrm{P}\};$$

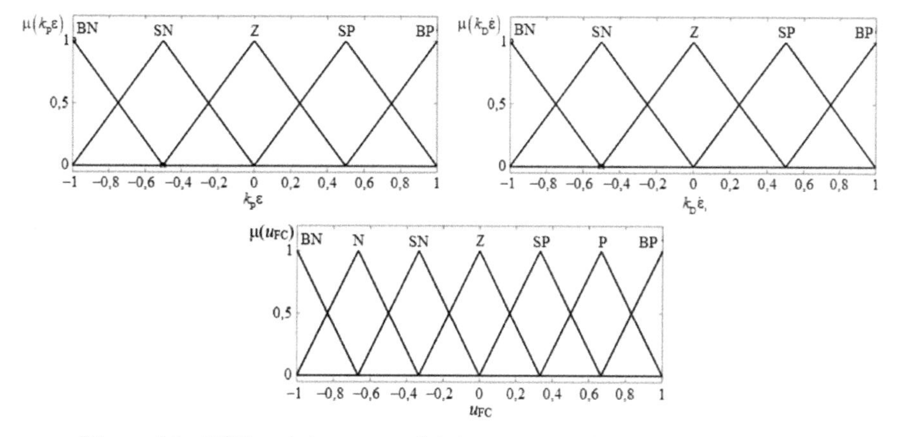


**Figure 4.3** FC linguistic terms and their parameters for the second structure.

$$\{\mathrm{BN}, \mathrm{SN}, \mathrm{Z}, \mathrm{SP}, \mathrm{BP}\};$$
$$\{\mathrm{BN}, \mathrm{N}, \mathrm{SN}, \mathrm{Z}, \mathrm{SP}, \mathrm{P}, \mathrm{BP}\};$$
$$\{\mathrm{VBN}, \mathrm{BN}, \mathrm{N}, \mathrm{SN}, \mathrm{Z}, \mathrm{SP}, \mathrm{P}, \mathrm{BP}, \mathrm{VBP}\},$$

where VBN is very big negative; BN is big negative; N is negative; SN is small negative; Z is zero; SP is small positive; P is positive; BP is big positive; VBP is very big positive.

In turn, LTs for each input and output variable are evenly distributed in the interval [–1; 1]. For example, the appearance of the terms with their parameters for the second structure (Table 4.1) of the fuzzy controller is shown in Figure 4.3.

In addition, for all structures (Table 4.1) of the MR's FC, the operation "min" is selected as an aggregation operation and the operation "max" is selected for both activation and accumulation operations. In turn, the gravity center method is chosen as the defuzzification method of the given fuzzy controller.

At the synthesis of the FC RBs when finding the consequents vector $\mathbf{R}_X$ using the presented algorithms, the objective function $J$ is chosen at the initialization stage (Step 1), which is the generalized integral deviation of the real transient of the FACS of the MR speed $V_{\mathrm{MR}}(t, \mathbf{R}_X)$ from the desired transient $V_{\mathrm{D}}(t)$ of its reference model (RM) [69, 74], which is represented by the transfer function

$$W_{\mathrm{RM}}(s) = \frac{V_{\mathrm{D}}(s)}{V_{\mathrm{S}}(s)} = \frac{1}{(T_{\mathrm{RM}}s + 1)^2}, \tag{4.33}$$

where $T_{\mathrm{RM}}$ is the time constant of the RM and $V_{\mathrm{S}}$ is the set value of the robot linear speed.

In turn, the used objective function is calculated on the basis of the equation

$$J(t, R_X) = \frac{1}{t_{\max}} \int_0^{t_{\max}} [(E_{\mathrm{V}})^2 + k_{J1}(\dot{E}_{\mathrm{V}})^2 + k_{J2}(\ddot{E}_{\mathrm{V}})^2]dt, \tag{4.34}$$

where $t_{\max}$ is the total time of the FACS transient of the MR speed; $E_{\mathrm{V}}$ is the deviation of $V_{\mathrm{MR}}(t, \mathbf{R}_X)$ from $V_{\mathrm{D}}(t)$, $E_{\mathrm{V}} = V_{\mathrm{D}}(t) - V_{\mathrm{MR}}(t, \mathbf{R}_X)$; $k_{J1}$, $k_{J2}$ are weighting coefficients that take into account the importance of deviation derivatives.

Since various indicators of control quality can be achieved at various structures of the MR's fuzzy controller, the following optimal values of the objective function $J_{\mathrm{opt}}$ are selected for the structures presented in Table 4.1: 1) $J_{\mathrm{opt}} = 0.22$; 2) $J_{\mathrm{opt}} = 0.2$; 3) $J_{\mathrm{opt}} = 0.18$; 4) $J_{\mathrm{opt}} = 0.1$; 5) $J_{\mathrm{opt}} = 0.09$; 6) $J_{\mathrm{opt}} = 0.075$.

By calculating the values of the objective function (4.34) for each tested algorithm on each $n$th iteration, the simulation of the MR speed FACS transients was carried out in all possible operating modes (under the action of various input and disturbing influences) to efficiently synthesize consequents of all the used RB rules. In turn, checking of the optimization completion was carried out under the condition of achieving the optimal value of the FACS objective function ($J(n) \leq J_{\mathrm{opt}}$).

To study the effectiveness and compare the presented algorithms, a synthesis of the FC RBs is carried out for six of its structures (Table 4.1) using three considered algorithms for each (18 experiments total). Also, each of these experiments is carried out five times with subsequent averaging of the results. The averaged experimental results obtained during the RB synthesis for the various structures (Table 4.1) of the speed FC using each of the presented algorithms are given in Table 4.2.

Here, for each of the presented algorithms, the values of the adjustable parameters are selected based on the experiments and recommendations obtained in previous studies [45, 69, 74]. In particular, for the ACO algorithm, the number of agents in the population $z_{\max} = 30$, the number of elite agents in the population $e = 10$, and the other adjustable parameters are: $\alpha = 2$; $\beta = 1$; $Q = 0.1$; $\rho = 0.5$. In turn, for GA, the number of individuals in the population $z_{\max} = 100$ and the values of the probabilities of crossover and mutation are: $P_{\mathrm{C}} = 0.25$; $P_M = 0.5$. As for the SSA, the number of rounds

**Table 4.2** Averaged experimental results obtained at the RB synthesis.

| Experiment number | Structure number | Algorithm | $J_{opt}$ | Obtained $J$ | $n_{J\text{opt}}$ | $\upsilon_{J\text{opt}}$ |
|---|---|---|---|---|---|---|
| 1 | 1 | ACO | 0.22 | 0.214 | 47 | 1410 |
| 2 | 1 | GA | 0.22 | 0.214 | 31 | 3100 |
| 3 | 1 | SSA | 0.22 | 0.207 | 202 | 202 |
| 4 | 2 | ACO | 0.2 | 0.192 | 52 | 1560 |
| 5 | 2 | GA | 0.2 | 0.197 | 34 | 3400 |
| 6 | 2 | SSA | 0.2 | 0.184 | 453 | 453 |
| 7 | 3 | ACO | 0.18 | 0.165 | 55 | 1650 |
| 8 | 3 | GA | 0.18 | 0.176 | 37 | 3700 |
| 9 | 3 | SSA | 0.18 | 0.17 | 885 | 885 |
| 10 | 4 | ACO | 0.1 | 0.99 | 61 | 1830 |
| 11 | 4 | GA | 0.1 | 0.99 | 44 | 4400 |
| 12 | 4 | SSA | 0.1 | 0.91 | 1804 | 1804 |
| 13 | 5 | ACO | 0.09 | 0.087 | 69 | 2070 |
| 14 | 5 | GA | 0.09 | 0.081 | 48 | 4800 |
| 15 | 5 | SSA | 0.09 | 0.087 | 3004 | 3004 |
| 16 | 6 | ACO | 0.075 | 0.061 | 77 | 2310 |
| 17 | 6 | GA | 0.075 | 0.071 | 53 | 5300 |
| 18 | 6 | SSA | 0.075 | 0.069 | 7005 | 7005 |

$l$ is set automatically as the consistent iterative procedures are carried out cyclically until the optimal value of the objective function is found ($J = J_{opt}$).

Figures 4.4 and 4.5 show the dynamics of change of the best values of objective function (4.34) during the synthesis of consequents of the FC RBs using the presented algorithms for the second and sixth structures, respectively.

The smoothed dependencies $\upsilon_{J\text{opt}} = f(C_{RB})$ and $K_{RPI} = f(C_{RB})$ are built for RB structures 1–6 and shown in Figures 4.6 and 4.7, respectively, for a comparative analysis of the investigated algorithms.

In addition, on these graphs, the distinctions are conditionally made depending on the dimension and complexity of the synthesized RBs: for RB of small dimension (SD), $C_{RB} \leq 370$; for RB of medium dimension (MD), $370 \leq C_{RB} \leq 700$; for RB of large dimension (LD), $C_{RB} > 700$.

As can be seen from Table 4.2 as well as from Figures 4.4–4.7, all three of the presented algorithms allow to efficiently synthesize and optimize RBs with the formation of optimal consequents $\mathbf{R}_{opt}$ to achieve the desired values of the objective function $J_{opt}$. Wherein, the computational and time costs of these algorithms are within acceptable limits. However, by synthesizing the RBs of small dimension (structures 1–3), the algorithm based on sequential

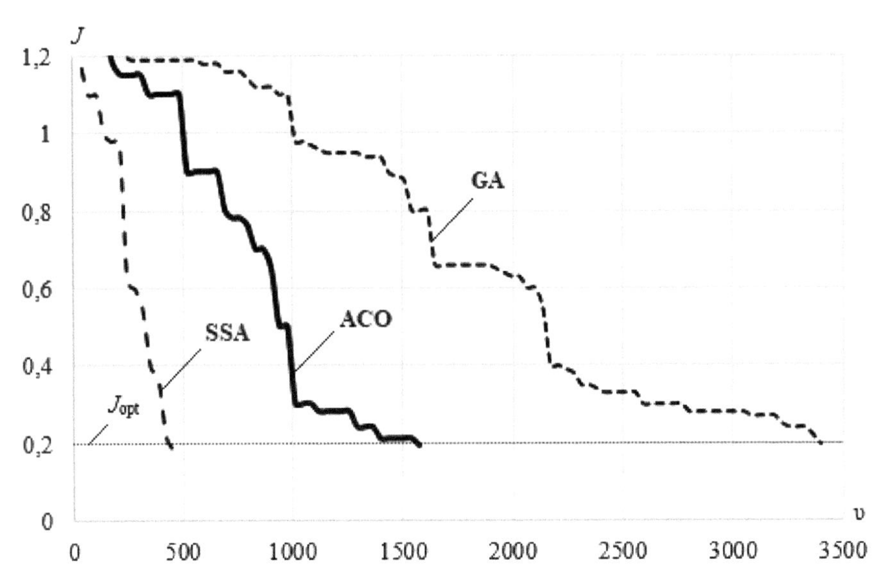


**Figure 4.4** Dynamics of change of the objective function's best values during the synthesis of consequents of the FC rule base for the second structure.

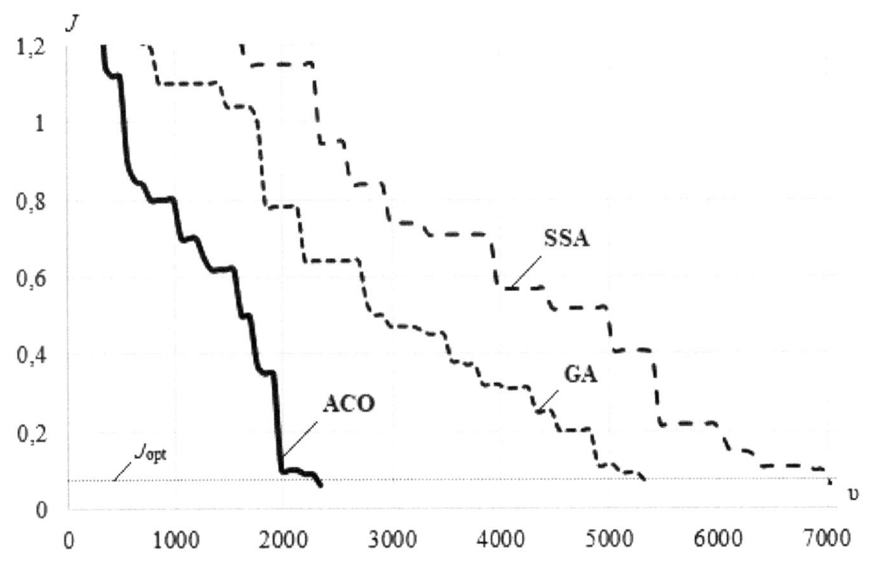


**Figure 4.5** Dynamics of change of the objective function's best values during the synthesis of consequents of the FC rule base for the sixth structure.

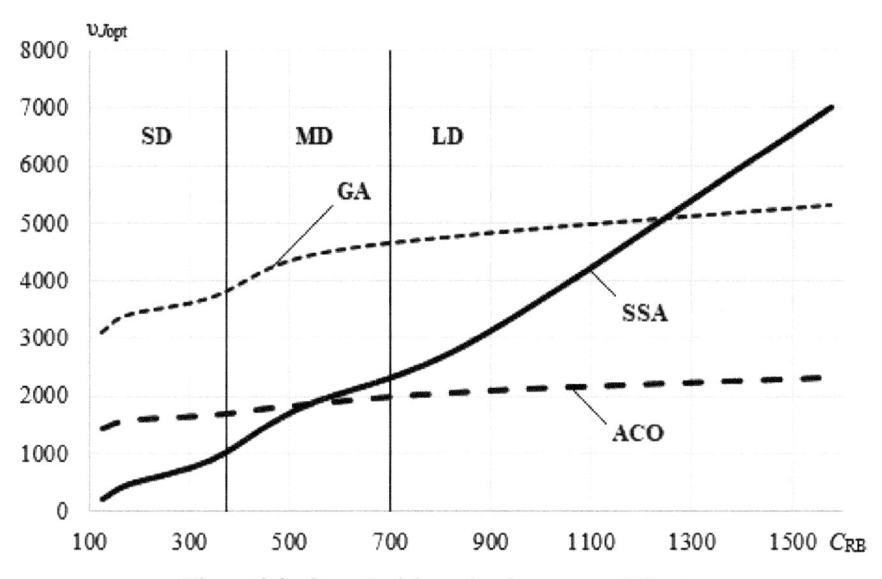


**Figure 4.6** Smoothed dependencies $\upsilon_{J\text{opt}} = f(C_{\text{RB}})$.

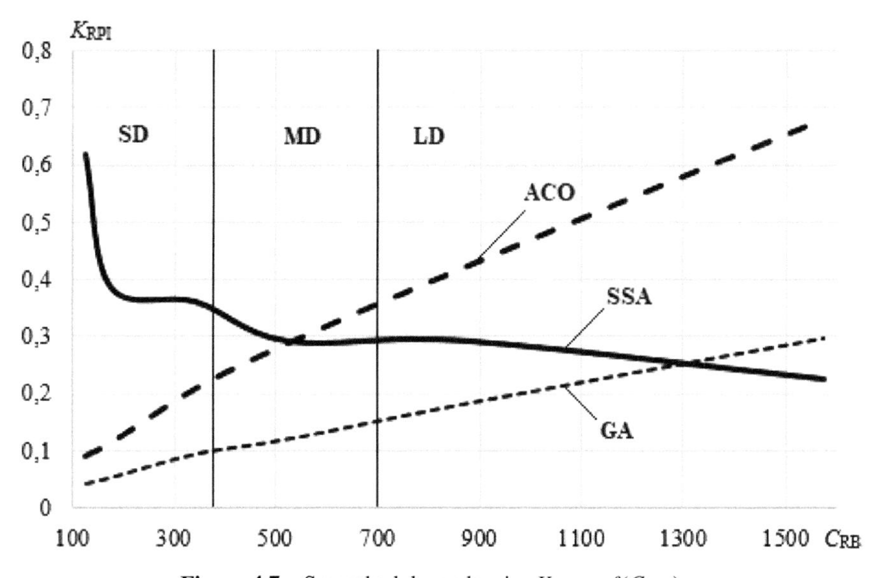


**Figure 4.7** Smoothed dependencies $K_{\text{RPI}} = f(C_{\text{RB}})$.

search is more efficient than bio-inspired algorithms (ACO and GA) since it allows to find optimal consequent vectors $\mathbf{R}_{opt}$, using significantly fewer calculations of the objective function $J$. In particular, for the first structure, SSA requires less by 1208 and 2898 calculations of the objective function than ACO and GA, respectively. The relative performance indicator $K_{RPI}$ is also significantly higher (0.618 for SSA, 0.088 for ACO, and 0.04 for GA). In turn, at the synthesis of the RBs of medium dimension (structure 4), SSA and ACO require almost the same amount of total number of calculations of the objective function $J$ for achieving its optimal value $J_{opt}$ (1804 and 1830). While for the GA, this parameter ($\upsilon_{Jopt}$) is much larger (4400). The relative performance indicator $K_{RPI}$ values are as follows: 0.291 for SSA, 0.287 for ACO, and 0.119 for GA. As for the RBs of large dimension (structures 5 and 6), for their synthesis, bio-inspired genetic and ACO algorithms are more efficient than SSA, as the parameter $\upsilon_{Jopt}$ is significantly less for them than for the SSA. The indicator $K_{RPI}$ is also significantly less (for structure 6: 0.224 for SSA, 0.681 for ACO, and 0.297 for GA). Also, an important feature of the SSA is that with an increase in the number of rules $s$ in the RB, the number of necessary rounds $l$ of consistent iterative procedures also increases to find the optimal vector of consequents $\mathbf{R}_{opt}$. So, for developing the RBs of the first structure (Table 4.1) it took two rounds ($l$ = 2), for developing the RBs of the second and third structures, three rounds ($l$ = 3), for developing the fourth and fifth structures, four rounds ($l$ = 4) and for sixth structures, five rounds ($l$ = 5).

Thus, the algorithm based on sequential search is quite efficient and of low cost (in terms of computational and time costs) at the synthesis of RBs of relatively small dimension ($C_{RB} \leq 370$). At the same time, when using this algorithm for the synthesis of RBs of medium and large dimension ($C_{RB} > 600$), the total amount of calculations increases significantly, the efficiency ratio $K_{RPI}$ decreases, and its application for solving this problem becomes unreasonable. In turn, the implementation of the ACO algorithm for the synthesis of small dimension RBs is inexpedient, as its coefficient $K_{RPI}$ is low enough and unnecessarily high computational and time costs are required. Here, with the increase of the complexity of synthesized RBs, the ratio $K_{RPI}$ significantly increases and it becomes the most effective for the synthesis of RBs of large dimension ($C_{RB} > 700$) in this case. As for GA, its use is also advisable only for systems of large dimension, although its efficiency indicator $K_{RPI}$ is somewhat lower than that of the ACO for all studied structures. Improving the efficiency of this algorithm can be achieved by using its various modifications by determining the optimal values of the tunable parameters, which is the subject of separate further studies.

Based on the analysis of the results obtained and all of the above, the following conclusions and recommendations can be made. At the synthesis of the RBs for the FACSs of small dimension ($C_{\mathrm{RB}} \leq 370$), it is advisable to implement the algorithm based on sequential search. At designing the RBs for the FACSs of medium dimension ($370 \leq C_{\mathrm{RB}} \leq 700$), the SSA and bio-inspired algorithms can be used. In turn, at developing and optimizing the RBs for the systems of large dimension ($C_{\mathrm{RB}} > 700$), it is reasonable to use only bio-inspired algorithms, in particular, the ACO algorithm.

The best variants of the RBs obtained in the synthesis process by means of presented algorithms are given in Tables 4.3 and 4.4 for the structures 1–3, respectively.

In turn, RBs presented in Table 4.3 are developed for structures 1 and 2 with the help of SSA, and RB presented in Table 4.4 is designed for structure 3 using the ACO algorithm.

The best variants of the RB fragments obtained in the synthesis process by means of presented algorithms are given in Tables 4.5–4.7 for structures 4–6, respectively. In turn, the RB fragment presented in Table 4.5 is developed for structure 4 with the help of SSA, the RB fragment given in Table 4.6 is

**Table 4.3** RBs synthesized for structures 1/2 with the help of SSA.

| $k_{\mathrm{P}}\varepsilon$ | $k_{\mathrm{D}}\dot{\varepsilon}$ | | | | |
|---|---|---|---|---|---|
| | **BN** | **SN** | **Z** | **SP** | **BP** |
| **BN** | BN | BN | BN | BN/N | SN |
| **SN** | BN | BN/N | SN | SN | SN |
| **Z** | BN/N | SN | Z | SP | BP/P |
| **SP** | SP | SP | SP | BP/P | BP |
| **BP** | SP | BP/P | BP | BP | BP |

**Table 4.4** RB synthesized for structure 3 with the help of ACO.

| $k_{\mathrm{P}}\varepsilon$ | $k_{\mathrm{D}}\dot{\varepsilon}$ | | | | | | |
|---|---|---|---|---|---|---|---|
| | **BN** | **N** | **SN** | **Z** | **SP** | **P** | **BP** |
| **BN** | BN | BN | BN | BN | N | N | SN |
| **N** | BN | BN | BN | N | N | SN | SN |
| **SN** | BN | BN | N | SN | SN | SN | SN |
| **Z** | BN | N | SN | Z | SP | P | BP |
| **SP** | SP | SP | SP | SP | P | BP | BP |
| **P** | SP | SP | P | P | BP | BP | BP |
| **BP** | SP | P | P | BP | BP | BP | BP |

**Table 4.5** Fragment of RB synthesized for structure 4 with the help of SSA.

| **Rule number** | **Input and output variables** | | | |
|---|---|---|---|---|
| | $k_P\varepsilon$ | $k_D\dot{\varepsilon}$ | $k_I\int\varepsilon\mathrm{dt}$ | $u_{FC}$ |
| 1 | BN | BN | N | BN |
| 12 | BN | SP | P | BN |
| 35 | Z | SN | Z | SN |
| 54 | SP | Z | P | SP |
| 75 | BP | BP | P | BP |

**Table 4.6** Fragment of RB synthesized for structure 5 with the help of GA.

| **Rule number** | **Input and output variables** | | | |
|---|---|---|---|---|
| | $k_P\varepsilon$ | $k_D\dot{\varepsilon}$ | $k_I\int\varepsilon\mathrm{dt}$ | $u_{FC}$ |
| 1 | BN | BN | BN | BN |
| 27 | SN | BN | SN | BN |
| 53 | Z | BN | Z | SN |
| 87 | SP | Z | SN | SP |
| 125 | BP | BP | BP | BP |

**Table 4.7** Fragment of RB synthesized for structure 6 with the help of ACO.

| **Rule number** | **Input and output variables** | | | |
|---|---|---|---|---|
| | $k_P\varepsilon$ | $k_D\dot{\varepsilon}$ | $k_I\int\varepsilon\mathrm{dt}$ | $u_{FC}$ |
| 1 | BN | BN | BN | VBN |
| 42 | N | SP | SN | N |
| 74 | SN | BP | SP | SN |
| 132 | P | SN | SN | P |
| 175 | BP | BP | BP | VBP |

designed for structure 5 using GA, and the RB fragment given in Table 4.7 obtained for the structure 6 by means of ACO.

Furthermore, the full vectors of consequents $\mathbf{R}_{\mathrm{opt}}$ for these structures (4–6) have the following forms.

For structure 4 $\mathbf{R}_{\mathrm{opt}}$ = (BN, BN, BN, BN, BN, BN, BN, BN, BN, BN, BN, BN, BN, BN, BN, BN, BN, BN, BN, BN, N, N, SN, SN, N, SN, SN, N, SN, SN, BN, N, N, N, SN, SN, SN, Z, SP, SP, SP, P, P, BP, BP, SP, SP, SP, SP, SP, P, SP, SP, SP, P, P, BP, P, BP, BP, BP, BP, BP, BP, BP, BP, BP, BP, BP, BP, BP, BP, BP, BP, BP).

For structure 5, $\mathbf{R}_{\mathrm{opt}}$ = (BN, BN, BN, BN, BN, BN, BN, BN, BN, BN, BN, BN, BN, BN, BN, BN, BN, BN, BN, BN, BN, BN, BN, BN, BN, BN, BN, BN, BN, BN, BN, BN, BN, BN, N, BN, N, SN, SN, SN, BN, N, SN, SN, SN, N, N, SN, SN, SN, N, N, SN, SN, Z, SN, SN, SN, SN, Z, SN, SN, Z, SP, SP, Z, SP, SP, SP, P, SP, P, P, BP, BP, SP, SP, SP, SP, P, SP, SP, SP, SP, P, SP,

SP, SP, SP, P, P, P, P, P, BP, BP, BP, BP, BP, BP, BP, BP, BP, BP, BP, BP, BP, BP, BP, BP, BP, BP, BP, BP, BP, BP, BP, BP, BP, BP, BP, BP, BP, BP).

For structure 6, $\mathbf{R}_{\text{opt}}$ = (VBN, VBN, VBN, VBN, VBN, VBN, VBN, VBN, VBN, VBN, VBN, VBN, VBN, VBN, VBN, VBN, VBN, VBN, VBN, VBN, VBN, VBN, VBN, VBN, VBN, VBN, VBN, VBN, VBN, VBN, VBN, BN, VBN, VBN, BN, BN, N, BN, BN, N, N, N, N, N, N, SN, SN, N, N, SN, SN, Z, VBN, VBN, BN, BN, BN, BN, N, N, N, SN, N, N, SN, SN, SN, SN, SN, SN, SN, Z, N, N, SN, SN, SN, BN, BN, N, N, SN, SN, SN, SN, SN, Z, SN, Z, Z, Z, SP, Z, SP, SP, SP, P, SP, SP, P, P, BP, SP, SP, SP, SP, P, SP, SP, SP, SP, SP, SP, SP, SP, SP, P, SP, P, P, BP, BP, P, BP, BP, BP, VBP, SP, SP, P, BP, BP, P, P, P, P, BP, P, P, P, P, BP, P, BP, BP, BP, VBP, BP, VBP, VBP, VBP, VBP, VBP, VBP, VBP, VBP, VBP, VBP, VBP, VBP, VBP, VBP, VBP, VBP, VBP, VBP, VBP, VBP, VBP, VBP, VBP, VBP, VBP, VBP, VBP, VBP).

Figure 4.8 shows the characteristic surfaces $u_{\text{FC}} = f_{\text{FC}}(k_{\text{P}}\varepsilon, k_{\text{D}}\dot{\varepsilon})$ of the speed fuzzy controller with the developed RBs for its different structures: *a* – for structure 2; *b* and *c* – for structures 4 and 6, respectively, at a fixed value $k_{\text{I}} \int \varepsilon dt = 0.5$.

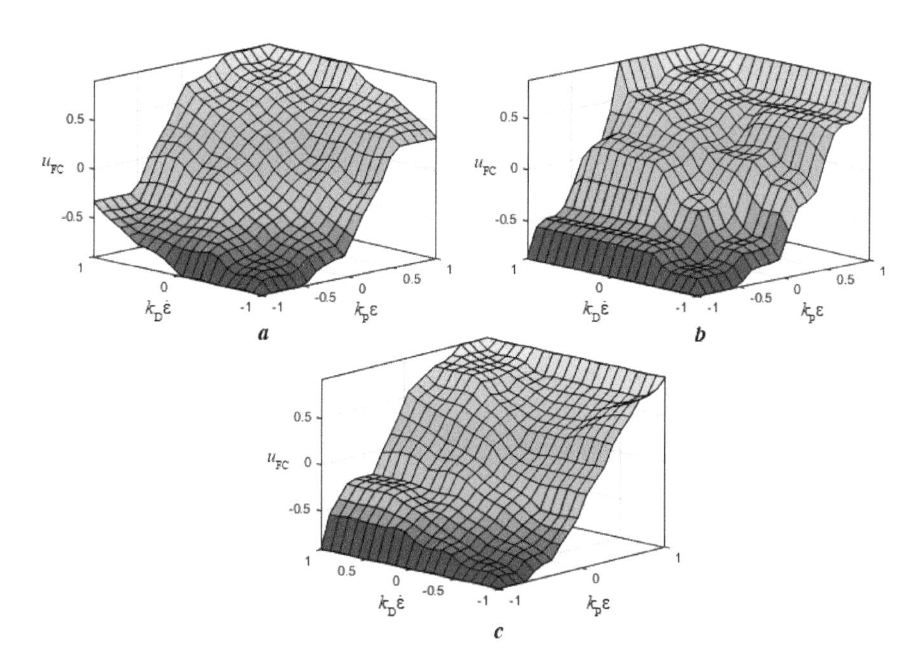


**Figure 4.8** Characteristic surfaces $u_{\text{FC}} = f_{\text{FC}}(k_{\text{P}}\varepsilon, k_{\text{D}}\dot{\varepsilon})$ of the speed fuzzy controller with the developed RBs.

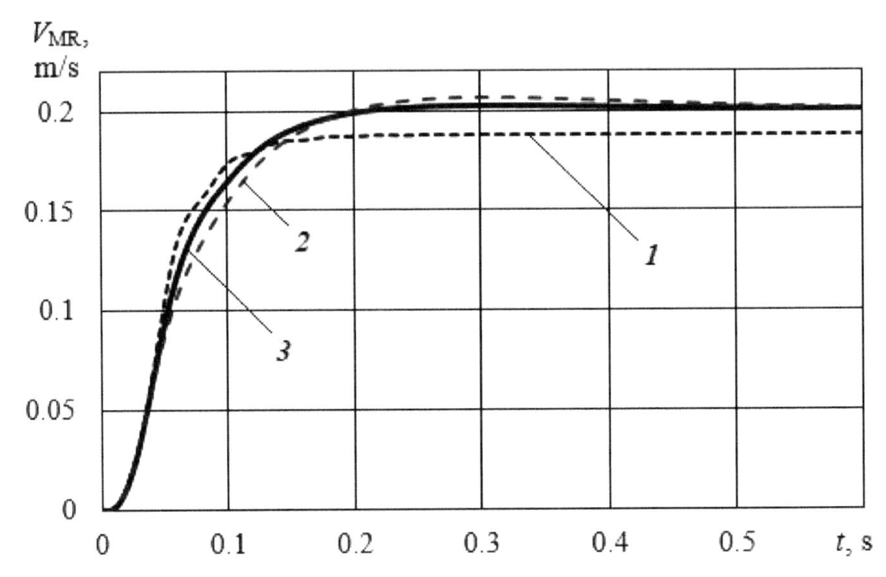


**Figure 4.9** Speed transients for the FACS of the mobile robot.

Figure 4.9 presents the accelerating characteristics of the multipurpose MR when moving along an inclined ferromagnetic surface at the following parameters: set value of the MR speed $V_S = 0.2$ m/s; permanent disturbance in the form of the load force of the technological operation $F_{TO} = 900$ N; angle of inclination of the working surface $\gamma = 60°$.

These transient graphs of MR speed change are obtained for the speed FACS with the developed FC for different structures of the RBs: 1 – structure 2; 2 – structure 4; 3 –structure 6.

Figure 4.9 shows that fuzzy control system of the MR speed with the developed RBs based on the presented algorithms have enough high control quality indicators, which confirms their high efficiency. In addition, the more complicated the structure of the developed FC, the higher the quality of control; however, the computational costs at the synthesis process also increases, as well as further software and hardware implementation of the controller.

In turn, if the problem arises for further improvement of the controller for the given MR speed FACS as well as simplification of its software and hardware implementation, after synthesis of the RB consequents vector $\mathbf{R}_X$ by means of the presented algorithms, it is possible to conduct additional structural-parametric optimization of the fuzzy controller, in particular, optimization of: the vector $\mathbf{X}_{\mathrm{IS}}$ that defines the FC input variables; the vector $\mathbf{X}_{\mathrm{IT}}$ that determines the number, types, and parameters of LTs of the FC input variables; the vector $\mathbf{X}_{\mathrm{FIE}}$ that defines the operations of aggregation, activation, and accumulation of the fuzzy inference engine; the vector $\mathbf{X}_{\mathrm{DF}}$ that defines the number, types, and parameters of LTs of the FC output variable $u_{\mathrm{FC}}$, as well as its defuzzification method.

## 4.5 Conclusion

The efficiency research and comparative analysis of the bio-inspired algorithms of automatic synthesis and optimization of RBs for the Mamdani-type FACSs is presented in this work. In particular, the ACO and GAs are investigated, which are adapted to the specifics of the task of automatic RBs synthesis with finding optimal consequents $\mathbf{R}_{\mathrm{opt}}$ at insufficient initial information (in conditions of a high degree of information uncertainty), at a sufficiently large number of rules for which the compilation of the FACS RB on the basis of expert knowledge is not always effective, and at the various levels of expert qualifications. Also, these bio-inspired algorithms are compared with an algorithm of automatic RB synthesis based on sequential search. To evaluate the effectiveness of the presented algorithms, the specialized criteria are introduced: the universal indicator for the quantitative assessment of the RB complexity $C_{\mathrm{RB}}$ and relative performance indicator $K_{\mathrm{RPI}}$. Moreover, the dependence of the relative performance indicator $K_{\mathrm{RPI}}$ on the complexity universal indicator $C_{\mathrm{RB}}$ is used for the comparison of the presented algorithms at different structures of the developed RBs.

The considered ACO algorithm, GA, and SSA are applied for the development of various RB configurations (six different structures) of the fuzzy control system for the multipurpose MR that is able to move on inclined and vertical ferromagnetic surfaces. The studies conducted on this particular example show that all three of the presented algorithms allow effectively synthesizing and optimizing RBs with the formation of optimal consequents $\mathbf{R}_{\mathrm{opt}}$ to achieve the desired values of the objective function that is confirmed by the simulation results of the MR control system. However, at the synthesizing of RBs of small dimension (structures 1–3), the SSA is more efficient than bio-inspired algorithms (ACO and GA) since it allows to find optimal consequent vectors $\mathbf{R}_{\mathrm{opt}}$, using significantly fewer calculations of the objective function. The relative performance indicator $K_{\mathrm{RPI}}$ for this algorithm is also significantly higher in this case (for example, for the first structure: 0.618 for SSA, 0.088 for ACO, and 0.04 for GA). In turn, at the synthesis of the RBs of medium dimension (structure 4), SSA and ACO require almost the same amount of calculations of the objective function for achieving its optimal value $J_{\text{opt}}$. While for the GA, this parameter is much larger. The values of the criterion $K_{\text{RPI}}$ for this structure are: 0.291 for SSA, 0.287 for ACO, and 0.119 for GA. As for the RBs of large dimension (structures 5 and 6), for their synthesis, bio-inspired ACO and GAs are more efficient than SSA, as the total number of the objective function calculations is significantly less for them than for the SSA. This is also confirmed by the indicator $K_{\text{RPI}}$ that is also significantly less (for example, for the sixth structure: 0.224 for SSA, 0.681 for ACO, and 0.297 for GA).

The obtained results of the efficiency research and comparative analysis provide an opportunity to formulate the following recommendations for rational using the investigated algorithms. The implementation of the bio-inspired algorithms of automatic synthesis and optimization of the FACSs RBs, in particular, ACO and GA, is expedient for the synthesis of predominantly medium and large dimension RBs ($C_{\text{RB}} > 600$). Moreover, the efficiency of these algorithms significantly increases with increase of the optimized system dimension. As for RBs of the small dimension ($C_{\text{RB}} \leq 370$), the SSA is the most efficient and of low cost for their synthesis and optimization. And the bio-inspired algorithms' application is unreasonable in this case. When comparing the ACO algorithm with GA, then in most cases, ACO has higher performance indicator $K_{\text{RPI}}$ than GA and requires less computational and time costs at the same complexity of the synthesized and optimized RB.

Further research should be conducted toward the studies of various modifications of ACO and GA as well as other bio-inspired algorithms (particle swarm optimization, bee colony optimization, biogeography-based optimization, bacterial foraging optimization, artificial immune systems, etc.) at solving the problem of the RBs synthesis and optimization of the different types of FACSs.

## References

[1] D. S. Huang, K. Li, G. W. Irwin, 'Intelligent Control and Automation', in Lecture Notes in Control and Information Sciences, Vol. 344. Springer, Berlin, Heidelberg, 1121 p., 2006.

[2] S. N. Vassilyev, A. Yu. Kelina, Y. I. Kudinov, F. F. Pashchenko, 'Intelligent Control Systems', in Procedia Computer Science, Vol. 103, pp. 623-628, 2017.

[3] Y. P. Kondratenko, O. V. Kozlov, 'Mathematic Modeling of Reactor's Temperature Mode of Multiloop Pyrolysis Plant', in book: Lecture Notes in Business Information Processing: Modeling and Simulation in Engineering, Economics and Management, K. J. Engemann, A. M. Gil-Lafuente, J. M. Merigo, Eds., vol. 115. Berlin, Heidelberg: Springer-Verlag, pp. 178-187, 2012.

[4] G. Krivulya, I. Skarga-Bandurova, Z. Tatarchenko, O. Seredina, M. Shcherbakova, E. Shcherbakov, 'An intelligent functional diagnostics of wireless sensor network', in Proceedings of 2019 International Conference on Future Internet of Things and Cloud Workshops, FiCloudW 2019, pp. 135-139, 2019.

[5] Y. P. Kondratenko, O. V. Kozlov, O. S. Gerasin, A. M. Topalov, O. V. Korobko, 'Automation of Control Processes in Specialized Pyrolysis Complexes Based on Web SCADA Systems', in Proceedings of the $9^{th}$ IEEE International Conference on Intelligent Data Acquisition and Advanced Computing Systems: Technology and Applications (IDAACS), Vol. 1, Bucharest, Romania, pp. 107-112, 2017.

[6] Y. P. Kondratenko, O. V. Korobko, O. V. Kozlov, 'Synthesis and Optimization of Fuzzy Controller for Thermoacoustic Plant', in Recent Developments and New Direction in Soft-Computing Foundations and Applications, Studies in Fuzziness and Soft Computing 342, Lotfi A. Zadeh et al. (Eds.), Berlin, Heidelberg: Springer-Verlag, pp. 453-467, 2016.

[7] L. A. Zadeh, 'Fuzzy Sets', in Information & Control, 8, pp. 338-353, 1965.

[8] L. A. Zadeh, 'The role of fuzzy logic in modeling, identification and control', in Modeling Identification and Control, 15 (3), pp. 191-203, 1994.

[9] L. A. Zadeh, A. M. Abbasov, R. R. Yager, S. N. Shahbazova, M. Z .Reformat, Eds., 'Recent Developments and New Directions in Soft Computing', STUDFUZ 317, Cham: Springer, 2014.

[10] J. Kacprzyk, 'Multistage Fuzzy Control: A Prescriptive Approach', John Wiley & Sons, Inc., New York, NY, USA, 1997.

[11] J. M .Mendel, 'Uncertain Rule-Based Fuzzy Systems, Introduction and New Directions', Second Edition, Springer International Publishing, 684 p.,2017.

[12] M. Solesvik, Y. Kondratenko, G. Kondratenko, I. Sidenko, V. Kharchenko, A. Boyarchuk, 'Fuzzy decision support systems in marine

practice', in: Fuzzy Systems (FUZZ-IEEE), 2017 IEEE International Conference on Fuzzy Systems, pp. 1-6, 2017.

[13] E. H. Mamdani, 'Application of fuzzy algorithms for control of simple dynamic plant', in Proceedings of IEEE, vol. 121, pp. 1585-1588, 1974.

[14] E. H. Mamdani, S. Assilian, 'An experiment in linguistic synthesis with a fuzzy logic controller', in International journal of man-machine studies, No. 7(1), pp. 1-13, 1975.

[15] A. Piegat, 'Fuzzy Modeling and Control', New York, Heidelberg: Physica-Verlag, 2001.

[16] B. Kosko, 'Fuzzy Systems as Universal Approximators', in IEEE Trans. on Computers, Vol. 43, Iss. 11, pp. 1329-1333, 1994.

[17] G. V. Kondratenko, Y. P. Kondratenko, D. O. Romanov, 'Fuzzy Models for Capacitive Vehicle Routing Problem in Uncertainty', in Proc. 17th International DAAAM Symposium "Intelligent Manufacturing and Automation: Focus on Mechatronics & Robotics", Vienna, Austria, pp. 205-206, 2006.

[18] Y. Kondratenko, G. Kondratenko, I. Sidenko, 'Two-stage method of fuzzy rule base correction for variable structure of input vector', in 2017 IEEE First Ukraine Conference on Electrical and Computer Engineering (UKRCON), Kiev, Ukraine, pp. 1043-1049, 2017.

[19] T. Takagi, M. Sugeno, 'Fuzzy identification of systems and its applications to modeling and control', in IEEE Transactions on Systems, Man, and Cybernetics, SMC–15, N 1, P. 116-132, 1985.

[20] Y. P. Kondratenko, O. V. Kozlov, G. V. Kondratenko, I. P. Atamanyuk, 'Mathematical Model and Parametrical Identification of Ecopyrogenesis Plant Based on Soft Computing Techniques', in Complex Systems: Solutions and Challenges in Economics, Management and Engineering, Christian Berger-Vachon, Anna María Gil Lafuente, Janusz Kacprzyk, Yuriy Kondratenko, José M. Merigó, Carlo Francesco Morabito (Eds.), Book Series: Studies in Systems, Decision and Control, Vol. 125, Berlin, Heidelberg: Springer International Publishing, pp. 201-233, 2018.

[21] Y. P. Kondratenko, O. V. Kozlov, 'Mathematical Model of Ecopyrogenesis Reactor with Fuzzy Parametrical Identification', in: Recent Developments and New Direction in Soft-Computing Foundations and Applications, Studies in Fuzziness and Soft Computing 342, Lotfi A. Zadeh et al. (Eds.). Berlin, Heidelberg: Springer-Verlag, pp. 439-451, 2016.

[22] Y. P. Kondratenko, N. Y. Kondratenko, 'Synthesis of Analytic Models for Subtraction of Fuzzy Numbers with Various Membership Function's Shapes', in: Gil-Lafuente A., Merigó J., Dass B., Verma R. (eds) Applied Mathematics and Computational Intelligence, FIM 2015, Advances in Intelligent Systems and Computing, Vol. 730, Springer, Cham, pp. 87-100, 2018.

[23] J.-S. R. Jang, 'ANFIS: Adaptive-Network-based Fuzzy Inference Systems', in IEEE Transactions on Systems, Man, and Cybernetics, Vol. 23, No. 3, pp. 665-685, 1993.

[24] J.-S. R. Jang, C.-T. Sun, E. Mizutani, 'Neuro-Fuzzy and Soft Computing: A Computational Approach to Learning and Machine Intelligence', Prentice Hall, 1996.

[25] M. Jamshidi, V. Kreinovich, J. Kacprzyk, Eds., 'Advance Trends in Soft Computing', Cham: Springer-Verlag, 2013.

[26] Z. Xiao, J. Guo, H. Zeng, P. Zhou, S. Wang, 'Application of Fuzzy Neural Network Controller in Hydropower Generator Unit', in J. Kybernetes, Vol. 38, No. 10, pp. 1709-1717, 2009.

[27] O. Kozlov, G. Kondratenko, Z. Gomolka, Y. Kondratenko, 'Synthesis and Optimization of Green Fuzzy Controllers for the Reactors of the Specialized Pyrolysis Plants', in: Kharchenko V., Kondratenko Y., Kacprzyk J. (eds) Green IT Engineering: Social, Business and Industrial Applications, Studies in Systems, Decision and Control, Vol 171, Springer, Cham, pp. 373-396, 2019.

[28] O. Kosheleva, V. Kreinovich, 'Why Bellman-Zadeh approach to fuzzy optimization', in Appl. Math. Sci. Vol. 12, pp. 517-522, 2018.

[29] Y. P. Kondratenko, T. A .Altameem, E. Y. M. Al Zubi, 'The optimization of digital controllers for fuzzy systems design', in Advances in Modelling and Analysis, AMSE Periodicals, Series A 47, pp. 19-29, 2010.

[30] W. A. Lodwick, J. Kacprzhyk, Eds., 'Fuzzy Optimization', STUDFUZ 254, Berlin, Heidelberg: Springer-Verlag, 2010.

[31] D. Simon, 'H$\infty$ estimation for fuzzy membership function optimization', in International Journal of Approximate Reasoning, 40, pp. 224-242, 2005.

[32] J. Zhu, F. Lauri, A. Koukam, V. Hilaire, 'Fuzzy Logic Control Optimized by Artificial Immune System for Building Thermal Condition', in: Siarry P., Idoumghar L., Lepagnot J. (eds) Swarm Intelligence Based Optimization, ICSIBO 2014, Lecture Notes in Computer Science, Vol. 8472. Springer, Cham, pp. 42-49, 2014.

[33] A. Melendez, O. Castillo, 'Evolutionary optimization of the fuzzy integrator in a navigation system for a mobile robot', in: Recent Advances on Hybrid Intelligent Systems, pp. 21-31, 2013.
[34] S. Khan, et al., 'Design and Implementation of an Optimal Fuzzy Logic Controller Using Genetic Algorithm', in Journal of Computer Science, Vol. 4, No. 10, pp. 799-806, 2008.
[35] R. Hampel, M. Wagenknecht, N. Chaker, 'Fuzzy Control: Theory and Practice, New York: Physika-Verlag, Heidelberg, 410 p., 2000
[36] K. Tanaka, H. O. Wang, 'Fuzzy Control Systems Design and Analysis: A Linear Matrix Inequality Approach', John Wiley & Sons, New York, USA, 2001.
[37] V. Opanasenko, S. Kryvyi, 'Synthesis of multilevel structures with multiple outputs', in $10^{th}$ International Conference on Programming, UkrPROG, Kyiv, Ukraine, pp. 32-37, 2016.
[38] D. Driankov, H. Hellendoorn, M. Reinfrank, 'An introduction to fuzzy control', Springer Science & Business Media, 2013.
[39] Q. Suna, R. Li, P. Zhang, 'Stable and Optimal Adaptive Fuzzy Control of Complex Systems Using Fuzzy Dynamic Model', in J. Fuzzy Sets and Systems, Vol. 133, pp. 1-17, 2003.
[40] B. Jayaram, 'Rule reduction for efficient inferencing in similarity based reasoning', in International Journal of Approximate Reasoning 48, no. 1, pp. 156-173, 2008.
[41] Y. Yam, P. Baranyi, C.- T. Yang, 'Reduction of fuzzy rule base via singular value decomposition', in IEEE Transactions on Fuzzy Systems, 7, no. 2, pp. 120-132, 1999.
[42] Y. P. Kondratenko, D. Simon, 'Structural and parametric optimization of fuzzy control and decision making systems', in: Zadeh L., Yager R., Shahbazova S., Reformat M., Kreinovich V. (eds), Recent Developments and the New Direction in Soft-Computing Foundations and Applications, Studies in Fuzziness and Soft Computing, Vol. 361, Springer, Cham, pp. 273-289, 2018.
[43] Y. Kondratenko, P. Khalaf, H. Richter, D. Simon, 'Fuzzy Real-Time Multiobjective Optimization of a Prosthesis Test Robot Control System', in Yuriy P. Kondratenko, Arkadii A. Chikrii, Vyacheslav F. Gubarev, Janusz Kacprzyk (Eds) Advanced Control Techniques in Complex Engineering Systems: Theory and Applications, Dedicated to Professor Vsevolod M. Kuntsevich, Studies in Systems, Decision and Control, Vol. 203. Cham: Springer Nature Switzerland AG, pp. 165-185, 2019.

[44] Y. P. Kondratenko, E. Y. M. Al Zubi, 'The Optimisation Approach for Increasing Efficiency of Digital Fuzzy Controllers', in Annals of DAAAM for 2009 & Proceeding of the 20th Int. DAAAM Symp. "Intelligent Manufacturing and Automation", Published by DAAAM International, Vienna, Austria, pp. 1589-1591, 2009.
[45] D. Simon, 'Evolutionary Optimization Algorithms: Biologically Inspired and Population-Based Approaches to Computer Intelligence', John Wiley & Sons, 2013.
[46] S. K. Oh, W. Pedrycz, 'The Design of Hybrid Fuzzy Controllers Based on Genetic Algorithms and Estimation Techniques', in J. Kybernetes, Vol. 31, No. 6, pp. 909-917, 2002.
[47] H. Ishibuchi, T. Yamamoto, 'Fuzzy rule selection by multi-objective genetic local search algorithms and rule evaluation measures in data mining', in Fuzzy Sets and Systems 141, no. 1, pp. 59-88, 2004.
[48] R. Alcalá, J. Alcalá-Fdez, M. J. Gacto, F. Herrera, 'Rule base reduction and genetic tuning of fuzzy systems based on the linguistic 3-tuples representation', in Soft Computing 11, no. 5, pp. 401-419, 2007.
[49] W. Pedrycz, K. Li, M. Reformat, 'Evolutionary reduction of fuzzy rule-based models', in Fifty Years of Fuzzy Logic and its Applications, STUDFUZ 326, Cham: Springer, pp. 459-481, 2015.
[50] R. Haupt, S. Haupt, 'Practical genetic algorithms', New Jersey: John Wiley & Sons, 261 p., 2004.
[51] A. Prakash, S. G. Deshmukh, 'A multi-criteria customer allocation problem in supply chain environment: An artificial immune system with fuzzy logic controller based approach', in Expert Systems with Applications, Volume 38, Issue 4, pp. 3199-3208, 2011.
[52] A. Nabi, N.A. Singh, 'Particle swarm optimization of fuzzy logic controller for voltage sag improvement', in Proceedings of 2016 3rd International Conference on Advanced Computing and Communication Systems (ICACCS), vol. 01, pp. 1-5, 2016.
[53] A. Engelbrecht, 'A study of particle swarm optimization particle trajectories', in Information Sciences, No 176(8), pp. 937-971, 2006.
[54] R. Menon, S. Menon, D. Srinivasan, L. Jain, 'Fuzzy logic decision-making in multi-agent systems for smart grids', in: Computational Intelligence Applications in Smart Grid (CIASG), 2013 IEEE Symposium, pp. 44-50, 2013.
[55] H. Ashby, R. V. Yampolskiy, 'Genetic Algorithm and Wisdom of Artificial Crowds Algorithm Applied to Light Up', in Proceedings of 16th International Conference on Computer Games, AI, Animation,

Mobile, Interactive Multimedia, Educational & Serious Games, Louisville, KY, USA, pp. 27-30, 2011.

[56] O. Cordon, F. Gomide, F. Herrera, F. Hoffmann, L. Magdalena, 'Ten Years of Genetic Fuzzy Systems: Current Framework and New trends', in Fuzzy Sets and Systems, Vol. 141, Iss. 1, pp. 5-31, 2004.

[57] M. Dorigo, M. Birattari, 'Ant Colony Optimization', in Encyclopedia of Machine Learning, Sammut C., Webb G.I. (eds.), Springer, Boston, MA, 2011.

[58] N. Quijano, K. M. Passino, 'Honey bee social foraging algorithms for resource allocation: theory and application', in Columbus: Publishing house of the Ohio State University, 39 p., 2007.

[59] S. Vaneshani, H. Jazayeri-Rad, 'Optimized Fuzzy Control by Particle Swarm Optimization Technique for Control of CSTR', in International Journal of Electrical and Computer Engineering, Vol. 5, No:11, pp. 1243-1248, 2011.

[60] D. H. Kim, C. H. Cho, 'Bacterial foraging based neural network fuzzy learning', Proceedings of the $2^{nd}$ Indian International Conference on Artificial Intelligence (IICAI – 2005), Pune: IICAI, pp. 2030-2036, 2005.

[61] L. Fan, E. M. Joo, 'Design for auto-tuning PID controller based on genetic algorithms', in Industrial Electronics and Applications, 2009, ICIEA 2009, 4th IEEE Conference on IEEE, pp. 1924-1928, 2009.

[62] F. L. Minku, T. Ludermir, 'Evolutionary strategies and genetic algorithms for dynamic parameter optimization of evolving fuzzy neural networks', in Evolutionary Computation, The 2005 IEEE Congress, Vol. 3, pp. 1951-1958, 2005.

[63] O. Castillo, P. Ochoa, J. Soria, 'Differential Evolution with Fuzzy Logic for Dynamic Adaptation of Parameters in Mathematical Function Optimization'. in: Angelov P., Sotirov S. (eds) Imprecision and Uncertainty in Information Representation and Processing. Studies in Fuzziness and Soft Computing, Vol. 332. Springer, Cham, pp. 361-374, 2016.

[64] R. T. Alves, M. R. Delgado, H. S. Lopes, A. A. Freitas, 'An Artificial Immune System for Fuzzy-Rule Induction in Data Mining', in: Yao X. et al. (eds) Parallel Problem Solving from Nature - PPSN VIII. PPSN 2004. Lecture Notes in Computer Science, Vol 3242. Springer, Berlin, Heidelberg, pp. 1011-1020, 2004.

[65] D. Simon, 'Biogeography-Based Optimization', in IEEE Transactions on Evolutionary Computation, 12(6), pp. 702-713, 2008.

[66] D. Simon, 'Sum Normal Optimization of Fuzzy Membership Functions', in International Journal of Uncertainty: Fuzziness and Knowledge-Based Systems, 10, pp. 363-384, 2002.

[67] J. Zhao, L. Han, L. Wang, Z. Yu, 'The fuzzy PID control optimized by genetic algorithm for trajectory tracking of robot arm', in 2016 12th World Congress on Intelligent Control and Automation (WCICA), Guilin, China, pp. 556-559, 2016.

[68] G. Thomas, P. Lozovyy, D. Simon, 'Fuzzy Robot Controller Tuning with Biogeography-Based Optimization', in Modern Approaches in Applied Intelligence: 24th International Conference on Industrial Engineering and Other Applications of Applied Intelligent Systems, IEA/AIE 2011, Syracuse, NY, USA, June 28 – July 1, Proceedings, Part II, pp. 319-327, 2011.

[69] Y. P. Kondratenko, A. V. Kozlov, 'Parametric optimization of fuzzy control systems based on hybrid particle swarm algorithms with elite strategy', in Journal of Automation and Information Sciences, Vol. 51, Issue 12, New York: Begel House Inc., 25-45, 2019.

[70] L. T. Koczy, K. Hirota, 'Size reduction by interpolation in fuzzy rule bases', in IEEE Transactions on Systems, Man, and Cybernetics, Part B: Cybernetics, 27, no. 1, pp. 14-25, 1997.

[71] D. Simon, 'Design and rule base reduction of a fuzzy filter for the estimation of motor currents', in International Journal of Approximate Reasoning, 25, pp. 145-167, 2000.

[72] Y. P. Kondratenko, L. P. Klymenko, E. Y. M. Al Zu'bi, 'Structural Optimization of Fuzzy Systems' Rules Base and Aggregation Models', in Kybernetes, Vol. 42, Iss. 5, pp. 831-843, 2013.

[73] Y. P. Kondratenko, O. V. Kozlov, O. V. Korobko, 'Two Modifications of the Automatic Rule Base Synthesis for Fuzzy Control and Decision Making Systems', in: J. Medina et al. (Eds), Information Processing and Management of Uncertainty in Knowledge-Based Systems: Theory and Foundations, 17th International Conference, IPMU 2018, Cadiz, Spain, Proceedings, Part II,, CCIS 854, Springer International Publishing AG, pp. 570-582, 2018.

[74] Y. P. Kondratenko, A. V. Kozlov, 'Generation of Rule Bases of Fuzzy Systems Based on Modified Ant Colony Algorithms', in Journal of Automation and Information Sciences, Vol. 51, Issue 3, New York: Begel House Inc., pp. 4-25, 2019.

[75] A. P. Rotshtein, H. B. Rakytyanska, 'Fuzzy evidence in identification, forecasting and diagnosis', Vol. 275, Heidelberg: Springer, 2012.

[76] C. Von Altrock, 'Applying fuzzy logic to business and finance', in Optimus, 2, pp. 38-39, 2002.
[77] A. Gozhyj, I. Kalinina, V. Vysotska, V. Gozhyj, 'The method of web-resources management under conditions of uncertainty based on fuzzy logic', in Proceedings of the 2018 IEEE 13th International Scientific and Technical Conference on Computer Sciences and Information Technologies, (CSIT 2018), Vol. 1., "Vega and Ko", Lviv, pp. 347-352, 2018.
[78] M. Pasieka, N. Grzesik, K. Kuźma, 'Simulation modeling of fuzzy logic controller for aircraft engines', in International Journal of Computing, 16(1), pp. 27-33, 2017.
[79] B. Benhala, A. Ahaitouf, M. Fakhfakh, A. Mechaqrane, 'New Adaptation of the ACO Algorithm for the Analog Circuits Design Optimization' in International Journal of Computer Science (IJCSI), Vol. 9, no. 3, pp. 360-367, 2012.
[80] J.- Y. Potvin, 'Genetic algorithms for the traveling salesman problem', in Annals of Operations Research, Vol. 63, pp. 339-370, 1996.
[81] Y. Nagata, D. Soler, 'A new genetic algorithm for the asymmetric traveling salesman problem', in Expert Systems with Applications, Vol. 39, No. 10, pp. 8947-8953, 2012.
[82] R. Gan, Q. Guo, H. Chang, Y. Yi, 'Improved Ant Colony Optimization Algorithm for the Traveling Salesman Problems', in Journal of Systems Engineering and Electronics, pp. 329-333, 2010.
[83] A. Piwonska, 'Genetic algorithm finds routes in travelling salesman problem with profits', in Zeszyty Naukowe Politechniki Bia lostockiej, Informatyka, pp. 51-65, 2010.
[84] R.-M. Chen, Y.-M. Shen, C.-T. Wang, 'Ant Colony Optimization Inspired Swarm Optimization for Grid Task Scheduling', in 2016 International Symposium on Computer, Consumer and Control (IS3C), pp. 461-464, 2016.
[85] A. Sadegheih, 'Scheduling problem using genetic algorithm, simulated annealing and the effects of parameter values on GA performance', in Applied Mathematical Modelling, Vol. 30, Iss. 2, pp. 147-154, 2006.
[86] A. Sadegheih, P. R. Drake, 'Network optimization using linear programming and genetic algorithm, Neural Network World', in Int. J. Non-Stand. Comput. Artif. Intell., Vol. 11, Iss. 3, pp. 223-233, 2001.
[87] Q. Chengming, 'Vehicle Routing Optimization in Logistics Distribution Using Hybrid Ant Colony Algorithm', in TELKOMNIKA Indonesian Journal of Electrical Engineering, vol. 11, no. 9, pp. 5308-5315, 2013.

[88] X. Sun, J. Wang, 'Routing design and fleet allocation optimization of freeway service patrol: Improved results using genetic algorithm', in Phys. A Stat. Mech. Appl., Vol. 501, pp. 205-216, 2018.
[89] R. K. Arakaki, F. L. Usberti, 'Hybrid genetic algorithm for the open capacitated arc routing problem', in Comput. Oper. Res., Vol. 90, pp. 221-231, 2018.
[90] R. Abbasian, M. Mouhoub, 'An efficient hierarchical parallel genetic algorithm for graph coloring problem', in Proceedings of The 13th annual conference on Genetic and evolutionary computation, ACM, Dublin, Ireland, pp. 521-528, 2011.
[91] C. Croitoru, H. Luchian, O. Gheorghies, A. Apetrei, 'A New Genetic Graph Coloring Heuristic', in Proceedings of The Computational Symposium on Graph Coloring and its Generalizations, Ithaca, New York, USA, pp. 63-74, 2002.
[92] C. A. Glass, A. Prugel-Bennett, 'Genetic algorithm for graph coloring: Exploration of Galinier and Hao's algorithm', in Journal of Combinatorial Optimization, Vol. 3, pp. 229-236, 2003.
[93] Y. Khaluf, S. Gullipalli, 'An Efficient Ant Colony System for Edge Detection in Image Processing', in Proceedings of the European Conference on Artificial Life, pp. 398-405, 2015.
[94] R. Putha, L. Quadrifoglio, E. Zechman, 'Comparing ant colony optimization and genetic algorithm approaches for solving traffic signal coordination under oversaturation conditions', in Comput. Aided Civ. Infrastruct. Eng. Vol. 27, pp. 14-28, 2012.
[95] G. Narvydas, R. Simutis, V. Raudonis, 'Autonomous Mobile Robot Control Using Fuzzy Logic and Genetic Algorithm', in IEEE International Workshop on Intelligent Data Acquisition and Advanced Computing Systems: Technology and Applications, Dortmund, Germany, pp. 460-464, 2007.
[96] Y. P. Kondratenko, J. Rudolph, O. V. Kozlov, Y. M. Zaporozhets, O. S. Gerasin, 'Neuro-fuzzy observers of clamping force for magnetically operated movers of mobile robots', in Technical Electrodynamics, № 5, pp. 53-61, 2017 (in Ukrainian).
[97] Y. P. Kondratenko, Y. M. Zaporozhets, J. Rudolph, O. S. Gerasin, A. M. Topalov, O. V. Kozlov, 'Features of clamping electromagnets using in wheel mobile robots and modeling of their interaction with ferromagnetic plate', in Proc. of the 9th IEEE International Conference on Intelligent Data Acquisition and Advanced Computing Systems:

Technology and Applications (IDAACS), Vol. 1, Bucharest, Romania, pp. 453-458, 2017.

[98] Y. Kondratenko, Y. Zaporozhets, J. Rudolph, O. Gerasin, A. Topalov, O. Kozlov, 'Modeling of clamping magnets interaction with ferromagnetic surface for wheel mobile robots', in International Journal of Computing, Vol. 17, Iss. 1, pp. 33-46, 2018.

[99] Y. Kondratenko, O. Kozlov, O. Korobko, A. Topalov, 'Complex Industrial Systems Automation Based on the Internet of Things Implementation', in: Bassiliades N. et al. (eds) Information and Communication Technologies in Education, Research, and Industrial Applications. ICTERI 2017. Communications in Computer and Information Science, vol 826. Springer, Cham, pp. 164-187, 2018.

# 5

# Inverse Model Approach to Disturbance Rejection Problem

**L. Lyubchyk**

National Technical University “Kharkiv Polytechnic Institute”,
Kirpitcheva str., 2, Kharkiv, 61002, Ukraine
E-mail: Leonid.Liubchyk@khpi.edu.ua

## Abstract

Disturbance rejection and output tracking problem in linear multivariable control systems is considered. Using inverse dynamic models approach, controller structures including disturbance observer and compensator are proposed for disturbance estimation and compensation. The method for inverse dynamic models synthesis based on unknown input observer theory is justified. The properties of closed-loop control system with inverse model-based controller have been investigated for the purpose of attainable disturbance rejection accuracy assessment. Moreover, non-singularity conditions of controlled plant structure are established, allowing eliminating disturbance estimates from decoupling compensator equation. For this case, when such conditions are violated, the realizable form of disturbance decoupling compensator includes internal dynamic filter with small time constant being proposed. It was shown that slow motion in two-time-scale system coincides with the processes in the system with ideal compensator, if the fast motion in closed-loop system is stable. A generalization of inverse model method for systems with variable structure is also considered. Sliding mode equivalence principle is proposed to design non-stationary sliding surfaces with integral parts, in that, inverse model-based equivalent control ensures mismatched disturbances compensation. Variable structure disturbance observer and compensator are also developed and sliding mode conditions for proposed control scheme are established.

Peculiarities of applying inverse model approach for discrete-time systems are also considered and procedures for structural–parametric synthesis of discrete disturbance observers and compensators are suggested.

**Keywords**. Inverse model, discrete systems, disturbance decoupling problem, disturbance rejection, equivalent control, output control, sliding mode, unknown input observer, variable structure system.

## 5.1 Introduction

One of the most important problems in control theory is disturbance rejection along with reference signal tracking, known as output control problem [1, 2]. As far as uncertainties of controlled plant model may be considered as a parametric disturbance, these tasks are closely related with general problems of control systems invariance and robustness [3, 4].

There are two main approaches to such a problem. First, namely, disturbance attenuation methods use available *a priori* information concern disturbances in statistical or set-membership form. In that, the solution is obtained in a class of classical feedback control structures and formalized as an optimization problem with the averaged or guaranteed cost function; moreover, it additionally takes into account requirements of controller's internal stability. Cost functions as a norm of closed-loop transfer function are widely used, and solution may be obtained using $\mathbf{H}_2$ or $\mathbf{H}^\infty$ optimal control methods [5, 6].

Recently, different methods for describing uncertain disturbances using interval or set-membership models [7, 8] have become widespread, and an approach based on invariant ellipsoids [9] has been found to be effective in solving disturbance rejection problems [10], wherein, *linear matrix inequalities* (LMI) [11] method is used to solve the corresponding optimization problems.

Another approach uses current actual disturbance information obtained by its direct or indirect measurements. Such an approach realizes in control structures known as "two degree of freedom controllers" [12] in the form of combined feedback/feedforward control systems and is now widely used in active disturbance rejection approach [1, 13]. Appropriate design methods, which use different types of plants and disturbances models in control loop and known as *internal model control* (IMC) [3], are very popular in robust process control. In [14], it was shown that the IMC approach ensures selective invariance properties of closed-loop system, i.e. rejection of a certain class of disturbances.

As far as the idea of selective invariance was initially developed for single-input single-output (SISO) systems with scalar disturbance, its development and generalization for multivariable systems are of the great interest. Recently, in such a way, a number of model-based control methods in IMC framework have been developed for disturbance rejection in multivariable systems taking into account requirements of accuracy, dynamic performance, stability, and robustness.

These problems are connected with general theoretical *disturbance decoupling problem* (DDP) [15] which is eliminating the influence of unknown and unmeasurable disturbances to systems output. In the framework of geometric approach, DDP solution via static-state feedback has been obtained using fundamental concept of ($A$,$B$)-invariant and controllability sub-spaces [16, 17]. The requirement of closed-loop system stability along with disturbance decoupling leads to the *disturbance decoupling problem with stability* (DDPS). Moreover, in many practical situations, system state vector cannot be measured directly and *disturbances decoupling problem by measurement feedback* (DDPM) can be defined. The geometric approach was extended to DDPM solution, wherein dynamic output feedback compensator design is considered in geometric framework [18].

Solvability conditions for problems mentioned above are well known [19], but in spite of the existence of general solution of DDP in terms of invariant subspaces, finding the analytical description for controllers that solve DDP is quite complex and a difficult task. Besides, for complete characterization of the solution, the most important step of the design procedure is parameterization of corresponding controllers and disturbance dynamic compensators equations [20, 21].

From a practical point of view, it is desirable to decompose DDPM into structural synthesis of designed controller that renders the fixed and free parameters and parametric synthesis based on appropriate parameters tuning methods in order to satisfy design goals, such as internal and general stability, performance optimization, and so on.

Practical methods development for disturbance decoupling leads to further IMC elaboration, including *inverse model control* (InvMC), which is well adapted for solving this class of problems. The idea of *inverse model* (IM) utilization for both disturbance estimation and rejection was proposed in [22]. The InvMC approach includes disturbance estimation, output controlled object reaction prediction, and disturbances compensation.

The appropriate control structure consists of *disturbance observer* (DO), feedback/feedforward controller, and *disturbance compensator* (DC). Such an approach ensures not only closed-loop system stabilization but also reference signal tracking and immeasurable arbitrary disturbance rejection. Various approaches to solving control problems using inverse models were considered in [23–26].

The basic of InvMC approach for multivariable control is IM state-space representation (structural synthesis) and IM parametric design in order to ensure the desired dynamic properties. If the invertibility conditions take place [22, 27], structure inversion algorithm [28] may be applied. In that, IM structure and parameter models are strictly determined by plant parameters; so for a non-minimum phase system, IM will be unstable. Thus, such an approach may be directly applied only to systems with stable transmission zeros.

Thus, for practical implementation of InvMC method, it is necessary to use tuning inverse models; moreover, the IM design method should include the suitable parameterization of its equations, whereupon free tuning parameters are selected from the simultaneous conditions of stability and desired dynamic properties. The development of InvMC method using synthesized inverse models was considered in [29, 30], while the theory of *unknown input observer* (UIO) [31–33] was used to solve the IM design problem. In that, UIO equation, combined with unknown input signal estimate, may be treated as designed IM. Furthermore, DO and DC parametric synthesis comes to be relevant modal control problems.

The basic point of the proposed method is its purely algebraic essence. In such a way, a simple algebraic condition for the DDPM solvability is obtained; so the existence of solutions is easily verified. If for a controlled plant, some conditions, called *conditions of structural non-singularity*, are fulfilled, disturbance estimate can be eliminated from the control law and DD controller equations are obtained in the explicit form. If conditions mentioned above are violated, the realizable form of DDC can be obtained using additional internal dynamic filter with small time constant. So we get two-time-scale system, which can be analyzed by singular perturbation method, and if fast motion in the closed-loop system is stable, the slow one coincides with the processes in the system with ideal compensator.

An important direction of InvMC development is its implementation in *variable structure system* (VSS), which has been recognized as an extremely effective control scheme due to its robustness properties with respect to plant parameters variations, unmodeled dynamics, and external disturbances

[34, 35]. A lot of VSS control schemes have been developed using both state and output feedback [36, 37]; in the last case, state observer is used for state estimate, what implements dynamic output feedback sliding mode control [38].

One of the most important problems in VSS design is unmeasurable persistent disturbances rejection along with the reference signal output tracking [39]. Disturbances influence not only degrades the control accuracy but also leads to the some undesirable effects, including increasing of required amplitude of discontinuous control and also chattering phenomena. However, though traditional VSS design method ensures sliding mode stability for a broad range of bounded disturbances, the disturbance uncertain term does not appear in sliding mode dynamics only for special class of disturbances, satisfying the so-called matching or invariance conditions, initially specified in [40]. A similar problem arises in VSS design under sliding surface implementation based on a state observer in feedback loop; in such a case, state estimation error is an uncertain term in sliding mode.

Various approaches for VSS control under mismatched disturbance are considered in [41, 42]. Prediction error approach for VSS control of uncertain plants was justified in [43, 44], with proper modification of sliding surface using prediction error, obtained by reference model and averaging filter. The method is proved to ensure reference model following and chattering elimination for SISO plants with unmodeled dynamics. Another approach, referred to as *integral variable structure control* (IVSC), has been developed in [45, 46] and uses DO in relay control loop in order to improve control robustness. VSS with dynamic feedback including integral term are considered in [47] for SISO model following problem and in [48] for output tracking systems design. A perspective approach for state estimation in VSS under mismatched disbranches is based on UIO using [49, 50].

The key idea of InvMC method implementation in VSS is *sliding mode equivalence* (SME) principle proposed in [51]. According to this, sliding surfaces are selected so that equivalent control in the sense of [34] will coincide with the "ideal" control derived by InvMC. For this purpose, an additional time-dependent term of integral type is included into sliding surface equations. Consequently, system dynamics during sliding mode has required properties. In such a way, variable structures of DO and DC can be designed since slow components of discontinuous signals are proved proportional to disturbance estimate and compensative control signal. Finally, a novel cascade VSS control scheme arises; in that, the discontinuous signals generated by preceding variable structure blocks are transformed by the integral part of following ones.

The chapter is organized as follows. Section 5.1 provides an overview of disturbance rejection methods and InvMC approach formulation. In Section 5.2, an unified approach to DO and DC design problems is considered based on IM of controlled plant channels and structural–parametric IM synthesis problem solution based on UIO theory. Further, *disturbance decoupling compensator* (DDC) design method sets out, and conditions of DDC existence are given in the form of structural non-singularity criterion; moreover, an implemented form of DDC is proposed in violation of these conditions. In Section 5.3, output control problem for multivariable VSS under mismatched disturbances is considered and InvMC in a VSS framework applied for both disturbances observer (DO) and feedforward compensator (FC) design. In Section 5.4, the features of InvMC implementation and application for discrete-time system are considered.

## 5.2 Disturbance Rejection via Inverse Model Control

### 5.2.1 Inverse Model Control Principle

Consider the linear multivariable control system

$$\begin{aligned} &\dot{x}(t) = Ax(t) + B_1 u(t) + B_2 w(t), \quad x(t) = x_0, \\ &y_1(t) = C_1 x(t), \quad y_2(t) = C_2 x(t), \end{aligned} \tag{5.1}$$

where $x(t) \in \mathbf{R}^n$ is the system state vector, $x_0$ is the initial state, $\|x_0\| \leq c_0$, $u(t) \in \mathbf{R}^{m_1}$ is the input control, $w(t) \in \mathbf{R}^{m_2}$ is the unknown disturbance, $\|w(t)\| \leq c_w$, $y_1(t) \in \mathbf{R}^{q_1}$, $y_2(t) \in \mathbf{R}^{q_2}$ are the output controlled and measured variables, respectively, $m_1 = q_1, \quad m_2 \leq q_2$.

We introduce matrix transfer functions for "*i–j*" channels of multivariable system (5.1)

$$G_{ij}(s) = C_i (sI_n - A)^{-1} B_j, \quad i, j = 1, 2. \tag{5.2}$$

Let$S_{ij}(\alpha) = C_i A^{\alpha-1} B_j$ for integer $0 < \alpha \leq n$ be Markov parameters of system (5.1), and $\alpha_{ij}$ is a relative degree of corresponding channels, i.e. the minimal integer, so that $S_{ij}(\alpha) \neq 0$, and, moreover, the following assumptions take place:

$$\begin{aligned} &\text{(a)} \quad \operatorname{rank} B_1 = \operatorname{rank} S_{11}(\alpha_{11}) = m_1, \\ &\text{(b)} \quad \operatorname{rank} B_2 = \operatorname{rank} S_{22}(\alpha_{22}) = m_2. \end{aligned} \tag{5.3}$$

Assumptions (5.3) are equivalent to invertibility conditions of system (5.1) channels

$$\mathrm{Rank}\left[G_{11}\left(s\right)\right] = m_1, \quad \mathrm{Rank}\left[G_{22}\left(s\right)\right] = m_2, \tag{5.4}$$

where $\mathrm{Rank}\left[\cdot\right]$ denotes the rank over the field of rational functions.

For simplicity reason, we will assume that $\alpha_{ij} = 1$ and use the notation $S_{ij}\left(1\right) = S_{ij}$.

Consider DDPM along with the problem of input reference signal $y_1^*(t)$ tracking, $\|y_1{}^*(t)\| \le c_0{}^*, \quad \|\dot{y}_1{}^*(t)\| \le c_1{}^*$. Control goal is to determine control law $u(t) = U\left(Y(t)\right)$, depending on vector of measured variables $Y(t)$ so that disturbance will be eliminated from control error $e(t) = y_1^*(t) - y_1(t)$, which should have desired dynamic properties.

Depending on the accessible measurement information, we will distinguish

- Direct DDPM (*feedforward controller* (FFC) design), if $Y(t) = \{w(t), y_1(t)\}$, i.e. disturbance and controlled variables can be measured directly.
- Indirect DDPM, if $Y(t) = y_2(t)$, when only part of output variables is available to measurement (measured output feedback).

InvMC approach to DDPM in terms of Laplace transform includes following steps:

- Disturbance estimation using of "2-2"channel IM

$$\hat{w}(s) = G_{22}^{+}(s)\left[y_2(s) - G_{21}(s)\cdot u(s)\right], \tag{5.5}$$

where "+" denotes the Moore–Penrose matrix inverse,

- Control signal formation using of "1-1"channel IM

$$u^*(s) = G_{11}^{-1}(s)\cdot\left[y_1^*(s) - G_{12}(s)\hat{w}(s)\right]. \tag{5.6}$$

Predictive model (PM) $G_{12}\left(s\right)$ is used for controlled output prediction, initialized by input disturbance estimate.

If structural non-singularity condition for system (5.1) in matrix form

$$\det\left[I_{m_1} - G(s)\right] \ne 0, \quad G(s) = G_{11}^{-1}(s)G_{12}(s)G_{22}^{+}(s)G_{21}(s) \tag{5.7}$$

is fulfilled, then for particular case $m_1 = q_1$

$$u^*(s) = \left[I_{m_1} - G(s)\right]^{-1}\left[G_{11}^{-1}(s)y_1^*(s) - G_{22}^{+}(s)y_2(s)\right], \tag{5.8}$$

and $y_1(s) = y_1^*(s)$ if $u(s) = u^*(s)$, i.e. the DDPM problem is solved.

It should be noted that such a "naive" approach is practically impossible to implement not only due to difficulty of matrix transfer function inverting but also owing to possible instability and lack of robustness of obtained control law.

Finally, control law includes feedforward component $u_{\mathrm{FF}}(s) = u^*(s)$, solving the DDPM, and feedback component $u_{\mathrm{FB}}(s)$ ensures the closed-loop system stability. Practical realization of InvMC method is admissible if IMs are stable and have robust properties.

### 5.2.2 Inverse Model Design

The basis of InvMC approach is state-space representation of inverse models. If system (5.1) invertibility conditions take place [30, 31], the structure inversion algorithm [28] may be applied; in this case, the IM structure and parameters are strictly determined by parameters of corresponding channels. Inverse model design method must include suitable parameterization of its equations defining free tuning parameters, which are selected from simultaneous conditions of stability and desired dynamic properties. The most general way for such parameterization is the use of UIO theory [31, 32], wherein the observer equation combined with unknown input signal estimate may be treated as the designed IM.

*Full-order inverse model:* Consider the problem of system (5.1) "1-1"channel inversion. For this purpose, suppose that$w(t) \equiv 0$, in that $u(t)$ and $y_1(t)$ will be treated as the unknown input and measured output, respectively. In such a case, using UIO observer

$$\begin{aligned} \dot{\tilde{x}}(t) &= F_1\tilde{x}(t) + G_{11}y_1(t), \quad \tilde{x}(0) = \tilde{x}_0, \\ \hat{x}(t) &= \tilde{x}(t) + H_{11}y_1(t), \end{aligned} \tag{5.9}$$

where $\hat{x}(t) \in \mathbf{R}^n$ is the state vector estimate. We can obtain the IM equation in the form

$$\begin{aligned} \dot{x}_{I_1}(t) &= F_1 x_{I_1}(t) + (G_{11} - F_1 H_{11})\, u_{I_1}(t) + H_{11}\dot{u}_{I_1}(t) \\ y_{I_1}(t) &= B_1^+\left[\dot{x}_{I_1}(t) - A x_{I_1}(t)\right], \quad x_{I_1}(0) = \tilde{x}_0 + H_{11}C_1 x_0, \end{aligned} \tag{5.10}$$

where $x_{I_1}(t) = \hat{x}(t) \in \mathbf{R}^n$, $u_{I_1}(t) \in \mathbf{R}^{q_1}$, $y_{I_1}(t) \in \mathbf{R}^{m_1}$ is the IM state vector, input signal, and output signal, respectively, and $u_{I_1}(t) = y_1(t)$.

If parameters of UIO (5.9) satisfy the so-called "invariance conditions" [30, 31]

$$(I_n - H_{11}C_1)\,F_1 - F_1\,(I_n - H_{11}C_1) = G_{11}C_1, \quad B_1 - H_{11}C_1B_1 = 0, \tag{5.11}$$

then the unknown input $u(t)$ will be eliminated from deviation vectors $e_x(t) = x(t) - x_{I_1}(t)$, $e_u(t) = u(t) - y_{I_1}(t)$ which will be given by the following equations:

$$\dot{e}_x(t) = F_1 \cdot e_x(t), \quad e_u(t) = -B_1^+ (F_1 - A)\, e_x(t). \tag{5.12}$$

As it has been shown in [30], if condition (5.3a) takes place, then in the case when $m_1 \leq q_1$, linear algebraic equation system (5.11) has a matrix solution

$$\begin{array}{l} F_1 (L_1) = \Pi_1 A - L_1 C_1, \quad H_1 = B_1 S_{11}^+, \\ G_{11} (L_1) = \Pi_1 A H_1 + L_1 \Omega_1, \end{array} \tag{5.13}$$

where matrices $\Pi_1 = I_n - B_1 S_{11}^+ C_1$, $\Omega_1 = I_n - S_{11} S_{11}^+$, $S_{11}^+ = \left(S_{11}^T S_{11}\right)^{-1} S_{11}^T$, and $L_1$ is the arbitrary $(n \times q_1)$ matrix of free tuning parameters.

Therefore, if pair matrix$(\Pi_1 A,\ C_1)$ is observable (input observability conditions), the eigenvalues of $F_1 (L_1)$ may be assigned by tuning matrix $L_1$ selection via any pole-placement method.

Finally, parameterized state-space IM representation are obtained in the form

$$\begin{array}{l} x_{I_1}(t) = F_1 (L_1)\, x_{I_1}(t) + L_1 u_{I_1}(t) + H_1 \dot{u}_{I_1}(t), \\ y_{I_1}(t) = -C_1 (L_1)\, x_{I_1}(t) + B_1^+ L_1 u_{I_1}(t) + S_{11}^+ \dot{u}_{I_1}(t), \\ C_1 (L_1) = S_{11}^+ C_1 A + B_1^+ L_1 C_1. \end{array} \tag{5.14}$$

Using the special form of system (5.1), obtained by non-singular state-space transformation,

$$A = \begin{pmatrix} A_{11} & A_{12} \\ A_{21} & A_{22} \end{pmatrix}, B_1 = \begin{pmatrix} B_{11} \\ B_{12} \end{pmatrix}, C_1 = \begin{pmatrix} I_{q_1} \mid 0_{q_1, n-q_1} \end{pmatrix}, \tag{5.15}$$

equations of IM matrices may be represented as

$$\begin{array}{l} \Pi_1 = \left( \begin{array}{c:c} \Omega_{B_{11}} & 0_{q_1, n-q_1} \\ \hdashline -B_{12} B_{11}^+ & I_{n-q} \end{array} \right), \quad F_1 (L_1) = \left( \begin{array}{c:c} \Omega_{B_{11}} A_{11} - L_{11} & \Omega_{B_{11}} A_{12} \\ \hdashline \tilde{A}_{21} - L_{12} & \tilde{A}_{22} \end{array} \right), \\ \Omega_{B_{11}} = I_{q_1} - B_{11} B_{11}^+, \quad \tilde{A}_{1i} = A_{2i} - B_{12} B_{11}^+ A_{1i}, \quad i = 1, 2, \end{array} \tag{5.16}$$

where $L_1^T = \left( L_{11}^T \mid L_{12}^T \right)$, $L_{11} \in \mathbf{R}^{q_1 \times q_1}$, and $L_{12} \in \mathbf{R}^{n-q_1 \times q_1}$ are arbitrary tuning matrices, selected by pole-placement methods for IM design.

*Reduced-order inverse model:* The minimal state-space IM realization may be obtained by means of reduced order UIO. Let $z(t) = R_1 x(t) \in \mathbf{R}^{n-q_1}$ be an aggregated auxiliary variables, where $R_1$ is an aggregate matrix such that $\mathrm{rank}\left( C_1^T \,\vdots\, R_1^T \right) = n$.

Then state vector estimate can be obtained as follows:

$$\hat{x}(t) = P_1 y_1(t) + Q_1 \hat{z}(t), \tag{5.17}$$

where $\hat{z}(t)$ is given by minimal-order UIO

$$\begin{aligned} &\dot{\bar{x}}(t) = \bar{F}_1 \cdot \bar{x}(t) + \bar{G}_{11} y_1(t), \quad \bar{x}(0) = \bar{x}_0, \\ &\hat{z}(t) = \bar{x}(t) + \bar{H}_{11} y_1(t) \end{aligned} \tag{5.18}$$

and matrices $P_1 \in \mathbf{R}^{n \times q_1}$ and $Q_1 \in \mathbf{R}^{n \times n - q_1}$ are defined as

$$\begin{aligned} &\left( P_1 \,\vdots\, Q_1 \right) = \begin{pmatrix} C_1 \\ R_1 \end{pmatrix}^{-1}, \quad C_1 P_1 = I_{q_1}, \; R_1 Q_1 = I_{q_1}, \\ &P_1 C_1 + Q_1 R_1 = I_n \quad C_1 Q_1 = 0_{q_1, n-q_1}, \; R_1 P_1 = 0_{n-q_1, q_1}. \end{aligned} \tag{5.19}$$

The "invariance conditions" in such a case take on the form of matrix equations system

$$\begin{aligned} &\left(R_1 - \bar{H}_1 C_1\right) A - \bar{F}_1 \left(R_1 - \bar{H}_1 C_1\right) = \bar{G}_{11} C_1, \\ &R_1 B_1 - \bar{H}_{11} C_1 B_1 = 0, \end{aligned} \tag{5.20}$$

and a corresponding solution of Equation (5.20) may be obtained as

$$\begin{aligned} &\bar{F}_1 \left(R_1\right) = R_1 \Pi_1 A Q_1, \quad \bar{H}_{11} = R_1 B_1 S_{11}^{+} = R_1 H_1, \\ &\bar{G}_{11} \left(R_1\right) = R_1 \Pi_1 A \left(\bar{H}_{11} + P_1 \Omega_1\right), \end{aligned} \tag{5.21}$$

where matrices $P_1$ and $Q_1$ are uniquely determined by $R_1$ selection.

Therefore, minimal-order IM is given by equations

$$\begin{aligned} &\dot{\bar{x}}_{I_1}(t) = \bar{F}_1 \left(R_1\right) \bar{x}_{I_1}(t) + R_1 \Pi_1 A P_1 u_{I_1}(t) + R_1 H_{11} \dot{u}_{I_1}(t), \\ &y_{I_1}(t) = -\bar{C}_1 \left(P_1\right) \left[C_1 A Q_1 x_{I_1}(t) + C_1 A P_1 u_{I_1}(t) - \dot{u}_{I_1}(t)\right], \end{aligned} \tag{5.22}$$

where $x_{I_1}(t) = z(t) \in \mathbf{R}^{n-q_1}$ is the IM state vector, and $\bar{C}_1 \left(P_1\right) = S_{11}^{+} + B_1^{+} P_1 \Omega_1$.

Deviation vectors$\bar{e}_x(t) = R_1 x(t) - x_{I_1}(t)$,$e_u\left(t\right)$ also are invariant with respect to$u(t)$:

$$\dot{\bar{e}}_x(t) = \bar{F}_1 \left(R_1\right) e_x(t), \quad e_u = -C_1 \left(P_1\right) C_1 A Q_1 \bar{e}_x(t), \tag{5.23}$$

and its dynamic properties are determined by tuning the matrix $R_1$ selection.

Concretely defining the matrices $P_1$ and $Q_1$ choice, one can admit

$$\left( P_1 \,\vdots\, Q_1 \right) = \begin{pmatrix} P_{11} & Q_{11} \\ P_{12} & Q_{12} \end{pmatrix}, \quad P_{11} = I_q, \quad Q_{11} = 0_{q,n-q}, \tag{5.24}$$

and in such a case, $R_1 = Q_{12}^{-1} \left( -P_{12} \,\vdots\, I_{n-q} \right)$ and $P_{12}$ and $Q_{12}$ are arbitrary matrices with $\det Q_{12} \neq 0$. For system representation (5.15) from Equation (5.16), Equation (5.21) follows that

$$\begin{aligned} &\bar{F}_1 (R_1) = Q_{12}^{-1} \left( \bar{A}_{22} - P_{12}\bar{A}_{12} \right) Q_{12}, \\ &\bar{A}_{12} = \Omega_{B_{11}} A_{12}, \quad \bar{A}_{22} = A_{22} - B_{12}B_{11}^{+} A_{12}. \end{aligned} \tag{5.25}$$

Thus, matrix $Q_{12}$ defines the similarity transformation and does not change spectrum of $\bar{F}_1 (R_1)$, which is determined by arbitrary matrix $P_2 \in \mathbf{R}^{n-q_1 \times q_1}$. The last may be chosen by pole-placement method if pair $\left( \bar{A}_{22}, \bar{A}_{12} \right)$ is observable; moreover, aggregate matrix $R_1$ is determined up to an arbitrary non-singular matrix $Q_{12}$.

*Regularized inverse model:* The problem of IM parametric design is solvable, if matrix pair $(\Pi_1 A, C_1)$ is observable. This condition is violated if $m_1 = q_1$. In fact, using representation (5.15), one can obtain that $\Omega_{B_{11}} = 0$ and expression (5.16) becomes

$$\Pi_1 = \begin{pmatrix} 0_{m_1,m_2} & 0_{m_1,n-m_1} \\ -B_{12}B_{11}^{+} & I_{n-m} \end{pmatrix}, \quad F_1(L_1) = \begin{pmatrix} -L_{11} & 0_{m_1,n-m_1} \\ \tilde{A}_{21} - L_{12} & \tilde{A}_{22} \end{pmatrix}. \tag{5.26}$$

Non-observable subsystem of Equation (5.14) with matrix $\tilde{A}_{22} = A_{12} - B_2 B_{11}^{-1} A_{12}$ will be unstable for non-minimum phase transfer function $G_{11}(s)$ and cannot be stabilized by $L_1$ choosing.

In order to maintain the conditions of input observability, the regularized inverse model based on the "sub-invariant" UIO may be utilized [28, 29]. Using "sub-invariance" conditions for deviation vector $e_x(t)$ with respect to the unknown input

$$B_1 - H_{11}C_1B_1 = \varepsilon_1 B_1, \quad 0 < \varepsilon_1 < 1, \tag{5.27}$$

where $\varepsilon_1$ is a small parameter, and regularized IM matrix parameters are obtained as follows:

$$\begin{aligned} &F_1 (\varepsilon_1, L_1) = \Pi_1 (\varepsilon_1) A - L_1 C_1, \\ &G_{11} (\varepsilon_1, L_{11}) = \Pi_1 (\varepsilon_1) A H_1 (\varepsilon_1) - L_1 \Omega_1 (\varepsilon_1), \\ &H_1 (\varepsilon_1) = (1 - \varepsilon_1) B_1 S_{11}{}^{-1} = (1 - \varepsilon_1) H_{11}, \\ &\Pi_1 (\varepsilon_1) = I_n - (1 - \varepsilon_1) B_1 S_{11}^{-1} C_1, \quad \Omega_1 (\varepsilon_1) = \varepsilon_1 I_{q_1}. \end{aligned} \tag{5.28}$$

For $\varepsilon_1 \neq 0$ $\det \Pi_1(\varepsilon_1) \neq 0$ and pair matrix $(\Pi_1(\varepsilon_1) A,\ C)$ will be observable, if $(A,\ C_1)$ is observable.

It should be also noted that for stable non-minimum phase transfer function $G_{11}(s)$, there exists $\varepsilon^* > 0$ so that $\Pi_1(\varepsilon_1) A$ will be stable for any $\varepsilon_1 > \varepsilon_1^*$ because $\Pi_1(\varepsilon_1 = 1) = I_n$, and boundary value $\varepsilon_1^*$ may be easily found by root-locus method. As a result, tuning matrix may be obtained for any $\varepsilon_1 > 0$ by pole-placement methods using representation

$$\Pi_1(\varepsilon_1) = \left( \begin{array}{c|c} \varepsilon_1 I_{q_1} & 0_{q_1, n-q_1} \\ \hline -(1-\varepsilon_1) B_{12} B_{11}^{-1} & I_{n-q_1} \end{array} \right),$$

$$F_1(\varepsilon_1, L_1) = \left( \begin{array}{c|c} \tilde{A}_{11}(\varepsilon_1) - L_{11} & \tilde{A}_{12}(\varepsilon_1) \\ \hline \tilde{A}_{21}(\varepsilon_1) - L_{12} & \tilde{A}_{22}(\varepsilon_1) \end{array} \right), \tag{5.29}$$

where $\tilde{A}_{1i}(\varepsilon_1) = \varepsilon_1 A_{1i}$, $\tilde{A}_{2i}(\varepsilon_1) = A_{2i} - (1-\varepsilon_1) B_{12} B_1^{-1} A_{1i}$, $i = 1, 2$.

The corresponding deviation vectors are given by equations

$$\begin{aligned} \dot{e}_x(t) &= F_1(\varepsilon_1, L_1) e_x(t) + \varepsilon_1 B_1 u(t), \\ e_u(t) &= -C_1(\varepsilon_1, L_1) e_x(t) + \varepsilon_1 u(t), \end{aligned} \tag{5.30}$$

where $C_1(\varepsilon_1, L_1) = (1-\varepsilon_1) S_{11}^{-1} + B_1^{+} L_1 C_1$.

Then steady-state deviations include biases, which are proportional to the regularization parameter $\varepsilon_1$ and may be considered as a payoff for the parametric design problem solvability. Actual value of $\varepsilon_1$ may be found upon tradeoff between steady-state deviation and IM transition performance using *a priori* information about input signal$u(t)$.

For example, in low frequency, domain bias in $e_u(t)$ may be partially eliminated by multiplication of the $u_{I_1}(t)$ on the correction matrix

$$K_u = \left[ (1-\varepsilon_1) I_{q_1} + \varepsilon_1 C_1(\varepsilon_1, L_1) \left[ I_n - F_1(\varepsilon_1, L_1) \right]^{-1} B_1 \right]^{-1}. \tag{5.31}$$

Reduced-order regularized inverse model may be developed in a similar way, using corresponding "sub-invariance" conditions

$$R_1 B_1 - \bar{H}_{11} C_1 B_1 = \varepsilon_1 R_1 B_1, \quad 0 < \varepsilon_1 < 1; \tag{5.32}$$

it that, reduced-order IM matrices are defined as

$$\begin{aligned} &\bar{F}_1(\varepsilon_1, R_1) = R_1 \Pi_1(\varepsilon_1) A Q_1, \\ &\bar{G}_{11}(\varepsilon_1, R_1) = R_1 \Pi_1(\varepsilon_1) A \left[ H_{11}(\varepsilon_1) + P_1 \Omega_1(\varepsilon_1) \right], \\ &\bar{H}_{11}(\varepsilon_1) = (1-\varepsilon_1) R_1 B_1 S_{11}^{-1}. \end{aligned} \tag{5.33}$$

Taking into account parameterization $R_1(Q_{12}, P_{12})$ (5.24), IM dynamic matrix $\bar{F}_1(\varepsilon_1, R_1)$ with regard to (5.29) may be represented as

$$\bar{F}_1(\varepsilon_1 R_1) = Q_{12}^{-1}\left[\tilde{A}_{22}(\varepsilon_1) - P_{12}\tilde{A}_{12}(\varepsilon_1)\right]Q_{12}. \tag{5.34}$$

It may be shown that if pair matrix $(A_{22}, A_{12})$ is observable, parametric design problem of reduced-order IM is solvable for any$\varepsilon_1 > 0$.

Finally the deviation equations are

$$\begin{aligned} \dot{\bar{e}}_x(t) &= \bar{F}_1(\varepsilon_1, R_1)\bar{e}_x(t) + \varepsilon_1 R_1 B_1 u(t), \\ e_x(t) &= Q_1\bar{e}_x(t), \\ e_u(t) &= -\bar{C}_1(\varepsilon_1, R_1)C_1 A Q_1 e_x(t) + \varepsilon_1\left(I_{m_1} - B_1^+ P_1 S_{11}\right)u(t), \end{aligned} \tag{5.35}$$

where $\bar{C}_1(\varepsilon_1, R_1) = \left[(1-\varepsilon_1)S_{11}^{-1} + \varepsilon_1 B_1^+ P_1\right]$ and also have a sub-invariance properties up to the regularization parameter $\varepsilon_1$.

### 5.2.3 Inverse Model Based Feedforward Control

Consider direct DDPM when both disturbance and controlled variable can be measured directly and condition $m_1 = q_1$ takes place.

In such a case, control law includes output feedback and feedforward components

$$u(t) = -K\left[y_1 * (t) - y_1(t)\right] + u * (t), \tag{5.36}$$

where $K$ is gain matrix and $u * (t)$ is compensative component, formed in accordance with the InvMC principle as output signal $u * (t) = y_{I_1}(t)$ of IM (5.14) with input signal $u_{I_1}(t)$obtained from PM with state vector $x_P(t) \in \mathbf{R}^n$:

$$\begin{aligned} \dot{x}_p(t) &= Ax_p(t) + B_2 w(t), \quad x_p(0) = x_{po}, \\ u_{I_1}(t) &= y_1 * (t) - C_1 x_p(t). \end{aligned} \tag{5.37}$$

From Equations (5.14) and (5.37), IM-based FFC equation follows

$$\begin{aligned} \dot{x}_C(t) =& F_1(\varepsilon_1, L_1)x_C(t) \\ &+ L_1 y_1 * (t) + H_{11}(\varepsilon_1)\dot{y}_1 * (t) + \Pi_1(\varepsilon_1)B_2 w(t), \\ u * (t) =& -C_1(\varepsilon_1, L_1)x_C(t) \\ &+ B_1^+ L_1 y_1 * (t) + (1-\varepsilon_1)S_{11}^{-1}\left[\dot{y}_1 * (t) - S_{12}w(t)\right], \end{aligned} \tag{5.38}$$

where $x_C(t) = x_{I_1}(t) + x_p(t)$ is compensator's state vector.

From Equations (5.36) and (5.38), it follows that control error equation for closed-loop system is given by

$$\begin{aligned}
&\dot{\Theta}_1(t) = (A - B_1 K C_1)\,\Theta_1(t) + (B_1 K_1 - \Omega_{B_1} L_1)\,\Theta_2(t),\\
&\dot{\Theta}_2(t) = -\,(C_1 L_1)\,\Theta_2(t) + \varepsilon_1 f_1(t),\\
&e(t) = \Theta_2\,(t) - C_1\Theta_1(t), \quad f_1(t) = \dot{y}_1 * (t) - C_1 A x_C(t) - S_{12} w(t),
\end{aligned} \tag{5.39}$$

where $\Theta_1(t) = x\,(t) - x_C(t)$, $\Theta_2(t) = e(t) - C_1\Theta_1(t)$ is state vectors of error dynamics equation, $\Omega_{B_1} = I_n - B_1 B_1^{+}$.

If reduced-order IM is used, the corresponding equation of FFC may be derived in a similar way in the following form:

$$\begin{aligned}
\dot{\bar{x}}_C(t) =& \bar{F}_1\,(\varepsilon_1, R_1)\,\bar{x}_C(t) + R_1\Pi_1\,(\varepsilon_1)\,AP_1 y_1^*(t)\\
&+ R_1 H_1\,(\varepsilon_1)\,\dot{y}_1^*(t) + R_1\Pi_1\,(\varepsilon_1)\,B_2 w(t),\\
u^*(t) =& -\,\bar{C}_1\,(\varepsilon_1, P_1)\,[C_{\,1}\,AQ_1\bar{x}_C(t)\\
&+ C_1 A P_1 y_1^* - \dot{y}_1^*(t) + S_{12}\,w(t)]\,,
\end{aligned} \tag{5.40}$$

where $\bar{x}_C(t) = \bar{x}_{I_1}(t) + R_1 x_p(t)$ is the reduced-order FFC state vector.

By elementary algebraic transformations, it is easy to show that control error dynamic for system (5.1) with control law (5.36) is given by

$$\begin{aligned}
&\dot{\Theta}\,(t) = (A - B_1 K C_1)\,\Theta\,(t) + \varepsilon_1 P_1 f_1\,(t)\,,\\
&e\,(t) = -C_1\Theta\,(t)\,,\\
&f_1\,(t) = \dot{y}_1^* - C_1 A P_1 y_1^* - C_1 A Q_1 \bar{x}_c\,(t) - S_{12} w\,(t)\,,
\end{aligned} \tag{5.41}$$

when $\Theta(t) = Q_1\,[R_1 x(t) - \bar{x}_c(t)] - P_1 e(t)$ -is the generalized deviation vector.

As a result, DDPM in the presence of complete measurements $Y(t) = \{w(t), y_1(t)\}$ is reduced to state modal control problem for FFC design using tuning parameters $\varepsilon_1$, $L_1$, or $P_1$ and output modal control problem for closed-loop system design using gain matrix $K$; so well-developed methods may be applied.

Thus, the upper bound of the attainable control accuracy $\|e\,(\infty)\| \leq \delta^*\,(c_0^*,\ c_1^*,\ c_w)$ may be easily obtained from Equations (5.38) and (5.39) or Equations (5.40) and (5.41). It is evident that steady-state control error is determined by parameters, obtained under FFC design problem solution, namely,$\varepsilon_1{}^*$,$L_1{}^*$, or$P_1{}^*$. In the minimum phase case, one can admit that $\varepsilon_1 = 0$ and, theoretically, full disturbance decoupling is accessible. In general case, attainable accuracy of InvMC depends on mutual placement of "1-1"channel transmission zeroes and desired poles location of FFC dynamic matrix.

### 5.2.4 Inverse Model Based Disturbance Observer

In the case when$w(t)$,$y_1(t)$ cannot be measured directly, and only $y_2(t)$are measured variables, corresponding disturbance estimates should be used in control law (5.36); in that, estimates are obtained by means of the "2-2" channel IM; wherein, UIO approach leads to following state observer:

$$\begin{aligned}&\dot{\tilde{x}}(t) = F_2\tilde{x}(t) + G_{21}u(t) + G_{22}y_2(t), \quad \tilde{x}(0) = \tilde{x}_0,\\&\hat{x}(t) = \tilde{x}(t) + H_{21}u(t) + G_{22}y_2(t).\end{aligned} \tag{5.42}$$

The corresponding "invariance" condition for estimation error with respect to$w(t)$is

$$\begin{aligned}&(I_n - H_{22}G_2)\,F_2 - F_2\,(I_n - H_{22}G_2) = G_{22}C_2,\\&(I_n - H_{22}G_2)\,B_2 = 0,\\&(I_n - H_{22}G_2)\,B_1 - G_{21} - F_2H_{21} = 0, \quad H_{21} = 0.\end{aligned} \tag{5.43}$$

If invertibility condition (5.3b) takes place and $q_2 > m_2$, then the solution of matrix linear equations (5.43) may be obtained in parametric form

$$\begin{aligned}&F_2\,(L_2) = \Pi_2 A - L_2 C_2,\ H_{21} = 0,\ H_{22} = B_2 {S_{22}}^{+},\\&G_{21} = \Pi_2 B_1,\ G_{22} = \Pi_2 A H_{22} - L_2\Omega_2,\end{aligned} \tag{5.44}$$

where $\Pi_2 = I_n - B_2 {S_{22}}^{+} C_2$, $\Omega_2 = I_q - S_{22}{S_{22}}^{+}$, and $L_2$ is $(n \times q_2)$ tuning matrix.

Finally, IM of "2-2" channel has the following representation:

$$\begin{aligned}&\dot{x}_{I_2}(t) = F_2\,(L_2)\,x_{I_2}(t) + L_2 u_{I_2}(t) + H_{22}\dot{u}_{I_2}(t) + \Pi_2 B_1 u(t),\\&y_{I_2}(t) = -C_2\,(L_2)\,x_{I_2}(t) + {B_2}^{+} L_2 u_{I_2}(t) + {S_{22}}^{+}\dot{u}_{I_2}(t) - {S_{22}}^{+} S_{21} u(t),\end{aligned} \tag{5.45}$$

where $x_{I_2}(t) = \hat{x}(t)$, $C_2\,(L_2) = {S_{22}}^{+} C_2 A + {B_2}^{+} L_2 C_2$.

If measured output variables are used as input signal of IM (5.45) $u_{I_2}(t) = y_2(t)$, then state and output vectors of IM may be treated as corresponding state and disturbance estimates $x_{I_2}(t) = \hat{x}(t)$, $y_{I_2}(t) = \hat{w}(t)$; so Equation (5.45) represents DO.

In fact, estimation errors$e_x(t) = x(t) - \hat{x}(t)$ and $e_w(t) = w(t) - \hat{w}(t)$ are given by

$$\dot{e}_x(t) = F_2\,(L_2)\,e_x(t), \quad e_w(t) = -C_2\,(L_2)\,e_x(t), \tag{5.46}$$

and its desired dynamic properties may be attached by pole-placement method if pair $(\Pi_2 A, C_2)$ is observable, i.e. input observability condition of "2-2" channel takes place.

In the same way, reduced-order DO may be derived:

$$\begin{aligned}\dot{\bar{x}}_{I_2}(t) &= \bar{F}_2(R_2)\,\bar{x}_{I_2}(t) + R_2\Pi_2 AP_2 y_2(t) + R_2 H_{22}\dot{y}_2(t) + R_2\Pi_2 B_1 u(t),\\ \hat{w}(t) &= -\bar{C}_2(P_2)\left[C_2 AQ_2\bar{x}_{I_2}(t) + C_2 AP_2 y_2(t) - \dot{y}_2(t) + S_{21}u(t)\right],\end{aligned} \tag{5.47}$$

where $\bar{C}_2(P_2) = S_{22}^{+} + B_2{}^{+}P_2\Omega_2$, $\bar{F}_2(R_2) = R_2\Pi_2 AQ_2$, $R_2$ is arbitrary aggregate matrix, and $P_2$ and $Q_2$ are defined as before.

From Equation (5.47), estimation error equation for reduced-order observer is the following:

$$\dot{\bar{e}}_x(t) = \bar{F}_2(R_2)\,\bar{e}_x(t), \quad e_w(t) = -C_2(P_2)\,C_2 AQ_2\bar{e}_x(t). \tag{5.48}$$

### 5.2.5 Disturbance Decoupling Compensator Design

Implementation of the proposed approach for solving DDPM involves the use of control law

$$u(t) = K_1\left[y_1{}^{*}(t) - C_1\hat{x}(t)\right] + \hat{u}^{*}(t), \tag{5.49}$$

where compensative signal $\hat{u}^{*}(t)$ is formed by FFC (5.38) or (5.40), where $w(t)$ is replaced by its estimate, obtained by DO (5.45) or (5.47).

Construction of DDC requires the exclusion of intermediate estimates of unknown disturbance from its equations. The possibility of DDC design depends on system (5.1) structural properties. We introduce the concept of structural non-singularity, which for full-order FFC and DO have the form of following conditions:

$$\operatorname{rank}\left(I_{m_1} - S\right) = m_1, \quad S = (1-\varepsilon_1)\,S_{11}{}^{-1}S_{12}S_{22}{}^{+}S_{21}. \tag{5.50}$$

When conditions (5.50) are met, disturbance estimate may be eliminated from the FFC and DO equations and control signal $u(t)$ may be directly expressed through dynamic transformation of measured variables; so DDC exists and its equations in the form of two degree of freedom controller may be obtained from Equations (5.38) and (5.45).

Consider reduced-order DDC design for reference signal $y^{*}(t)$ given by reference model

$$\dot{y}^{*}(t) = A^{*}y^{*}(t) + y_{\text{ref}}(t). \tag{5.51}$$

Then for "square plant" ($q_1 = m_1$) under assumption (5.3a)

$$\begin{aligned}u^{*}(t) &= S_{11}^{-1}(y_{\text{ref}}(t) + C_A\hat{x}(t) - S_{12}\hat{f}(t)),\\ C_A &= A^{*}C_1 - C_1 A.\end{aligned} \tag{5.52}$$

For reduced-order DDC design problem, structural non-singularity condition is

$$\begin{aligned}&\text{rank } \bar{S} = m + q, \quad \bar{S} = \begin{pmatrix} I_m & S_{11}^{-1}S_{12} \\ C_1S_{21} & I_q \end{pmatrix} \quad \text{or } \det \Phi \neq 0, \\ &\Phi = I_q - C_N S_{21} S_{11}^{-1} S_{12}, \quad C_N = S_{21}^{+} + B_2{}^{+} P\Omega_2, \quad \Omega_2 = I_p - S_{22}S_{22}^{+}.\end{aligned} \tag{5.53}$$

In this case, using considered methodology, DDC equations can be obtained in explicit form by eliminating intermediate disturbance estimates. For example, in particular case $S_{12} = 0$, DDC takes the form

$$\begin{aligned}&\dot{\bar{x}}(t) = F^0\bar{x}(t) + R\Pi_2 A^0(P\Omega_2 + H_2)y_m(t) + R\Pi_2 H_1 y_{\text{ref}}(t), \\ &u^*(t) = S_{11}^{-1}(y_{\text{ref}}(t) + C_A Q\bar{x}(t)) + S_{11}^{-1} C_A(P\Omega_2 + H_2)y_m(t)), \\ &F^0 = R\Pi_2 A^0 Q, \quad A^0 = A + H_1 C_A, \quad \Pi_2 = I_n - B_2 S_{22}^{+} C_2, \\ &H_1 = B_1 S_{11}^{-1}, \quad H_2 = B_2 S_{22}^{+}.\end{aligned} \tag{5.54}$$

In many practical applications, conditions (5.53) are usually violated. In such a case, realizable DDC may be obtained using disturbance estimations, which are dynamically transformed by additional internal low-pass filter with small time constant$\varepsilon$:

$$\begin{aligned}&u^*(t) = S_{11}^{-1}(y^*(t) + C_A\hat{x}(t) - S_{12}\tilde{w}(t), \\ &\varepsilon \cdot \dot{\tilde{w}}(t) = -\tilde{w}(t) + (1 - \mu) \cdot \hat{w}(t),\end{aligned} \tag{5.55}$$

where $0 < \varepsilon \ll 1, \ \ 0 < \mu \ll 1$, and $\mu/\varepsilon \leq c < \infty$ are filter parameters.

Then disturbance compensative component of control law is described as

$$\begin{aligned}&u^*(t) = \tilde{u}(t) + \varphi_1(t), \\ &\varepsilon\dot{\tilde{u}}(t) = -\mu\tilde{u}(t) + (1 - \mu)(\varphi_1(t) + S_{11}^{-1} S_{12}\varphi_2(t)), \\ &\varphi_1(t) = S_{11}^{-1}(y_{ref}(t) + C_A\hat{x}(t)), \\ &\varphi_2(t) = \bar{C}_2(\dot{y}_m(t) - !_2 AQ\bar{x}(t) - !_2 APy_m(t)), \\ &\bar{C}_2 = S_{22}^{+} + B_2{}^{+} P\Omega_2.\end{aligned} \tag{5.56}$$

Consider the properties of closed-loop system with controls (5.49) and (5.56)

$$\begin{aligned}&\dot{x}(t) = A^0 x(t) + \Pi_1 B_2 \mathrm{w}(t) + H_1 y_{\text{ref}}(t) + Le_x(t), \\ &A^0 = A + H_1 C_A = \Pi_1 A + H_1 A^* C_1,\end{aligned} \tag{5.57}$$

where $L$ is a certain matrix.

For non-minimum phase system, matrix$A^0$is unstable; so the problem of closed-loop system stabilizing arises. Moreover, classical state feedback

$u(t) = u^*(t) - K\hat{x}(t)$ does not change closed-loop matrix spectrum because$\Pi_1(A + BK) = 0$.

In such a case, *local optimal control* (LOC) method [52] may be applied. In accordance with LOC, control should be found from local criteria minimization problem

$$\left\| y_{\text{ref}}(t) + C_A A\hat{x}(t) - S_{11}u(t) - S_{12}\hat{f}(t) \right\|^2 + \beta \left\| u(t) \right\|^2 \to \min_u, \quad (5.58)$$

where $\beta$ is a weight coefficient. The corresponding control law is given by

$$\begin{aligned} u^*_\beta(t) &= D_1(\beta)\left(y_{\text{ref}}(t) + C_A A\hat{x}(t) - S_{12}\hat{w}(t)\right), \\ D_1(\beta) &= \left(\beta I_m + S_{11}^{\mathrm{T}} S_{11}\right)^{-1} S_{11}^{\mathrm{T}}. \end{aligned} \quad (5.59)$$

From Equation (5.59), the closed-loop system equation follows

$$\begin{aligned} &\dot{x}(t) = A_0(\beta, K)\, x(t) + BD(\beta)\, y_{\text{ref}}(t) + \Pi_1(\beta)\, B_2 w(t) + L_\beta e_x(t), \\ &A_0(\beta, K) = A_0(\beta) - B_\beta K, \quad B_\beta = \beta B_1 \left(\beta I_m + S_{CB}^{\mathrm{T}} S_{CB}\right)^{-1}, \\ &A_0(\beta) = A + BD(\beta)\, C_A = \Pi_1(\beta)\, A + BD(\beta)\, A^* C, \end{aligned} \quad (5.60)$$

where $\Pi_1(\beta) = I_n - B_1 D(\beta)\, C_1$,and system (5.60) may be stabilized if matrix pair $(A_0(\beta),\ B_\beta)$is controllable. Finally, control error is given by

$$\dot{e}_c(t) = A^* e_c(t) - \beta S_{11}\left(\beta I_m + S_{11}^{\mathrm{T}} S_{11}\right)^{-1} u^*(t) \quad (5.61)$$

and control goal is achieved with $\varepsilon^*(\beta)$.

For a system with structural singularity, when DDC with internal filter (5.56) is used, closed-loop system equations are given by

$$\begin{aligned} &\dot{x}(t) = A^0 x(t) + B_2 w(t) - H_1 S_{12}\tilde{w}(t) + H_1 y_{\text{ref}}(t) + Le_x(t), \\ &\varepsilon\dot{\tilde{w}}(t) = -\tilde{w}(t) + (1-\mu)w(t) - (1-\mu)e_f(t). \end{aligned} \quad (5.62)$$

System (5.62) is two-time-scale system [53], in which slow motion under $\varepsilon = 0$coincides with the process in system with FFC using directly measured disturbances, and the fast one satisfies dynamic equation

$$\begin{aligned} &E(\varepsilon)\dot{\tilde{x}}(t) = \tilde{A}^0 \tilde{x}(t) + \tilde{B}^0 w(t), \\ &E(\varepsilon) = \begin{pmatrix} I_n & 0 \\ 0 & \varepsilon I_q \end{pmatrix}, \ \tilde{A}^0 = \begin{pmatrix} A^0 & -H_B S_{CN} \\ 0_{q,n} & -I_q \end{pmatrix}, \ \tilde{B}^0 = \begin{pmatrix} N \\ (1-\mu) I_q \end{pmatrix}. \end{aligned} \quad (5.63)$$

Thus, the feasibility of the proposed DDC with internal dynamic filter for structurally singular systems is determined by ensuring stability of fast movements.

## 5.3 Sliding Mode Inverse Model Control

### 5.3.1 Sliding Mode Equivalence Principle

Consider output tracking and disturbance rejection problem for linear multivariable system (to simplify writing, we will omit argument *t*)

$$\begin{aligned} \dot{x} &= Ax + B_1 u + B_2 w, \\ y_1 &= C_1 x, \quad y_2 = C_2 x. \end{aligned} \tag{5.64}$$

For this purpose, taking advantage of InvMC implementation within the framework of VSS paradigm approach, we will construct a discontinuous control that ensures sliding mode implementation in system (5.64), which, in a certain sense, is close to processes in system (5.64) under InvMC.

The implementation of VSSs is usually based on discontinuous control

$$\begin{aligned} &u^{\sigma} = -U(x) \cdot \text{sign}\,[\sigma(x)], \quad \sigma = (\sigma_1, \ldots, \sigma_{m_1})^{\mathrm{T}}, \\ &\text{sign}\,(\sigma) = [\text{sign}(\sigma_1), \ldots, \text{sign}(\sigma_{m_1})]^{\mathrm{T}}, \end{aligned} \tag{5.65}$$

ensuring a sliding mode motion on the intersection of discontinuity surfaces$\sigma = 0$.

The equation of sliding mode on the manifold $\sigma = 0$ may be derived using *equivalent control* method [34]; in that, equivalent continuous control $u^{\text{eq}}$ should be found such that$\dot{\sigma}(u^{\text{eq}}) = 0$. The design procedure includes selection the discontinuity surfaces $\sigma(x)$; so controlled plant behavior during sliding motion described by system (5.64) equation with $u = u^{\text{eq}}$ has desired dynamic properties. A proper choice of continuous modulating control function $U(x)$ is in a way that sliding motion is asymptotically stable and any state trajectories hit the sliding surface.

In that, sliding mode existing condition [34, 35] are usually used in the following form:

$$\dot{\sigma}_i \sigma_i < 0, \qquad |u_i^{eq}| \leq U(x), \qquad i = \overline{1, m}. \tag{5.66}$$

The basic idea of SME principle, proposed in [51], is sliding surfaces $\sigma(x)$ selection so that equivalent control will coincide with control function, obtained by InvMC approach.

Assume, at first, that disturbance are accessible to direct measurement and, using SME principle, we will construct a control law equivalent to that obtained by InvMC based on DC with regularized feedforward IM:

$$\begin{aligned} &u^{\text{eq}} = K_1 e + u^*, \\ &\dot{x}_c = F_1(\varepsilon_1, L_1) x_c + L_1 y_1^* + H_1(\varepsilon_1) \dot{y}_1^* + \Pi_1(\varepsilon_1) B_2 w, \\ &u^* = \bar{C}_1(L_1) x_c + B_1^+ L_1 y_1^* + (1 - \varepsilon_1) S_{11}^{-1} (\dot{y}_1^* - S_{12} w). \end{aligned} \tag{5.67}$$

Suppose that desired switching surface is defined as

$$\sigma = Gx + \sigma_0 = 0\,, \tag{5.68}$$

where $G$ arbitrary matrix such that matrix $GB$ has full rank, and an auxiliary time-dependent function $\sigma_0$ is chosen from the equivalence condition (5.67).

Accept $G = S_{11}{}^{-1}C_1$, in such a case

$$\dot{\sigma}_0 = (S_{11}^{-1}C_1 A - K_1 C_1)x + S_{11}^{-1}S_{12}w + u^* + K_1 y_1{}^*. \tag{5.69}$$

Consequently, conditions (5.68) and (5.69) lead to IVSC with state feedback. Control law (5.65) with the switching surface (5.68) may be treated as the variable structure converter of continuous combined control.

In order to obtain sliding mode conditions consider two types of control function:

- **(A)** State-dependent amplitude modulated relay $U\,(e) = \alpha \cdot \|e\| + \beta\,, \alpha, \beta > 0\,.$
- **(B)** Bounded relay control $U\,(e) = U^M > 0.$

Stability conditions for VSS with InvMC may be derived using the following result.

Let system $\dot{x} = Ax + Bu\,, y = c^{\mathrm{T}}x + d^{\mathrm{T}}u, x \in \mathbf{R}^n,\ x_0 = 0,\ y \in \mathbf{R}^1$ be stable, and $\|u\| \le c_u\,.$

Then the following estimate holds:

$$\begin{aligned} &|y| \le |y\,(\infty)| \le \left[\delta\left(c^{\mathrm{T}}V^{-1}c\right)^{1/2} \cdot \|B\| + \|d\|\right] c_u, \\ &\delta = \mathrm{cond}\,(V) = \lambda_m\,(V)\,/\lambda_M\,(V)\,, \end{aligned} \tag{5.70}$$

where $\delta$ is the degree of conditionality of matrix $V$, which is a positive-definite solution of Lyapunov equation $VA + A^{\mathrm{T}}V = -W,\ W > 0$, and $\lambda_m,\ \ \lambda_M$ are minimal and maximal eigenvalues of matrix $V$.

To prove the validity of Equation (5.70), define suitable Lyapunov function in the form [53]: $v(x) = (x^{\mathrm{T}}Vx)^{1/2},\ VA + A^{\mathrm{T}}V = -W,\ W > 0\,,$ for which

$$\begin{aligned} &\dot{v}(x) \le \eta\,\|x\|\,, \qquad \|\mathrm{grad}(x)\| \le \mu x, \\ &\eta = (1/2)\cdot\lambda_N^{1/2}(V)\lambda_m(W),\quad \mu = \lambda_m^{-1/2}(V)\cdot\lambda_M(V). \end{aligned} \tag{5.71}$$

Following the technique proposed in [52], we will obtain

$$\dot{v} \le -\eta v + \mu\,\|B\|\,c_u, \qquad v(0) = 0. \tag{5.72}$$

Using known properties of convex sets support functions [54]

$$\max_{x^{T}Qx\leq 1} z^{T}x = +(z^{T}Q^{-1}z)^{1/2}, \qquad \min_{x^{T}Qx\leq 1} z^{T}x = -(z^{T}Q^{-1}z)^{1/2}, \quad (5.73)$$

it is easy to verify correctness of Equation (5.70).

Note that it is desirable to choose matrix $W$ so that

$$\eta = 0.5 \cdot \lambda_{M}^{1/2}(V)\lambda_{m}(W) = \eta_{0}(A), \quad (5.74)$$

where $\eta_{0}(A) = -\max\{\mathrm{Re}\,\lambda_{i}(A)\}$ is degree of stability of system (5.64).

In such a case

$$\delta(A,V) = \eta_{0}{}^{-1}(A)\,\lambda_{M}^{1/2}(V)\,[\mathrm{cond}\,(V)]^{1/2}. \quad (5.75)$$

Using this result, the equivalent sliding mode existence conditions may be obtained.

From Equation (5.68), it follows that $\dot{\sigma} = u^{\sigma} - u^{*} - K_{1}e$.

Then inequality (5.66) fulfillment conditions for modulating control law (**A**) are

$$\begin{aligned} &\alpha > \max_{i}\left\{ \left\|E_{i}^{T}K_{1}\right\|\right\}, \quad i=\overline{1,m_{1}}, \\ &\beta > \max_{i}\left\{ \delta_{1}(d_{i}{}^{T}V_{1}{}^{-1}d_{i})^{1/2}\chi_{0}+\chi_{i}\right\}, \end{aligned} \quad (5.76)$$

where $E_{i}^{T}$ denotes the $i$th unit vector,

$$\begin{aligned} \chi_{0} &= \left\|L_{1}\right\|c_{0}^{*} + (1-\varepsilon_{1})\left\|B_{1}S_{11}{}^{-1}\right\|c_{1}^{*} + \left\|\Pi_{1}(\varepsilon_{1})B_{2}\right\|c_{w}, \\ d_{i}{}^{T} &= E_{i}{}^{T}\bar{C}_{1}, \quad \chi_{i} = \left\|E_{i}{}^{T}B_{1}^{+}L_{1}\right\|c_{0}^{*} \\ &\quad + (1-\varepsilon_{1})\left\|E_{i}{}^{T}S_{11}{}^{-1}\right\|c_{1}^{*} + (1-\varepsilon_{1})\left\|S_{11}{}^{-1}S_{11}\right\|c_{w}. \end{aligned} \quad (5.77)$$

In accordance with Equation (5.75)

$$\delta_{1} = \delta(F_{1},V_{1}), \qquad V_{1}F_{1} + F_{1}{}^{T}V_{1} = -W_{1}, \quad (5.78)$$

where $W_{1} \geq 0$ is chosen so that $\eta\,(V_{1},W_{1}) = \eta_{0}\,(F_{1})$.

For control function (**B**), equivalent sliding mode existence conditions may be obtained in a similar way, but only the local stability may be guaranteed.

The VSS control laws (5.65), (5.68), and (5.69) may be directly applied in the only case when the full state vector $x$ and disturbances $w$ measurements are accessible. In the other case, it may be replaced by its estimates obtained by means of inverse model-based state/disturbances observer (IVSC with output feedback).

### 5.3.2 5.3.2. Variable Structure Feedforward Compensator

Let us choose *variable structure feedforward compensator* (VSFC) in the form

$$\begin{aligned} \dot{x}_c &= A\dot{x}_c + L_1 r_1^* + B_1 u_1^\sigma + B_2 \hat{w}, \\ r_1^* &= y_1^* - C_1 x_c, \end{aligned} \tag{5.79}$$

where discontinuous controlis defined as

$$u_1^\sigma = -U_1\left(x_c\right) \cdot \text{sign}\left(\sigma_1\right), \sigma_1 = -S_{11}^{-1}\left(r_1^* + \sigma_1^0 - \varepsilon_1 y_1^*\right), \tag{5.80}$$

with modulating control function.

In accordance with SME principle, the switching surfaces should be selected so that sliding mode equation of system (5.64) with controls (5.65) and (5.68) will coincide with IM-based FFC equation.

Using equivalent control

$$u_1^{\text{eq}} = S_{11}^{-1}[(1-\varepsilon_1)\dot{y}_1^* - C_1 A x_c - C_1 L_1 r_1^* - S_{12}\hat{w} + \dot{\sigma}_1^0], \tag{5.81}$$

equation for integral term$\sigma_1^0$may be obtained as

$$\dot{\sigma}_1^0 = C_1 L_1 r_1^* + \varepsilon_1 C_1 (A x_c + B_2 \hat{w}), \tag{5.82}$$

where $\varepsilon_1$ and $L_1$ are VSFC tuning parameters.

From Equations (5.81) and (5.82), it follows that

$$u_1^{\text{eq}} = (1-\varepsilon_1) \cdot S_{11}^{-1}(\dot{y}_1^* - C_1 A x_c - S_{12}\hat{w}). \tag{5.83}$$

As a result, VSFC equation in sliding mode regime

$$\dot{x}_c = \Pi_1(\varepsilon_1) A x_c + L_1 r_1^* + H_1(\varepsilon_1)\dot{y}_1^* + \Pi_1(\varepsilon_1) B_2 \hat{w} \tag{5.84}$$

will be identical to IM-based FC. Note that in such a way, there is no necessity in the reference signal$y_1^*$ differentiation.

The output signal of VSFC should be taken in the form

$$u^* = u_1^\sigma + B_1^+ L_1 r_1^*. \tag{5.85}$$

Because the slow component of will coincide with the averaging value of may be used as a compensative control. In turn, it can be obtained by means of multivariable averaging filter consisting of low-pass filters with high enough cutoff frequency. As far as variable structure converter is used, then such a filter may be eliminated from control scheme. Indeed, discontinuous compensative control (5.85) is transformed by integral part

(5.69) of control laws (5.65) and (5.68), which eliminates the need of high-frequency component averaging.

Using the equation for switching function derivative

$$\dot{\sigma}_1 = u_1^\sigma + (1 - \varepsilon_1) \cdot S_{11}^{-1} \cdot \left( C_1^0 A^0 x_c + S_{12} \hat{w} - \dot{y}_1^* \right), \tag{5.86}$$

we obtain VSFC sliding modes existence conditions.

- **(A)** Control law$U(x_c) = \alpha_1 \|x_c\| + \beta_1$:

$$\alpha_1 > (1 - \varepsilon_1) \cdot \max_i \left\{ \|d_i\| \right\}, \quad \beta_1 > \max_i \left\{ \left\| E_i^{\mathrm{T}} S_{12} \right\| \right\} c_w + c_1^*. \tag{5.87}$$

- **(B)** Control law$U(x_c) = U_1^M$:

$$U_1^M > \max_i \left\{ \delta_1 \left( d_i^{\mathrm{T}} A V_1^{-1} A^{\mathrm{T}} d_i \right)^{1/2} \chi_0 + \psi_i^{(1)} c_w \right\} + c_1^*, \tag{5.88}$$

where $d_i^{\mathrm{T}} = E_i^{\mathrm{T}} S_{11} C_1, \psi_i^{(1)} = \left\| E_i^{\mathrm{T}} S_{12} \right\|$, and $\delta_1 = \delta(F_1, V_1)$ is defined similarly to Equation (5.78).

### 5.3.3 Variable Structure Disturbance Observer

*Variable structure disturbance observer* (VSDO) may be designed as a dynamic system with auxiliary discontinuous control and input signal, which involves output plant's measurement in the form of DO state feedback:

$$\dot{\hat{x}} = A\hat{x} + B_1 u + B_2^\sigma u_2^\sigma + L_2 r_2, \quad r_2 = y_2 - C_2 \hat{x}, \tag{5.89}$$

where $u_2^\sigma \in \mathbf{R}^{q_2}$ is discontinuous signal.

Let matrix $B_2^\sigma$ be chosen so that$\operatorname{rank}(C_2 B_2^\sigma) = q_2$. Define the discontinuous function as

$$u_2^\sigma = -U_2 \operatorname{sign}(\sigma_2), \sigma_2 = -\left( C_2 B_2^\sigma \right)^{-1} \left( r_2 - \varepsilon_2 y_2 + \sigma_2^0 \right), \tag{5.90}$$

where in accordance with SME principle

$$\dot{\sigma}_2^0 = C_2 L_2 r_2 + \varepsilon_2 C_2 (A\hat{x} + B_1 u). \tag{5.91}$$

From Equations (5.89) and (5.91), it follows that equivalent control

$$u_2^{\mathrm{eq}} = (1 - \varepsilon_2) \left( C_2 B_2^\sigma \right)^{-1} C_2 (A e_x + B_2 w), \tag{5.92}$$

and consequently depends on unmeasurable disturbances.

In such a case, the state estimation error is given by

$$\begin{aligned} \dot{e}_x &= \left[\Pi_2^\sigma\left(\varepsilon_2\right) A - L_2 C_2\right] e_x + \left[\varepsilon_2 B_2 + \left(1-\varepsilon_2\right) \Pi_2^\sigma B_2\right] w \\ \Pi_2^\sigma\left(\varepsilon_2\right) &= I_n - \left(1-\varepsilon_2\right) B_2^\sigma \left(C_2 B_2^\sigma\right)^{-1} C_2, \quad \Pi_2^\sigma = \Pi_2^\sigma(0) \end{aligned} \tag{5.93}$$

Let us choose matrix $B_2^\sigma$ so that $\Pi_2^\sigma B_2 = 0$ in order to ensure the state estimation error "$\varepsilon$-invariance" toward $w$. Such conditions are fulfilled, provided $B_2^\sigma = \left(I_n - \Pi_2^+ \Pi_2\right) L_2^\sigma$ with $\Pi_2 = I_n - B_2 S_{22}^+ C_2$, where $L_2^\sigma$ is arbitrary tuning matrix such that $\det\left(C_2 B_2^\sigma\right) \neq 0$; tuning parameters $\varepsilon_2$ and $L_2$ are corresponding tuning parameters.

As it follows from Equations (5.27) and (5.29), the unmeasurable disturbance estimate is determined by linear transformation of averaging discontinuous function

$$\hat{w} = S_{22}^+ \cdot \left(C_2 B_2^\sigma\right) \cdot \hat{u}_2^\sigma. \tag{5.94}$$

Thus, the estimate may be obtained without averaging filter by using the integral part of VSFC switching surface, in such a way thatmay be directly admitted as.

Using the switching derivative equation

$$\dot{\sigma}_2 = u_2^\sigma - \left(1-\varepsilon_2\right)\left(C_2 B_2^\sigma\right)^{-1} C_2\left(A e_x + B_2 w\right), \tag{5.95}$$

it is easy to obtain the VSDO sliding mode conditions.

Note that only control law (**B**) may be used because the estimation error cannot be directly inserted in control function.

Consequently, sliding mode conditions for variable structure observer are the following:

$$\begin{aligned} U_2^M &> \left(1-\varepsilon_2\right) \cdot \max\left\{\, \gamma_1(c_0, c_w), \;\; \gamma_2(c_w)\right\}, \\ \gamma_1(c_0, c_w) &= \max_i \left\{\, \left\|d_i^{\mathrm{T}} A^0\right\| c_0 + \psi_i^{(2)} c_w \right\}, \\ \gamma_2(c_w) &= \max_i \left\{\, \varepsilon_2 \delta_2 (d_i^{\mathrm{T}} A V_2^{-1} A^{\mathrm{T}} d_i)^{1/2} \left\|B_2\right\| + \psi_i^{(2)} \right\} c_w, \end{aligned} \tag{5.96}$$

where $d_i^{\mathrm{T}} = E_i^{\mathrm{T}} (C_2 B_2^\sigma)^{-1} C_2 \,, \psi_i^{(2)} = \left\|d_i^{\mathrm{T}} B_2\right\|, \delta_2 = \delta\left(F_2, V_2\right),$

$$V_2 F_2 + F_2^{\mathrm{T}} V_2 = -W_2, \quad W_2 > 0, \; \eta\left(V_2, W_2\right) = \eta_0\left(F_2\right). \tag{5.97}$$

It is clear that only local stability conditions may be obtained and the stability domain is determined by upper bound of initial state$c_0$.

Therefore, we obtained the proper multivariable modification of cascade VSS control scheme, recently proposed for robustness improvement [55]. Averaging of discontinuous functions is realized by integral parts of VSS control blocks so that the special filters may be eliminated from the control scheme. Moreover, characteristic feature of proposed approach is the possibility of order reduction for measured signals derivatives used in control law. For plants with relative degree not greater than 2, the differentiators may be eliminated; in general case, the approximate differentiators should be used [56, 57].

## 5.4 Discrete Inverse Model Control

### 5.4.1 Problem Statement

Disturbance rejection and output tracking methods based on InvMC approach are also applicable to discrete-time systems. The synthesis of control laws, as well as DC and DO with inverse models, remains unchanged; the peculiarity is that the approximate differentiation of measured variables, ensuring the controller's physical feasibility, is replaced by taking their differences, which leads to a delay in disturbance estimate.

Consider output control problem for discrete multivariable control system described by linear state-space nominal model with input and parametric disturbances:

$$\begin{aligned} x_{k+1} &= (A + \delta A_k)x_k + B_1 u_k + D_1 f_k, \\ y_k^1 &= C_1 x_k, \quad y_k^2 = C_2 x_k, \end{aligned} \tag{5.98}$$

where $x_k \in \mathbf{R}^{\mathrm{n}}$ is state vector at time instant $k$, $u_k \in \mathbf{R}^{\mathrm{m}_1}$ is control vector, $y_k^1 \in \mathbf{R}^{\mathrm{q}_1}$ is output controlled variables, $y_k^2 \in \mathbf{R}^{\mathrm{q}_2}$ is vector of measured variables, $f_k \in \mathbf{R}^{\mathrm{d}}$ is input disturbance, and $\delta A_k$ is parametric disturbance.

Let only $(p < n)$ rows of system dynamic matrix $A$ be subject to parametric disturbance, which may be factorized as $\delta A_k = D_2 \cdot \Delta$, where matrix $\Delta$ is composed of object parameters deviation regarding nominal values, and $(n \times p)$ matrix $D_2$ describes the parametric disturbance structure. In that, the controlled system model (5.98) is

$$\begin{aligned} x_{k+1} &= A x_k + B_1 u_k + B_2 w_k, \\ y_k^1 &= C_1 x_k, \quad y_k^2 = C_2 x_k, \end{aligned} \tag{5.99}$$

where $B_2 = (D_1, D_2)$,$w_k^{\mathrm{T}} = (f_k, \Delta_k x_k) \in \mathbf{R}^{m_2}$ is equivalent disturbance, $m_2 = p + d$.

Output control problem is to find the control sequence $\{u_k\}$, depending on the measured output variables $y_k^2$, which ensure reference signal $y_k^*$ tracking and disturbances $w_k$ rejection. The requirement of closed-loop system stabilization along with the disturbance rejection leads to the disturbance rejection problem with stability.

Following InvMC approach, control structure for disturbance rejection and closed-loop system stabilization are selected as a combination of DO and DC, both based on the designed IMs of the respective controlled system channels.

From a practical point of view, it is desirable to decompose the design procedure into two steps, namely, *structural synthesis* of DO and DC that renders the fixed and free parameters and *parametric synthesis* based on the appropriate parameters tuning methods in order to satisfy the design goals, such as internal stability, robustness, performance optimization, and so on. To solve these control problems, the theory of discrete UIO in conjunction with pole-placement methods turned out to be effective.

### 5.4.2 Discrete Disturbance Observer

By analogy with the dynamic observers theory, discrete dynamic system

$$\begin{aligned} \bar{x}_{k+1} &= A^I\bar{x}_k + B_1^I y_k + B_2^I y_{k+d} + B_0^I u_k, \\ \hat{w}_k &= C^I\bar{x}_k + D_1^I y_k + D_2^I y_{k+d} \end{aligned} \tag{5.100}$$

with state vector $\bar{x}_k \in \mathbf{R}^{n-q_2}$ will be referred to as asymptotic reduced-order DO, if dynamic system (4.3) is asymptotically stable and the following conditions take place: $\|\bar{x}_k - Rx_k\| \to 0$, $\|\hat{w}_k - w_k\| \to 0$, $k \to \infty$, where $R \in \mathbf{R}^{n-q_2\times n}$ is the appropriate aggregate matrix such that $\operatorname{rank}\left(C_2^{\mathrm{T}} \vdots R^{\mathrm{T}}\right) = n$, $\operatorname{rank} R = n - q$.

Vector $\hat{w}_k$ may be treated as input disturbance estimation, obtained by IM (5.100). In such a case, integer $r \geq 1$ determines the minimal disturbance estimation delay, which concise with control channel relative order, namely, minimal integer, such as matrix of Markov parameter $S_{22}(r) = C_2A^{r-1}B_2 \neq 0$.

The minimal state-space realization of IM is defined by means of reduced order UIO. For the purpose of simplicity, consider only the case of relative order 1, under the assumption $S_{22}(1) = S_{22} = C_2B_2 \neq 0$. Treated $w_k$as an unknown input signal, state vector estimate $\hat{x}_k \in \mathbf{R}^n$ is determined by minimal-order UIO as follows:

$$\begin{aligned} &\tilde{x}_{k+1} = \bar{F}\tilde{x}_k + \bar{G}y_k + \bar{G}_0 u_k, \quad \bar{x}_k = \tilde{x}_k + \bar{H}y_k^2, \\ &\hat{x}_k = Py_k^2 + Q\bar{x}_k, \end{aligned} \tag{5.101}$$

where $\bar{x}_k \in \mathbf{R}^{n-q_2}$ is aggregated state vector $Rx_k$ estimation and $\tilde{x}_k$ is observer state vector.

Matrices$P_1 \in \mathbf{R}^{n\times q_1}$ and $Q_1 \in \mathbf{R}^{n\times n-q_1}$ are uniquely determined by aggregate matrix$R$:

$$\left( \begin{array}{cc} P & Q \end{array} \right) = \left( \begin{array}{c} C \\ R \end{array} \right)^{-1}, \begin{array}{l} CP = I_q, \quad RQ = I_q, \quad PC + QR = I_n, \\ CQ = 0_{q\times n-q}, \quad RP = 0_{n-q\times q}. \end{array} \tag{5.102}$$

Equations of observer (5.101) dynamic part may be transformed to the equivalent form

$$\bar{x}_{k+1} = \bar{F}\bar{x}_k + \left(\bar{G} - \bar{F}\bar{H}\right) y_k^2 + Hy_{k+1}^2 + \bar{G}_0 u_k. \tag{5.103}$$

In that, state vector estimation error $e_k^x = x_k - \hat{x}_k$ and aggregated variables deviation $\bar{e}_k^x = Rx_k - \bar{x}_k$ are connected by linear transformation$e_k^x = Q\bar{e}_k^x$; thus $\bar{e}_k^x$ will be given by

$$\begin{aligned} &\bar{e}_{k+1}^x = \bar{F}\bar{e}_k^x + \left[RA - \bar{F}R - (\bar{G} - \bar{F}\bar{H})C_2 - \bar{H}C_2A\right] x_k + \\ &+(RB_2 - HC_2)w_k + (RB_1 - \bar{G}_0)u_k. \end{aligned} \tag{5.104}$$

From Equation (5.104), observer design conditions, namely, state estimation independence from unknown input disturbance (so-called "unknown input invariance conditions") follows:

$$\begin{aligned} &(R - \bar{H}C_2)A - \bar{F}(R - \bar{H}C_2) = \bar{G}C_2, \\ &RB = HC_2B_2, \quad RB_1 - \bar{G}_0 = 0. \end{aligned} \tag{5.105}$$

When invertibility conditions $\operatorname{rank} S_{22} = q_2,\ q_2 \geq m_2$take place, which define UIO structural synthesis problem solvability, linear matrix equation (5.105) solution is

$$\begin{aligned} &\bar{F}_2 = R\Pi_2 AQ, \quad \bar{G}_2 = R\Pi_2 A(H + P\Omega_2), \\ &\bar{H}_2 = RB_2 S_{22}{}^+, \quad \bar{G}_0 = RB_1, \end{aligned} \tag{5.106}$$

where "+" denotes Moore–Penrouze generalized inverse, and projection matrices$\Pi_2 = I_n - B_2 S_{22}{}^+ C_2$, $\Omega_2 = I_q - S_{22}S_{22}{}^+$ while$C_2\Pi_2 = \Omega_2 C_2$.

Taking an input vector $w_k$ estimate as $\hat{w}_k = B^+(\hat{x}_{k+1} - A\hat{x}_k)$ and considering it as IM output vector, then channel "D-M" IM equations may be obtained from Equations (5.103) and (5.106):

$$\begin{aligned}\bar{x}_{k+1} &= R\Pi_2 AQ\bar{x}_k + R\Pi_2 APy_k^2 + RB_2 S_{22}^+ y_{k+1}^2 + RB_1 u_k,\\ \hat{w}_k &= (S_{22}^+ + B_2^+ PR)(y_{k+1}^2 - C_2 AQ\bar{x}_k - C_2 APy_k^2).\end{aligned} \quad (5.107)$$

From Equation (5.107), it follows that IM dynamics matrix $A^I = \bar{F}_2 = R\Pi_2 AQ$ depends on the arbitrary aggregate matrix $R$ of given rank $n - q_2$, which is DO tuning matrix.

### 5.4.3 Disturbance Observer Parameterization

Using the block form of system (5.98) (for notation simplicity, indices at $q$ are omitted),

$$A = \begin{pmatrix} A_{11} & A_{12} \\ A_{21} & A_{22} \end{pmatrix}, \quad C_2 = \begin{pmatrix} I_q & 0_{n-q \times q} \end{pmatrix}, \quad B_2 = \begin{pmatrix} B_{21} \\ B_{22} \end{pmatrix}\begin{matrix} q \\ n-q \end{matrix}, \quad (5.108)$$

which may be obtained by non-singular state-space transformation, and concretely defining matrices $P$, $Q$ choice, one can admit

$$\begin{pmatrix} P & Q \end{pmatrix} = \begin{pmatrix} P_1 & Q_1 \\ P_2 & Q_2 \end{pmatrix}\begin{matrix} q \\ n-q \end{matrix}, \quad P_1 = I_q, \; Q_1 = 0_{q\times n-q}. \quad (5.109)$$

In such a case, for any $Q_2$ such that $\det Q_2 \neq 0$, aggregate matrix may be obtained in the form $R = Q_2^{-1}\begin{pmatrix} -P_2 & I_{n-q} \end{pmatrix}$, and, consequently, IM dynamic matrix is the following:

$$R\Pi AQ = Q_2^{-1}\left(\tilde{A}_{22} - P_2\Omega_{B_{21}} A_{12}\right) Q_2, \quad (5.110)$$

where $\tilde{A}_{22} = A_{22} - B_{22}B_{21}{}^+ A_{12}$, $\Omega_{B_{21}} = I_q - B_{21}B_{21}{}^+$.

Thus, in fact, non-singular matrix $Q_2$ specifies the similarity transformation and does not change matrix $A^I$ spectrum, which, as following from Equation (5.110), is completely determined by only arbitrary tuning matrix $P_2$. Any type of pole-placement method may be used for matrix $A^I$ stabilization, and appropriate modal control problem will be solvable if matrix pair $(\tilde{A}_{22}, \; \tilde{A}_{12})$ is observable. Such a condition is equivalent to the well-known UIO design solvability condition, namely, input observability [32, 33]. Therefore, aggregate matrix $R$ is determined up to an arbitrary non-singular matrix $Q_2$.

Input observability condition is obviously violated in the case when $m_2 = q_2$. In that, $\Omega_{B_1} = 0$ and $A^I$ does not depend on $P_2$. In such a case for the tuning properties guarantee, it is possible to use regularized UIO [54], which ensures the approximately invariance with respect to the unknown disturbance:

$$\|RB_2 - HC_2B_2\|^2 + \varepsilon \|H\|^2 \to \min_H, \tag{5.111}$$

where $\varepsilon > 0$ is regularization parameter.

Optimization problem (5.111) solution is $\bar{H}_2(\varepsilon) = RB_2{S_{22}}^{\mathrm{T}}(\varepsilon I_q + S_{22}{S_{22}}^{\mathrm{T}})^{-1}$, and, accordingly, the results of regularized IM structural design problem are the following:

$$\begin{aligned}
&\bar{F}_2(\varepsilon) = \tilde{A}_{22}(\varepsilon)P_2\Omega_{B_{21}}(\varepsilon)A_{12},\ \tilde{A}_{22}(\varepsilon) = A_{22} - B_2\Psi_{B_{21}}(\varepsilon)A_{12},\\
&\Psi_{B_{21}}(\varepsilon) = B_{21}^{\mathrm{T}}(\varepsilon I_q + B_{21}B_{21}^{\mathrm{T}}),\\
&\Omega_{B_{21}}(\varepsilon) = I_q - B_{21}B_{21}^{\mathrm{T}}(\varepsilon I_q + B_{21}B_{21}^{\mathrm{T}})^{-1} = \varepsilon(\varepsilon I_q + B_{21}B_{21}^{\mathrm{T}})^{-1}.
\end{aligned} \tag{5.112}$$

It is obvious that, provided $q_2 = m_2$, matrix $\Omega_{B_1}(\varepsilon) \neq 0$ for any $\varepsilon > 0$ and corresponding pole-placement problem may be solved if observability conditions for matrix pair $\{\tilde{A}_{22}(\varepsilon),\tilde{A}_{11}(\varepsilon)\}$, $\tilde{A}_{12}(\varepsilon) = \Omega_{B_1}(\varepsilon)A_{12}$ are met.

Estimation errors for designed DO based on regularized IM are given by

$$\begin{aligned}
\bar{e}_{k+1}^x &= \bar{F}_2(\varepsilon)\bar{e}_k^x + \varepsilon RB_2(\varepsilon I_q + {S_{22}}^{\mathrm{T}}S_{22})^{-1}w_k,\\
e_k^w &= -{B_2}^+\left(H_2(\varepsilon) + P\Omega(\varepsilon)\right)C_2AQ\bar{e}_k^x +\\
&\quad +\varepsilon {B_2}^+(I - PC_2)(\varepsilon I_q + {S_{22}}^{\mathrm{T}}S_{22})^{-1}w_k.
\end{aligned} \tag{5.113}$$

The proposed IM regularization ensures desired dynamic properties of DO, but estimation errors depend on the real disturbance, though the factor level is determined by regularization parameter $\varepsilon > 0$, which, in turn, should be specified by the tradeoff between DO dynamic performance and additional estimation error, component caused by inverse model regularization.

### 5.4.4 Disturbance Compensator Structural Synthesis

Disturbance estimate is used for disturbance rejection control law construction. Consider the combined control law as the sum of stabilizing and compensation components $u_k = u_k^s + u_k^c$, though control-stabilizing component $u_k^s$ is taken as $u_k^s = K \cdot e_k^y$, where $e_k^y = (y_k^* - y_k)$ is control error $K$ is linear output feedback matrix.

Compensative control component $u_k^c$ is formed by multivariable dynamic DC constructed by IM control channel with special input signal $r_k$, generated by PM

$$\begin{aligned} x_{k+1}^{p} &= Ax_{k}^{p} + B_{f}\hat{w}_{k}, \\ y_{k}^{p} &= Cx_{k}^{p}, \quad r_{k} = y_{k}^{*} - y_{k}^{p}, \end{aligned} \tag{5.114}$$

where $x_{k}^{p} \in \mathbf{R}^{n}$ is PM state vector and $y_{k}^{p} \in \mathbf{R}^{q_1}$ is predicted reaction on input disturbance.

Using the approach described above, multivariable DC equation is obtained as a combination of control channel IM and disturbance channel PM.

Considering the one-step PM input signal prediction $r_{k+1} = y_{k+1}^{*} - C_1Ax_{k}^{p} - C_1B_2\hat{w}_k$, it is easy to obtain DC equation in the form:

$$\begin{aligned} \bar{x}_{k+1}^{c} &= \bar{F}_1(\varepsilon, P_1)\,\bar{x}_{k}^{c} + R(P_1)\Pi_1(\varepsilon)B_2\hat{w}_k, \\ u_{k}^{c} &= -\bar{B}(\varepsilon)\left[C_1AQ\bar{x}_{k}^{c} + C_1B_2\hat{w}_k\right], \end{aligned} \tag{5.115}$$

where$x_{k}^{c} = \bar{x}_k + Rx_{k}^{p}$is dynamic compensator state vector.

As a result, compensator matrices

$$\begin{aligned} &\bar{F}_1(\varepsilon){=}R\Pi_1(\varepsilon)AQ, \bar{H}_1(\varepsilon){=}RB_1{S_{11}}^{+}(\varepsilon), \bar{B}(\varepsilon){=}{B_1}^{+}\left(H_1(\varepsilon) + P\Omega_1(\varepsilon)\right), \\ &\Pi_1(\varepsilon) = I_n - H_1(\varepsilon)C_1, \Omega_1(\varepsilon) = I_{q_1} - S_{11}{S_{11}}^{+}(\varepsilon), \end{aligned} \tag{5.116}$$

depend on the appropriate IM tuning parameters$\varepsilon$, $P_1$.

Matrix ${S_{11}}^{+}(\varepsilon) = {S_{11}}^{\mathrm{T}}(\varepsilon I_q + S_{11}{S_{11}}^{\mathrm{T}})^{-1}$ is generalized Moore–Penrouze pseudo-inverse of$S_{11}$, though ${S_{11}}^{+}(\varepsilon)\left|_{\varepsilon=0}\right. = {S_{11}}^{-1}$ and ${S_{11}}^{+}(\varepsilon)\left|_{\varepsilon=\infty}\right. = 0$.

Consider dynamic properties of closed-loop feedback–feedforward control system with designed DO (5.107) and DC (5.115) and (5.116).

In order to obtain control error equation, consider, at first, equation for output controlled signal for system with disturbance feedforward controller (DFC). For the purpose of simplicity, consider the case when reference signal $y_{k}^{*} = 0$ and $m_1 = q_1$

$$\begin{aligned} y_{k+1} &= C_1Ax_k + C_1B_1u_k + C_1B_2w_k \\ &= Ax_k + S_{11}(-Ky_k + \bar{u}_{k}^{c}) + C_1B_2w_k = \\ &= C_1Ax_k + S_{11}Ky_k - S_{11}\bar{B}(\varepsilon)C_1AQ\bar{x}_{k}^{c} \\ &\quad - (I_m - S_{11}\bar{B}(\varepsilon))C_1B_2w_k. \end{aligned} \tag{5.117}$$

Using matrix identities

$$\begin{aligned} &I_q - S_{11}\bar{B}(\varepsilon) = (I_q - S_{11}{B_1}^{+}P)\Omega_1(\varepsilon), \\ &S\bar{B}(\varepsilon) = I_q - (I_q - S_{11}{B_1}^{+}P)\Omega_1(\varepsilon), \\ &B_1\bar{B}(\varepsilon) = B_1{S_{11}}^{-1}(I_q - (I_q - S_{11}B^{+}P)\Omega_1(\varepsilon)), \end{aligned} \tag{5.118}$$

and taking into account that $x_k - Q\bar{x}_k^c = Q\bar{\theta}_k - Pe_k$, where $\bar{\theta}_k = Rx_k - \bar{x}_k^c$, after some algebraic transformation, it is possible to get the output control error equation:

$$\begin{aligned} e_{k+1} &= C_1\left(AP - B_1K\right)e_k - C_1AQ\bar{\theta}_k + \Lambda_1(\varepsilon)\bar{w}_k, \\ \bar{\theta}_{k+1} &= R\left(B_1K - AP\right)e_k + RAQ\bar{\theta}_k + \Lambda_2(\varepsilon)\bar{w}_k, \\ \bar{w}_k &= -C_1B_w w_k + C_1AQ\bar{x}_k^c, \end{aligned} \quad (5.119)$$

where $\Lambda_1(\varepsilon) = (I_q - SB_1^+P)\Omega_1(\varepsilon)$, $\Lambda_2(\varepsilon) = RB_1B_1^+P\Omega_1(\varepsilon)$.

For introduced matrices, it follows that $\Omega_1\left(\varepsilon = 0\right) = 0$ and $S_{11}\bar{B}_u\left(\varepsilon = 0\right) = I_q$, $S_{11}\bar{B}_u\left(\varepsilon = \infty\right) = S_{11}B_1{}^+P$. Using notation $x_k^0 = Q\bar{\theta}_k - Pe_k$, equation for control error (4.22) may be represented in equivalent form:

$$x_{k+1}^0 = \left(A - B_1KC_1\right)x_k^0 + \bar{\Lambda}(\varepsilon)\bar{w}_k, \quad e_k = -C_1x_k^0, \quad (5.120)$$

where $\bar{\Lambda}(\varepsilon, P_2) = (I_m - B_1B_1^+)P\Omega_1(\varepsilon)$, though $\bar{\Lambda}\left(\varepsilon = 0\right) = 0$.

Note that $x_k^0$ is a closed-loop system state vector, and $\bar{w}_k$ may be considered as some equivalent disturbance. From Equation (5.120), it follows that including in control law additional compensation component, formed by the DC, leads to the effect, which is equivalent to initial disturbance transformation by some dynamic filter. In that, equivalent disturbance $\bar{w}_k$ influence on control error is determined by matrix$\bar{\Lambda}\left(\varepsilon, P_2\right)$, which depends on the DC tuning parameters.

### 5.4.5 Disturbance Compensator Parametric Synthesis

DC parametric design problem is to find the compensator tuning parameters $\varepsilon,\ P_1$ based on the tradeoff between the requirements to the compensator dynamic properties, namely, degree of stability, dynamic performance, and so on, and degree of disturbance compensation.

In the case when disturbance is immeasurable, the DC input signal should be formed by the real-time disturbance estimates$\hat{w}_k$, obtained by DO (5.107). This situation may be considered as indirect disturbance measurement with some measurement noise$\hat{w}_k = w_k + \xi_k$, which must be taken into account under DC parametric design.

Using the proposed technique, the equations for control error dynamics for control system with DO and DC in the presence of measurement noise have been obtained as

$$\begin{aligned} e_{k+1} &= C_1 (AP - B_1 K) e_k - C_1 AQ\bar{\theta}_k + \Lambda_1(\varepsilon)\bar{w}_k - S\bar{B}C_1 B_2 \xi_k, \\ \bar{\theta}_{k+1} &= R(B_1 K - AP) e_k + RAQ\bar{\theta}_k + \Lambda_2(\varepsilon)\bar{w}_k - R\left(\Pi_1(\varepsilon) + B_1\bar{B}C_1\right) B_2 \xi_k, \\ \bar{w}_k &= -S_{11} w_k + C_1 AQ\bar{x}_k^c, \end{aligned} \tag{5.121}$$

where$\Lambda_1(\varepsilon, P_2) = (I_q - S_1 B_1^+ P)\Omega_1(\varepsilon)$,$\Lambda_2(\varepsilon, P_2) = RB_1 B_1^+ P\Omega_1(\varepsilon)$.

From Equation (5.121), it follows the equation of closed-loop feedback–feedforward system with indirect disturbance measurement, realized by DO:

$$x_{k+1}^0 = (A - B_1 K C_1) x_k^0 + \bar{\Lambda}(\varepsilon)\bar{w}_k - \bar{\Gamma}(\varepsilon)\xi_k, \quad e_k = -C_1 x^0, \tag{5.122}$$

where$\bar{\Gamma}(\varepsilon) = (I_n - (I_n - B_1 B_1^+)P\Omega_1(\varepsilon)C_1)B_2$,$\bar{\Gamma}(\varepsilon = 0) = B_2$.

From Equation (5.122), it follows that for non-minimum phase object, there exists extremely achievable control accuracy level that depends on both object and disturbance characteristics and compensator parameters.

To formulate the problem of DC parametric synthesis, stated at first relationship between discrete $Z$–transform of control error $e(z)$ and disturbance and measurement noise$w(z)$, $\xi(z)$ :

$$e(z) = G_w(z) w(z) + G_\xi(z)\xi(z), \tag{5.123}$$

where discrete matrix transfer functions$G_f(z)$ and $G_\xi(z)$ are the following:

$$\begin{aligned} G_w(z) &= -C_1 (zI_n - A + B_1 K C_1)^{-1} \bar{\Lambda}(\varepsilon) C_1 \\ &\quad \cdot (AQ\left(zI_{n-q} - \bar{F}\right)^{-1} R\Pi_1 - I_n) B_2, \\ G_\xi(z) &= C_1 (zI_n - A + B_1 K C_1)^{-1} \bar{\Gamma}. \end{aligned} \tag{5.124}$$

The DC design problem formalization should be used as *a priori* information concerning disturbance estimation distortion. Considering problem solution under non-statistic *a priori* information, namely, restricted energy signals are used as a model of disturbance and measurement noise:

$$\|w\|_2 = \sqrt{\sum_{k=0}^{\infty} |w_k|_2^2} \le E_w < \infty, \quad \|\xi\|_2 = \sqrt{\sum_{k=0}^{\infty} |\xi_k|_2^2} \le E_\xi < \infty. \tag{5.125}$$

Disturbance compensation level characterized by control error energy $E_e$with upper bond

$$E_e \le \bar{E}_e = \|G_w(z)\|_\infty E_w + \|G_\xi(z)\|_\infty E_\xi, \tag{5.126}$$

as it follows from norm relationship theorem [6], where $\mathbf{H}^{\infty}$ is a norm of matrix transfer function, is defined as follows:

$$\|G(z)\|_{\infty} = \sup_{|z|<1} \|G(z)\|_2 = \sup_{0\leq\omega\leq 2\pi} \lambda_{\max}^{1/2}[G^{\mathrm{T}}(e^{-j\omega})G(e^{j\omega})]. \quad (5.127)$$

In accordance with properties of $\mathbf{H}^{\infty}$ norm, $\bar{E}_u$ criterion minimization is equivalent to control error energy minimization for worst input signals of restricted energy [5]. Therefore, DDC parametric $\mathbf{H}^{\infty}$ optimization problem with additional performance constraint, namely, transfer process duration and variation, is stated as

$$\begin{aligned} &\bar{E}_e(\varepsilon, P_2) \to \min, \\ &T(\varepsilon) \leq \left(\ln\left(\lambda^{-1}(\varepsilon)\right)\right)^{-1} \ln\left(\delta^{-1}\right), \\ &\sigma(\varepsilon) \leq (\beta - 1)/(\beta + 1), \quad \beta = \lambda_1(\varepsilon)\left(1 - \lambda(\varepsilon)\right), \end{aligned} \quad (5.128)$$

where$\lambda(\varepsilon) = \left\|\bar{F}(\varepsilon)\right\|_2 = \lambda_{\max}^{1/2}\left(\bar{F}^{\mathrm{T}}(\varepsilon)\bar{F}(\varepsilon)\right)$, $\delta$ is predetermined level, which characterized DC transfer processes duration.

Hereby, the proposed DC design method ensures its internal stability and desired dynamic properties and robustness with respect to uncertain inaccuracy of disturbance indirect measurement. Thus, reachable control accuracy level depends on disturbance and measurement noise properties defined by Equations (5.126) and (5.127).

## 5.5 Conclusion

A significant increase of control system performance under disturbance rejection and output tracking can be achieved by complexity of controller's structure, which provides great opportunities in comparison with classical feedback. The structures of multivariable controllers with dynamic models in control loop have a number of advantages, providing not only high control accuracy but also robustness. The control method based on inverse dynamic models provides a unified approach to solving the problem of structural–parametric synthesis of compensators for measured disturbances and decoupling controllers using indirect measurements. The use of synthesized inverse models using the theory of observers for systems with unknown input makes it possible to ensure stability and specified dynamic properties of control structures and control system as a whole. The advantage of inverse model method is the possibility of formalizing the choice of multivariable controller structure in the form of analytical design procedures and using well-known methods for parameters tuning or optimization. In this case, it becomes possible, even at the stage of control system designing, to establish the solvability of synthesis problems and potentially achievable accuracy of control and disturbance suppression. Further inverse model method development is advisable on the way of ensuring robustness with respect to deviations of controlled plant parameters and parametric disturbances compensating methods development.

## References

[1] H. Sira-Ramírez, A. Luviano-Juárez, M. Ramírez-Neria and E. W. Zurita-Bustamante, 'Active Disturbance Rejection Control of Dynamic Systems. A Flatness Based Approach', Elsevier, 2017.

[2] Z. Chen, D. Xu, 'Output Regulation and Active Disturbance Rejection Control: Unified Formulation and Comparison', Asian Journal of Control, vol. 18(5), 2016, pp. 1668-1678.

[3] V. Morari, E. Zafirov, 'Robust processes control', New Jersey, Prentice Hall, 1989.

[4] Yu. Kondratenko, A. Chikrii, V. Gubarev and J. Kacprzyk, (Eds), 'Advanced Control Techniques in Complex Engineering Systems: Theory and Applications', Dedicated to Professor Vsevolod M. Kuntsevich. Studies in Systems, Decision and Control, vol. 203, Cham: Springer Nature Switzerland AG, 2019.

[5] K. Zhou, J. Doyle, K. Glover, 'Robust and Optimal Control', Prentice-Hall, NJ, 1996.

[6] B. Francis, 'A course of $H^{\infty}$ control theory', Lecture Notes in Control and Information Sciences, Berlin, Springer-Verlag, 1987.

[7] V. Kuntsevich, 'Estimation of Impact of Bounded Perturbations on Nonlinear Discrete Systems', In Kuntsevich, V.M., Gubarev, V.F., Kondratenko, Y.P., Lebedev, D.V., Lysenko, V.P. (Eds)., Control Systems: Theory and Applications, Series in Automation, Control and Robotics, River Publishers, Gistrup, Delft, 2018, pp. 3-15.

[8] A. Kurzhanski, P. Varaiya, 'Dynamics and Control of Trajectory Tubes', In Theory and Computation, Systems & Control, Birkhäuser, 2014.

[9] A. Poznyak, A. Polyakov and V. Azhmyakov, 'Attractive Ellipsoids in Robust Control, Systems & Control: Foundations & Applications', Birkhäuser, 2014.

[10] M. Khlebnikov, B. Polyak and V. Kuntsevich, 'Optimization of linear systems subject to bounded exogenous disturbances: The invariant

ellipsoid technique', Automation and Remote Control, vol. 72, no. 11, 2011, pp. 2227-2275.

[11] C. Scherer, S. Weiland, 'Linear Matrix Inequalities in Control', Delft University of Technology, Netherlands, 2004.

[12] H. Taguchi, M. Araki, 'Two-Degree-of-Freedom PID Controllers, Their Functions and Optimal Tuning,' IFAC Proc. Volumes, vol. 33(4), 2000, pp. 91-96.

[13] Yi. Huang, W. Xue, 'Active disturbance rejection control: Methodology and theoretical analyses', ISA Transactions, vol. 53(4), 2014, pp. 963-976.

[14] Ya. Tsypkin, U. Holmberg, 'Robust stochastic control and internal model control', Int. Journal of Control, vol. 61, no. 4, 1995, pp. 809-822.

[15] D. Cheng, H. Qi, Z. Li, 'Disturbance Decoupling', In Analysis and Control of Boolean Networks, Springer-Verlag, London, 2011, pp. 275-296.

[16] G. Basile, G. Marro, 'Controlled and conditioned invariants in linear system theory', Edgewood Cliffs, NJ, Prentice-Hall, 1992.

[17] A. Kaldmäe, Ü. Kotta, 'Disturbance decoupling by measurement feedback', IFAC Proceedings, vol. 47(3), 2014, pp. 7735-7740.

[18] Y. Hou, R. Huang, Q. Cheng, L. Hou and X. Wang, 'Fault detection and isolation for output feedback system based on space geometry method', Cluster Computing, vol. 22, 2019, pp. 9313-9321.

[19] W. Wonham, 'Linear Multivariable Control: A Geometric Approach', Springer-Verlag, 1985.

[20] A. Tsuchiya, T. Murakami, 'Characteristic analysis of feedback control system with simplified disturbance compensator', In 29th Annual Conference of IEEE Industrial Electronics Society IECON'03, 2003.

[21] Ch. Jeang-Lin, Wu. Tsui-Chou, 'Dynamic Compensator-Based Output Feedback Controller Design for Uncertain Systems with Adjustable Robustness', Journal of Control Science and Engineering, vol. 2018, 2018.

[22] H. Seraji, 'Minimal inversion, command tracking and disturbances decoupling in multivariable systems', Int. Journal of Control, vol. 49, 1989, pp. 2093-2191.

[23] N. Adhikary, C. Mahanta, 'Inverse dynamics based robust control method for position commanded servo actuators in robot manipulators', Control Engineering Practice, vol. 66, 2017, pp. 146-155.

[24] C. Liu, H. Peng, 'Inverse-Dynamics Based State and Disturbance Observers for Linear Time-Invariant Systems', Journal of Dynamic Systems, Measurement, and Control, vol. 124, 2002, pp. 376-381.
[25] L. Zhiteckii, K. Solovchuk, 'Pseudoinversion in the problems of robust stabilizing multivariable discrete-time control systems of linear and nonlinear static objects under bounded disturbances', Journal of Automation and Information Sciences, vol. 49, no. 5, 2017.
[26] L. Zhiteckii, K. Solovchuk, 'Robust adaptive pseudoinverse model-based control of an uncertain SIMO memoryless system with bounded disturbances', In IEEE 2nd Ukraine Conf. on Electrical and Computer Engineering, 2019, pp. 621-627.
[27] L. Silverman, 'Inversion of multivariable linear systems', IEEE Transaction on Automatic Control, vol. 14, 1969, pp. 270-276.
[28] S. Pushkov, 'Inversion of Linear Systems on the Basis of State Space Realization', Systems Theory and General Control Theory, vol. 57, 2018, pp.7-17.
[29] L. Lyubchik, 'Inverse model control and sub-invariance in linear discrete multivariable systems', In Proc. of 3-rd European Control Conf., vol. 4, 1995, pp. 3654-3659.
[30] L. Lyubchyk, 'Disturbance rejection in linear discrete multivariable systems: inverse model approach', IFAC Proceedings Volumes, 2011.
[31] C. Commault, J-M. Dion, O. Sename and R. Motyeian,' Unknown input observer - A structural approach', In 2001 European Control Conference (ECC), 2001.
[32] J. Chen, R. Patton, 'Design of unknown input observers and robust fault detection filters', Int. Journal of Control, vol. 63(1), pp. 85-105, 2007
[33] S. Zak, 'Observer design for systems with unknown inputs', Int. Journal of Applied Mathematics and Computer Science, vol. 15, no. 4, 2008, pp. 431-446.
[34] V. Utkin, 'Sliding mode in control and optimisation', Springer, 1992.
[35] Yu. Shtessel, Ch. Edwards, L. Fridman and A. Levant, 'Sliding mode control and observation', 2014, Springer, New York, pp. 1-42.
[36] C. Edwards, S. Spurgeon and A. Akoachere, 'Sliding mode output feedback controller design using linear matrix inequalities', IEEE Transaction on Automatic Control, vol. 46(1), 2001, pp. 115–119.
[37] C. Edwards, S. Spurgeon and R. Hebden, 'On the design of sliding mode output feedback controllers', Int. Journal of Control, vol. 76, 003, pp. 893–905.

[38] V. Acary, B. Brogliato and Yu. Orlov, 'Chattering-free digital sliding-mode control with state observer and disturbance rejection', IEEE Transaction on Automatic Control, vol. 57(5), 2012, pp. 1087–1101.
[39] M. Rubagotti, A. Estrada and F. Castanos, A. Ferrara and L. Fridman, 'Integral sliding mode control for nonlinear systems with matched and unmatched perturbations', IEEE Transaction on Automatic Control, vol. 56(11), 2011, pp. 2699–2704.
[40] B. Drazenovic, 'The invariance conditions in variable structure systems', Automatica, vol. 5, 1969, pp. 287-297.
[41] B.-K. Lee, L.-W. Mau, 'An output tracking VSS control in the presence of a class of mismatched uncertainties', Asian Journal of Control, vol. 4, no. 2, 2002, pp. 206-216.
[42] H. Choi, 'Variable structure control of dynamical systems with mismatched norm-bounded uncertainties: an LMI approach', Int. Journal of Control, vol. 74, 2001, pp. 1324-1334.
[43] M. Basin, A. Ferreira and L. Fridman, 'Sliding mode identification and control for linear uncertain stochastic systems', In Proc. of 45th IEEE Conference on Decision and Control, 2006, pp. 1339-1344.
[44] M. Steinberger, I. Castillo, M. Horn and L. Fridman, 'Robust output tracking of constrained perturbed linear systems via model predictive sliding mode control', Int. Journal of Robust and Nonlinear Control, vol. 30(3), 2020, pp. 1258-1274.
[45] F. Castanos, J. Xu and L. Fridman, 'Integral sliding modes for systems with matched and unmatched uncertainties', Edwards, C., Colet, E.F., Fridman, L. (eds.), Advances in Variable Structure and Sliding Mode Control, Lecture Notes in Control and Inform.n Sciences, vol. 334, Springer, Berlin, 2006, pp. 227-246.
[46] J. Bejarano, L. Fridman and A. Poznyak, 'Output integral sliding mode control based algebraic hierarchical observer', Int. Journal of Control, 80(3), 2007, pp. 443-453.
[47] M. Hamayun, C. Edwards and H. Alwi, 'Design and analysis of an integral sliding mode fault-tolerant control scheme', IEEE Trans. Automatic Control, vol. 57, 2012, pp. 1783–1789.
[48] F. Castanos, L. Fridman, 'Analysis and design of integral sliding manifolds for systems with unmatched perturbations', IEEE Trans. Automatic Control, vol. 55(5), 2006, pp. 853–858.
[49] T. Floquet, C. Edwards and S. Spurgeon, 'On sliding mode observers for systems with unknown inputs', Int. Journal of Adaptive Control and Signal Processing, vol. 21, 2007, pp. 63-66.

[50] K. Kalsi, J. Lian, S. Hui and S. Zak, 'Sliding-mode observers for systems with unknown inputs: A high-gain approach', Automatica, vol. 46, 2010, pp. 347-353.

[51] L. Lyubchyk, 'Output tracking and mismatched disturbances rejection using inverse model based equivalent sliding mode control', In Proc. of 8th IEEE International Conference on Electrical Engineering, Computing Science and Automatic Control, Merida - Yucatan, Mexico, 2011.

[52] G. Kelmans, A. Poznyak and A. Chernitser, 'Adaptive locally optimal control', Int. Journal of System Science, 1981, vol. 12, no. 2, pp. 235-254.

[53] D. Shiljak, 'Large-scale dynamic systems: stability and structure', North-Holland, 1978.

[54] R. Rokafeler, 'Convex analysis', Princeton, 1970.

[55] C. Bonivento, A. Nersisian and R. Zanasi, 'A cascade structure for robust control design', IEEE Trans. Autom. Control, vol. 39, no. 4, 2002, pp. 846-849.

[56] E. Cruz-Zavala, J. Moreno and L. Fridman, 'Uniform robust exact differentiator', IEEE Transaction on Automatic Control, vol. 56 (11), 2011, pp. 2727–2733.

[57] A. Levant, 'Higher-order sliding modes, differentiation and output feedback control', Int. Journal of Control, vol. 76, no. 9, 2003, pp. 924-941.

# 6

# Invariant Relations in the Theory of Optimally Controlled Systems

**B. Kiforenko**[1], **S. Kiforenko**[2]

[1]S.P. Timoshenko Institute of mechanics of NASU, P. Nesterov str., 3, Kiev, 03057, Ukraine
[2]International Research and Training Center for Information Technologies and Systems of the National Academy of Sciences of Ukraine and Ministry of Education and Science of Ukraine, 40 Glushkov ave.,Kiev, 03680 GSP, Ukraine
E-mail: bkifor@ukr.net, skifor@ukr.net

## Abstract

The article introduces a new concept of price-target invariance of dynamical system optimal controls. Several types of optimization problems have been discovered in which, in the process of analyzing the necessary conditions, it is possible to obtain relations between control functions that are independent of the initial and final conditions for maneuver and criterion for assessing the quality of control. These relationships appear to violate the Leibniz sufficient reason principle. The relevance of determining the nature of these relationships is due to the need to solve the problem of the possibility of their practical use, which greatly simplifies the structure of the facility's control system.

Below are the results of the study of invariant relations of both regular and singular controls of the motion of dynamical systems. The latter turned out to be especially effective in controlling the movement of aircraft in the atmosphere as well as in controlling devices with propulsion systems that use engines with various methods of thrust generating. In the case of controlled objects, the motion of which is described by systems of linear equations with constant coefficients, there are the invariant relations indicated by the

Feldbaum between the moments of control functions switching. An analysis of these relations showed that oscillatory systems possess, in addition to the known time constant, the energy time constant. This article establishes this constant dynamic meaning. The new principle is proposed as a form of the universal minimum energy dissipation principle, specific for describing processes in living nature.

**Keywords:** Dynamical systems, optimal control, necessary conditions, control system structure, regular and singular control, invariant relations of control functions, minimum energy dissipation principle.

## 6.1 Introduction

The wording of any variational problem contains information about the functioning of the controlled object, represented by its equations of motion. If the goal of a particular maneuver is achievable, the control functions that ensure its achievement in the best way depend on the prescribed boundary conditions and on the conditions of transversality. At the same time, there are several classes of variational problems whose necessary optimality conditions contain relations between control functions that are invariant with respect to a change in the conditions specified in the statement of the problem [88, 46, 48]. The invariance of these relations seems to violate the universal principle of causation of the resulting solution to each specific problem. The apparent contradiction requires an analysis of the causes of this invariance and the properties of the managed object, which these relations express. The lack of reliable answers to these questions causes a natural distrust of the solution, justified by principle of a sufficient base Leibnitz [100]: "Not a single phenomenon can be true or valid, not a single statement is true, without sufficient reason why this is the case, and not otherwise ... ." When analyzing optimal control in some practically interesting Mayer problems, the authors obtained relations, when recording which adjoined functions are not used [47, 48]. Therefore, they are not dependent on specific teleological information. Moreover, it is possible to write down the indicated formulas without formulating any variational problem at all, and it is necessary to have only the equations of motion of the object. Therefore, they express some special property of the object itself, regardless of the subjective component of the formulation of the variational problem, which expresses any desires of the controlling subject.

Below we consider three types of variational problems announced in Section 6.1, which allow one to obtain price-target invariants for both regular and singular optimal trajectories.

Section 6.2 analyzes the nature of invariant relations. In Section 6.3, we analyze the problems of using singular controls in rocket flight mechanics since the invariant relations between the control functions in this case significantly increase the efficiency of motion control, although the calculation of the corresponding trajectories becomes more complicated. The author of this paper proposes a method of overcoming this difficulty, which made it possible to obtain solutions to practically interesting problems of rocket dynamics [44].

The invariance problems of dynamical systems with redundant control are studied in Section 6.4. This control method is especially effective for rocket movement in the atmosphere. An understanding of the nature of the occurrence of invariance was made possible by the analysis of their motion in the space of the hodograph of phase velocity. The information obtained made it possible to formulate the variational principle of the minimum dissipation of control energy caused by the possible inconsistency of excess controls. A relation is established between the relations of this principle with formulas describing the Pareto set in the theory of analytic games. The Anokhin–Pareto principle is proposed as a form of the universal principle of minimum energy dissipation, specific for describing processes in living nature.

Section 6.5 shows that in the case of linear controlled systems with constant coefficients, there are invariant relations between the moments of switching control functions. These relations order the set of switching times indicated by the Feldbaum theorem. An analysis of these relations showed that oscillatory systems possess, in addition to the known time constant, the energy time constant. In a second-order system, this constant is equal to the duration of the transition of the system from a position with a maximum potential energy to a position with a maximum kinetic energy.

Section 6.6 presents the results of a study of invariance in modeling functioning in living nature by methods of the theory of dynamical systems. An analysis of the necessary optimality conditions in the variational problem of the controlled movements of biological objects confirmed Anokhin about the specific result of activity, as a system-forming factor, ensuring its expedient functioning. The Anokhin–Pareto principle is formulated as a form of the universal experimental principle of minimum energy dissipation, specific to living nature.

## 6.2 The Problems of Price–Target Invariance in the Theory of Optimal Control

Below we consider three types of variational problems announced in Section 6.1, which allow one to obtain price-target invariants for both regular and singular optimal trajectories. The relevance of determining the nature of these relationships is due to the need to solve the problem of the possibility of their practical use, which greatly simplifies the structure of the facility's control system. In this part of the article, we investigate the nature of the invariant relations between control functions in the theory of dynamical systems. Let for an object whose operation is described by a system of differential equations (6.1)

$$\frac{dx_i}{dt} = f_i(x_1, ..., x_n, u_1, ..., u_r), \quad i = \overline{1, n}; \quad u(t) \in U . \tag{6.1}$$

Here, $U$ is the set of admissible controls. Let for this system be formulated Mayer's variational problem of transition from a certain initial position $x(t_0) = x_0$ to a final state $x(t_f)$ with minimum functional value:

$$J[u(t)] = \Phi(x(t_f)). \tag{6.2}$$

Here, the symbol $\Phi$ indicates the functional optimization, which is provided in the ongoing study. Note that there may be a more general case: $J[u(t)] = \Phi(x(t_f), t_f)$, for a non-autonomous system.

If a solution to any variational problem exists, it satisfies necessary optimality conditions. These conditions *select optimally controlled motions* of the dynamic system from the set of admissible motions (the set of solutions of the system of differential equations (6.1) under any control $u(t) \in U$). Let the functions $f_i(x_1, ..., x_n, u_1, ..., u_r)$ for $i = \overline{1, n}$,$\Phi(x(t_f))$, initial state $x(t_0)$,and final state $x(t_f)$ for the controlled object be defined in such a way that the optimal control analysis can be performed using Pontryagin's maximum principle.

The principle of maximum Pontryagin [72] was chosen as a necessary condition for optimality since it is widely used for the analysis of control processes in the theory of dynamical systems. An important feature of this method of analysis is the preservation of the structure of the resulting formulas in the study of the dynamics of both deterministic and stochastic systems. The only necessary change is to replace in the formulas all the determinate parameters with their mathematical expectations [87].

In accordance with the principle procedure, the Hamilton–Pontryagin $H$-function is formed:

$$H = \psi_1 f_1(x, u) + \psi_2 f_2(x, u) + \cdots + \psi_n f_n(x, u) \tag{6.3}$$

where $\psi_i$, $i = \overline{1, n}$ are conjugate variables satisfying the system of differential equations:

$$\frac{d\psi_i}{dt} = -\frac{\partial H(x, u)}{\partial x_i}. \tag{6.4}$$

A prerequisite for optimality: if a solution to the problem for $\min J\left[u(t)\right]$exists, then there is a non-trivial solution to the adjoint system (6.4). At each time point, the *optimal control* $u(t)$ is derived from the condition:

$$u(t) = \underset{u \in U}{\operatorname{argmax}} H(x(t), \psi(t), u). \tag{6.5}$$

Construction of the optimal solution of the Mayer problem under consideration was reduced to solving a two-point boundary value problem for the system of $2n$ differential equations (6.1) and (6.4), when selecting $r$ control functions from condition (6.5). The necessary $2n$ conditions for determining the system integration constants for Equations (6.1) and (6.4) consist of initial and final conditions: $x(t_0) = x_0$ and $x(t_f) = x_f$, as well as the transversality conditions $\psi_i(t_f) = -\frac{\partial \Phi(x(t_f))}{\partial x_i}$, $i = \overline{1, n}$. This ensures dependence of the vector function $\psi(t)$ on the purpose of motion $x(t_f) = x_f$ and the transversally conditions. It follows from condition (6.5) that the control depends both on the current state of system $x(t)$ at time point $t$ and the conjugate vector $\psi(t)$. This is how *control is performed in accordance with a specific goal*$x(t_f) = x_f$*, taking into account the chosen quality assessment*$\Phi(x(t_f))$. Therefore, it is natural to name $\psi(t)$ *the guidance vector.*

Next, we consider the features of expediently functioning objects with excessive control $r \geq n$.

Taking inequality $r \geq n$ into account when analyzing the necessary optimality conditions leads to an unexpected result – *reduction in the number of degrees of freedom* of the object controlling *with efficient functioning*. As known, the number of variational problem degrees of freedom is equal to number $r$ of control functions. When calculating optimal controls by the formula, controlling functions should additionally $(r - n + 1)$ meet the consistency conditions of controlling actions: det $A_j = 0$ $j = \overline{n, r}$, where

$$A_j = \begin{pmatrix} \frac{\partial f_1}{\partial u_1} & \frac{\partial f_2}{\partial u_1} & \cdots\cdots & \frac{\partial f_n}{\partial u_1} \\ \cdots\cdots & \cdots\cdots & \cdots\cdots & \cdots\cdots \\ \frac{\partial f_1}{\partial u_{n-1}} & \frac{\partial f_2}{\partial u_{n-1}} & \cdots\cdots & \frac{\partial f_n}{\partial u_{n-1}} \\ \frac{\partial f_1}{\partial u_j} & \frac{\partial f_2}{\partial u_j} & \cdots\cdots & \frac{\partial f_n}{\partial u_j} \end{pmatrix}. \tag{6.6}$$

The principal feature of these conditions is their independence from coordinates of the guidance vector. Therefore, they are *invariant in relation to the teleological aspects of any Mayer variational problem* for a given object. Moreover, these conditions can be deduced *even prior to formulation of any variational problem.* It is only necessary to possess information about the nature of the object in the form of system (6.1). Relations (6.6) are valid if controlling impacts do not reach their boundary values:$u(t) \in intU$. The amount of these relations is $r - n + 1$.

If some controls reach boundary values, the invariant relations take a different form: let the set of admissible controls $U$ be closed and some $q < r$ from the set of controlling functions $u_1, u_2, ..., u_r$ reach the boundary of their admissible values, and, in this case,$r - q > n - 1$. Equations of the corresponding boundary surfaces will be given as

$$g_s(x_1, \ldots, x_n, u_1, \ldots, u_r) = 0, s = \overline{1, q}. \tag{6.7}$$

In this case, $r - n + 1$ of the relations (Appendix 2) are replaced by the relations: det $A_j = 0 \ j = \overline{n, r}$, where

$$A_j = \begin{pmatrix} \frac{\partial f_1}{\partial u_1} & \frac{\partial f_2}{\partial u_1} & \cdots\cdots & \frac{\partial f_n}{\partial u_1} & \frac{\partial g_1}{\partial u_1} & \frac{\partial g_2}{\partial u_1} & \cdots\cdots & \frac{\partial g_q}{\partial u_1} \\ \cdots\cdots & \cdots\cdots & \cdots\cdots & \cdots\cdots & \cdots\cdots & \cdots\cdots & \cdots\cdots & \cdots\cdots \\ \frac{\partial f_1}{\partial u_{n-1}} & \frac{\partial f_2}{\partial u_{n-1}} & \cdots\cdots & \frac{\partial f_n}{\partial u_{n-1}} & \frac{\partial g_1}{\partial u_{n-1}} & \frac{\partial g_2}{\partial u_{n-1}} & \cdots\cdots & \frac{\partial g_q}{\partial u_{n-1}} \\ \frac{\partial f_1}{\partial u_j} & \frac{\partial f_2}{\partial u_j} & \cdots\cdots & \frac{\partial f_n}{\partial u_j} & \frac{\partial g_1}{\partial u_j} & \frac{\partial g_2}{\partial u_j} & \cdots\cdots & \frac{\partial g_q}{\partial u_j} \end{pmatrix}. \tag{6.8}$$

Invariant constraints for the choice of rational controlling are still $r - n + 1$: $(r - n + 1) - q$ of relations (6.8) plus $q$ equations of boundary surfaces:

$$g_s(x_1, \ldots, x_n, u_1, \ldots, u_r) = 0, s = \overline{1, q}.$$

The concept of "invariance" as a scientific term was introduced into the theory of automatic control by Shchipanov [78]. In the work of Luzin [57] and in the monograph of Kuhtenko [52] were obtained and analyzed the conditions of invariance, i.e. independence of one or several controlled values from external disturbances. In our paper, a new concept is introduced into consideration – *a system-wide invariance as a way of coordinating* controlling actions *that do not depend on objective and assessment of the quality* of controlling in each case. We note that the influence of invariance on the methodology of controlling dynamic systems is posed by Smolyaninov [80].

## 6.3 The Problems of Using Singular Controls in Rocket Flight Mechanics

Historians of science will have to recognize as one of the main factors of its development in the 20th century the urgent need to solve unusually complex scientific and technical problems, which was dictated by the space race. Many branches of modern mechanics are brought to life by the needs of rocket and space technology. The need to achieve cosmic speeds that are many times greater than the values that are usual for land and air vehicles, and even under conditions due to the impossibility of energetic power from the Earth, makes the problem of rationality of control unprecedentedly acute [44].

In this regard, the problems of flight mechanics of rockets and spacecraft from the very first works of the scientific stage of the study are considered as variational. Apparently, the first paper containing an analysis of the variational problem of rocket motion was an article by Hamel [33] published in 1927. However, this publication was premature, without provoking the response of researchers for almost two decades. An important role in consolidating the optimization approach to the problems of flight mechanics was played by the publications of Okhotsimsky [22, 66]. Very soon, the inefficiency of the classical calculus of variations was found to solve problems arising in the practice of controlling the movement of rockets [22]. This stimulated the creation of a modern mathematical theory of optimal processes. One of the most difficult sections here is the theory of special (singular) optimal controls [27].

### 6.3.1 Power Consumption in Degeneracy of the Optimal Control of Rocket Thrust in Atmosphere

The first problem related to the development of rocketry, the attempts to solve which demonstrated the inefficiency of classical calculus of variations, was formulated by Goddard in 1919 [28]. The results of the analysis of the optimal control of the thrust of the engine during vertical rocket motion in the atmosphere, presented in the works of Okhotsimsky [22, 66], indicated a feature of the problem caused by its degeneracy: the Euler equation turned out to be the final relation between phase variables. It became clear that in the general case, it was impossible to satisfy the conditions of the problem at the ends of the trajectory using only the Euler extremal. These solutions to the problem were obtained by methods of direct study of variation long before the introduction of the Rozonoer [77] concepts of special optimal control. The study of the optimality conditions for arcs of variable traction caused widespread discussion and stimulated the development of necessary conditions of a higher order than the Euler equations and the maximum principle of Pontryagin. The analysis of flight control problems confirmed the relevance of the study of degenerate variational problems since typical flight control problems of aircraft, both atmospheric and space, turn out to be degenerate [51, 58].

### 6.3.2 Necessary Conditions for the Optimality of a Singular Control

The possibility of including singular arcs in the optimal trajectory complicates the solution of the corresponding variational problems, which are incorrect according to Tikhonov. The well-known difficulties of solving ill-posed problems in constructing optimal trajectories [23] are supplemented with specific features associated with the "overdetermination" of special sections [27]. Let us illustrate what has been said with the example of a terminal control problem with scalar control typical of flight mechanics:

$$\begin{aligned} &\frac{dx}{dt} = f_0(x) + uf_1(x), x = (x_1, ..., x_n\}^T u(t) \in (0, 1), \\ &J(u) = \Phi(x(t_f)) \to \min . \end{aligned} \tag{6.9}$$

Here $x$ is an $n$-dimensional phase – vector, $u(t)$ control function. The Hamiltonian for system (6.7) can be represented as

$$\begin{aligned} &H\,(x\,,\psi\,,u = H_0\,(x\,,\psi) + u\,H_1\,(x\,,\psi)\,,\;\; H_0\,(x\,,\psi\,) = \psi^T f_0(x\,)\,, \\ &H_1\,(x\,,\psi) = \psi^T f_1(x\,). \end{aligned} \tag{6.10}$$

The attached system is written as

$$\frac{d\psi}{dt} = -\frac{\partial f_0^T}{\partial x}\psi - \frac{\partial f_1^T}{\partial x}\psi. \tag{6.11}$$

The control on regular arcs of the optimal trajectory takes boundary values: $u(t) = 1$ for $H_1\,(x, \psi) > 0$, $u(t) = 0$ and $u(t) = 0$ at $H_1\,(x, \psi) < 0$.

The necessary optimality condition for a singular control $u(t)$ belonging to the open kernel of the set of admissible controls is reduced to the form:

$$\frac{\partial H\,(x,\psi,u)}{\partial u} = H_1\,(x(t),\,\psi\,(t)) = 0. \tag{6.12}$$

If condition (6.12) is satisfied identically on an interval of non-zero duration, the corresponding section of the optimal trajectory is called special and it is impossible to calculate the special control from this condition directly. To calculate this control, differentiation of relation (6.12) with respect to time is performed [27]. From the fulfillment of condition (6.12) along a singular arc of non-zero duration, it follows that its derivative is equal to 0:

$$\frac{d}{dt}H_1(x\,(t),\,\psi(t)) = -\,\{H_0,\,H_1\} + u(t)\{H_1,\,H_1\} = 0, \tag{6.13}$$

where $\{X,\,Y\}$ is the Poisson bracket of the functions $X\,(x,\,\psi)$ and $Y\,(x,\,\psi)$. The Poisson bracket of identical functions $\{H_1,\,H_1\}$ is identically equal to zero. Therefore, relation (6.13) is not an equation for determining control $u(t)$, but a new first integral of the system of differential equations (6.9) and (6.11), having the form: $\{H_0,\,H_1\} = 0$. Calculating the second time derivative of the switching function $H_1\,(x(t),\,\psi(t))$ and equating it to 0, we obtain an equation for calculating the singular control:

$$\frac{d^2}{dt^2}H_1 = \{\,H_0,\{H_0\,,\,\{H_1\}\} + u(t)\,\{\,H_1,\{H_0\,,\,\{H_1\,\}\} = 0, \tag{6.14}$$

If condition (6.14) is satisfied identically on an interval of non-zero duration, the corresponding section of the optimal trajectory is called special and it is possible to calculate the singular control from this condition directly (if $\{\,H_1,\{H_0\,,\,\{H_1\,\}\} \neq 0$). The necessary condition for its optimality obtained by Kelley [42] is written as

$$\frac{\partial}{\partial u}\,(\frac{d^2}{dt^2}\,(\frac{\partial H}{\partial u})) \geq 0. \tag{6.15}$$

It should be noted that in a number of practically interesting problems of modern and classical aerodynamics and rocket dynamics, the ratios for calculating the magnitude of the special thrust control of the rocket engine and the surface of the special control are invariant with respect to the change in the boundary conditions of the optimization problem and the choice of a functional that evaluates the quality of motion control. However, the authors of this work are not aware of publications discussing this amazing property [44, 88].

### 6.3.3 The Problem of Calculating Optimal Trajectories With Singular Arcs

The calculation of optimal trajectories with singular arcs turned out to be a very difficult task of computational mathematics [44, 95] in connection with the overdetermination of singular arcs since in this sections, in addition to the equations of motion and the associated system, the relations $H_1 = 0$ and $\dot{H}_1 = 0$ are fulfilled, representing first integrals of the indicated optimal motion equations with zero values of the integration constants. The need for the simultaneous fulfillment of these two relations at the starting point of a singular arc $t = \tau$ imposes a restriction $(\vec{x}(0), \vec{\psi}(0)) \in \Phi_0$ on the initial values of the components of the phase $\vec{x}\,(0)$ and associated $\vec{\psi}\,(0)$ vectors, and the variety is not known before solving the problem.

Access to this variety in the search process seems unlikely, and $\vec{\psi}\,(0)$ varying without going off $\Phi_0$is almost impossible. Even if in the search process of one of the samples the condition $(\vec{x}(0), \vec{\psi}(0)) \in \Phi_0$ is satisfied, subsequent variation of the values $\vec{\psi}\,(0)$, leading to the departure from the manifold $\Phi_0$ significantly changes the nature of the trajectory. Illustration is given in the work [96]. To calculate the optimal trajectories with singular arcs, a method for calculating the optimal trajectories with singular arcs by integrating from the surface of a special control [45] was developed and was then developed together with Zlatsky to the integration method from a particular variety [96]. The essence of the proposed method consists in the transition from searching for a vector $\vec{\psi}\,(0)$ on an unknown manifold $\Phi_0$ to searching "under the lantern" [11]. Initially, on a set of points satisfying all the necessary optimality conditions for a singular section, a certain intermediate point $X_\psi\,(\vec{x}^*, \vec{\psi}^*)$ is assigned, starting from which the system of equations of optimal motion is integrated with the reverse flow of time. Then a numerical search is made for such a point $\tilde{X}_\psi$, the result of integration from which will be a trajectory that passes in phase space through a given initial position (with a reasonably chosen degree of accuracy). The found point is taken as the starting point of the singular arc. Then, a singular section is constructed and the moment of descent from it is determined from the condition of getting into the final state and satisfying the transversality conditions with satisfactory accuracy. So what does the promised lantern illuminate?

Let us consider in detail the special sections of the first order of degeneracy [27] described above. At the same time, two conditions in the form of equalities $H_1 = 0$and $\dot{H}_1 = 0$ are fulfilled in a singular section. Using these equalities, we can exclude from the remaining relations the two components of the adjoint vector $\vec{\psi}$ so that in $2n$-dimensional space, a singular manifold is -dimensional. On it, the search for the above point is performed. If the equations of motion of the object are autonomous, in problems with a free value of the execution time of the maneuver, the indicated relations, due to the need to fulfill the transversality condition, add an integral and a special variety becomes $(2n - 2)$-dimensional. In this case, the relations $H = 0$, $H_1 = 0$, and $\dot{H}_1 = 0$represent a system of three linear homogeneous algebraic equations with respect to the three components of the vector $\vec{\psi}$. Determinant of this system should be 0 due to these components' non-triviality. The zero determinant of this system is the surface equation of a singular control $S(x) = 0$. The solution of this algebraic system associates with each point of the surface $S(x) = 0$ the value of the conjugate vector $\psi = \psi(x)$. Two regular arcs of optimal trajectories satisfying the maximum principle can go out from any point $(x, \psi(x))$ of this surface: one starts at the maximum and the second at the minimum allowable control value $u$.

The conjugation of each of the indicated trajectories with a singular site lying on the surface $\Sigma$ does not contradict the conditions of optimal conjugation [27]. Thus, in the case of $n = 3$ and maneuvers and with non-fixed flight durations $t_f$, each selected point on the surface of the singular control can serve as the beginning (or end) point of the optimal trajectory regular arc. It is on this surface $\Sigma$ of singular control that it is proposed to search for the end point $\tilde{X}_{\psi}$ of the initial regular arc of the optimal trajectory starting from the given starting point of the maneuver and continuing until it reaches the surface $S(x) = 0$. If on this arc the singular control condition for reaching the surface $S(x) = 0$ is not fulfilled, then singular control is not optimal.

The proposed method, of course, has no (and cannot have!) advantages in terms of the number of parameters to be determined. Either look for two initial values of the vector $\vec{\psi}(0)$ (the third is determined from the first integral $H(x(0), \psi(0), u(0)) = 0$) on an *a priori* unknown manifold $\Phi_0$ or two coordinates of the point $X_s$ in the three-dimensional phase space of the problem (the third is determined from the surface control equation $S(x(0)) = 0$). Note that, in contrast to the search of $\vec{\psi}(0)$, when searching for a point on the surface of a special control, precisely (analytically!), all information on singular arcs, which can be obtained even before the start of calculations, at the stage of analysis of the necessary optimality conditions, is completely and directly used.

Indeed, the choice of the point $X_s$ coordinates from the condition $S(x(0)) = 0$ is the consequence of the exact fulfillment of the necessary conditions $H_1 = 0$ and $\dot{H}_1 = 0$ of the optimality of the singular control. Otherwise, the possibility of establishing the fact of simultaneous fulfillment of these two conditions in the process of numerical solution of any problem is determined by a subjective factor – an agreement on which small number can be considered a machine zero.

A very important advantage of the proposed approach is that the search is carried out among the values of the phase coordinates that have a direct and understandable physical meaning, which is by no means important for choosing the initial approximation and tactics of the numerical search procedure [58, 64, 84]. An advantage of the proposed method is the possibility of efficient use of hypothetical *a priori* information about the nature of the initial portion of the optimal trajectory, followed by *a posteriori* verification of the hypotheses made [92].

The reaction to the publication in 1983 of an article in which our method for calculating optimal trajectories with singular arcs by integration from a particular manifold was proposed was a report by one of the developers of the modern mathematical theory of optimal control Kelley in collaboration with Kumar at *Fifth AIAA Conference on Guidance, Navigation and Control* (Monterey, California, August 17–19, 1987), and then published in *Journal of Guidance, Control and Dynamics* [53]. In the introduction to the text of the report and article, it is noted that the numerical solution of problems of singular optimal control is fraught with great difficulties when using both direct and indirect methods. The approach to solving such problems, proposed in the article [96], was rated as effective and recompiled in [53] due to the inaccessibility of this work in the US scientific press. A detailed analysis of the proposed approach is presented in this article, in which the problem of ascending and accelerating an aircraft in a vertical plane is considered and a detailed study is made of the dependence of the structure of the optimal trajectory on the boundary conditions of the maneuver. Using this approach, the Goddard problem with a limited rise time was studied [86]. Studies of traction control features when reaching the maximum height are continued taking into account phase restrictions; for example, on the value of the velocity head of the incoming flow, the impact of the resistance crisis on traction control in the Goddard problem is estimated.

The Goddard problem also tests numerical methods for solving optimization problems, for example, the direct method of variation in the control space. One commonly used approach involves the use of quadratic regularization. A regularizing term of the form $(\lambda - 1)u^2$ is added to the right side of the differential equation for changing the functional. This eliminates the linearity of the right-hand sides of the equations of the dynamic system with respect to control $u(t)$. Namely, this property of the controlled system is a formal sign of the possibility of including a singular arc in an optimal trajectory. Next, a sequence of non-singular two-point boundary value problems is solved for a regularized dynamical system with a decreasing value of the regularizing factor $(\lambda - 1)$. Theoretically, for $\lambda \to 1$, the solution to the original problem should be obtained. A good example of using such a procedure for the Ariane 5 launcher is described in detail in [59].

The practical convergence of this procedure is clearly illustrated by the curves in Figures 6.5 and 6.6 given on page 45 of this article. In addition, the authors propose an alternative approach to the calculation of singular control, oriented directly to the numerical solution. At each integration step, a numerical search is performed in the space $(x, \psi)$ of points close to a particular manifold at which the function $Y = H_1^2(x, \psi) + \dot{H}_1^2(x, \psi)$ does not exceed a certain number $\tilde{a}_{\text{sign}}$. A comparison of the thrust value obtained by this method in the Goddard problem with the analytical expression (3.5) shown in Figure 3 of this article looks satisfactory. However, we will comment on the authors' message that the choice of 0.1 or 0.9 as the initial value $\tilde{a}_{\text{sign}}$ gives similar results, which, of course, is reassuring.

The author's assessment of the results as "soothing" should be taken into account considering the remark of Fedorenko [23] that the so-called practical convergence of minimizing discrepancies can be a consequence of both achieving a minimum and the ineffectiveness of the minimization method. It is good when there is an analytical solution to the problem with which you can compare the result, as was the case with the test account of the Goddard problem. But the confidence in the correctness of the solution to the main problem of the article on the control of the Ariane 5 launch vehicle lift, illustrated in Figure 6.5 (p. 45), can be based only on the no more than plausible conclusion of [15] on the practical convergence of the minimization process.

In the survey works [14, 92], the state of the problem of the numerical solution of optimal control problems in space flight mechanics is analyzed with an emphasis on the possibility of overcoming the difficulties of choosing the initial conditions for the shooting method, and geometric optimal methods control and homotope method [14], as well as some approaches of the theory of dynamical systems successfully applied in astrodynamics. Overcoming the difficulties of the solution largely depends on the availability and reliability of information about the structure of the desired trajectory, the number and location of special sections before the start. Information of this kind can be obtained by careful analysis of the necessary conditions for optimality, considering the characteristics of a particular task. It was the results of such an analysis performed by various researchers mentioned in this article that provided the calculators with such reliable information that would ensure the convergence of the numerical procedures used. The authors of the works, who solved the Goddard problem by the method of variation in the control space without preliminary analysis of the necessary conditions, faced various difficulties to overcome [23].

Summing up the above, we focus on the permanent relevance of the generalization of the accumulated experience of numerical solution of problems of optimization of trajectories with singular arcs. As for the further development of the theory of singular optimal control, it is appropriate to point out the remark by Gabasov and Kirillova [27] that the condition of a singularity is not a sign of exceptional situation, rather, it is a signal that the task is quite complicated and it is impossible to fully explore only the maximum principle (a necessary condition of the first order). And the transition to higher conditions is dictated by the general level of development of science since work with first approximations is already recognized as insufficient.

## 6.4 Addition to the Feldbaum Theorem on Number of Switching

We turn to the case of dynamical systems described by differential equations, the right-hand sides of which linearly depend on the phase coordinates and controls:

$$\frac{dx}{dt} = Ax + Bu. \tag{6.16}$$

Here, $A$ and $B$ are matrices of sizes $n \times n$ and $r \times n$; accordingly, $x$ is an $n$-dimensional phase vector column and $u$ is an $r$-dimensional control vector. Matrices $A$ and B are constants, and the set of admissible control $U$ is a dimensional box. Let the roots of the characteristic equation of the associated system be

$$\dot{\psi} = -A^T \psi. \tag{6.17}$$

In this case, by the theorem of Feldbaum on the number of switching, any of the controls $r$ takes only its boundary values and has no more than $(n - 1)$

switching. We illustrate this with the example of the problem of translating an object whose movements are described by a system of differential equations:

$$\begin{aligned} \frac{dx_1}{dt} &= a_{11}x_1 + a_{12}x_2 + b_1 u, \quad u \in [u_1,\, u_2], \\ \frac{dx_2}{dt} &= a_{21}x_1 + a_{22}x_2 + b_2 v, \quad v \in [v_1,\, v_2]. \end{aligned} \tag{6.18}$$

Let the roots of the characteristic equation of the system be real. By the Feldbaum theorem, each of the controls of this system on an optimal trajectory has no more than one switch. We denote by $\tau$ and $\vartheta$ the moments of switching controls $u$ and $v$, respectively and by $\alpha_1$and $\alpha_2$ the real different roots of the characteristic equation of the adjoint system. It was possible to show [47] that both $\tau$ and $\vartheta$ are related by

$$\vartheta = \tau + \frac{1}{\alpha_1 - \alpha_2} \ln \frac{a_{11} + \alpha_1}{a_{11} + \alpha_2}. \tag{6.19}$$

Therefore, in the general case, the control interval is divided by points $\tau$ and $\vartheta$ into three sub-intervals, and the length $\Omega$ of the segment $[\tau,\, \vartheta]$ depends not only on the boundary conditions and functional but also on the boundary values of the control actions $u_1$, $u_2$, $v_1$, and $v_2$, i.e. on the size of the set of permissible controls. When considering, as an example, the problem of controlling the oscillations of a pendulum in a medium with resistance, the interval $\Omega$ is the time during which all potential energies of an uncontrolled pendulum at the maximum deviation from the equilibrium position pass into kinetic when passing through the equilibrium position. Thus, we are talking about ***a new concept – the energy constant of the time of the oscillatory system.***

In the general case, out of the whole set allowed by the Feldbaum theorem, the moments of only $(n-1)$ switching are selected based on the need to satisfy the boundary conditions and the transversality conditions of the Mayer variational problem. The remaining moments are determined using the formulas given in [43], which include only the matrix elements and, therefore, are invariant with respect to the boundary conditions and the functional of the variational problem under consideration.

***Comment***. The number of selected moments is equal to the number of unknown integration constants of the adjoint system, which is necessary to uniquely determine the optimal trajectory: although there are only $n$ constants, it is sufficient to know $(n-1)$ constants from them in connection with the homogeneity of the function $H$ to calculate the solution to the variational problem.

## 6.5 Investigation of the Invariance in the Modeling of Functioning in Living Nature

The main difficulty in the theoretical analysis of processes in living nature is that general principles in systematic biology cannot be derived from physical or chemical laws [61] since when only these laws are used, it is impossible even to simply distinguish between living and non-living. The complexity of the exploration of processes in living nature is so great that, at present, the relevant selection principles are only being formed. The development of a system of fundamental principles from which follows the entire continuum of observable phenomena is the central issue of theoretical biology [56]. The application of mathematical modeling methods in this paper for studying the problems of finding the principles that determine functioning of living matter are stimulated by the fact that only the interplay between mathematics and biology will help to accomplish one of the greatest scientific revolutions, namely to implement mathematical formalization of biology [19]. In this part of article, we introduce the definition for the new principle obtained in the mathematical analysis of the problem of finding a system-forming factor explaining the nature of the interaction of a variety of active components in the process of forming the functional systems of the organism, thus addressing the problem posed by Anokhin [5] in 1971.

### 6.5.1 Statement of the Anokhin Problem

Consider control of a dynamic system whose operation is described by a system of differential equations (6.1). Here, $U$ is the set of admissible controls. Let for this system be formulated Mayer's variational problem of transition from a certain initial position $x(t_0) = x_0$ to a final state $x(t_f)$ with minimum functional value (6.2). Symbol $\Phi$ indicates the functional optimization, which is provided in the ongoing study. Note that there may be a more general case: $J[u(t)] = \Phi(x(t_f), t_f)$, for a non-autonomous system. This variational problem represents one of the possible formalizations of Anokhin's problem in terms of the theory of controllable systems. *The first basic assumption* adopted in this formulation is *finiteness n.* The choice of the number of variables defining the state of an object is determined by the measure of the adequacy of the formulated mathematical model to a specific object. The adequacy of the model to the problem for the study of which it is being formed appears to be the only rational measure. We are talking about an agreement on the essential $n$ characteristics of the object while neglecting, within the framework of the conducted research, all the other characteristics.

*The second basic assumption* is the *predictability of change in the state of the object.* It is assumed that knowledge of the $n$ of the object phase coordinates at a time moment $t = t_0$: $x(t_0) = \{x_1(t_0),\ x_2(t_0),\ \dots, x_n(t_0)\}$ allows, using the chosen mathematical model structure, to trace the change in the state of the object during a certain time interval, i.e. at $t > t_0$. Formalized representation of the process of this change is performed in this paper in terms of the theory of controllable dynamic systems (6.1).

The choice of this representation is explained by the fact that for modeling of the manifestations that are found only in living systems, it is necessary to use new concepts [32] that are not found in inorganic sciences. The concept, which is redundant when describing *naturally proceeding processes* in the inert matter but without which it is impossible to describe any behavioral act in the living nature, is *control* [60]. The use of this concept in the description of technical devices is an exception that proves the rule: technology is "nature" created by man. In his aforementioned work, Anokhin formulated the problem of choosing the research algorithm as follows: it is impossible to formalize a biological system without reflecting at the same time the most important feature of a living system – the need to obtain a certain result in the best possible way. Aspects pointed out by Anokhin exactly relate to the problems of the theory of controllable dynamic systems.

*The third basic assumption* used in the further analysis is *the redundant controllability of the dynamic system* (6.1) which is formally represented by the inequality $r \geq n$. Illustrating this feature of a living organism, Anokhin points to the redundant number of degrees of freedom in brain. This redundancy makes it impossible to imagine the organized behavior of the whole organism without orderly interaction in the nervous system. The same applies to the muscle system and many chemical constellations [5].

### 6.5.2 Solution of the Anokhin Problem

If a solution to any variational problem exists, it satisfies necessary optimality conditions. These conditions *select optimally controlled motions* of the dynamic system from the set of admissible motions (the set of solutions of the system of differential equations (6.2) under any control $u(t) \in U$). Let the functions $f_i(x_1,\ \dots,\ x_n,\ u_1,\ \dots,\ u_r)$ for $i = \overline{1, n}$,$\Phi(x(t_f))$, initial state $x(t_0)$,and final state $x(t_f)$ for the controlled object be defined in such a way that the optimal control analysis can be performed using Pontryagin's maximum principle.

The principle of maximum Pontryagin [72] was chosen as a necessary condition for optimality since it is widely used for the analysis of control processes in the theory of dynamical systems. An important feature of this method of analysis is the preservation of the structure of the resulting formulas in the study of the dynamics of both deterministic and stochastic systems. The only necessary change is to replace in the formulas all the determinate parameters with their mathematical expectations [87]. This feature of the derived relations is especially important in the study of control in living nature since these processes are always stochastic.

In accordance with the principle procedure, the Hamilton–Pontryagin *H*-function (6.3) is formed.

A prerequisite for optimality: if a solution to the problem for $\min J\left[u(t)\right]$exists, then there is a non-trivial solution to the adjoint system (6.4). At each time point, the *optimal control* $u(t)$ is derived from the condition (6.5).

Construction of the optimal solution of the Mayer problem under consideration was reduced to solving a two-point boundary value problem for the system of $2n$ differential equations (6.1) and (6.4), when selecting $r$ control functions from condition (6.5). The necessary $2n$ conditions for determining the system integration constants for (6.1) and (6.4) consist of initial and final conditions: $x(t_0) = x_0$ and $x(t_f) = x_f$, as well as the transversality conditions $\psi_i(t_f) = -\frac{\partial \Phi(x(t_f))}{\partial x_i}$, $i = \overline{1,n}$. This ensures dependence of the vector function $\psi(t)$ on the purpose of motion $x(t_f) = x_f$ and the transversality conditions. It follows from condition (6.5) that the control depends both on the current state of system $x(t)$ at time point $t$ and the conjugate vector $\psi(t)$. This is how *control is performed in accordance with a specific goal*$x(t_f) = x_f$, *taking into account the chosen quality assessment*$\Phi(x(t_f))$. Therefore, it is natural to name $\psi(t)$ *the guidance vector*. Look at features of expediently functioning objects with redundant control. This is the basic assumption used when formulating Anokhin's problem about functioning of living nature in terms of controllable systems theory. Taking inequality $r \geq n$ into account when analyzing the necessary optimality conditions leads to an unexpected result –*reduction in the number of degrees of freedom* of the object controlling *with efficient functioning.* In variational problem, the number of degrees of freedom is equal to number $r$ of control functions. When functioning at the maximum point, controlling functions should additionally $(r - n + 1)$ meet the consistency conditions (6.6) .

The principal feature of these conditions is their independence from coordinates of the guidance vector. Therefore, they are *invariant in relation to the teleological aspects of any Mayer variational problem* for a given object. Moreover, these conditions can be deduced *even prior to formulation of any variational problem.* It is only necessary to possess information about the nature of the object in the form of a system (6.1). Relations (6.6) are valid if controlling impacts do not reach their boundary values:$u(t) \in intU$. If some of the controls reach boundary values, the invariant relations take a different form. In this case, $r - n + 1$ formulas (6.20) are replaced by the relations (6.8). Invariant constraints for the choice of rational controlling are still $r - n + 1$: $(r - n + 1) - q$ formulas (6.20) plus $q$ equations of boundary surfaces:

$$g_s(x_1, \ \dots \ , \ x_n, \ u_1, \ \dots \ , \ u_r) = 0, s = \overline{1, q}. \tag{6.20}$$

### 6.5.3 Features of Expediently Functioning Objects with Redundant Control

The above results have been obtained without regard to basic assumption used when formulating Anokhin's problem in terms of the theory of controllable systems. Taking inequality $r \geq n$ into account when analyzing the necessary optimality conditions leads to an unexpected result – *reduction in the number of degrees of freedom* of the object controlling *with efficient functioning*. As stated in paragraph 2 of the variational problem, the number of degrees of freedom is equal to number $r$ of control functions. When functioning expediently (see Appendix 1) at the maximum point, controlling functions should additionally $(r - n + 1)$ meet the consistency conditions of controlling actions: $\det A_j \ = \ 0 \ j \ = \ \overline{n, r}$ (see Equation (6.6) or Equation (6.8)).

Having solved these equations, we obtain the following relations between the control functions:

$$\begin{aligned}
&u_n = u_n(x_1, \ \dots, \ x_n, \ u_1, \ u_2, \ \dots, \ u_{n-1}) \\
&u_{n+1} = u_n(x_1, \ \dots, \ x_n, \ u_1, \ u_2, \ \dots, \ u_{n-1}) \\
&\dots\dots\dots\dots\dots\dots\dots\dots\dots\dots\dots \\
&u_r = u_r(x_1, \ \dots, \ x_n, \ u_1, \ u_2, \ \dots, \ u_{n-1}).
\end{aligned} \tag{6.21}$$

Substitution of these expressions into a system of equations of motion transforms it into a form:

$$\frac{dx_i}{dt} = f_i(x_1, ..., x_n, u_1, ..., u_{n-1}, u_n(x_1, ..., x_n, u_1, ..., u_{n-1}), ..., u_r(x_1, ..., x_n, u_1, ..., u_{n-1})), (i = \overline{1, n}).$$

Dynamics of an object whose functioning is described by a system of equations (6.22) can only be influenced by choice of $(n - 1)$ the controlling impact $u_1, ..., u_{n-1}$. The other controlling actions $u_n, u_{n+1}, ..., u_r$ are selected by invariant rules (6.21). Transformation of system of equations of motion (6.1) into system (6.22) is true on a single condition – *on expedient of intended control.* Specification of the purpose and assessment of quality control is not even required. Therefore, system (6.22), unlike more general system (6.1), is called *expediently controlled.*

The principal result of the transition from arbitrarily admissible control of the object $u(t) \in U$ system (6.1) to *expedient* control system (6.22) is a reduction in number of degrees of freedom by $r - n + 1$. In this case, system (6.22), unlike system (6.1), *is no longer redundantly controlled*, it *is expediently controlled.* It should be noted that solution of any particular variational problem for system (6.9) simplifies search for optimal controls. When solving the problem of controlling system (6.1), it is necessary to investigate function *H* for by *r* variables $u_1, ..., u_r$. When substituting system (6.1) for system (6.22), a search for maximum of function *H* must be performed only by variables $u_1, ..., u_{n-1}$. The other controls are determined by relations $\det A_j = 0 \; j = \overline{n, r}$ (see Equation (6.6) or Equation (6.8)). This reduces influence on procedure of search for a classical trouble – curse of dimensionality [12]. If, after all, $r < n$, choice of all controls is determined by the need to satisfy problem boundary conditions and transversality conditions, leaving no room for additional coordination. Thus, our theoretical analysis of the variational problem formulated in Section 6.5.1 confirmed the assumption, stated by Anokhin, that the *advantageous effect of the system* is the determinant factor that exempts components of the system from redundant degrees of freedom.

Interesting is Anokhin's comment regarding Pavlov's assessment of the said independence. Thus, in his fundamental work [5], Anokhin writes that genius of Pavlov daringly ventured into a most subtle and most intimate aspect of work of human brain – *the goal of behavior.* At the Third Congress on Experimental Pedagogy in Petrograd on January 2, 1916, Pavlov called his famous speech on this issue "The Reflex of Goal." Analyzing the activities of animals and humans, he concludes that it is necessary to introduce *a special reflex – a reflex of the goal*, in addition to known reflexes. In fact, here Pavlov had already expressed Anokhin's hypothesis about the advantageous effect as a system-forming factor. But Anokhin does not focus on the following observation, from our point of view the most important one. Pavlov argued that the act of striving to achieve the goal should be separated from the meaning and value of the goal since *the essence of the matter lies in the striving itself, whereas the goal itself is unimportant*. This remark can be interpreted as verbalization of Pavlov's anticipation of the *system-wide invariance value*, discovered in the process of our mathematical analysis of Anokhin's hypothesis [5].

### 6.5.4 Structure of the Controlling System of an Expediently Functioning Object

It is this invariance of relations $\det A_j = 0 \; j = \overline{n, r}$ (see Equation (6.6) or Equation (6.8)), with regard to the teleological aspects of any particular problem, that allows to significantly simplify structural configuration of controlling system of an efficiently functioning object. Extremely important is the significant difference between the problems of defining controls $u_1, ..., u_{n-1}$ and controls $u_n, \; u_{n+1}, \; ..., \; u_r$. To calculate values of the controls of the first group, it is necessary to solve the traditional problem of optimal performing of a given transition from $x(t_0)$ state to final state $x(t_f)$ with minimum value of functional: $J[u(t)] = \Phi(x(t_f))$. Controlling functions of the second group are calculated according to the rules (6.6), (6.8), and (6.21) which are known prior to solving any particular problem. *Invariance* of these formulas, their *immutability*, and *their a priori knowledge* allow to constructively isolate the program implementation of second problem into a separate unit. Program implementation of solution of the first problem, which is fundamentally much more complex, constitutes the main controlling module. The sequence of these sub-problem solutions in the process of solving any particular problem makes it expedient to constructively combine these modules in a hierarchical structure. In the upper (external) level of such a hierarchy, a specific task of dynamic system controlling (6.9) is solved. The lower (internal) level is only intended for calculation of controls $u_n, \; u_{n+1}, \; ..., \; u_r$, according $\det A_j = 0 \; j = \overline{n, r}$ (see Equation (6.6) or Equation (6.8)), to *that fully ensure the object functioning when solving a specific controlling problem.* We find the result of this section a positive response to Herrero's question on the possibility of using mathematical methods for reducing perceived complexity of living systems by splitting them at a formalized description into suitable subsystems possessing a much simpler structure [37].

### 6.5.5 Hierarchy and Invariance of Expediently Controlled System

Transferring the results of Section 6.5.4 to any living being, we note that the control system of a rationally controlled object has at least a two-level hierarchical structure. The task of the upper level is to assess life situation $x(t_0)$, the choice of the control objective $x(t_f)$, and criterion $\Phi(x(t_f))$, consciously or unconsciously exercised by the subject. Also, controls $u_1$, ..., $u_{n-1}$ are chosen at the same level in accordance with the subject's motivation and his personal experience in solving rather difficult problems of choice. The task of the lower, internal level is to develop controls $u_n$, $u_{n+1}$, ..., $u_r$ according to rules $\det A_j = 0 \; j = \overline{n, r}$ (see Equation (6.6) or Equation (6.8)), whose invariance enables them to be stored compactly in the control system memory in the form of stable algorithms. Here the *principle of hierarchy of control*, its universality, is important. Assertion of Patte [69] that if there is to be any theory of general biology, it must explain the origin and operation of hierarchical constraints that provide the performance of coherent functions seems to be relevant to the evaluation of theoretical result obtained. An indication of hierarchical support of internally consistent functions, corresponding to the nature of the $\det A_j = 0 \; j = \overline{n, r}$ (see Equation (6.6) or Equation (6.8)) is considered to be particularly important here.

The article presents the results of the analysis of invariant relations between control functions in the theory of optimally controlled systems. We analyze the relationships related to a new type of invariance: of the price-target invariance. The concept of "invariance" as a scientific term was introduced into the theory of automatic control by Shchipanov [78]. In the work of Luzin [57] and in the monograph of Kuhtenko [52] were obtained and analyzed the conditions of invariance, i.e. independence of one or several controlled values from external disturbances. It can also be about independence from perturbations from a wider class of causes up to a purely algorithmic type [96]. It also seems interesting to test the feasibility of applying the ideology of game problems to analyze the structure of control circuits which ensures the invariance of the functioning of dynamic systems with respect to perturbations of various natures.

In the particular case of optimizing the movement of objects whose functioning is described by differential equations with constant coefficients, an addition to the Feldbaum theorem on the number of switching is obtained. At the same time, a new concept was introduced into consideration: the energy time constant of the oscillatory system, and to find out its dynamic meaning.

Next, we dwell in detail on the study of the results of using invariance in mathematical modeling of processes in living nature. The fact is that mathematical modeling, as a tool for studying the principles of theoretical biology occupies a position between Scylla and Charybdis of two Amosov's statements [4].On the one hand, modern level of knowledge of the material world provides a natural scientific approach to the explanation of all things in existence. On the other hand, the use of simulation modeling as a method of cognition does not in any way approximate the researcher to understanding the laws of the function of the living brain due to the multivariance of the possible mind simulation models. Therefore, the above results of the analysis, obtained from the basic assumptions formulated in Section 6.5.1, should be commented in this section following the classic Isaac Newton's maxim: "Explain the maximum possible number of facts using the minimum possible number of basic assumptions."

## 6.6 Investigation Analysis of Results

As a result of any mathematical model analysis using a mathematical tool, as a rule, a whole number of facts that were encrypted in it turns out to be decoded [61]. After all, applied mathematics is a branch of science which precisely develops the tools for such decoding for a researcher. The principal issue here is how the decoded facts relate to continuum of observed phenomena. The following are considerations about whether these facts are consistent with fundamental assumption that future theoretical biology would probably be a hybrid theory of judgment and an algorithmic explanation [50].

### 6.6.1 Mathematical Modeling – A Tool for Research of Complex Systems

We evaluate the result precisely as a hybrid of "parsimonious reasoning" by Anokhin and the above algorithmic optimization analysis. The hybrid, on whose urgency Leonard Euler had persistently focused the attention of the researchers: "There is absolutely no doubt that every effect in the universe can be explained from final causes, by the aid of the method of maxima and minima, as it can from the effective causes themselves... But one ought to make a special effort to see that both ways of approach to the solution of the problem be laid open; for thus not only is one solution greatly strengthened by the other but, more than that, from the agreement between the two solutions we secure the very highest satisfaction." [21].

It is shown that, in general case, rationally controlled systems differ from arbitrarily controlled ones by the fact that variational principle of minimum controlling energy loss is true for the former due to control components inconsistency.

Presented results are valid for any controlled objects whose functioning is described by the system of equations (6.1) when $r \geq n$. The above results relate to redundantly controlled dynamic objects of any nature. Moreover, the system of equations of the form (6.1) can also be used in description of stochastic processes. In this case, both deterministic and stochastic controlling systems can be considered as using the same analytical method [12] with substitution of corresponding values by their mathematical anticipations. Relations (6.20) and (6.21) are the rules that arrange, as required by Anokhin, a *more ordered set of rational motions* from the chaos of an object's motions, with a possible random interaction of control actions from the initial set $U$. At $r < n$, optimality conditions are satisfied without additional constraints (6.20). It should be noted that the relationship between the control functions, independent of the goal and the assessment of maneuvering quality, are found in publications related to controls of technical objects movement (see, e.g. [46, 88]). However, these articles do not express surprise at such an unexpected property of controls; the reasons for its invariant optimality are not investigated. There are also no indications of the possibility of using invariant relations when designing technical controlling systems, although formulas (6.8) imply a solution to the synthesis problem of redundant controls.

### 6.6.2 Optimality and Evolution Selection

Evaluation of evolutionary selected living beings, both optimal and optimally controlled ones, is fairly common in theoretical biology (see, e.g. [1, 9,17, 29, 42, 47, 54, 55, 60, 61, 75, 82, 88]) and dates back to ancient times. Aristotle considered optimality of biological systems to be the cornerstone of the biological phenomena understanding. At the same time, a rather convincing criticism of the optimality concept in biology is given in the monograph [63]. It is hard not to recognize the relevance of the thesis that the concept of the biosystem optimality does not correspond to that of the optimal systems problem in the theory of control. A perfection of biological control systems, *a priori* unattainable in engineering systems, is the direct evidence of its validity. Moreover, according to the recent publication [30], many philosophers and biologists question the very possibility of formulating any laws in biology due to the contingency and variation of living systems. At the same time, general principles in systems biology though not derived from physical and chemical laws are *additional* to them. The last statement almost textually coincides with the complementarity principle of Niels Bohr. The monograph [38] states that Niels Bohr defined *complementarity relation* between the physicochemical aspect of life processes which is controlled by causality and their functional features, dominated by a different kind of causality – teleological. Anokhin's hypothesis about the advantageous effect as a system-forming imperative factor fully complies with this principle.

### 6.6.3 Hierarchy and Invariance of Expediently Controlled System

Transferring the results of Sections 6.5.2–6.5.5 to any living being, we note that the control system of a rationally controlled object has at least a two-level hierarchical structure. The task of the upper level is to assess life situation $x(t_0)$, the choice of the control objective $x(t_f)$ and criterion $\Phi(x(t_f))$, consciously or unconsciously exercised by the subject. Also, controls $u_1$, ..., $u_{n-1}$ are chosen at the same level in accordance with the subject's motivation and his personal experience in solving rather difficult problems of choice. The task of the lower, internal level is to develop controls $u_n$, $u_{n+1}$, ..., $u_r$ according to rules (6.7) (or (6.8)), whose invariance enables them to be stored compactly in the control system memory in the form of stable algorithms. Here the *principle of hierarchy of control*, its universality, is important. Assertion of Patte [69] that if there is to be any theory of general biology, it must explain the origin and operation of hierarchical constraints that provide the performance of coherent functions seems to be relevant to the evaluation of theoretical result obtained. An indication of hierarchical support of internally consistent functions, corresponding to the nature of the rules (6.7) and (6.8), is considered to be particularly important here.

Invariance of coordination rules ensuring their immutability when performing any vital act allows to concluding that it is expedient to keep them lifelong in the structure of object control system. Information about these rules is transmitted genetically and accumulates in the process of individual development (formation of conditioned reflexes, training, sports training, etc.). It is important that in the process of learning, “lowering” answer to a recurring question from the sphere of consciousness to subconscious [7] is characteristic of the work of brain. In this way, a hierarchy is being gradually formed, providing both reflexive and conscious support of vital activity.

Under normal conditions, not only information processes that ensure work of the lower level do not reach consciousness, but also all information about these rules is reliably protected from its interference. Possibly, unblocking of such communication channels only occurs in the presence of pathology whose correction is not guaranteed without involving consciousness. Moreover, the upper-level intervention in the debugged work of the lower one can lead to the false functioning of the latter. Such is the etiology of psychosomatic diseases, neuroses, and certain types of hypertension [79]. However, targeted intervention of upper layer can be quite an effective way of treatment, stabilization, and improvement of the entire system – meditation, for example.

## 6.7 Optimal Control Theory as a Tool for Cognition

Mathematical analysis of Anokhin’s problem as an *optimization task* allowed to define not obvious *a priori* features of the *expedient rational* functioning of redundantly controllable objects. The ability to write relations (6.20) *prior to (and even without!) the wording* of any optimization task seems an undoubted testimony of *immanent optimality of the universe* granted by divine design – “so vivid a manifestation of this truth” noted by Euler. These relations are the result of applying the maximum principle to a *specific problem* with certain teleological conditions. Their invariance with respect to these conditions and the rationality of the control system hierarchy arising from it are the universal characteristics of all living things. It is important that the theoretical confirmation of these universal features is obtained by analyzing the optimality conditions. Without this, there is no reasonable motivation to writing formulas (6.20) or (6.21) and attaching them to system motion as constraints on the controlling influences choice from the $U$ set. The analysis of such an unexpected property of rational functioning allows to interpret it as a form of the universal experimental principle of minimum energy dissipation specific for living nature. This universal principle is referred to as experimental because, strictly speaking, it still has not been proved logically, although we do not know of any examples contradicting it [60].

As for its specific biological interpretation, it certainly includes the principle of adequate design of Rashevsky [74], interpreted by Rosen [76], an indication of the minimum hemodynamic vascular resistance, the minimum of the mechanical power developed by the muscles [24], a minimum energy consumption for biological structure formation and its functioning [25], optimal regulation of the respiratory air flow [34, 67, 94], etc. The possible discrepancy between theoretical results and the experimental data is naturally due to the insufficiently complete formulation of optimality criterion. In addition, the reason for this discrepancy may be any physiological factor effect that was not taken into account in the mathematical model [42].

The statement of Anokhin's problem as a variation task is so general that neither the specific formulation of the optimality criterion $J[u(t)] = \Phi(x(t_f))$ nor taking into consideration any particular physiological factor when writing Equation (6.1) was needed.

The features of the functioning of rationally controlled systems described in the above paragraph 5 leave open an extremely important question: what property of an object is expressed in rules (6.20), if they depend only on its nature, formalized by Equation (6.1). Essential only is the *desire to control its functioning rationally*, whatever the objective and the assessment of the quality of control may be.

It was possible to show that with the rational control of excessively controlled systems, the energy loss caused by the unbalancing of the control actions is minimal. It turned out that formulated by Anokhin, the problem of reducing the number of degrees of freedom is solved by the Pareto method.

*Formal coincidence of the rules* (6.6) and (6.20) with the mathematical definition of the Pareto set and *coincidence of the semantic meaning of belonging* to this set with the consistency of controls, as well as the hierarchy of the structure for providing expedient functioning, seems to be a reasonable basis for the following *Anokhin–Pareto principle (AP principle)* formulation:

***A solution of Anokhin's problem on the exemption of the system components from redundant degrees of freedom is achieved in accordance with Pareto's rule, which provides the maximum possible level of internal consistency of controlling influences to minimize the loss of controlling energy when functioning expediently in living nature.***

In his time, academician Moiseev [61] wrote about the relevance of the Pareto principle for understanding the processes occurring in living nature. He argued that the diversity of life forms is connected in a certain way with the multitude of possible compromises between provision of their own homeostasis and the pursuit of accomplishing the generalized minimum dissipation principle.

## 6.8 Is Teleology Theological?

The significance of the system-forming factor isomorphism is highlighted by Anokhin in the paper [5] by raising the result of system functioning to the level of the progress driving factor for all living things on the planet. The emotional coloring of this generalizing conclusion does not in any way diminish the relevance of the question of how this isomorphism was established. Confirmation of Anokhin's conclusion, obtained "from the properties of a living organism," using the results of a theoretical analysis of the variational problem of paragraph 2 testifies to effectiveness of the optimal control theory as a method of cognition of the material world. Indeed, it could not be otherwise since, according to Euler, "nothing at all takes place in the universe in which some rule of maximum or minimum does not appear." From a cognitive point of view, *the universality of variational principles* for those forms of motion for which they have already been (and may still be!) formulated is the key result of the optimization research. Conscious awareness of the subordination of *all* world events to variational principles seems to be convincing evidence of the *purposeful creation of the world* [60]. In this regard, the discovery of the variational calculus in the 17th century seems to be quite an expected consequence of the world's godly roots: since man did not arise as a result of the self-development of the biosphere on planet Earth but was instead created purposefully in the image and likeness of the Creator, it means that the world was created for man, according to apologists of the anthropic principle of the universe [19].

One of the authors of the present work is impressed by the debate on the above issues which has continued since the days of Maupertuis. Another one finds true the following statement by Heraclitus resembling the scenario of a pulsating universe: "The universe, which is the same for all, has not been made by any god or man, but it always has been, is, and will be - an ever-living fire, kindling itself by regular measures and going out by regular measures." This famous reasoning by Heraclitus is given in the collection of Diels [100] under number 30.

A theological solution of this antinomy is impossible since, as the Holy Luka Krymsky (Voyno-Yasenetsky) recognized, even the Holy Scripture is nothing more than our human concept of God [16]. At the same time, Caecilius, an ancient critic of Christianity, points out that all things in human are doubtful, unknown, incorrect, and generally rather plausible than true [73]. No better is the situation with the attempts of scientific proof of the existence of a Supreme Being. Asserting anthropomorphism and, therefore, rejecting idea of a personal God, Albert Einstein recognizes his understanding of the problem as being close to the views of Spinoza, who humbly admired the beauty and believed in the logical simplicity of order in the universe [39].

This disagreement, typical for researchers of living nature, is in our view, precisely the inner contradiction that urges further search. An effective tool for resolving such contradictions is the *fundamental* interaction of scientists, the idea of which was expressed by Meyen, a prominent expert in the field of methodology of biological research, who asserted that it is impossible to agree with the opponent's opinion without having to believe in the truth of the postulates he accepted, without *sympathy*, without what could be called mutual intuition [40].

Convergence of views as a goal is not a rejection of one's own research principles but as a possibility of fundamental joint clarification of the achieved milestones, starting positions, and direction of efforts for further cognition of the universe. This is especially important in view of scientific insolubility of the problem because the existing scientific concepts only apply to a very limited area of reality, while the unknown area remains infinite [36]. We can either accept belief in finality of the theological version or, following Spinoza and confessing admiration for the beauty and belief in the logical simplicity of order and harmony, continue the search, improving methods of research *in vivo*, *in vitro*, in model, *in principium*, and following the wise Japanese principle: "Moving to a goal is more important than achieving it."

## 6.9 Acknowledgment

The authors consider it their duty to express gratitude to Vsevolod Mikhailovich Kuntsevich for constant support and fruitful discussion of the results presented in this paper.

## References

[1] G. Ackland, 'Maximization principles and daisy world', Journal of Theoretical Biology, 227, 1, pp. 121-128, 2004.

[2] M. Ajmone, M. Bellomo, M. Egidi, 'Towards a mathematical theory of com plex socio-economical systems by functional subsystems representation', American Institute of Mathematical Sciences Models, **1,** 2 , pp. 249–278, 2008.

[3] G. Albi, N. Bellomo, L. Fermo et al., 'Traffic, crowds, and swarms. From kinetic theory and multiscale methods to applications and research perspectives', Math. Models Methods Appl Sci, **29,** 10, pp. 1901–2005, 2019.

[4] N. Amosov, 'Mind. Consciousness. True', Glushkov Institute of Cybernetics of the NAS of Ukraine, Kiev [in Russian], 1993.

[5] P. Anokhin, 'Fundamental questions of the general theory of functional systems', USSR Academy of Sciences, Department of Physiology, Moscow, [in Russian], 1971.

[6] P. Anokhin, 'Ocherki po fiziologii funktsional'nykh system' [Essays on the Physiology of Functional Systems], Meditsina Publ, Moscow [In Russian], 1975.

[7] Yu. Antomonov, 'Systems, complexity, dynamics', Naukova Dumka, Kiev, [In Russian], 1969.

[8] Yu. Antomonov, 'Principles of neurodynamics', Naukova Dumka, Kiev, [In Russian], 1974.

[9] Aristotle, 'On the parts of animals', Biomed Giz, Moscow, [In Russian], 1937.

[10] J. Banga, 'Optimization in computational systems biology', BMC Systems Biology, **2(1),** 47.

[11] E.Bauer, 'Theoretical biology', All-Union Institute of Experimental Medicine (VIEM), Publishing House Berezin and Zidkov, Moscow-Leningrad, [In Russian], 1935.

[12] R. Bellman, 'Dynamic programming', Publishing foreign lit, Moscow, [In Russian], 1960.

[13] A. Bellouquid, N.Chouhad, 'Kinetic models of chemotaxis towards the diffusive limit: asymptotic analysis', Math. Models Methods Appl Sci, 39, pp. 3136-3151, 2016.

[14] A.Bellouquid, M. Delitala, , 'Mathematical Modeling of Complex Biological Systems A Kinetic Theory Approach', *Birkhäuser*-Springer, *Boston,* 2006.

[15] J. Betts, 'Survey of Numerical Methods for Trajectory Optimization', Journal of Guidance, Control and Dynamics, 21, 2, pp. 193 – 207, 1998.
[16] I. Blekhman, A. Myshkis, Ya. Panovko, 'Mechanics and Applied Mathematics. Logic and Features of Mathematics Applications. Science', Moscow, [In Russian], 1983.
[17] D. Bolotina, O.Golosova, 'On the representation of God in human form', Lepta Kniga, Moscow, [in Russian], 2013.
[18] E. Bukvareva, G. Aleshchenko, 'The principle of optimal diversity of biosystems', KMK Scientific Partnership, MTposcow, [in Russian], 2013.
[19] B. Carter, 'Large number coincidences and the anthropic principle in cosmology' In: Confrontation of cosmological theories with observational data, Springer Dordrecht, pp. 291-298, 1974.
[20] L. Coscia, V. Fermo, N. Bellomo, 'On the mathematical theory of living systems II: The interplay between mathematics and system biology', Computers & Mathematics with Applications, 62, 10, pp. 3902-3911, 2011.
[21] M. Delitala,'On the Mathematical Modelling of Complex Systems' , Biological and Kinetic Theory Approach, Birkhäuser, Appl Sci, *17, pp.* 1647-1673, 2007.
[22] L. Euler, 'Methodus inveniendi lineas curvas maximi minimive proprietate gaudentes. apud Marcum-Michaelem Bousquet', 1744.
[23] V. Egorov, 'On the solution of one inherent variational problem and the optimal rise of a space rocket', Adj Mat. and mehan, 22, 1, pp.16-26, [in Russian], 1958.
[24] R. Fedorenko, 'Approximate Solution of the Problems of Optimal Control', Nauka, Moscow, [In Russian], 1978.
[25] W. Fehn, H. Rahn, 'Mechanics breathing in man', J Appl Physiol, 2, pp. 592– 607, 1950.
[26] R. Fisher, '*The Genetical Theory of Natural Selection*', *Edited with a foreword and notes by J. Bennett (A complete variorum ed), University Press Oxford, Oxford, 1999.*
[27] P. Fursova, A. Levich, V. Alekseev, 'Extreme Principles in Mathematical Biology' Successes of Modern Biology, 123, 2, pp. 115-137, [in Russian], 2003.
[28] R. Gabasov, F. Kirillova, 'Special optimal controls', Nauka, Moscow, [in Russian], 1973.
[29] R. Goddard, 'A method of reaching extreme altitudes', Smithsonian Inst. Publs. Misc. Collections, 171, 2, 1919.

[30] A. Gorban, R. Khlebopros, 'Demon Darwin: The idea of optimality and natural selection', Nauka, Moscow, [In Russian], 1988.
[31] M. Green, J. Piel, 'Theories of human development: A comparative approach' Psychology Press, 2015.
[32] S. Green, O. Wolkenhauer, 'Tracing Organizing Principles: Learning from the History of Systems Biology', History *and Philosophy of the Life Sciences,* 35, 4, pp. 553–576, 2013.
[33] Gurvich, 'Principles of analytical biology and the theory of cell field', Science, Moskow, [In Russian], 1991.
[34] G. Hamel, 'Uber eine mit dem Problem der Rakete zuzammenhangendeAufgabederVariationsrechnung', Zeitschrift fur angewandte Mathematik und Mechanik, 7, 6, pp. 451- 452, 1927.
[35] R. Hämäläinen, A. Viljanen, 'Modeling the Respiratory Airflow Pattern by Optimization Criteria', Cybernetics, 29, pp. 143–149, 1978.
[36] M. Haruno, D. Wolpert, 'Optimal control of redundant muscles in step-tracking wrist movements', Journal of Neurophysiology, 94, 6, pp. 4244-4255, 2005.
[37] V. Heisenberg, 'Physics and Philosophy Part and Whole' Science, Moscow, [In Russian], 1989.
[38] M. Herrero, 'On the role of mathematics in biology' J Math Biol, 54, pp. 887–889, 2007.
[39] G. Holton, 'Thematic origins of scientific thought Kepler to Einstein' Harvard, University Press, 1988.
[40] M. Jammer, 'Einstein and Religion: Physics and Theology', Princeton University Press, Princeton, 2000.
[41] R. Karpinskaya, 'Why does a biologist need a methodologist? Methodology of biology: new ideas (synergetics, semiotics, co-evolution)', Ed. Baksansky. Editorial URSS, Moscow, [In Russian], 2001.
[42] H. Kelley, 'Singular Extremals', [In book 'Topics in Optimization', Acad. Press, pp. 63 – 101, 1967.
[43] M. Khanin, N. Dorfman, I.Bukharov et al., 'Extreme principles in biology and physiology', Science, Moscow, [In Russian], 1978.
[44] B. Kiforenko, 'Optimal Relations in the Theory of Optimal Systems', Institute of Mathematics, Academy of Sciences of the Ukrainian SSR, Kiev, [In Russian], 1989.
[45] V. Zlatsky, 'Investigation of degenerate problems of flight mechanics', dis. Candidate of physical and mathematical sciences, special 01.02.01,1982.

[46] B. Kiforenko, 'Singular Optimal Controls in the Mechanics of Space Flight', Nau kova Dumka, Kiev. [In Russian], 2017.

[47] B. Kiforenko, 'Optimal Trajectories with Areas of Special Control'. In the book 'Complex Control Systems'. Institute of Cybernetics, Academy of Sciences of the Ukrainian SSR, Kiev, pp. 45 – 55, [In Russian], 1974.

[48] B. Kiforenko, V. Kuzmenko, V. Berezenko, 'Some problems of mechanics with low flying flyers', Bulletin of the University of Kiev, Serie Mat. and Mechanics, 13, pp. 44-47, [In Russian], 1971.

[49] B. Kiforenko, S. Kiforenko, 'The Anokhin-Pareto order in the theory of optimal systems', Izv AN SSSR, Techn Cybernetics, 5, pp., 187–191, [in Russian], 1991.

[50] H. Kitano, 'Perspectives on systems biology', New Generation Computing, 18, 3, pp. 199-216, 2000.

[51] D. Krakauer, J. Collins, D. Erwin et al., 'The challenges and scope of theoretical biology', Journal of theoretical biology, 276, 1, pp. 269-276, 2011.

[52] V. Krotov, V. Gourman, 'Methods and Tasks of Optimal Control', Science, Moscow, [in Russian], 1973.

[53] A. Kukhtenko, 'Das Problem der Invarianz in der Automatisierung', Gostekhizdat der Ukrainischen SSR, Kiew, [in Russian], 1963.

[54] R. Kumar, H. Kelley, 'Singular Optimal Atmospheric Rocket Trajectories', Jou nal of Guidance, Control, and Dynamics, 11, 4, pp. 305 – 312, 1988.

[55] A. Levich, 'The principle of maximum entropy and the theorem of variational modeling in community ecology', Successes of Modern Biology, 124, 6, pp. 515-533, [In Russian], 2004.

[56] W. Li, 'Optimal control for biological movement systems', Doctoral dissertation, UC San Diego, 2006.

[57] V. Lischuk, 'Mathematical theory of blood circulation', Medicine, Moscow, [In Russian], 1991.

[58] N. Luzin, 'A study of the matrix theory of differential equations', Avtomat i Telemekh 5, pp. 4- 66, [in Russian], 1940.

[59] C. Marshal, P. Kontensou, 'Singularities in optimization of deterministic dynamic systems', Journal of Guidance and Control, 4, 3, pp. 633–666, 1981.

[60] P. Martinon, et al., 'Numerical Study of Optimal trajectories for an Arian 5 Launcher', Journal of Guidance, Control and Dynamics, 32, 1, pp. 51-55, 2009.

[61] N. Moiseev, 'Man environment society: problem of formalized of the description', Nauka, Moscow, [In Russian], 1982.
[62] N. Moiseev, 'Algorithms of development', Nauka, Moscow, [In Russian], 1987.
[63] M. Mossio, M. Montévil, G. Longo, 'Theoretical principles for biology: Organization', Progress in Biophysics and Molecular Biology, 122, 1, pp. 24-35, 2016.
[64] V. Novoseltsev, 'Theory of management and biosystems', Nauka, Moscow, [In Russian], 1978.
[65] H. Oberle, 'Numerical Computation of Singular Control Functions in Trajectory Optimization Problems', Journal of Guidance and Control, 13, 1, pp. 153 – 159, 1990.
[66] I. Obraztsov, M. Khanin, 'Optimal biological systems', Medicine, Moskow, [In Russian], 1989.
[67] D. Okhotsimsky, 'To the theory of rocket movement', App. Mat. and Mehan., 10, 2, pp. 251 – 272, [in Russian] , 1946.
[68] Y. Onopchuk, 'On one general scheme of regulation of external respiration regimes minute blood volume and tissue blood flow by oxygen request', Cybernetics, pp. 110–115, [In Russian], 1980.
[69] V. Pareto, 'Manuel d/économie politique', Paris, 1909.
[70] H. Pattee, 'The nature of hierarchical controls in living matter', Foundations of mathematical biology, 1, pp. 1-22, 1972.
[71] I. Pavlov, 'Reflex of goal', MSS, 3, pp. 306-313, 1916, [in Russian].
[72] I. Petrov, 'Mathematical modeling in medicine and biology based on models of mechanics of continuous media', Proceedings of the Moscow Institute of Physics and Technology, Mathematics, 1, 1, pp. 5-16, [in Russian], 2009.
[73] L. Pontryagin et al. 'Mathematical theory of optimal processes', Nauka, Moscow, [in Russian], 1961.
[74] A. Ranovich, 'Primary sources in the history of early Christianity Antique critics of Chrstianity', Politizdat, Moscow, [in Russian], 1990.
[75] N. Rashevsky, 'Mathematical biophysics', The Univ. of Chicago Press, Chicago, 1948.
[76] M. Reed, 'Why is mathematical biology so hard', Notices of the MS, 3, pp. 338-342, 2004.
[77] R. Rosen, 'Optimality Principles in Biology',Plenum Press, New York. 1969.
[78] L. Rozonoer, 'The Pontryagin maximum principle in the theory of optimal systems' 20, 11, pp. 1442-1458, [in Russian], 1959.

[79] G. Shchipanov, 'Theory and methods of designing automatic controllers', Autoatic and Telemekh, 1, pp. 49–66, [in Russian], 1939.
[80] G. Sidorenko, G. Borisova, E. Ageenkova, 'Psychophysiological aspects of cardiological research', Belarus, Minsk, [In Russian], 1982.
[81] V. Smolyaninov, 'From geometry invariants to control invariants' in: E. Velikhov, A. Chernavsky (ed) 'Intelligent processes and them modeling', Nauka, Moscow, pp. 66-110, 1987.
[82] H. Spencer, 'The principles of biology', v. 2, D Appleton and Company, New-York, 1875.
[83] E. Todorov, 'Efficient computation of optimal actions', Proceedings of the national academy of sciences, Washington DC, 106, 28, pp. 11478 – 11483, 2009.
[84] E. Todorov, W. Li, X. Pan, 'From task parameters to motor synergies: A hierarchical framework for approximately optimal control of redundant manipulators', J. Robot Syst, 22, 11, pp. 691–710, 2005.
[85] E. Tre'lat , 'Optimal Control and Applications to Aerospace: Some Results and Challenges', Journal of Application Theory and Applications, 154**,** 3, pp. 713-758, 2012.
[86] M. Tsetlin, 'Investigations on the automata theory and modeling of biological systems', Nauka, Moscow, [in Russian], 1969.
[87] P. Tsiotras, H. Kelley, 'Goddard Problem with Constrained Time of Flight', Journal of Guidance and Control, 27, 3, pp. 289-296, 1992.
[88] I. Tsypkin, Y.Tsypkin, 'Relay control systems', CUP Archive, 1984.
[89] N. Vinh, 'Integrals of the motion for optimal trajectories in atmospheric flight', AIAA Journal, 11, 5, pp. 700-703, 1973.
[90] L. Von Bertalanfy, 'Modern theories of development: An introduction to theoretical biology', Oxford Univ. Press, New-York, 1933.
[91] J. Von Uexküll, D. Mackinnon, 'Theoretical biology', K Paul, Trench, Trubner and Company Limited, 1933.
[92] P. Wellstead, et al., 'The role of control and system theory in systems biology', Annual reviews in control, 32**, 1**, pp. 33-47, 2008.
[93] J. Willems, A. Kitapci, L. Silverman, 'Singular Optimal Control: a Geometric Approach', SIAM Journal of Control and Optimization, 24, 2, pp.323-337, 1986.
[94] J. Woodger, 'Biological principles. A Critical Study', Routledge, New York, 1926.
[95] S. Yamashiro, F. Grodins, 'Optimal regulation of respiratory air flow', J Appl Physiol, 30**,** pp. 597–602, 1971

[96] V. Zlatsky, B. Kiforenko, 'About Calculation of Optimal Trajectories with Areas of Special Control', Questions of Compute and Applied Mathematics, 33, pp. 55- 62 , [in Russian].

[97] V. Kuntsevich, 'Estimation of Impact of Bounded Perturbations on Nonlinear Descrete Systems', In: V. Kuntsevich, V. Gubarev, Y. Kondratenko, D. Lebedev, V. Lysenko, (Eds). 'Control Systems: Theory and Applications', Series in Automation, Control and Robotics, River Publishers, Gistrup, Delft, 2018, pp. 3-15.

[98] Chikrii, A., at *all.* Method of Resolving Functions in the Theory of Conflict—Controlled Processes. In Advanced Control Techniques in Complex Engineering Systems: Theory and Applications Springer, Cham, 2013, pp. 3-33.

[99] Diels, H. Die Fragmente der Vorsokratiker, Berlin, Orpheus, fr, 17, 1988).

[100] McDonough, Jeffrey K., "Leibniz's Philosophy of Physics", The Stanford Encyclopedia of Philosophy (Fall 2019 Edition), Edward N. Zalta (ed.).

# 7

# Robust Adaptive Controls for a Class of Nonsquare Memoryless Systems

**L. Zhiteckii[1], K. Solovchuk[2]**

[1]International Research and Training Center for Information Technologies and Systems of the National Academy of Science of Ukraine and Ministry of Education and Sciences of Ukraine, Acad. Glushkova av., 40, Kiev, 03187, Ukraine
[2]Poltava Scientific Research Forensic Center of the MIA of Ukraine, Rybalskiy lane, 8, Poltava, 36011, Ukraine
E-mail: leonid_zhiteckii@i.ua; solovchuk_ok@ukr.net

## Abstract

The discrete-time robust adaptive control for a class of uncertain multivariable memoryless (static) plants with arbitrary unmeasurable bounded disturbances is addressed in this chapter. The main feature of their gain matrices is that they have the full rank. The cases of nonsquare systems, where the number of the control inputs is less than the number of the outputs, are studied. It is assumed that the plant parameters defining the elements of its gain matrix are unknown. Again, bounds on the disturbances may also be unknown. The control problem is here to design robust adaptive controllers to be able to ensure the boundedness of all the input and output system signals in the presence of parametric and of possibly nonparametric uncertainties. To solve this problem, the so-called pseudoinversion concept is utilized. The asymptotic properties of the feedback control systems containing these controllers are established. Simulation results are also presented to support the theoretic study.

**Keywords:** Boundedness, discrete time, estimation algorithm, feedback control system, robust adaptive controller, pseudoinverse model-based control, uncertainty.

## 7.1 Introduction

The problem of efficient control of multivariable systems with arbitrary unmeasurable external disturbances stated several decades ago remains important both from theoretical and also practical points of view until recently. Novel results in this scientific area have been reported in numerous papers and generalized in several books including [1–4]. Last time, the problem of a perfect performance of the so-called multi-input multi-output (MIMO) and single-input multi-output (SIMO) systems attracted attention of many researches [5–7]. Similar class of multivariable control systems is also studied in [8, chap. 3] dealing with uncertain memoryless (static) multi-input single-output (MISO) systems in the presence of arbitrary bounded disturbances.

Among other methods advanced in the modern control theory, the inverse model-based method that is an extension of the well-known internal model principle seems to be perspective in order to cope with arbitrary unmeasurable disturbances and to optimize some classes of multivariable control systems. It turned out that this method first intuitively advanced in [9] makes it possible to optimize the closed-loop control system containing the MIMO memoryless plants whose gain matrices are square and nonsingular. Since the beginning of the 21st century, a significant progress has been achieved, utilizing the inverse model-based approach; see, e.g. [10] and other works. Nevertheless, it is quite unacceptable if the MIMO plants to be controlled have singular square or any nonsquare gain matrices because they are noninvertible.

In order to optimize the closed-loop control system containing an arbitrary MIMO memoryless plant, the so-called pseudoinverse model-based approach has been proposed and substantiated in [11]. Its gain matrix must be known to implement this approach. In practice, however, the plant parameters defining the elements of gain matrices may not be known *a priori*. In this case, the problem of designing the robust multivariable control system may be stated. The books [8, 12, 13] give a fairly full picture concerning the results achieved in the robust control theory to the beginning of the 2000s. Novel results regarding its practical application can be found in [14]. To achieve the robustness of the closed-loop system with unmeasurable disturbances, they can be indirectly measured, by employing the inverse model-based approach [10]. A new method to deal with the estimation of unknown bounds on unmeasurable disturbances is advanced in recent work [15]. Within the framework of the robust control theory, the pseudoinverse model-based method has been modified in [11, 16, 17] to stabilize some classes of uncertain linear and nonlinear MIMO systems whose gain matrices are arbitrary. (Note that the problem of robust control of some nonlinear one-dimensional static plant has before been solved in [18].) Unfortunately, the pseudoinverse model-based controller having fixed parameters may not be suitable if the parametric uncertainty is great enough.

An adaptation concept plays a role of some universal tool to cope with uncertain systems [8, 19–25], and others. This concept has been employed in the papers [26–28] in which adaptive controllers for the time-invariant linear and nonlinear MIMO and MISO memoryless plants are designed and studied. The latest methods for the robust adaptive control of linear and some nonlinear MISO memoryless plants can be found in [8, chap. 3]. Difficulties that take place when adaptive control uses the point estimation algorithms are how to guarantee the stability (the boundedness) of the closed-loop control system [29]; see also [19, 20]. To overcome these difficulties, the so-called frequency theorem given in [19, theorem 4.П.3] and the key technical lemma given in [20, subsection 6.2] are usually utilized to establish the boundedness properties of adaptive control systems. However, they become not applicable if the number of system outputs exceeds the number of their control inputs; see [19, p. 242] and [20, p. 202]. Nevertheless, the problem of the adaptive stabilization of SIMO memoryless plants with bounded disturbances has been solved in the recent paper [30]. Again, some fruitful ideas have before been advanced in [31] to deal with the adaptive control of nonsquare MIMO memoryless plants whose gain matrices have the full rank.

This chapter generalizes the pseudoinverse model-based approach reported in [11, 16, 17] and extends the results obtained in [30, 31].

## 7.2 Problem Formulation

Let

$$y_n = Bu_{n-1} + v_{n-1} \tag{7.1}$$

be the equation describing a linear discrete-time multivariable memoryless plant whose measurable $m$-dimensional output vector, $r$-dimensional control vector, and unmeasured $m$-dimensional disturbance vector at the $n$th time instant $(n = 1, 2, \ldots)$ are defined as $y_n = [y_n^{(1)}, ..., y_n^{(m)}]^T$, $u_n = [u_n^{(1)}, ..., u_n^{(r)}]^T$, and $v_n = [v_n^{(1)}, ..., v_n^{(m)}]^T$, respectively. $B$ is some time-invariant $m \times r$ gain matrix given by

$$B = \begin{pmatrix} b^{(11)} & \dots & b^{(1r)} \\ \dots & \dots & \dots \\ b^{(m1)} & \dots & b^{(mr)} \end{pmatrix}. \tag{7.2}$$

Consider the class of nonsquare multivariable plants, where the number $m$ of the outputs exceeds the number $r$ of its control inputs, i.e. $1 \le r < m \ (m \ge 2)$.

The following assumptions with respect to the gain matrix $B$ in Equation (7.2) and the sequences $\{v_n^{(i)}\} = v_0^{(i)},\ v_1^{(i)},\ \dots\ (i = 1,\ \dots,\ m)$ are made.

A1) The elements of $B$ are all unknown. However, there are some interval estimates defined as

$$\underline{b}^{(ij)} \le b^{(ij)} \le \overline{b}^{(ij)}, \qquad i = 1, \dots, m, \quad j = 1, \dots, r \tag{7.3}$$

where the upper and lower bounds $\underline{b}^{(ij)}$ and $\overline{b}^{(ij)}$, respectively, are assumed to be known.

A2) The rank of $B$ satisfies

$$1 \le \ \text{rank}\ B = r \tag{7.4}$$

implying that $B$ has the full rank because $\text{rank}\ B \le \min\{r,\ m\}$; see, e.g. [32, item 4.41].

A3) $\{v_n^{(i)}\}\ (i = 1, \dots,\ m)$ are all the arbitrary scalar nonstochastic sequences bounded in modulus according to

$$|v_n^{(i)}| \le \varepsilon^{(i)} < \infty \tag{7.5}$$

where $\varepsilon^{(i)}$s are constant. It is assumed that these upper bounds may be unknown, in general.

Denote by $y^0 = [y^{0(1)}, \dots, y^{0(m)}]^T$ the desired $m$-dimensional output vector $(y^{0(i)} \equiv \text{const}\ \forall i = 1,\ \dots,\ m)$. Without loss of generality, suppose $0 < \|y^0\| < \infty$. This means that, at least, one $y^{0(i)}$ of $y^{0(1)}, \dots, y^{0(m)}$ is nonzero: $|y^{0(1)}| + \dots + |y^{0(m)}| \ne 0$.

Define the output error vector

$$e_n = y^0 - y_n \tag{7.6}$$

with the components $e_n^{(i)} = y^{0(i)} - y_n^{(i)}$ giving $e_n = [e_n^{(1)}, \dots, e_n^{(m)}]^T$.

As in the case of the scalar memoryless system studied in [18], admissible control strategies will be defined by the feedback controls

$$u_n = U_n(u_{n-1},\ e_n) \tag{7.7}$$

with a time-varying operator $U_n\ :\ \mathbf{R}^r \times \mathbf{R}^m \to \mathbf{R}^r$.

The problem stated below is formulated as follows. Within the assumptions A1)–A3), it is required to design the feedback controller guaranteeing the ultimate boundedness of the sequences $\{e_n\} = e_1,\ e_2,\ \ldots$ and $\{u_n\} = u_0,\ u_1,\ u_2,\ \ldots$ giving

$$\limsup_{n\to\infty} ||e_n|| < \infty \tag{7.8}$$

$$\limsup_{n\to\infty} ||u_n|| < \infty. \tag{7.9}$$

## 7.3 Background on Pseudoinverse Model-Based Method

Using Equation (7.6), rewrite Equation (7.1) in the form

$$e_n = e_{n-1} - B\nabla u_{n-1} - \nabla v_{n-1} \tag{7.10}$$

where the notations

$$\nabla u_n := u_n - u_{n-1} \tag{7.11}$$

$$\nabla v_n := v_n - v_{n-1} \tag{7.12}$$

are introduced.

If follows from Equation (7.10) that for given $e_n$, the output error vector at $(n + 1)$th time instant, $e_{n+1}$, depends only on the control increment, $\nabla u_n$, given by Equation (7.11), at previous $n$th time instant and also on the unmeasurable variable $\nabla v_n$ defined by Equation (7.12). Taking into account this fact together with the boundedness assumptions

$$|\nabla v_n^{(i)}| \le 2\varepsilon_i \quad (i = 1,\ \ldots,\ m) \tag{7.13}$$

on the components $\nabla v_n^{(i)}$ of $\nabla v_n = [\nabla v_n^{(1)}, ..., \nabla v_n^{(m)}]^T$ in Equation (7.12) caused by Equation (7.5), introduce the local performance index

$$J_{n+1} := \sup_{\nabla \bar{v}:\ 0\le|\nabla \bar{v}^{(i)}|\le 2\varepsilon_i} ||e_{n+1}||_2 \tag{7.14}$$

where $\nabla \bar{v}_n := [\nabla \bar{v}_n^{(1)}, ..., \nabla \bar{v}_n^{(m)}]^T$.(In view of Equation (7.10), this quality criterion depends only on $u_n$ to be chosen at $n$th time instant and evaluates the capability of a controller to rejecting any bounded disturbances $v_n^{(1)}, ..., v_n^{(m)}$,which satisfy Equation (7.5) in the worst case.)

Based on the representation (7.7), the one-step iterative procedure of the form

$$u_n = u_{n-1} + \chi(e_n), \tag{7.15}$$

which generalizes the one proposed before in [33], will then be chosen as the control law. In this expression, $\chi : \mathbf{R}^m \to \mathbf{R}^r$ is an operator to be determined via minimizing $J_{n+1}$ of the form (7.14) in all possible $\nabla u_n = \chi(e_n)$ to achieve

$$J_{n+1} \longrightarrow \inf_{\chi(e_n)} J_{n+1} \tag{7.16}$$

In [16], it has been shown that if $B$ is known, then the optimization task (7.16) is solved by the choice

$$\chi(e_n) = B^+ e_n \tag{7.17}$$

where $B^+$ denotes the pseudoinverse matrix specified as follows [34, Theorem 3.4]:

$$B^+ = \lim_{\delta \to 0} (B^T B + \delta I_r)^{-1} B^T. \tag{7.18}$$

(Note that if $\mathrm{rank}\ P = r$, for some $m \times r$ matrix $P$, then $P^+$ is determined by the formula [32, item 7.46]

$$P^+ = (P^T P)^{-1} P^T \tag{7.19}$$

which is simpler than Equation (7.18) after setting $B = P$.)

Substituting Equation (7.17) into Equation (7.15) yields

$$u_n = u_{n-1} + B^+ e_n. \tag{7.20}$$

Thus, Equation (7.20), together with Equation (7.6), defines the optimal control which may be called the pseudoinverse model-based control [16].

It has been shown in [16] that, subject to Assumption A3), the model-based control algorithm (7.20), (7.6) when applied to the system (7.1) guarantees the boundedness of the signal sequences $\{y_n\}$ and $\{u_n\}$ written as

$$\{y_n\} \in \underbrace{\ell_\infty \times \ldots \times \ell_\infty}_{m}$$

$$\{u_n\} \in \underbrace{\ell_\infty \times \ldots \times \ell_\infty}_{r}$$

for any $B = (b^{(ij)}) \in \mathbf{R}^{m \times r}$. Furthermore, if there are no disturbances ($v_n \equiv 0_m$), and Equation (7.4) takes place, then for given $y^0$, there exists a stable feedback control system equilibrium defined by the pair $(u^{\mathrm{e}},\ y^{\mathrm{e}})$, where $u^{\mathrm{e}}$ represents the solution of the equation

$$B^{+}Bu = B^{+}y^{0}$$

with respect to $u$, and $y^{\mathrm{e}} = Bu^{\mathrm{e}}$. More certainly, $u^{\mathrm{e}} = B^{+}y^{0}$ and $y^{\mathrm{e}} = BB^{+}y^{0}$.

*Remark 7.1.* If rank $B = r$, then the control law (7.20) takes the form

$$u_n = u_{n-1} + (B^T B)^{-1} B^T e_n \tag{7.21}$$

because formula (7.19) is here applicable. After setting $C_{\mathrm{opt}} = (B^T B)^{-1}$in Equation (7.21), this leads directly to the optimal control law

$$u_n = u_{n-1} + C_{\mathrm{opt}} B^T e_n$$

derived before in [33]. Moreover, if *B* is a square nonsingular matrix, then the control law (7.20) becomes

$$u_n = u_{n-1} + B^{-1} e_n$$

representing the usual inverse model-based control law followed from results of [9].

Since, in practice, *B* is unknown, it needs to be replaced by an appropriate estimate $B_0$ remaining the fixed matrix (in the nonadaptive case). Then, instead of Equation (7.20), another control law

$$u_n = u_{n-1} + B_0^{+} e_n \tag{7.22}$$

yielding

$$\nabla u_n = B_0^{+} e_n \tag{7.23}$$

has been proposed in [16] for controlling the system (7.1).

The structure of the closed-loop control system described by Equations (7.1), (7.6), and (7.22) is shown in Figure 7.1. It contains the plant having the gain matrix $B$, and the controller which consists of a pseudoinverse model whose gain matrix is $B_0^{+}$, and of the discrete integrator summing the increments $\nabla u_n$ given by Equation (7.23) to produce

$$u_n = \sum_{k}^{n} \nabla u_k.$$

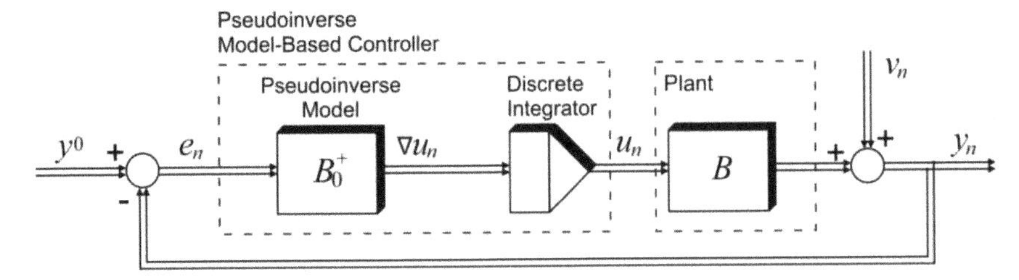


**Figure 7.1** Configuration of the pseudoinverse model-based feedback control system.

Recalling Assumption A1), define the interval matrix family (set) as

$$\Xi := \{(b^{(ij)}) \ : \ \underline{b}^{(ij)} \leq b^{(ij)} \leq \overline{b}^{(ij)}, \ \ i = 1, \ldots, m, \ j = 1, \ldots, r\}$$

and choose $B_0$ from $\Xi$ to satisfy

$$\text{rank } B_0 = r. \tag{7.24}$$

It can easily be shown that the equilibrium point, $u^{\mathrm{e}}$, of the closed-loop control systems (7.1), (7.6), and (7.22) has now to satisfy the equation

$$B_0^+ B u = B_0^+ y^0 \tag{7.25}$$

by setting $u = u^{\mathrm{e}}$. Unfortunately, this equation may have no solution with respect to $u$. Therefore, the point $u^{\mathrm{e}}$ may not exist, in general. It turns out, however, that under certain requirements given below, the system equilibrium will exist. To show this and to analyze the properties of the closed-loop system (7.1), (7.6), (7.22), introduce the matrix

$$\Delta := B_0 - B \tag{7.26}$$

which, due to $b^{(ij)} \in [\underline{b}^{(ij)}, \ \overline{b}^{(ij)}]$ and $b_0^{(ij)} \in [\underline{b}^{(ij)}, \ \overline{b}^{(ij)}]$, is specified as

$$\Delta = \{(\delta^{(ij)}) : \quad \underline{\delta}^{(ij)} \leq \delta^{(ij)} \leq \overline{\delta}^{(ij)}\} \tag{7.27}$$

where $\underline{\delta}^{(ij)} = b_0^{(ij)} - \overline{b}^{(ij)}$ and $\overline{\delta}^{(ij)} = b_0^{(ij)} - \underline{b}^{(ij)}$.

The following preliminary results can be shown to be valid.

**Lemma 7.1.** Let A1) to A3) hold. If Equation (7.24) takes place and the condition

$$\max_{\Delta \,:\, \delta^{(ij)} \in [\,\underline{\delta}^{(ij)}, \ \overline{\delta}^{(ij)}]} ||B_0^+ \Delta|| < 1 \tag{7.28}$$

with any matrix norm $||\cdot||$ is satisfied, then we have the following:

(a) the equilibrium of the closed-loop control systems (7.1), (7.6), and (7.22) exists;
(b) the boundedness of $\{u_n\}$ and $\{e_n\}$ is achieved.

*Proof.* To prove part (a), consider Equation (7.25). It is well known that this equation has always a solution if

$$\det B_0^+ B \neq 0. \tag{7.29}$$

Next, by the definition $||P||_2 = \alpha_{\max}(P)$ of the Euclidean matrix norm for any *P* (see, e.g. [32, item 14.48]), where $\alpha_{\max}(P)$ denotes the maximum singular value of *P*, we have

$$||B_0^+ \Delta||_2 = \alpha_{\max}(B_0^+ \Delta).$$

Substituting Equation (7.26) into the right-hand side of this expression gives

$$||B_0^+ \Delta||_2 = \alpha_{\max}(I_r - B_0^+ B) \tag{7.30}$$

where the fact that Equation (7.24) produces $B_0^+ B_0 = I_r$ [34, Theorem 3.12, result (b)] has been used.

Now utilizing the Browne inequality [35, part III, item 1.5.1],

$$|\lambda_i(P)| \leq \alpha_{\max}(P)$$

in which $\lambda_i(P)$ represents any *i*th eigenvalue of *P*, due to Equation (7.30), we obtain

$$|\lambda_i(I_r - B_0^+ B)| \leq ||B_0^+ \Delta||_2 \quad \forall i = 1, \ldots, r. \tag{7.31}$$

On the other hand, the property mentioned above, together with Brauer inequalities [35, part III, item 1.6.5] rewritten as $|\lambda_i(P)| \leq \min\{||P||_1, ||P||_\infty\}$, leads to

$$|\lambda_i(\underbrace{I_r - B_0^+ B}_{B_0^+ \Delta})| \leq ||B_0^+ \Delta||_1,$$

$$|\lambda_i(\underbrace{I_r - B_0^+ B}_{B_0^+ \Delta})| \leq ||B_0^+ \Delta||_\infty \quad \forall i = 1, \ldots, r$$

where $||P||_1$ and $||P||_\infty$ denote the so-called 1-norm and $\infty$-norm of *P*, respectively [32, item 14.48]. Combining these inequalities with Equation (7.31) yields

$$|\lambda_i(I_r - B_0^+ B)| \leq ||B_0^+ \Delta|| \quad \forall i = 1, \ldots, r \tag{7.32}$$

for any matrix norm $||\cdot||$. Further, by virtue of the property $\lambda_i(I_r + P) = 1 + \lambda_i(P)$ taken from [35, part I, item 2.15.3], the inequalities (7.32) finally produce

$$|1 - \lambda_i(B_0^+ B)| \leq ||B_0^+ \Delta|| \quad \forall i = 1, \ldots, r. \tag{7.33}$$

Since condition (7.28) has to be satisfied for all $\Delta$s defined by Equation (7.27), it follows from Equation (7.33) that there are no matrices $B$s from $\Xi$ such that $\lambda_i(B_0^+ B) = 0$, meaning $\det B_0^+ B = 0$. Consequently, condition (7.28) together with Equation (7.24) guarantees that requirement (7.29) will be satisfied. This fact establishes the existence of the point $u^{\mathrm{e}}$.

The proof of part (b) is given in [16, Theorem 2]. □

Thus, Lemma 7.1 establishes the conditions under which the robustness property of the feedback controller (7.6), (7.22) will be ensured. Although these conditions, which are the sufficient condition on the robust stability of the control system (7.1), (7.6), (7.22) (if $v_n^{(i)} \equiv 0, \; i = 1, \ldots, m$), can simply be verified using the linear programing tool [17], it may not be satisfied when the lengths of the intervals $[\underline{b}^{(ij)}, \; \overline{b}^{(ij)}]$ are large enough. In this case, the pseudoinverse model-based control law (7.22) will be unacceptable since the ultimate boundedness of $\{e_n\}$ and $\{u_n\}$ as $n \to \infty$ is not guaranteed. Therefore, the adaptive version of the pseudoinverse model-based control will be proposed to cope with arbitrary uncertain SIMO and nonsquare MIMO memoryless systems.

## 7.4 Robust Adaptive Pseudoinverse Model-Based Controllers for SIMO systems

Consider a linear discrete-time memoryless SIMO system described by the difference equation (7.1) in which *B* represents the time invariant gain matrix of the nonzero elements given by

$$B = \begin{pmatrix} b^{(1)} \\ \vdots \\ b^{(m)} \end{pmatrix}. \tag{7.34}$$

Suppose (for the time being) that *B* is known. In this ideal case, the feedback controller that ensures the boundedness of $\{e_n\}$ and $\{u_n\}$ may be designed as the pseudoinverse model-based control law of the form (7.20). Since $b^{(i)}$s are nonzero, it can be concluded that *B* defined in Equation (7.34) is the matrix of full rank $(\mathrm{rank}\ B = 1)$. This gives that $B^+$ is determined by

$$B^+ = \frac{1}{[b^{(1)}]^2 + \cdots + [b^{(m)}]^2} B^{\mathrm{T}} \tag{7.35}$$

because, in this case, the formula (7.19) holds.

Following the certainty equivalence principle which is well known in adaptive control literature (see, e.g. [20, p. 180], we will design a pseudoinverse model-based robust adaptive controller replacing unknown *B* in Equation (7.20) by its suitable estimate $B_n$ obtained at each *n*th time instant via some adaptive identification procedure. Then the control law takes the form

$$u_n = u_{n-1} + B_n^+ e_n. \tag{7.36}$$

Similar to Equation (7.35), $B_n^+$ is calculated as

$$B_n^+ = \frac{1}{[b_n^{(1)}]^2 + \ldots + [b_n^{(m)}]^2} B_n^{\mathrm{T}}$$

using the matrix

$$B_n = \begin{pmatrix} b_n^{(1)} \\ \vdots \\ b_n^{(m)} \end{pmatrix}$$

whose elements $b_n^{(1)}, \ \ldots, \ b_n^{(m)}$ represent the current estimates of unknown $b^{(1)}, \ \ldots, \ b^{(m)}$, respectively.

Let $\varepsilon^{(i)}$s be known. It can be concluded from Equation (7.1) together with Equations (7.5) and (7.6) that $b^{(1)}, \ \ldots, \ b^{(m)}$ satisfy the infinite sets of inequalities

$$|\tilde{b}^{(i)} \nabla u_{n-1} + e_n^{(i)} - e_{n-1}^{(i)}| \leq 2\varepsilon^{(i)}, \quad n = 1, \ 2, \ \ldots \tag{7.37}$$

$$|\tilde{b}^{(i)} u_{n-1} - y_n^{(i)}| \leq \varepsilon^{(i)}, \quad n = 1, \ 2, \ \ldots \tag{7.38}$$

with respect to some $\tilde{b}^{(i)}$s for each $i = 1, \ \ldots, \ m$. In other words, $\tilde{b}^{(i)} = b^{(i)}$ $(i = 1, \ \ldots, \ m)$ are the solutions of these sets of inequalities.

Inequalities (7.37) and (7.38) yield the sets that are the segments

$$\beta_n^{*(i)} = [\underline{b}_n^{*(i)}, \ \bar{b}_n^{*(i)}] \tag{7.39}$$

$$\beta_n^{**(i)} = [\underline{b}_n^{**(i)}, \ \bar{b}_n^{**(i)}] \tag{7.40}$$

to which $\tilde{b}^{(i)}$s belong. It follows from Equations (7.37) and (7.38) that the bounds on $\beta_n^{*(i)}$ and $\beta_n^{**(i)}$ are given by

$$\underline{b}_n^{*(i)} = \begin{cases} (e_{n-1}^{(i)} - e_n^{(i)} - 2\varepsilon^{(i)})/\nabla u_{n-1} & \text{if } \nabla u_{n-1} > 0 \\ (e_{n-1}^{(i)} - e_n^{(i)} + 2\varepsilon^{(i)})/\nabla u_{n-1} & \text{if } \nabla u_{n-1} < 0 \end{cases} \tag{7.41}$$

$$\bar{b}_n^{*(i)} = \begin{cases} (e_{n-1}^{(i)} - e_n^{(i)} + 2\varepsilon^{(i)})/\nabla u_{n-1} & \text{if } \nabla u_{n-1} > 0 \\ (e_{n-1}^{(i)} - e_n^{(i)} - 2\varepsilon^{(i)})/\nabla u_{n-1} & \text{if } \nabla u_{n-1} < 0 \end{cases} \tag{7.42}$$

and

$$\underline{b}_n^{**(i)} = \begin{cases} (y_n^{(i)} - \varepsilon^{(i)})/u_{n-1} & \text{if } u_{n-1} > 0 \\ (y_n^{(i)} + \varepsilon^{(i)})/u_{n-1} & \text{if } u_{n-1} < 0 \end{cases} \tag{7.43}$$

$$\bar{b}_n^{**(i)} = \begin{cases} (y_n^{(i)} + \varepsilon^{(i)})/u_{n-1} & \text{if } u_{n-1} > 0 \\ (y_n^{(i)} - \varepsilon^{(i)})/u_{n-1} & \text{if } u_{n-1} < 0 \end{cases}. \tag{7.44}$$

Next, introduce the intersections

$$\Omega_n^{*(i)} := \bigcap_{\mu=1}^{n} \beta_\mu^{*(i)}, \quad \Omega_n^{**(i)} := \bigcap_{\mu=1}^{n} \beta_\mu^{**(i)} \tag{7.45}$$

which can recursively be designed as follows:

$$\Omega_n^{*(i)} = \Omega_{n-1}^{*(i)} \bigcap \beta_n^{*(i)}, \quad \Omega_n^{**(i)} = \Omega_{n-1}^{**(i)} \bigcap \beta_n^{**(i)} \tag{7.46}$$

with $\beta_n^{*(i)}$ and $\beta_n^{**(i)}$ given by Equations (7.39) and (7.40), respectively. Further, define

$$\Omega_n^{(i)} = \Omega_n^{*(i)} \bigcap \Omega_n^{**(i)} \tag{7.47}$$

having the property that $b^{(i)} \in \Omega_n^{(i)}$ for any $n = 1,\ 2,\ \ldots$ . This property results from Equation (7.46) together with the fact that $b^{(i)} \in \beta_n^{*(i)}$ and $b^{(i)} \in \beta_n^{**(i)}$.

To derive the adaptive identification algorithm in the case where $\varepsilon^{(i)}$s are known, we will use the variables

$$\tilde{e}_n^{(i)} = e_n^{(i)} - e_{n-1}^{(i)} + b_{n-1}^{(i)} \nabla u_{n-1} \quad (i = 1,\ \ldots,\ m) \tag{7.48}$$

which play a role of some estimation errors (as in [36]). This algorithm is designed as a recursive procedure for solving the inequalities (7.37) with respect to unknown $\tilde{b}^{(i)}$ in the following form:

$$b_n^{(i)} = \begin{cases} b_{n-1}^{(i)} & \text{if } |\tilde{e}_n^{(i)}| \leq 2\,\varepsilon^{(i)} \\ P_{\Omega_n^{(i)}}\{b_{n-1}^{(i)} - \gamma_n^{(i)}[\tilde{e}_n^{(i)} & \\ \quad -\varepsilon^{(i)}\operatorname{sign}\tilde{e}_n^{(i)}]/\nabla u_{n-1}\} & \text{otherwise} \quad (i = 1,\ \ldots,\ m) \end{cases} . \tag{7.49}$$

In this algorithm, $P_{\Omega_n^{(i)}}\{w\}$ denotes the projection of $w$ onto $\Omega_n^{(i)}$ given by Equation (7.47), and $\gamma_n^{(i)}$ is a scalar multiplier satisfying

$$0 < \gamma' \leq \gamma_n^{(i)} \leq \gamma'' < 2 \tag{7.50}$$

*Remark* 7.2. Note that the recursive adaptive estimation procedure (7.49) is derived by combining the standard identification algorithm giving a point solution $b^{(i)} = b_n^{(i)}$ of inequalities (7.37) and (7.38) (as in [19, chap. 4]) and the membership set identification algorithm (7.45)–(7.47) determining the lower and upper bound on admissible $b^{(i)}$s (as in [8]).

The following preliminary result is needed to establish the ultimate properties of the adaptive closed-loop control system (7.1), (7.36), (7.49) together with Equations (7.6), (7.48) and (7.39)–(7.50).

**Lemma 7.2.** Let $\varepsilon'(B,\ \tilde{B})$ denote an $\varepsilon'$-neighborhood of $B$ given by $\varepsilon'(B,\ \tilde{B}) = \{\tilde{B} :\ ||B - \tilde{B}|| \ < \varepsilon'\}$.Then there is a sufficiently small positive $\varepsilon'$ such that if matrix estimate sequence $\{B_n\}$ caused by Equation (7.49) produces $B_n \in \varepsilon'(B,\ \tilde{B})$ for all $n_* \leq n < \infty$ with a finite $n_*$, then the sequences $\{u_n\}$ and $\{y_n\}$ will be bounded for all time.

*Proof.* The proof follows the lines of the proof of Theorem 1 in [16]. □

With this lemma, the following basic result can be shown to be valid.

**Theorem 7.1.** Subject to Assumptions A2) and A3), if the adaptive controller described in Equations (7.36) and (7.49) together with Equations (7.6) and (7.48), and Equations (7.39)–(7.44) together with Equation (7.50) is applied to the system (7.34), then the requirements (7.8) and (7.9) are satisfied.

*Proof.* Assume that, under the conditions of Theorem 7.1, the sequence $\{y_n\}$ is unbounded, implying

$$\lim_{n\to\infty} \sup ||y_n|| = \infty$$

This produces

$$\lim_{n\to\infty} \sup|u_n| = \infty \tag{7.51}$$

because Equation (7.34) together with the property that $\text{rank } B = 1$ causes

$$u_{n-1} = B^{+}y_n - B^{+}v_{n-1}$$

If Equations (7.5) and (7.51) are satisfied, then there is a subsequence $\{n_j\}$ $(n_1 < n_2 < \cdots)$ such that

$$\lim_{n_j\to\infty} |u_{n_j}| = \infty \tag{7.52}$$

Taking into account that due to Equations (7.43) and (7.44) the length $d_n^{**(i)}$ of the segment $\beta_n^{**(i)}$ is equal to $d_n^{**(i)} = 2\varepsilon^{(i)}/u_{n-1}$, it follows that Equation (7.52) yields

$$\lim_{n_j\to\infty} d_{n_j}^{**(i)} = 0 \tag{7.53}$$

From definitions (7.45) together with (7.47), it can be directly seen that $\{\Omega_n^{(i)}\}$ is a sequence of the one-dimensional nonincreasing sets in *n*: $\Omega_n^{(i)} \supseteq \Omega_{n+1}^{(i)} \quad \forall n = 1, 2, \ldots$ and $i = 1, \ldots, m$. Since $b^{(i)} \in \Omega_n^{(i)}$, it gives that if Equation (7.53) takes place, then there exists a number $n_+$ such that

$$\Omega_n^{(1)} \times \cdots \times \Omega_n^{(m)} \subseteq \varepsilon'(B, \tilde{B}) \quad \forall n \geq n_+$$

and any $\varepsilon' > 0$, where $\varepsilon'(B, \tilde{B})$ is some $\varepsilon'$-neighborhood of $B$ determined in Lemma 7.2. By this lemma, we have

$$\lim_{n\to\infty} \sup|u_n| < \infty$$

But this contradicts Equation (7.51). Hence, the assumption that $\{u_n\}$ is unbounded is false and the boundedness of $\{u_n\}$ holds.

By virtue of Equation (7.1) and due to Equation (7.5), the boundedness of $\{u_n\}$ causes

$$\lim_{n\to\infty} \sup||y_n|| < \infty$$

This completes the proof of Theorem 7.1. □

Now, let the bounds $\varepsilon^{(i)}$s in Equation (7.5) be unknown. Then the adaptive estimation algorithm described above cannot be implemented. In this case, the procedure for estimating the elements of $B$ is added by the estimation of $\varepsilon^{(i)}$s. Namely, instead of Equation (7.49), the adaptation algorithm becomes

$$b_n^{(i)} = \begin{cases} b_{n-1}^{(i)} & \text{if } |\tilde{e}_n^{(i)}| \leq 2\,\varepsilon_{n-1}^{(i)} \\ P_{\Omega_n^{(i)}}\{b_{n-1}^{(i)} - \gamma_n^{(i)}[\tilde{e}_n^{(i)} \\ \quad -\varepsilon_{n-1}^{(i)}\text{sign}\,\tilde{e}_n^{(i)}]/\nabla u_{n-1}\} & \text{otherwise} \end{cases} \tag{7.54}$$

and

$$\varepsilon_n^{(i)} = \varepsilon_{n-1}^{(i)}, \quad i = 1, \ldots, m \tag{7.55}$$

when $\Omega_n^{(i)} \neq \emptyset$; else

$$b_n^{(i)} = b_0^{(i)} \tag{7.56}$$

$$\varepsilon_n^{(i)} = \varepsilon_{n-1}^{(i)} + d_\varepsilon^{(i)}, \quad i = 1, \ldots, m \tag{7.57}$$

where $d_\varepsilon^{(i)}$ are arbitrarily sufficiently small positive numbers. In this algorithm, $\Omega_n^{(i)}$ is designed according to Equation (7.47) in which

$$\Omega_n^{*(i)} := \bigcap_{v=n_i(j)}^{n} \beta_v^{*(i)}, \quad \Omega_n^{**(i)} := \bigcap_{v=n_i(j)}^{n} \beta_v^{**(i)} \tag{7.58}$$

(instead of Equation (7.47)), where $n_i(j)$ denotes the latest time instant, when either $\Omega_{n_i(j)}^{*(i)}$ or $\Omega_{n_i(j)}^{**(i)}$ becomes empty ($n_i(1) < n_i(2) < \cdots < n_i(j) < n$). The lower and upper bounds on $\beta_n^{*(i)}$ and $\beta_n^{**(i)}$ given by Equations (7.39) and (7.40), respectively, are specified from Equations (7.41)–(7.44) after replacing unknown $\varepsilon^{(i)}$ by $\varepsilon_{n-1}^{(i)}$, leading to

$$\underline{b}_n^{*(i)} = \begin{cases} (e_{n-1}^{(i)} - e_n^{(i)} - 2\varepsilon_{n-1}^{(i)})/\nabla u_{n-1} & \text{if } \nabla u_{n-1} > 0 \\ (e_{n-1}^{(i)} - e_n^{(i)} + 2\varepsilon_{n-1}^{(i)})/\nabla u_{n-1} & \text{if } \nabla u_{n-1} < 0 \end{cases} \tag{7.59}$$

$$\bar{b}_n^{*(i)} = \begin{cases} (e_{n-1}^{(i)} - e_n^{(i)} + 2\varepsilon_{n-1}^{(i)})/\nabla u_{n-1} & \text{if } \nabla u_{n-1} > 0 \\ (e_{n-1}^{(i)} - e_n^{(i)} - 2\varepsilon_{n-1}^{(i)})/\nabla u_{n-1} & \text{if } \nabla u_{n-1} < 0 \end{cases} \tag{7.60}$$

and

$$\underline{b}_n^{**(i)} = \begin{cases} (y_n^{(i)} - \varepsilon_{n-1}^{(i)})/u_{n-1} & \text{if } u_{n-1} > 0 \\ (y_n^{(i)} + \varepsilon_{n-1}^{(i)})/u_{n-1} & \text{if } u_{n-1} < 0 \end{cases} \tag{7.61}$$

$$\bar{b}_n^{**(i)} = \begin{cases} (y_n^{(i)} + \varepsilon_{n-1}^{(i)})/u_{n-1} & \text{if } u_{n-1} > 0 \\ (y_n^{(i)} - \varepsilon_{n-1}^{(i)})/u_{n-1} & \text{if } u_{n-1} < 0 \end{cases} \tag{7.62}$$

The numbers $d_{\varepsilon}^{(i)}$ $(i = 1, \ldots, m)$ arising in Equation (7.57) are some fixed and sufficiently small positive numbers chosen by the designer.

The asymptotic properties of the adaptive pseudoinverse model-based controller exploiting the adaptation algorithms described above are established in the following theorem.

**Theorem 7.2.** Subject to Assumptions A2) and A3), if the pseudoinverse model-based adaptive controller (7.36), (7.54)–(7.58) together with (7.6), (7.48), (7.47), and (7.59)–(7.62) is applied to the system (7.34), then we have the following:

(i) estimates $\varepsilon_n^{(1)}, \ldots, \varepsilon_n^{(m)}$ are upper bounded;
(ii) the boundedness of $\{u_n\}$ and $\{y_n\}$ is guaranteed.

*Proof.* Noting that Equations (7.55) and (7.57) yield the nondecreasing sequences $\{\varepsilon_n^{(i)}\}$ $(i = 1, \ldots, m)$, we first establish that these sequences cannot be unbounded. In fact, when they become such that $\varepsilon_n^{(i)} \geq \varepsilon^{(i)}$ for all sufficiently large $n_+$ and $i = 1, \ldots, m$, then the intersection (7.47) will be nonempty (by definition (7.45) of $\Omega_n^{*(i)}$ and $\Omega_n^{**(i)}$). This proves part (i).

To prove part (ii), we note that the property of $\{\varepsilon_n^{(i)}\}$ established in part (i) means that the sequences $\{n_i(j)\}$ $(i = 1, \ldots, m)$ are all finite. On the other hand, it can be shown that if either $\varepsilon_n^{(i)} < \lim_{n\to\infty} \sup\ v_n^{(i)}$ or $-\varepsilon_n^{(i)} > \lim_{n\to\infty} \inf\ v_n^{(i)}$, then there exists a finite time instant $n(j)$ such that the intersections (7.46) becomes empty. This gives that $\varepsilon_n^{(i)}$ will increase as long as these inequalities take place. It can be understood that when $\varepsilon_n^{(i)}$ stopped, then we will be in a position of Theorem 7.2. In view of this fact, part (ii) follows. □

Thus, the controllers described above are the robust adaptive pseudoinverse model-based controllers. This fact is established in Theorems 7.2 and 7.3.

Two simulation experiments were conducted to illustrate how the adaptive pseudoinverse model-based controller copes with the SIMO memoryless system having two inputs and three outputs in the presence of bounded disturbances with known and unknown bounds on these variables. In both cases, the true matrix gain, $B$, and its initial estimate, $B_0$, were chosen as follows: $B^T = (2,\ 14,\ 8)$ and $B_0^T = (0.5,\ 1.5,\ 1)$. The desired output vector was taken as $y^0 = [2,\ 7,\ 3]^T$. The disturbance sequences $\{v_n^{(1)}\}$, $\{v_n^{(2)}\}$, and $\{v_n^{(3)}\}$ were generated as the pseudorandom

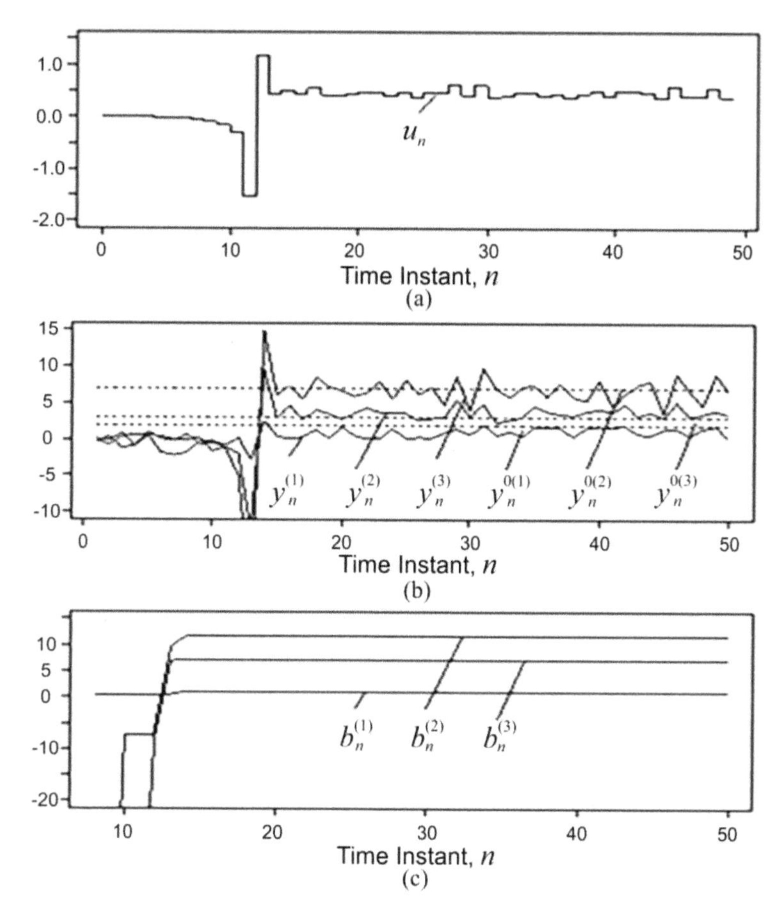


**Figure 7.2** Performance of adaptive controller in the first simulation experiment. (a) Control input. (b) Outputs. (c) Estimated system parameters. Dashed lines represent desired outputs.

i.i.d. sequences within the intervals $[-1,\ 1]$, $[-2,\ 2]$, and $[-0.8,\ 0.8]$, respectively.

The first simulation experiment was conducted assuming that the bounds on $v_n^{(i)}$ are known. Under such assumption, adaptation algorithm given by Equation (7.49) was used to adjust the controller parameters. The performance of the adaptive controller using adaptation algorithm (7.54)–(7.62) has been verified in the second simulation experiment. In this experiment, the initial values of the estimates $\varepsilon_n^{(1)}$, $\varepsilon_n^{(2)}$, and $\varepsilon_n^{(3)}$ were chosen

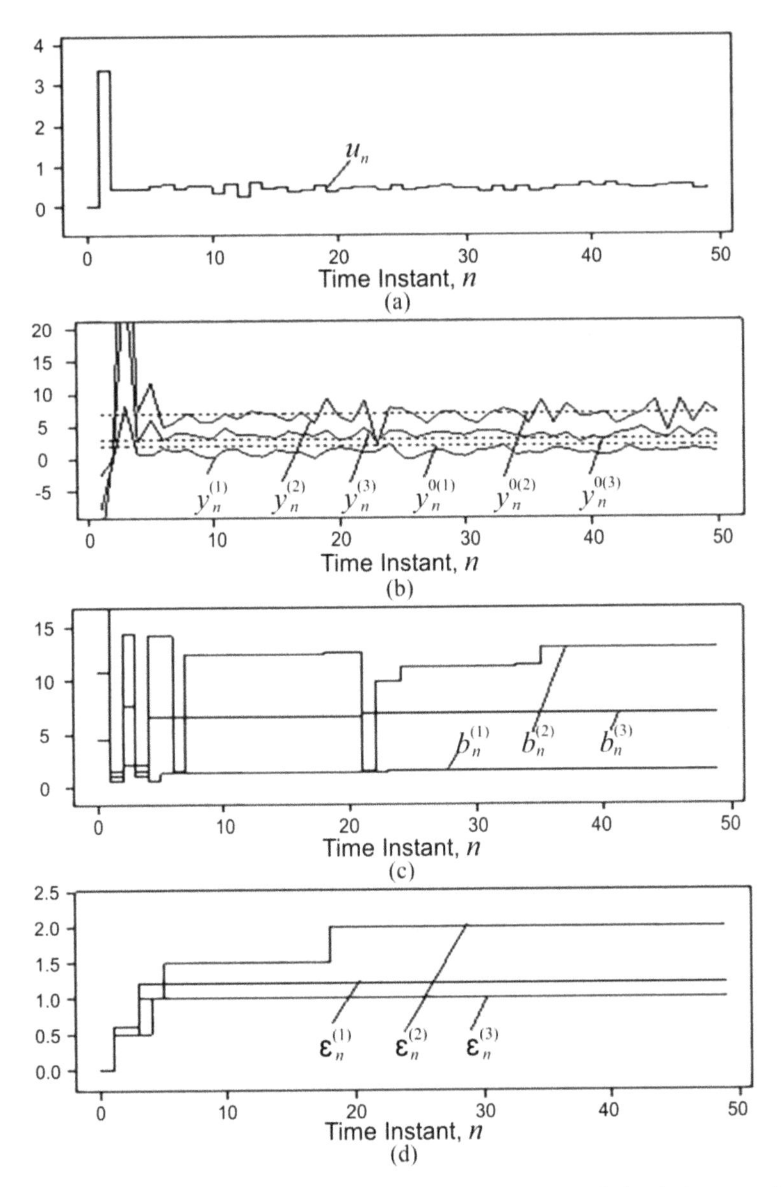


**Figure 7.3** Performance of adaptive control system in the second simulation experiment. (a) Control input. (b) Outputs. (c) Estimated system parameters. (d) Estimated disturbance bounds. Dashed lines represent desired outputs.

as $\varepsilon_0^{(1)} = \varepsilon_0^{(2)} = \varepsilon_0^{(3)} = 0$. $d_\varepsilon^{(i)}$s were taken as follows: $d_\varepsilon^{(1)} = 0.6$, $d_\varepsilon^{(2)} = d_\varepsilon^{(3)} = 0.5$.

The behavior of the closed-loop control systems containing the SIMO system (7.1) and the adaptive pseudoinverse model-based controllers, which use adaptation algorithms above, is shown in Figures 7.2 and 7.3. Figures 7.2(a) and (b) and 7.3(a) and (b) exhibit the robustness properties of these controllers. We observe that they are capable to ensure the ultimate boundedness of the outputs and the inputs. Figures 7.2(c) and 7.3(c) show the convergence of the adaptation procedures proposed above. Figure 7.3(d) demonstrates the boundedness of the estimated disturbance bounds.

## 7.5 Robust Adaptive Pseudoinverse Model-Based Control of MIMO System

Let Equation (7.1) be the equation of a nonsquare MIMO memoryless system $(2 \leq r < m)$ whose gain matrix, $B$, has full rank:

$$\text{rank } B = r. \tag{7.63}$$

Suppose that $\varepsilon^{(i)}$s are known. To control this system under the condition of the parametric uncertainty given by Equation (7.3), the adaptive controller will here be designed as the adaptive pseudoinverse model-based controller described by Equation (7.36) (as in Section 7.4) with $B_n^+$ defined by

$$B_n^+ = (B_n^T B_n)^{-1} B_n^T \tag{7.64}$$

(according to Equation (7.19)). Noting that the requirement

$$\text{rank } B_n = r \tag{7.65}$$

shall be satisfied to calculate $B_n^+$ by Equation (7.64), we will update the elements of $B_n = (b_n^{(ij)})$, exploiting the following standard adaptive estimation algorithm proposed in [36, sect. 4.2]:

$$b_n^{(i)} = \begin{cases} b_{n-1}^{(i)} & \text{if } |e_n^{(i)}| \leq \bar{\varepsilon}^{0(i)} \\ P_{\Xi^{(i)}}\{b_{n-1}^{(i)} - \gamma_n^{(i)} \frac{e_n^{(i)} - \bar{\varepsilon}^{(i)} \text{ sign} e_n^{(i)}}{||\nabla u_{n-1}||_2^2} \nabla u_{n-1}\} & \text{otherwise} \quad (i = 1, \ldots, m) \end{cases} . \tag{7.66}$$

In this algorithm, $b_n^{(i)} = [b_n^{(i1)}, ..., b_n^{(ir)}]^T$ is the $i$th row of $B_n$, $P_{\Xi^{(i)}}\{w\}$ represents the projection of a $w$ onto the $i$th set

$$\Xi^{(i)} = [\underline{b}^{(i1)}, \bar{b}^{(i1)}] \times \cdots \times [\underline{b}^{(ir)}, \bar{b}^{(ir)}] \tag{7.67}$$

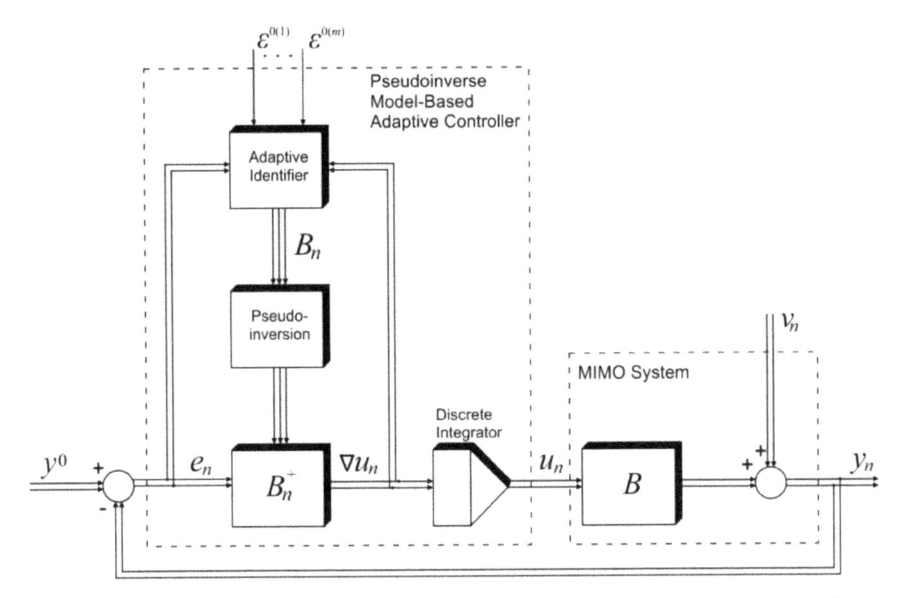


**Figure 7.4** Block diagram of the adaptive closed-loop MIMO pseudoinverse model-based control system.

and $\bar{\varepsilon}^{(i)} = 2\varepsilon^{(i)}$, $\bar{\varepsilon}^{0(i)} \geq \bar{\varepsilon}^{(i)}$. The coefficients $\gamma_n^{(i)}$s in Equation (7.66) are chosen from Equation (7.50) to satisfy Equation (7.65).

The recursive procedure (7.66) is the on-line identification algorithm needed to implement the adaptive pseudoinverse model-based controller (7.36). The adaptive closed-loop control system containing this controller is shown in Figure 7.4.

The convergence and robustness properties of the adaptive pseudoinverse model-based controller defined in Equation (7.36) together with Equation (7.66) are given below.

**Proposition 7.1.** Provided that Assumptions A1) to A3) are valid and condition (7.63) is satisfied, the adaptive control algorithm defined in Equations (7.36) and (7.66) and applied to the plant (7.1) gives the following:

(i) the estimate sequence $\{B_n\}$ converges in the sense of

$$B_n \underset{n \to \infty}{\longrightarrow} B_\infty \tag{7.68}$$

(ii) the ultimate boundedness (7.8) and (7.9) is achieved.

*Proof.* The proof is essentially based on the following important preliminary result.

**Lemma 7.3.** If A3) is valid, then there is a limit (7.68).

*Proof of Lemma 7.3.* The existence of $B_\infty$ follows from some early results found in the paper [37].

Due to space limitation, the proof of Proposition 7.1 is omitted. □

*Remark 7.3.* It does not mean that $B_\infty = B$. However, it is not necessary.

To verify how the adaptive controller proposed above performs, we give an illustrative example. In this example, the true $B$ and the initial $B_0$ were chosen as

$$B = \begin{pmatrix} 0.2 & 1.4 \\ 0.8 & 2.4 \\ 1.1 & 0.5 \end{pmatrix}, \qquad B_0 = \begin{pmatrix} 50 & 20 \\ 30 & 40 \\ 10 & 10 \end{pmatrix}$$

to ensure rank $B_0$ = rank $B = 2$. The desired output vector was taken as $y^0 = [2,\ 7,\ 3]^T$. $\{v_n^{(i)}\}$ were generated as pseudorandom i.i.d. sequences

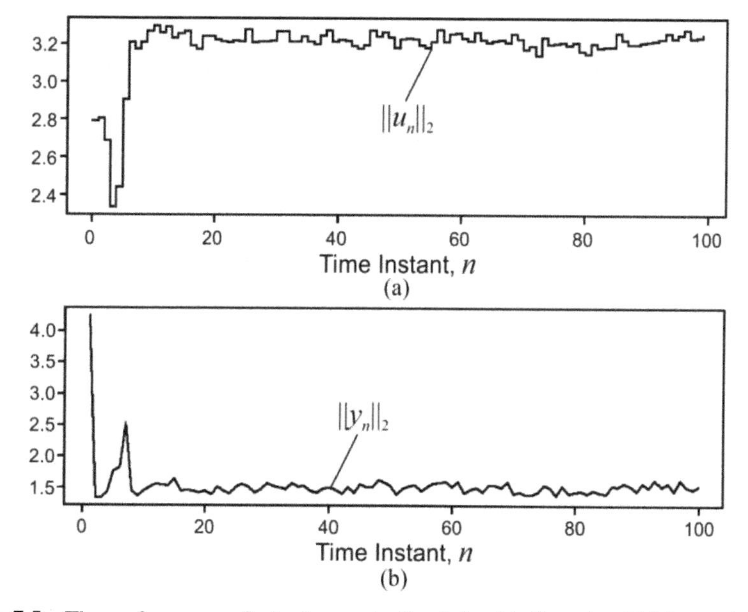


**Figure 7.5** The performance of adaptive controller defined in Equation (7.36) together with Equation (7.66) in a computer simulation. (a) Euclidean norm of control input vector. (b) Euclidean norm of output vector.

satisfying to $v_n^{(1)} \in [-0.1,\ 0.1]$, $v_n^{(2)} \in [-0.2,\ 0.2]$, and $v_n^{(3)} \in [-0.08,\ 0.08]$.

Computer simulations have been carried out to evaluate the behavior of the adaptive control system (7.1), (7.36), (7.66). This behavior is presented in Figure 7.5. It demonstrates that the closed-loop control systems containing the adaptive pseudoinverse model-based controller defined in Equation (7.36) together with Equation (7.66) are successful enough.

## 7.6 Conclusion

In this chapter, we dealt with the control of linear discrete-time SIMO and MIMO memoryless plants in the presence of parametric and nonparametric uncertainties. Their common feature is that the gain matrices of MIMO square and nonsquare systems have full ranks. We have assumed that the elements of these matrices are unknown. Nevertheless, it has been known that they belong to some interval sets. We have also supposed that the plants to be controlled are subjected to arbitrary bounded external unmeasurable disturbances whose bounds may be unknown.

To cope with the nonsquare uncertain systems, whose gain matrices have full rank, the adaptive pseudoinverse model-based approach has been proposed. This approach is directly applicable to the square MIMO systems provided that they are nonsingular (the case of singular square systems has before been studied in [30]).

The asymptotic properties of the adaptive pseudoinverse model-based controller have been derived in Sections 7.4 and 7.5. It has been established that these controllers are robust in the presence of both parametric and nonparametric uncertainties. Also, we have established that they guarantee the ultimate boundedness of all system signals.

The simulation examples have illustrated that within the pseudoinversion concept, it is possible to achieve a successful behavior of the feedback control systems containing uncertain SIMO and MIMO plants as predicted by the theory.

## References

[1] S. Skogestad and I. Postlethwaite, *Multivariable Feedback Control*, Wiley, Chichester, 1996.

[2] P. Albertos and A. Sala, *Multivariable Control Systems: an Engineering Approach*, Springer, London, 2006.

[3] T. Glad and L. Ljung, *Control Theory: Multivariable and Nonlinear Methods*, Taylor & Francis, New York, 2000.

[4] L. Tan, *A Generalized Framework of Linear Multivariable Control*, Elsevier, Oxford, 2017.

[5] V. F. Gubarev, M. D. Mishchenko, and B. M. Snizhko, 'Model predictive control for discrete MIMO linear systems', In: Kondratenko, Y.P., Chikrii, A.A., Gubarev, V.F., Kacprzyk, J. (Eds.). *Advanced Control Techniques in Complex Engineering Systems: Theory and Applications.* Dedicated to Prof. V.M. Kuntsevich. Studies in Systems, Decision and Control, pp. 63-81, vol. 203, Springer Nature Switzerland AG, Cham, 2019. DOI: https://doi.org/10.1007/978-3-030-21927-7_10

[6] G. Chen, J. Chen, and R. Middleton, 'Optimal tracking performance for SIMO systems', *IEEE Trans. Autom. Contr.*, pp. 1770–1775, no. 10, vol. 47, Oct. 2002.

[7] Sh. Hara, T. Bakhtiar, and M. Kanno, 'The best achievable $\mathcal{H}_2$ tracking performances for SIMO feedback control systems', *J. of Control Science and Engineering*, pp 1–12, no. 6, May 2007.

[8] V. M. Kuntsevich, *Control under Uncertainty: Guaranteed Results in Control and Identification Problems*, Nauk. dumka, Kiev, 2006 (in Russian).

[9] G. E. Pukhov and K. D. Zhuk, *Synthesis of Interconnected Control Systems via Inverse Operator Method*, Nauk. dumka, Kiev, 1966 (in Russian).

[10] L. M. Lyubchyk, 'Disturbance rejection in linear discrete multivariable systems: inverse model approach', In: *Proc. 18th IFAC World Congress*, pp. 7921–7926, Milano, Italy, 2011.

[11] L. S. Zhiteckii, et al., 'Discrete-time robust steady-state control of nonlinear multivariable systems: a unified approach', In: *Proc. 19th IFAC World Congress*, Cape Town, South Africa, pp. 8140–8145, 2014.

[12] B. T. Polyak and P. S. Shcherbakov, *Robust Stability and Control,* Nauka, Moscow, 2002 (in Russian).

[13] V. F. Sokolov, *Robust Control with Bounded Disturbances*, Komi Scientific Center, Ural Branch of the Russian Academy of Sciences, Syktyvkar, 2011 (in Russian).

[14] B. I. Kuznetsov, T. B. Nikitina, and I. V. Bovdui, 'Robust electromechanical servo system parametric synthesis as multi criteria game decision based on particles multi swarm optimization', In: *Proc. IEEE 5th International Conference "Actual Problems of Unmanned*

*Aerial Vehicles Development" (APUAVD-2019)*, Kyiv, Ukraine, pp. 206-209, 2019.

[15] V. M. Kuntsevich, 'Estimation of impact of bounded perturbations on nonlinear discrete systems', In: Kuntsevich, V. M., Gubarev, V. F., Kondratenko, Y. P., Lebedev, D. V., Lysenko, V. P. (Eds), *Control Systems: Theory and Applications. Series in Automation, Control and Robotics*, River Publishers, Gistrup, Delft, pp. 3-15, 2018.

[16] L. S. Zhiteckii and K. Yu. Solovchuk, 'Pseudoinversion in the problems of robust stabilizing multivariable discrete-time control systems of linear and nonlinear static objects under bounded disturbances', *J. of Automation and Information Sciences*, pp. 35–48, no. 5, vol. 49, 2017.

[17] L. S. Zhitetskii, V. I. Skurikhin, and K. Yu. Solovchuk, 'Stabilization of a nonlinear multivariable discrete-time time-invariant plant with uncertainty on a linear pseudoinverse model', *J. of Computer and Systems Sciences International*, pp. 759–773, no. 5, vol. 56, 2017.

[18] A. L. Bunich, 'On some nonstandard problems of the synthesis of discrete systems', *Autom. Remote Control*, pp. 994–1002, no. 6, 2000.

[19] V. N. Fomin, A. L. Fradkov, and V. A. Yakubovich, *Adaptive Control of Dynamic Plants*, Nauka, Moscow, 1981 (in Russian).

[20] G. C. Goodwin and K. S. Sin, *Adaptive Filtering, Prediction and Control*, Prentice-Hall, Engewood Cliffs, 1984.

[21] I. D. Landau, R. Lozano, and M. M'Saad, *Adaptive Control*, Springer, London, 1997.

[22] G. Tao, *Adaptive Control Design and Analysis*, John Wiley and Sons, New York, 2003.

[23] K. S. Narendra and A. M. Annaswamy, *Stable Adaptive Systems*, Dover Publications, New York, 2012.

[24] P. Ioannou and J. Sun, *Robust Adaptive Control*, Dover Publications, New York, 2013.

[25] K. J. Åström and B. Wittenmark, *Adaptive Control: 2nd Edition*, Dover Publications, New York, 2014.

[26] B. S. Lublinskii and A. L. Fradkov, 'Adaptive control of nonlinear statistical processes with an implicit characteristic', *Autom. Remote Control*, pp. 510–518, no. 4, 1983.

[27] G. M. Bakan, V. V. Volosov, and N. N. Salnikov, 'Adaptive control of a linear static plant by a model with unknown parameters', *Kibernetika*, pp. 63–68, no. 2, 1984.

[28] G. M. Bakan, 'Adaptive control of a multi-dimensional static process under nonstatistical uncertainty', *Autom. Remote Control*, pp. 76–88, no. 1, 1987.

[29] B. D. O. Anderson, et al., *Stability of Adaptive Systems: Passivity and Averaging Analysis*, MIT Press., Cambridge, 1986.

[30] L. S. Zhiteckii and K. Yu. Solovchuk, 'Robust adaptive pseudoinverse model-based control of an uncertain SIMO memoryless system with bounded disturbances', In: *Proc. IEEE 2nd Ukraine Conference on Electrical and Computer Engineering (UKRCON-2019),* Lviv, Ukraine, pp. 628–633, 2019.

[31] L. S. Zhiteckii and K. Yu. Solovchuk, 'Adaptive pseudoinverse model-based control of some memoryless SIMO and MIMO systems', In: *Proc. Int. conf. "Automatics–2018"*, Lviv, Ukraine, pp. 145–146, 2018.

[32] V. V. Voevodin and Yu. A. Kuznetsov, *Matrices and Calculations*, Nauka, Moscow, 1984 (in Russian).

[33] V. Ya. Katkovnik and A.A. Pervozvansky, 'Methods for the search of extremum and the synthesis problems of multivariable control systems', In: Medvedev, G. A. (Ed), *Adaptive Automatic Systems*, Sov. Radio, Moscow, pp. 17–42, 1973 (in Russian).

[34] A. Albert, *Regression and the Moore-Penrose Pseudoinverse*, Academic Press, New York, 1972.

[35] M. Marcus and H. Minc, *A Survey of Matrix Theory and Matrix Inequalities*, Aliyn and Bacon, Boston, 1964.

[36] L. S. Zhiteckii and V. I. Skurikhin, *Adaptive Control Systems with Parametric and Nonparametric Uncertainties*, Nauk. dumka, Kiev, 2010 (in Russian).

[37] S. Agmon, 'The relaxation method for linear inequalities', *Canad. J. Math.*, pp. 382–392, no 3, vol. 6, 1954.

# Part II

# Advances in Control Systems Applications

# 8

# Advanced Identification of Impulse Processes in Cognitive Maps

**Vyacheslav Gubarev[1], Victor Romanenko[2], and Yurii Miliavskyi[2,*]**

[1] Space Research Institute, National Academy of Sciences of Ukraine and State Space Agency of Ukraine, 40 Glushkov av., Kyiv, 03187, Ukraine
[2]"Institute for Applied System Analysis" of National Technical University of Ukraine "Igor Sikorsky Kyiv Polytechnic Institute", 37a Peremohy av., Kyiv, 03056, Ukraine
E-mail: yuriy.milyavsky@gmail.com

## Abstract

This chapter is devoted to the identification problem of a complex system using data obtained experimentally. Impulse processes in cognitive maps are very important and widespread class of such systems. Methods developed in this work take into account specific features of these systems. Model class that describes dynamic processes in cognitive maps is a state space discrete system of equations. So subspace identification approach is the most suitable method for solving the problem under consideration. Specific properties of cognitive maps are required to develop modified subspace method for this class of systems. Regularization procedure was introduced in standard subspace methods which allow finding an approximate solution consistent with errors in available data. Application of the developed methods is demonstrated in the example.

**Keywords:** Cognitive map, impulse processes, identification, subspace method, regularization, reduced model, regularized solution.

## 8.1 Introduction

In this chapter, new results on identification of a complex system are proposed, which are the further development of this problem earlier considered in [1]. A cognitive map (CM) is a weighted directed graph whose vertices (nodes) represent the coordinates (concepts, factors, and components) of complex systems of different nature, and the edges describe the relationships between these coordinates [2–6]. Description of interconnections between the components of the complex system, construction of a CM, and quantitative description of the coordinates' influence on each other are currently carried out by experts based on cause and effect relationships using the weight coefficients of the directed graph edges.

Under the influence of various perturbations, the coordinates of the CM vertices change in time during operation of the complex system. Each $i$th CM vertex takes on a value $x_i(k)$ at discrete time instants $k = 0, 1, 2, \ldots$, separated by a sampling period. According to [2], the change in the coordinate of the CM vertex at the moment $k$ is called impulse and is determined by the difference $P_i(k) = x_i(k) - x_i(k-1)$. The rule for changing the coordinates of the vertices in [2] is usually formulated as a first-order difference equation:

$$x_i(k+1) = x_i(k) + \sum_j^n a_{ij} P_j(k), i = 0, 1, 2, \ldots, n, \qquad (8.1)$$

where $n$ is the number of CM vertices (dimension of the model) and $a_{ij}$ is the weight coefficient of the directed graph edge which connects the $j$th vertex with the $i$th one. If there is no edge from node $l_j$ to node $l_i$, then the corresponding coefficient $a_{ij} = 0$. The impulse $P_i(k)$ arriving at one of the vertices $l_i$ will propagate along the CM chains to the remaining vertices being amplified or damped. The CM impulse process equation (8.1) is usually written in increments of variables

$$\Delta x_i(k+1) = \sum_j^n a_{ij} \Delta x_j(k), \qquad (8.2)$$

where $\Delta x_j(k) = x_j(k) - x_j(k-1), j = 0, 1, 2, \ldots, n.$

In vector-matrix form, expression (8.2) becomes

$$\Delta x(k+1) = A \Delta x(k), \qquad (8.3)$$

where $A$ is the incidence weight matrix ($n \times n$) and $\Delta x(k)$ is the vector of increments of the CM vertices coordinates $x_i$ at $i = 0, 1, 2, \ldots, n$.

The mathematical model (8.3), compiled by experts, will be adequate to the real processes of the complex system only for a short time since the coefficients $a_{ij}$ can vary widely during the system operating. In [7], to evaluate the matrix $A$ coefficients $a_{ij}$, the least squares recurrent method (LSRM) was used, which can only be correctly applied for a given CM structure, that is, for a known matrix $A$ dimension and given initial coefficients $a_{ij}$; moreover, all CM vertices should be completely measured. In [8], options for identifying matrix $A$ coefficients $a_{ij}$ were considered when all CM vertex coordinates were measured. Three identification methods that differ in areas of applicability and in the quality of the results obtained were considered and investigated. The first method was developed for a deterministic environment when all the CM vertices are measured accurately. The second method allows obtaining guaranteed intervals of estimates in the case of bounded measurement noise. However, it is applicable only at low noise levels or with a well-defined measurement matrix. The third identification method is the most common and is based on the least squares method. Theoretical and practical studies were carried out which revealed the dependence of the identification accuracy on the ratio of noise to the useful signal, on the duration of the observation interval, on the duration of testing exciting inputs, and on the number of CM nodes to which these inputs are applied. The regularization procedures proposed in these papers ensure the stability of the obtained solutions and increase the estimation accuracy in the case when additional information on zero connections between certain vertices of the CM is known.

However, it is most often in practice that not all CM vertices are measured. Moreover, even with a known structure, it is impossible to measure some coordinates of the CM vertices, and the identification methods proposed in [7, 8] cannot be applied.

Some questions of estimating state vector dimension for impulse process in a weakly structured CM with unknown model dimension were considered in [9].

Here we develop the identification methods for impulse processes in CM with regularization that allows us to consider a more general case when the system has very large and unknown dimension, with only inputs and outputs of arbitrary dimensions given.

## 8.2 Problem Statement

It is assumed that some of the CM vertices can be directly affected by applying external impulses $u_i$ (test signals or controls) to them.

Let the number of these external inputs be $r$, i.e. $\dim u = r$. Then the equation of the dynamic impulse process in the CM will have the form

$$\Delta x(k+1) = A\Delta x(k) + Bu(k), \tag{8.4}$$

where matrix $B$ $(n \times r)$ is composed of ones and zeros. The mathematical model (8.4) can be applied when the values of some coordinates of the CM vertices can be formed and controlled at discrete points in time by a decision maker via changing the available resources $u(k)$. In the general case for complex CM systems of different nature, these can be financial, energy, intellectual, information, economic, technological, administrative, defense, social, scientific, political, educational, environmental, and other resources that can be changed at each sampling instant as external controls acting on specific CM node. In this case, external controls $u_i$ must have the same physical nature with the vertices $x_i$ on which they act. For example, inputs for a complex system of socioeconomic type may be changes in financial costs or time for performing certain work, developing new types of products, varying prices for certain goods, and meeting the needs of employees in a field of activity presented in the form of the CM.

Besides controlled input, we suppose that CM is equipped with a measured system. Some of the coordinates $x_i$ may be considered as CM output. As a result, the measured subvector $y$ $(m \times 1)$ is represented in the form of the following measurement equation:

$$y(k) = C\Delta x(k), \tag{8.5}$$

where $C$ is a matrix $m \times n$.

Usually, instead of precise vector $y$, we have the output corrupted by noise. Real measurement vector $\hat{y}$ at any time $k$ can be represented as

$$\hat{y}_i = y_i(k) + \xi_i(k),$$

where $\xi_i(k)$ is a measurement noise, the nature of which is determined by the inaccuracy of the measurements. In this work, we suppose only that $\xi_i(k)$ is a random sequence but bounded in magnitude, i.e. $\xi(k)$ is a corresponding error component which meets the following inequality:

$$|\xi_i(k)| \leq \varepsilon_i, i = \overline{1,m}, k = 0, 1, 2, \ldots. \tag{8.6}$$

The identification problem for model class (8.4, 8.5) that describes dynamic CM is considered in this investigation. Initial information for this is input and output data corrupted by noise satisfying (8.6). In general case, model dimension and matrices $A$, $B$, $C$ should be found, i.e. the structure-parametric identification problem should be solved. A new method of dimension identification based on active experiment, i.e. on feeding input signal of a special type, will be suggested here.

## 8.3 CM Identification Features

A dynamic CM belongs to the class of complex systems since the number of directed graph vertices representing it can be quite large. Using a system analysis, the nature and meaning of some CM vertices can be established, and the coordinates of the others, which can also affect the system, often remain *a priori* unknown for various reasons. This leads to uncertainty in the representation of CM in the form of a directed graph, i.e. to the presence of hidden internal connections and consequently to its *a priori* unknown dimension. Then the structure of such a system is uncertain and has to be found when solving the identification problem. Even in the best case, when the CM dimension can be established, the question is how significant the contribution of all vertices to the process dynamics is. In addition, if the established dimension of the CM turns out to be large, then the presence of measurement errors and external disturbances can lead to the identification problem becoming ill-posed. Then one should find a regularized solution of this problem, in which the dimension is considered as a regularizing parameter. Any attempt to increase the dimension established by the regularization will lead to a practically unsuitable solution sensitive to the available errors. Small error variations produce significant distortion of the resulting solution. Maximum permissible dimension determined by the solution stability condition with respect to uncertainties will define the approximate solution of the structural-parametric identification problem with accuracy in agreement with the available errors. The method of its finding is the main goal of this work.

The mathematical model, which is usually used to describe processes in multidimensional multiconnected CM, is represented according to Equation (8.4) by a finite-difference system of equations. For the measured and controlled CM, the matrices $B$ and $C$ in state space (8.4) and (8.5) consist of zeros and ones. Therefore, *a priori*, only dimension $n$ and matrix $A$ remain unknown. From realization theory [10], it is well known that a model with a description in state space will not be unique. Therefore, the subspace method or 4SID [11–15] will be the most suitable for identification in this case. When the system dimension is known and the identification problem is well-posed, its solution can be found by this method using one of the known algorithms appropriate for the problem under consideration. A solution may be also found if there is sufficiently large volume of data and the conditions for asymptotic convergence of the solution to the exact one (with increasing data volume) are met. In practice, it is not always possible to fulfill these conditions for the correct solvability of the identification problem. So, it is proposed to modify the 4SID method so that it would be possible to find a regularized approximate solution, i.e. supplement the method with regularizing procedures allowing us to find maximum dimension of the model permissible by the solution stability condition. These procedures should be such that in the correct case, they get a standard solution that asymptotically tends to the exact one as the volume of data increases. Then the method of finding a regularized solution based on 4SID becomes universal with extended boundaries of its applicability. It can be used both in stochastic identification and in the case when the errors belong to sufficiently small bounded sets.

As to stochastic identification, the proposed method should be used when the corresponding multiple regression problem becomes ill-conditioned, namely, when the regressors are almost parallel [16, 17]. As a result, it may lead to great sensitivity of the identification problem solution to noise, especially with finite and not very large data volume. That is why several different methods based on 4SID were developed in order to improve the model parameters estimation. Some of them are described in [11–15].

The regularized solution that may be found by the proposed method has the maximum permissible dimension and should be consistent in accuracy with the error in the available data. This means that when the errors in the data tend to zero, the solution should asymptotically approach the one that corresponds the problem solution with accurate data and exact calculations, i.e. deterministic case. Moreover, if the exact solution is contained in the considered class of models and has a finite dimension, then the dimensions of the model and system must coincide. Otherwise, the dimension of the desired model will increase unlimitedly, and description of the system appears to be an expansion asymptotically approaching the exact one. The presence of errors in the data in most cases (at least for complex systems) leads to the fact that the dimension of the model corresponding to the regularized solution will be less than the dimension of the system generating the data.

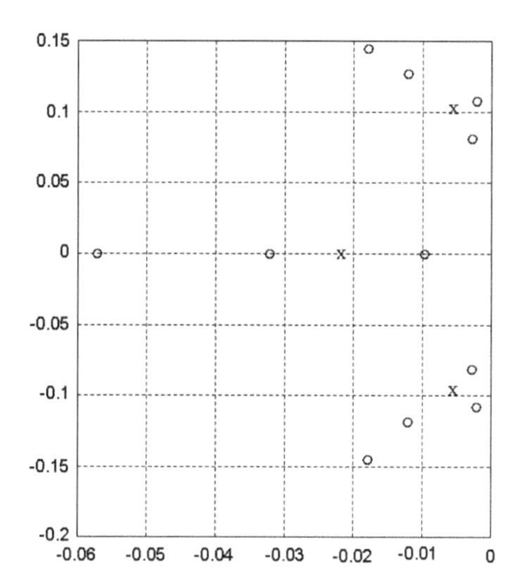


**Figure 8.1** Eigenvalues' location on complex plane for system – "o" and third-order approximate mode "x".

Therefore, there can be no talk of any unbiased parameter estimates. Figures 8.1 and 8.2 show the eigenvalues of the initial 11th-order system (circles) and approximating 3rd-order (Figure 8.1) and 4th-order (Figure 8.2) models (crosses), obtained as a result of the correct solution of identification problem. It is seen that each eigenvalue of the model corresponds to a certain cluster of eigenvalues of the system. Clustering is especially seen well in Figure 8.1. In more complex cases, clusters can intersect, i.e. be not to be well seen, but both sets of eigenvalues always belong to the same region.

## 8.4 Subspace Identification with Regularization

In this section, we consider how the well-known subspace method [11–13, 15] can be modified by integrating the regularization procedure into it, which allows one to find an error-consistent approximate solution of the identification problem. As already mentioned, the dimension of the model is taken as a regularizing parameter. With it increasing, the parametric identification problem starting from a certain dimension becomes ill-posed and its solution becomes highly sensitive to errors in the available data. The solution found for one error realization will be very different from that obtained for another error realization.

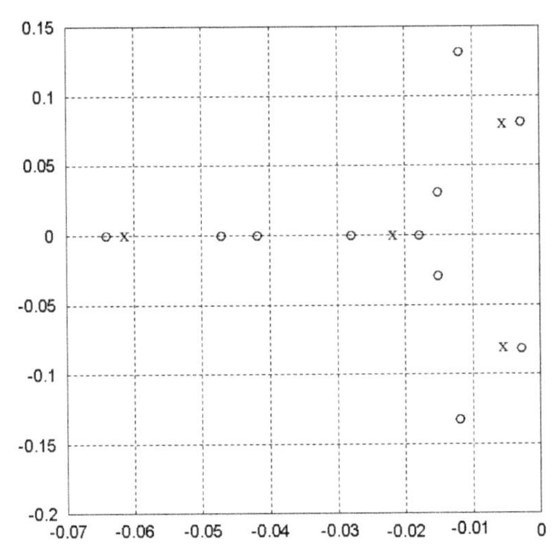


**Figure 8.2** Eigenvalues' location on complex plane for system – "o" and fourth-order approximate mode "x".

Then the regularization procedure consists in finding such a dimension of the system model that its further increase leads to instability of the obtained solution. It should be noted here that there is no sharp boundary between the stable and unstable regions. Therefore, different regularization procedures can give distinctive estimates of this boundary. Moreover, the result depends on the realized error. In a favorable case, an estimate of the maximum permissible dimension may turn out to be larger than that with an unfavorable one. This will not be critical when the data generating system is infinite-dimensional, at least in the considered class of models. A nontrivial situation may arise when in a given class of models the system has finite dimension. Then, as the simulation results show, when the difference between the dimensions of the system $n$ and the measurement vector $m$ is large enough, i.e. $n - m >> 1$, we obtain a reduced model of order less than that of the original system. However, when $n - m$ is small or equal to zero (full measurements), the regularization procedure can lead to a dimension exceeding the order of the system. It is assumed that all measurements are independent, i.e. the corresponding matrix $C$ in the measurement equation is well conditioned. In addition, dimensionality can be exceeded at a high noise level, i.e. at not very large signal-to-noise ratio. Therefore, it is advisable to consider several regularization procedures, which allow one to choose approximating model dimension more correctly, i.e. to find an appropriate regularized solution of the identification problem with their help.

As a result, the identification method with regularization is reduced to solving two problems: a) parametric identification at given dimensions; b) finding the model dimension corresponding to the regularized solution.

In this section, to solve problem a), it is proposed to use standard schemes of the 4SID method, which, for example, are given in [11, 12]. Its brief description is given in Section 8.4.1, and Section 8.4.2 is devoted to regularization procedures.

### 8.4.1 Identification for Given Model Dimension

The initial data for identification are vector sequences of input and output variables $\{\hat{y}(k)\}$ and $\{u(k)\}$, $k = \overline{1, N}$, where $N$ is the duration of the experiment, as well as the given dimension of the model. Then, in most of the existing direct 4SID methods, generalized Hankel matrices are formed from these data

$$Y = \begin{bmatrix} \hat{y}(1) & \hat{y}(2) & \dots & \hat{y}(N_1) \\ \hat{y}(2) & \hat{y}(3) & \dots & \hat{y}(N_1+1) \\ \vdots & \vdots & & \vdots \\ \hat{y}(K) & \hat{y}(K+1) & \dots & \hat{y}(K+N_1-1) \end{bmatrix},$$

$$U = \begin{bmatrix} u(1) & u(2) & \dots & u(N_1) \\ u(2) & u(3) & \dots & u(N_1+1) \\ \vdots & \vdots & & \vdots \\ u(K) & u(K+1) & \dots & u(K+N_1-1) \end{bmatrix}, \qquad (8.7)$$

where $K + N_1 = N$, $K$ must exceed the given model dimension $n$, $N_1 > K$.

Existing subspace methods are distinguished in various ways which make it possible to obtain from these matrices an estimate of the observability matrix for some realization. Using the matrices (8.7) and the Cauchy formula for the discrete case that describes the controlled and observed processes in the CM, the input–output relation may be written in the following form [11, 12, 14]:

$$Y = \Gamma_K X + \Phi_K U + \mathrm{E}, \qquad (8.8)$$

where

$$\Gamma_K = \begin{bmatrix} C \\ CA \\ \vdots \\ CA^{K-1} \end{bmatrix}, \Phi_K = \begin{bmatrix} 0 & 0 & \dots & 0 & 0 \\ CB & 0 & \cdots & 0 & 0 \\ \vdots & \vdots & & \vdots & \vdots \\ CA^{K-2}B & CA^{K-3}B & \dots & CB & 0 \end{bmatrix},$$

$$X = [\Delta x(1) \dots \Delta x(N_1)]\, \mathrm{E} = \Phi_K^n \cdot W + \mathrm{E}_e \,,$$

where $W$ is a matrix of disturbance inputs similar to $U$, and $\mathrm{E}_e$ is a matrix of measurement errors similar to $Y$. We assume that the system is stable, and at each time point $k$, the elements of the matrix $\mathrm{E}_e$ are limited in size so that for each element $e_{ij}$ of the matrix E, the inequality

$$|e_{ij}| \leq \varepsilon \tag{8.9}$$

holds. In Equation (8.9), $\varepsilon$ is a sufficiently small quantity for which the identification problem makes sense.

In the stochastic identification problem, the essential question is the consistency of the model parameters estimating, which means converging with probability almost one to exact values of these parameters while the data volume tends to infinity, i.e. possibility to obtain unbiased estimates. Therefore, the value $N_1$ should be as large as possible, ensuring the appropriateness of solution for practice. However, if the identification problems are ill-conditioned according to [16, 17], existing solution methods based on 4SID may not give the desired result. Nevertheless, when finding a regularized solution, it is also advisable, if allowed by the experimental conditions, to take a large value $N_1$.

Another important condition under which stochastic identification is consistent is the need for the input signal to be permanently exciting [11, 12]. In real situations, this is not always possible, especially in passive experiments. When finding a regularized solution, an approximate model of reduced dimension is reconstructed. This opens up new possibilities for identifying systems in a certain class of inputs. If the realized input signal allows factorization for some basis of independent functions, then as a result of reduced identification, we obtain a simplified model that gives an adequate response to any input, which is an arbitrary decomposition of informative basic signals. By "informative," we mean those terms of the expansion for which the ratio of their output signal to errors is quite large. This is especially important for identification using data obtained from passive experiments.

Then, in experiments with a large number of informative signals, it is possible to find a model that matches or is close to the one that corresponds to the regularized solution. Thus, finding a regularized solution of the identification problem allows us to construct approximate models suitable for practical use in a wide variety of systems including CM.

In all existing methods of stochastic identification, the signal subspace to which the matrix $\Gamma_K$ belongs is extracted from Equation (8.8) in various ways. For this, the second term is nullified in Equation (8.8), and when using the method of instrumental variables, the effect of the third term containing errors is also eliminated. After that, based on the singular value decomposition (SVD) of the subspace singled out in such a way, the matrix $\Gamma_K$ is estimated for some realization. Orthogonal projection into zero-space $U$ can be done by multiplying Equation (8.8) on the right by the matrix

$$\Pi_U^{\perp} = I - U^T(U \cdot U^T)^{-1}U \tag{8.10}$$

or by RQ decomposition of a composite matrix

$$\begin{bmatrix} U \\ Y \end{bmatrix} = \begin{bmatrix} R_{11} & 0 \\ R_{21} & R_{22} \end{bmatrix} \cdot \begin{bmatrix} Q_1^T \\ Q_2^T \end{bmatrix}. \tag{8.11}$$

Then the signal subspace of the matrix $\Gamma_K$ may be singled out by the SVD decomposition of the matrices $Y\Pi_U^{\perp}$ or $R_{22}$ obtained from Equations (8.8), (8.10), and (8.11), namely

$$Y\Pi_U^{\perp} = Q\Sigma V^T \text{ or } R_{22} = Q\Sigma V^T, \tag{8.12}$$

where $Q$, $V$ are the orthogonal matrices and $\Sigma$ is the matrix of singular numbers $\sigma_i$ located on the diagonal in nonincreasing order. In stochastic identification, a number of methods have been considered which are a development of the described approach based on Equation (8.8).

For a given model dimension in case (8.12), the following block decomposition of the matrices included in the SVD is performed:

$$Q\Sigma V^T = Q_1\Sigma_1 V_1^T + Q_2\Sigma_2 V_2^T, \tag{8.13}$$

where $Q = [Q_1\ Q_2]$, $\Sigma = \begin{bmatrix} \Sigma_1 & 0 \\ 0 & \Sigma_2 \end{bmatrix}$, $V = [V_1\ V_2]$, $\Sigma_1$ is a square matrix of a given dimension $n$, and blocks $Q_1$ and $V_1$ are composed of the first $n$ columns $Q$ and $V$. It is assumed that the input is informative and that for any values $n$, varied when finding a regularized solution, the matrix $\Sigma_1$ remains full-ranked. Then the first term in Equation (8.13) determines the reduced model of the system under study.

In accordance with the realization theory [10], the SVD decomposition allows one to single out the subspace which the extended observability matrix $\Gamma_K$ belongs to for some realization. It is common to take

$$\Gamma_K = Q_1. \quad (8.14)$$

From Equation (8.14) using well-known methods [11, 12, 14], it is quite easy to determine the matrices $A$ and $C$ for this realization. After finding matrices $A$ and $C$ for the same realization, one can find matrix $B$ from Equation (8.8) by the formulas that are given in [11–14]. Algorithms and standard matrices $A$, $B$, $C$ calculation software can be found, for example, in [15].

### 8.4.2 Model Dimension Determination

By means of varying dimension in expansion (8.13) and using the methods described in Section 8.4.1, one can obtain many approximate models, among which it is necessary to find the one that will be a regularized solution of the identification problem.

At its core, the task of determining the model dimension which is maximal permissible according to stability condition is close to the problem of model order reducing for a linear system of large dimension. In both cases, we look for an approximate model of a dimension smaller than that of the initial system such that its output response to any admissible input differs only slightly from the exact system [18]. Moreover, the problem of the model order reducing can be solved by the identification method. The variational method for solving the reduction problem described in [18] showed that, starting from a certain dimension $n_{\text{red}}$ of the reduced model, the norm deviation between the impulse response of the reduced model and of the system changes very slowly. This result is shown in Figure 8.3, where the norms of the difference of the impulse responses between the initial system of the 30th order with one measurement and reduced models of different dimensions are shown. It is seen that in this case, a significant improvement in the model quality, starting from the fifth order, is not observed. When identifying systems, this effect is exacerbated by the presence of errors in the measurement data.

We will use this property of reduced models when finding a regularized solution of the identification problem. In addition, important indicators of a dynamic system are the condition numbers of controllability and

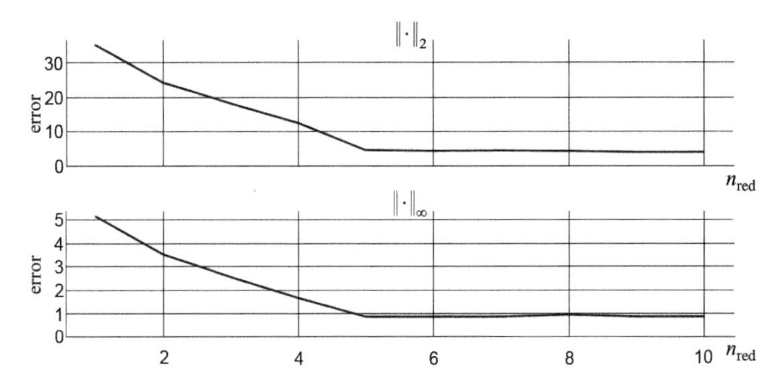


**Figure 8.3** Error dependence on reduced model order.

observability matrices, which characterize the system we are dealing with: well observable or controllable; poorly observable or controllable. If the model obtained as a result of identification is poorly observed or poorly controlled, then such a model is not appropriate for practice. State estimation and control problems to be solved by using such models will have the solution being very sensitive to noise in the data and disturbances. However, the conditionality of the observability and controllability matrices individually depends on the realization, i.e. on the chosen coordinate system of the state space. Under subspace identification, we get a model for an *a priori* unknown realization. Therefore, the model quality should be evaluated by the conditionality of the product of observability and controllability matrices, i.e. by the Hankel matrix, which is the impulse response matrix for discrete system. It can be considered as the identifiability matrix of the system, and its condition number allows to estimate the quality of the reduced model and its practical applicability because the identifiability matrix is invariant with respect to a nonsingular transformation.

In certain cases, a regularized solution can be found using an approach similar to the residual principle, which is widely used for solving ill-posed problems. For this, it is necessary to verify that among the singular numbers of the matrix $\Sigma$ in Equation (8.12), there is a value $\sigma_n$ such that $\sigma_n \geq \varepsilon$, but $\sigma_{n+1} < \varepsilon$. Then we select the partition into blocks (8.13) so that $\sigma_n$ is the last singular number of the matrix $\Sigma_1$ on its diagonal, and $\sigma_{n+1}$ is the first one of $\Sigma_2$ [19].

We denote the matrix $Q_2\Sigma_2V_2^T = \mathrm{E}$ and accept the following convention.

Suppose we have matrices $F$ and $L$ of the same dimension. We denote

$$F = |L| \Rightarrow f_{ij} = |l_{ij}|\,, \forall i \in \overline{1,m}, \quad \forall j \in \overline{1,s},$$

$$F \leq |L| \Rightarrow f_{ij} \leq |l_{ij}|\,, \forall i \in \overline{1,m}, \quad \forall j \in \overline{1,s},$$

where $|L|$ is a matrix of absolute values. So, for the selected value $n$, according to Equation (8.13), we check the feasibility of the inequality

$$|\mathrm{E}| = \left|Q_2\Sigma_2 V_2^T\right| \leq \mathrm{E}_0, \tag{8.15}$$

where $\mathrm{E}_0$ is a matrix with the same size as E. All elements of matrix $\mathrm{E}_0$ are equal to $\varepsilon$. If the inequality does not hold, then we increase the dimension $\Sigma_1$ by 1 and change the dimension of other blocks in Equation (8.13) accordingly. For a new partition, we check again the feasibility of Equation (8.15). If it is feasible, then the dimension of the desired model is equal to $n+1$. Otherwise, iterative search continues by increasing the dimension by 1 each time.

When the inequality is strictly satisfied for the chosen $n$ defined by $\sigma_n$, it is necessary to reduce the dimension by 1 and check Equation (8.15). Its violation means that the desired dimension of the model is equal to $n$. Otherwise, we continue to reduce $n$ until the sign changes in Equation (8.15), and as a result, the maximum permissible dimension is found.

For the model dimension obtained this way, the problem of parametric identification is solved by one of the methods described in Section 8.4.1.

This method is implemented when it is known that for the smallest singular number of the matrix $\Sigma$, inequality (8.15) is strictly satisfied. However, in practice, this does not happen always. In many cases, singular numbers quickly fall off first, and then, starting from some $n$, reach saturation, i.e. practically do not change. As a result, it is not possible to fulfill Equation (8.15) even for the smallest singular number and large$K$.

In given situation, another method is proposed for finding a regularized solution and the maximum permissible dimension of the model. In essence, it is similar to finding a stable solution with a quasi-optimal value of the regularization parameter [20]. It is based on the saturation effect, starting from a certain dimension of the reduced model (Figure 8.3). It is expected in this case that, starting from a certain dimension, the norm of the matrix $Q_2\Sigma_2 V_2^T$ stopped decreasing significantly. Then we introduce the following indicators. The first is the proportion or percentage of such matrix $\left|Q_2\Sigma_2 V_2^T\right|$ elements that do not exceed $\varepsilon$ in magnitude. The value $n$ starting from which this indicator ceases to decrease is taken as the dimension of the reduced model consistent in accuracy with the data error. This can be established interactively according to the chart of the indicator's dependence on the varied dimension.

As a second indicator, it is proposed to use the maximum value of the matrix $\left|Q_2\Sigma_2V_2^T\right|$ elements. According to its dependence on a varied dimension, i.e. on blocking Equation (8.13), one can also choose the dimension of the model corresponding to the regularized solution.

Finally, we will consider the condition number of the identifiability matrix as another indicator by which we can determine the reduced practically fitting CM model. The dimension, starting from which the conditioning of this matrix increases sharply, will determine the maximum permissible order of the model.

All considered regularized procedures except the last one (conditioning of identifiability matrix) determine the regularized solution. Performance of the identifiability matrix allows us to define a reduced solution of the identification problem since it does not depend on $\varepsilon$ and includes precise data.

This is the main difference between reduced and regularized solution, namely, the dimension of the reduced model order remains always smaller than that of the exact description. But for a given $\varepsilon$, their dimensions may be the same or close. Therefore, when identifying complex large-dimensional systems, such as CM, all proposed approaches are appropriate for using since, in practice, the more precise model obtained by identification may be useless for solving forecasting or control problems due to its sensitivity to errors.

## 8.5 Advanced Subspace Identification

When determining the model order as described in Section 8.4.2, we used the SVD decomposition of the $Y\Pi_U^{\perp}$ or $R_{22}$ matrices, which corresponds to some free motion extracted from the matrix $Y$. Therefore, comparing the signals corresponding to the error in the data using Equation (8.15) is not entirely adequate to the constraint (8.9). In other methods, inequality (8.15) is not used and they can also determine a regularized solution in the case when constraint (8.9) is satisfied, but the value $\varepsilon$ itself is *a priori* unknown. In order to determine the dimension from the behavior of the identifiability matrix, either knowledge of the Hankel matrix is required or it can be calculated directly from the matrices $A$, $B$, $C$ corresponding to an arbitrary but the same realization. And for this, a complete solution of the identification problem is required with variable model order $n$. This significantly complicates the procedure for finding a regularized (reduced) solution.

Therefore, in this section, we describe a method for solving the parametric identification problem, which is best suited for finding a regularized solution compared to Section 8.4.1. It can be implemented in active experiments, when it is possible to form relatively arbitrary input within the existing restrictions on the magnitude of the input.

Let the duration of the experiment $[0, N]$ be sufficiently large, allowing it to be divided into a sequence of intervals $[k_j, k_{j+1}]$ so that $k_0 = 0$ $(j = 0)$ corresponds to the first interval, and $j = K_1$ to the last one, i.e. point $k_{K_1+1}$ coincides with $N$. All even intervals $j = 0, 2, 4, \ldots$ are excitation intervals with a nonzero input, and all odd intervals are relaxation or free motion intervals. The number of points on the relaxation interval for all $j$ are chosen the same, which is equal to $N_2$. The excitation intervals can have different lengths$N_{3k}$, possibly random ones. Input sequences at excitation intervals should be formed in such a way that we have a wide variety of initial states at the moments $k_{j+1}$, $j = 0,\ 2,\ 4,\ \ldots$. If the input signals are limited in amplitude $|u_i(k)| \leq u_0$, $i \in \overline{1, r}$, and $m$ is not very small, it is enough to randomly assign the minimum and maximum values of the signal to each coordinate $u_i(k) = \pm u_0$, provided that the duration of the excitation interval varies in the required range.

At each time point, an output variable is measured, from which a temporary vector sequence $\{\hat{y}(k)\}$ is formed. In addition to the system response signal, each value $\hat{y}(k)$ contains measurement errors. From the values $\hat{y}(k)$ at the relaxation intervals, we form the following matrix:

$$Y_{relax} = \begin{bmatrix} \hat{y}^1(1) & \hat{y}^3(1) & \ldots & \hat{y}^{K_1}(1) \\ \hat{y}^1(2) & \hat{y}^3(2) & \ldots & \hat{y}^{K_1}(2) \\ \vdots & \vdots & \ddots & \vdots \\ \hat{y}^1(N_2) & \hat{y}^3(N_2) & \ldots & \hat{y}^{K_1}(N_2) \end{bmatrix}, \tag{8.16}$$

where $\hat{y}^s(k)$ is the measurement vector at the point $k$ for the $s$th relaxation interval $(s = 1, 3, \ldots)$; we assume that $K_1$ is even, corresponding to the last relaxation interval on the interval $[0, N]$.

According to the Cauchy formula for the discrete process, matrix (8.16) corresponds to the matrix equation

$$Y_{\text{relax}} = \Gamma_{N_2} \cdot X_0, \tag{8.17}$$

where $\Gamma_{N_2}$ is an expanded observability matrix,

$$X_0 = \begin{bmatrix} \Delta x_1^1(1) & \Delta x_1^3(1) & \ldots & \Delta x_1^{K_1}(1) \\ \Delta x_2^1(1) & \Delta x_2^3(1) & \ldots & \Delta x_2^{K_1}(1) \\ \vdots & \vdots & \ddots & \vdots \\ \Delta x_n^1(1) & \Delta x_n^3(1) & \ldots & \Delta x_n^{K_1}(1) \end{bmatrix}$$

is a matrix of initial states on relaxation intervals, and $n$ is the dimension of the state vector.

We assume that the dimension $n$ is given. Then using the methods described in Section 8.4.1, the problem of parametric estimation of matrix $A$, $B$, $C$ can be solved. In the case under consideration, we will also use the singular decomposition of the matrix $Y_{\text{relax}}$ as well as representation of the corresponding SVD matrices into blocks

$$Y_{\text{relax}} = Q_1 \Sigma_1 V_1^T + Q_2 \Sigma_2 V_2^T, \tag{8.18}$$

where the first term is an approximation of a noisy matrix $Y_{\text{relax}}$ by a matrix of incomplete rank $n$. It will determine the desired approximate model of the system.

In contrast to Section 8.4.1, we assume

$$\Gamma'_{N_2} = Q_1 \Sigma_1, X'_0 = V_1^T. \tag{8.19}$$

Representation (8.19) allows us to find matrices $A'$, $C'$, and $X'_0$ for some realization. The matrices $A'$ and $C'$ are found from the first equation in (8.19). The matrix $C'$ will be the first $m$ rows $Q_1\Sigma_1$, and the matrix $A'$ is found from the well-known [11, 12, 15, 21] overdetermined system of equations

$$\Gamma'_{N_2,1:N_2-m} A' \approx \Gamma_{N_1,m+1:N_2}, \tag{8.20}$$

where $\Gamma'_{N_2,1:N_2-m}$ is the matrix obtained from $\Gamma_{N_2}$ crossing out the last $m$ rows, and $\Gamma_{N_1,m+1:N_2}$ crossing out the first $m$ rows. System (8.20) is overdetermined and its solution is found using the ordinary or generalized least squares (LS) method.

To find the matrix $B'$, we write a vector-matrix equation similar to Equation (8.8) for the $k$th excitation interval $k = 2, 4, \ldots, K_1 - 1$

$$\hat{y}^k = \Gamma_{N_{3k}} A^{N_2-1} \Delta x^{k-1} + \Phi_{N_{3k}} u^k, k = 2,\ 4,\ \ldots,\ K_1 - 1, 0 \tag{8.21}$$

where $N_{3k}$ is the number of points on the $k$th excitation interval,

$$\hat{y}^k = \begin{pmatrix} \hat{y}^k(1) \\ \hat{y}^k(2) \\ \vdots \\ \hat{y}^k(N_{3k}) \end{pmatrix}, \Delta x^{k-1} = \begin{pmatrix} \Delta x_1^{k-1}(1) \\ \Delta x_2^{k-1}(1) \\ \vdots \\ \Delta x_n^{k-1}(1) \end{pmatrix}, u^k = \begin{pmatrix} u^k(1) \\ u^k(2) \\ \vdots \\ u^k(N_{3k}) \end{pmatrix},$$

$\hat{y}^k(j)$ and $u^k(j)$ are the output and input at the $j$th point of the $k$th excitation interval and $\Delta x^{k-1}$ is the $k-1$th column of the matrix $X_0$,

$$\Gamma_{N_{3k}} = \begin{bmatrix} C \\ CA \\ \\ CA^{N_{3k}-1} \end{bmatrix}, \Phi_{N_{3k}} = \begin{bmatrix} 0 & 0 & \dots & 0 & 0 \\ CB & 0 & \dots & 0 & 0 \\ \vdots & \vdots & & \vdots & \vdots \\ C^{N_{3k}-2}B & C^{N_{3k}-3} & \dots & CB & 0 \end{bmatrix}.$$

Let $\hat{y}^k - \Gamma_{N_{3k}-1}^{N_2-1}\Delta x^{k-1} = \bar{y}^k$, and since the matrix $X_0$ is determined by Equation (8.19), $\bar{y}^k$ is known. Then the matrix $B$ can be found from the overdetermined system of matrix equations

$$\Phi_{N_{3k}}u^k = \bar{y}^k, k = 2,\ 4,\ \dots,\ K_1 - 1, \tag{8.22}$$

using this ordinary or generalized LS method.

The attractiveness of this method is that it allows us to simply apply all the regularization procedures described in Section 8.4.2. From Equations (8.17) and (8.18), it follows that Equation (8.9) is fully consistent with these matrices and Equation (8.15) establishes a matrix of incomplete rank consistent in accuracy with Equation (8.9). In the case of singular numbers saturation, an estimate of the permissible dimension by two indicators characterizing the elements of the matrix $\left|Q_2\Sigma_2V_2^T\right|$ is similar to that described in Section 8.4.2.

It follows from Equation (8.17) that the product of the observability matrix and the matrix $X_0$ is invariant with respect to the basis in the state space. In accordance with the second relation in Equation (8.19), the matrix $X_0$ for any $n$ is orthogonal and its condition number is always equal to unity, i.e. corresponds to the same realization. Therefore, the properties of the invariant matrix $Y_{\text{relax}}$ are completely determined by the observability matrix, and the model quality can be analyzed based on dependence of its condition number on the order of the model. Indeed, as the modeling results show, starting with a certain $n$, condition number starts increasing rapidly, which indicates a poor conditionality of the observability matrix for dimensions larger than this value $n$.

Efficiency of the proposed method depends on the initial state formed at the end of each excitation interval. Their informative diversity provides maximum model dimension permissible by stability condition. Different iterative schemes may be considered which allow a consecutive approach to such signals.

## 8.6 Example

Consider the CM of an IT company (Figure 8.4)

This CM has both measurable and unmeasurable nodes. State vector of this CM is the following:

$x_1$ – project completion period, $x_2$ – expenditures for innovations, $x_3$ – salaries and bonuses, $x_4$ – project's budget, $x_5$ – profit, $x_6$ – expenditures for management staff, $x_7$ – marketing expenditures, $x_8$ – revenue from project sales, $x_9$ – expenditures for staff reassessment, $x_{10}$ – expenditures for skills upgrading, $x_{11}$ – technical control, $x_{12}$ – intelligent assets, $x_{13}$ – project's quality, $x_{14}$ – compatibility, $x_{15}$ – job satisfaction, and $x_{16}$ – experience exchange. The transposed incidence matrix of the CM has the form

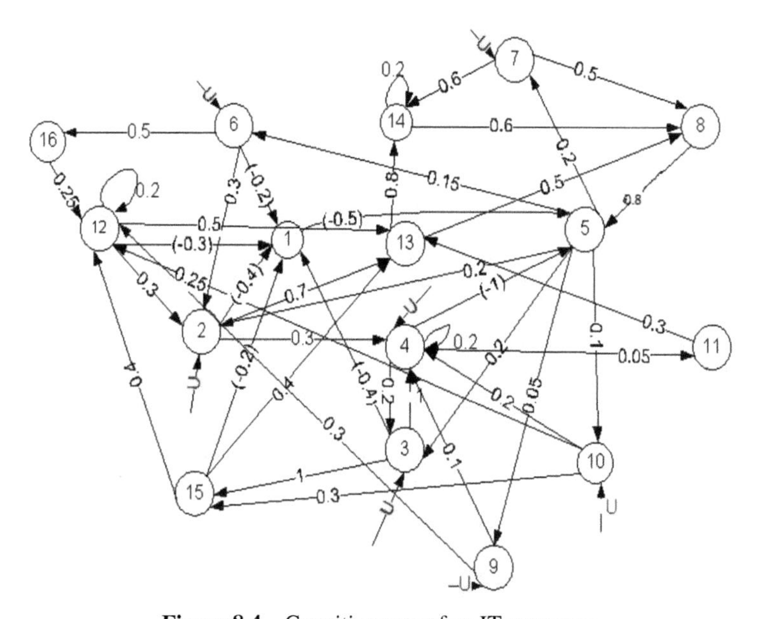

Figure 8.4 Cognitive map of an IT company.

$$
A = \begin{pmatrix}
0 & -0.4 & -0.4 & 0 & 0 & -0.2 & 0 & 0 & 0 & 0 & 0 & -0.3 & 0 & 0 & -0.2 & 0 \\
0 & 0 & 0 & 0 & 0.2 & 0.3 & 0 & 0 & 0 & 0 & 0 & 0.3 & 0 & 0 & 0 & 0 \\
0 & 0 & 0 & 0.2 & 0.2 & 0 & 0 & 0 & 0 & 0 & 0 & 0 & 0 & 0 & 0 & 0 \\
0 & 0.3 & 1 & 0.2 & 0 & 0 & 0 & 0 & 0.1 & 0.2 & 0 & 0 & 0 & 0 & 0 & 0 \\
-0.5 & 0 & 0 & -1 & 0 & 0 & 0 & 0.8 & 0 & 0 & 0 & 0 & 0 & 0 & 0 & 0 \\
0 & 0 & 0 & 0 & 0.15 & 0 & 0 & 0 & 0 & 0 & 0 & 0 & 0 & 0 & 0 & 0 \\
0 & 0 & 0 & 0 & 0.2 & 0 & 0 & 0 & 0 & 0 & 0 & 0 & 0 & 0 & 0 & 0 \\
0 & 0 & 0 & 0 & 0 & 0 & 0.5 & 0 & 0 & 0 & 0 & 0 & 0.5 & 0.6 & 0 & 0 \\
0 & 0 & 0 & 0 & 0.05 & 0 & 0 & 0 & 0 & 0 & 0 & 0 & 0 & 0 & 0 & 0 \\
0 & 0 & 0 & 0 & 0.1 & 0 & 0 & 0 & 0 & 0 & 0 & 0 & 0 & 0 & 0 & 0 \\
0 & 0 & 0 & 0.05 & 0 & 0 & 0 & 0 & 0 & 0 & 0 & 0 & 0 & 0 & 0 & 0 \\
0 & 0 & 0 & 0 & 0 & 0 & 0 & 0 & 0.3 & 0.25 & 0 & 0.2 & 0 & 0 & 0.4 & 0.25 \\
0 & 0.7 & 0 & 0 & 0 & 0 & 0 & 0 & 0 & 0 & 0.3 & 0.5 & 0 & 0 & 0.4 & 0 \\
0 & 0 & 0 & 0 & 0 & 0 & 0.6 & 0 & 0 & 0 & 0 & 0 & 0.8 & 0.2 & 0 & 0 \\
0 & 0 & 1 & 0 & 0 & 0 & 0 & 0 & 0 & 0.3 & 0 & 0 & 0 & 0 & 0 & 0 \\
0 & 0 & 0 & 0 & 0 & 0.5 & 0 & 0 & 0 & 0 & 0 & 0 & 0 & 0 & 0 & 0
\end{pmatrix}.
$$

For this CM, the identification method described in Section 8.5 was used to determine the approximate model, the dimension of which was established using the regularization procedures considered in Section 8.4.2. The number of measured vertices varied from 1 to 10. The exact output signal was noisy with a random sequence satisfying the constraint (8.9) with $\varepsilon = 0,01$. The main goal of the experimental studies was to evaluate the effectiveness of the advanced method described in Section 8.5. Therefore, only relaxation intervals with different initial conditions, which were randomly generated, were considered. In the experiments performed, the selected conditions and chosen system parameters did not allow inequality (8.15) to hold for all admissible partitions (8.13). Therefore, the assessment of the maximum permissible dimension of the model was carried out only by three indicators, namely, by the condition number of observability matrix, the percentage of matrix elements $\left|Q_2\Sigma_2V_2^T\right|$ not exceeding $\varepsilon$, and the dependence of the maximum element of this matrix on the dimension. The dependence of these indicators on the number of measured vertices $m$ and the model dimension $n$ are shown in Figures 8.5–8.7. Figure 8.5 shows the dependence on $n$ when one vertex of CM is measured, i.e. for $m = 1$. Similar dependences on $n$ are shown for $m = 4$ in Figure 8.6 and for $m = 10$ in Figure 8.7.

The results of interactive estimation of the model order based on the condition number of the observability matrix (reduced solution) are shown in the second row of Table 8.1, and the first row gives estimates of the model order based on the other two indicators.

The eigenvalues of the investigated CM (circles) and the eigenvalues of the identified model (crosses) placed on the complex plane are shown for $m = 1$ in Figure 8.8, for $m = 4$ in Figure 8.9, and for $m = 10$ in Figure 8.10.

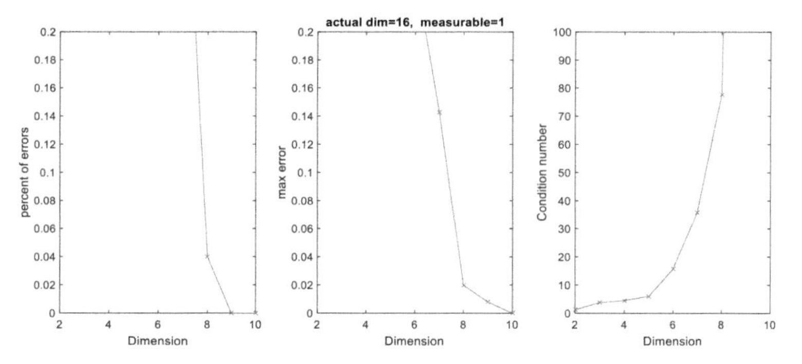


**Figure 8.5** Indicators dependence on model dimension for $m = 1$.

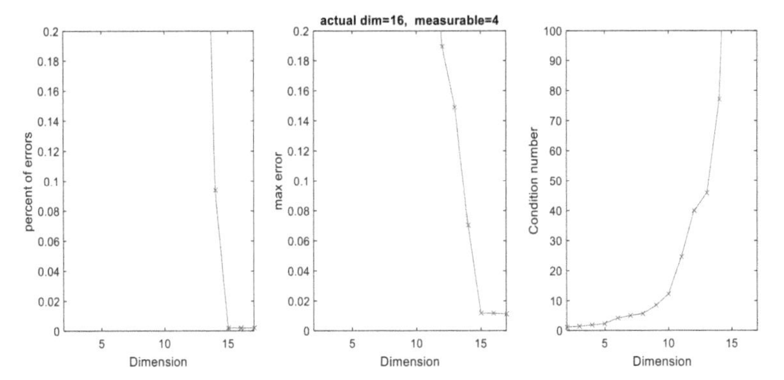


**Figure 8.6** Indicators dependence on model dimension for $m = 4$.

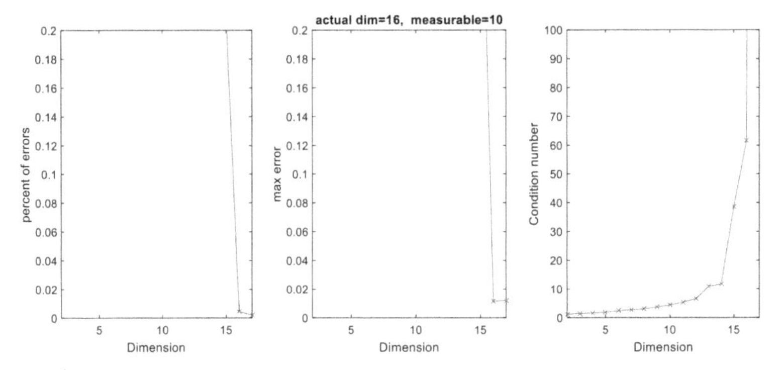


**Figure 8.7** Indicators dependence on model dimension for $m = 10$.

All of them, the CM and the models, are stable systems, i.e. the roots belong to the unit circle.

The presented results show that for a system with good controllability, characterized by condition number, it is possible to provide informative variety of initial states at the end of the excitation intervals, which allows improving the quality of estimation. The CM considered in the experiments

**Table 8.1** Model maximum possible dimension for different measurements.

| **Number of measurements** | **1** | **2** | **3** | **4** | **5** | **6** | **7** | **10** |
|---|---|---|---|---|---|---|---|---|
| **Regularized dimension** | 8 | 12 | 13 | 15 | 15 | 15 | 16 | 16 |
| **Reduced dimension** | 7 | 11 | 13 | 14 | 14 | 14 | 15 | 16 |

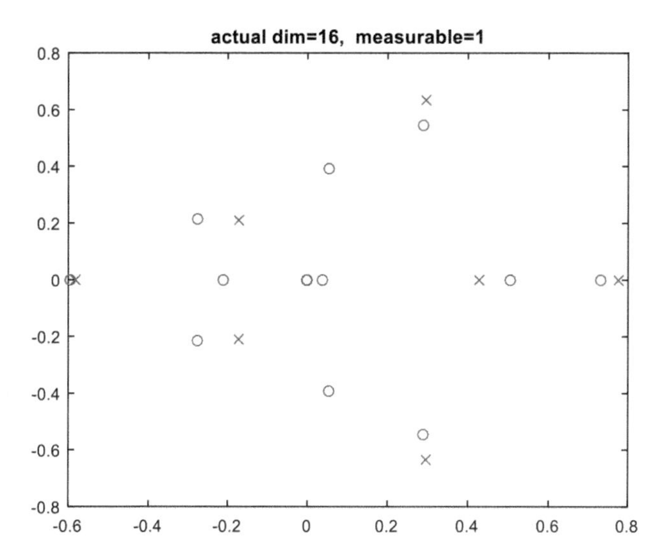


**Figure 8.8** Eigenvalues of system – "o" and model – "x" corresponding regularized solution for $m = 1$.

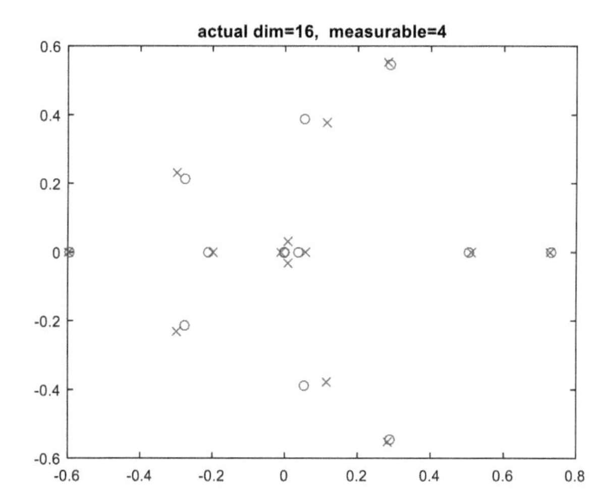


**Figure 8.9** Eigenvalues of system – "o" and model – "x" corresponding regularized solution for $m = 4$.

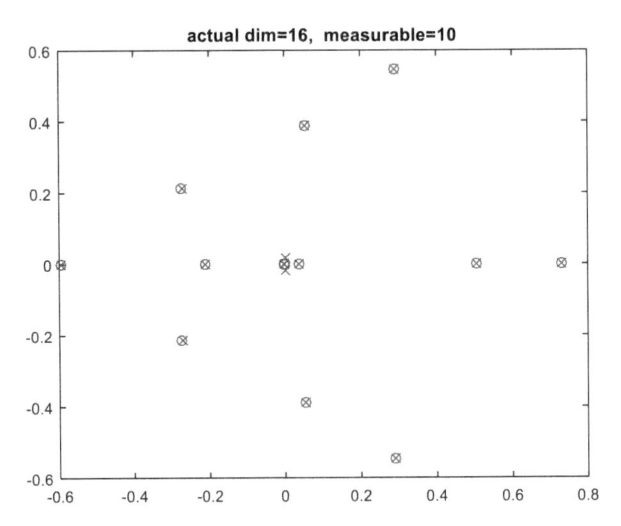


**Figure 8.10** Eigenvalues of system – "o" and model – "x" corresponding regularized solution for $m = 10$.

has an unfavorable feature for identification. Six eigenvalues are located in the neighborhood of zero, the signal of which rapidly decays in the relaxation interval which makes their identification difficult if errors exist. Nevertheless, with good observability for $m = 10$, it was possible to restore them all as well as all other eigenvalues, although approximate, but rather close to the exact model of the same dimension. In cases where the model of lower dimension was obtained, its eigenvalues represent clusters of the original system roots, as one can see from Figures 8.8 and 8.9. As a result, the responses of the approximate model and the system to any admissible input are close.

## 8.7 Conclusion

Advanced identification method and active experiments with controlled input open up the possibility to find high-quality model and to get data favorable for subspace identification. For example, in order to excite fast modes and not admit exciting slow ones, it is necessary to take a short length of the excitation interval. With increasing excitation interval, the signal of fast modes becomes small and the signal of slow ones grows. This allows a controlled way to generate an informative set of initial data for relaxation intervals and thus affect the quality of estimating the dimension and other parameters of the approximate model, which are found from the identification problem.

In passive experiments, standard subspace methods remain preferred. At the same time, identification with regularization allows one to construct approximate models oriented to a certain subset of input actions, i.e. to provide a model output close to real for a certain class of external influences.

## References

[1] V.M. Kuntsevich, V.F. Gubarev, Y.P. Kondratenko, D.V. Lebedev, V.P. Lysenko (Eds), *Control Systems: Theory and Applications.* Book Series in Automation, Control and Robotics, River Publishers, Gistrup, Delft, 2018, 329 p.

[2] F. Roberts, 'Discrete Mathematical Models with Application to Social, Biological, and Environmental Problems', Prentice Hall, Englewood Cliffs, 1976.

[3] E. Papageorgiou, J. Salmeron, 'A Review of Fuzzy Cognitive Maps ResearchDuring the Last Decade', IEEE transactions on fuzzy systems, vol. 21, no. 1, pp. 66-79, 2013.

[4] Y. Wang, W. Zhang, 'A Brief Survey on Fuzzy Cognitive Maps Research'. In: Huang DS., Han K. (eds), Advanced Intelligent Computing Theories and Applications, ICIC 2015, Lecture Notes in Computer Science, vol 9227, Springer, Cham, 2015.

[5] G. Gorelova, N. Pankratova, 'Innovational development of socio-economic systems based on methodology of foresight and cognitive modeling', Kyiv, Naukova dumka, 2015 (in Russian).

[6] G. Gorelova, E. Zakharova, S. Randchenko, 'Research of weekly-structured problems in socio-ecomonic systems', Rostov-na-Donu, RGU, 2006 (in Russian).

[7] V. Romanenko, Y. Miliavskyi, A. Reutov, 'Adaptive Control Method for Unstable Impulse Processes in Cognitive Maps Based on Reference Model', Journal of Automation and Information Sciences, vol. 47 (3), pp. 11-23, 2015.

[8] V. Gubarev, V. Romanenko, Y. Miliavskyi, 'Identification and Control Automation of Cognitive Maps in Impulse Process Mode'. In: Kuntsevich, V.M., Gubarev, V.F., Kondratenko, Y.P., Lebedev, D.V., Lysenko, V.P. (Eds). Control Systems: Theory and Applications. Series in Automation, Control and Robotics, River Publishers, Gistrup, Delft, pp. 43-64, 2018.

[9] V. Gubarev,, V. Romanenko, Y. Miliavskyi, 'Control and Identification in Cognitive Maps with Suppressing Constrained External and

Internal Disturbances in Impulse Processes', $10^{th}$ IEEE International Conference on Intelligent Data Acquisition and Advanced Computing Systems: Technology and Application (IDAACS), Metz, France, vol. 1, pp. 90-95, 2019.

[10] T. Kailath, 'Linear systems', Prentice Hall, Englewood Cliffs, N.J., 1980.

[11] M. Verhaegen, P. Dewilde, 'Subspace Model Identification. Part 1: The output-error state space model identification class of algorithms', International Journal of Control, Vol. 56, No 5, pp. 1187-1210, 1992.

[12] M. Verhaegen, P. Dewilde, 'Subspace model identification. Part 2. Analysis of the elementary output-error state space model identification algorithm', International Journal of Control, vol. 56, No 5, pp. 1211-1241, 1992.

[13] P. Van Overschee, B. De Moor, 'N4SID: Subspace algorithms for identification of combined deterministic-stochastic systems', Automatica, Vol. 30, No 1, pp. 75-93, 1994.

[14] M. Verhaegen, 'Identification of the Deterministic Part of MIMO State Space Models given in Innovations Form from Input-Output Data', Automatica, Vol. 30, No 1, pp. 61-74, 1994.

[15] P. Van Overschee, B.De Moor, 'Subspace Identification for Linear Systems', Kluwer Academic Publications, 1996.

[16] A. Chiuso, G. Picci, 'On the ill-conditioning of subspace identification with inputs', Automatica, vol. 40 (4), pp. 575-589, 2004.

[17] A. Chiuso, G. Picci, 'Numerical conditioning and asymptotic variance of susbspace estimates', Automatica, vol. 40 (4), pp. 677-683, 2004.

[18] V. F. Gubarev and V. V. Fatenko, 'Geometric and variational model order reduction methods. Comparative Analysis', J. Autom. Inform. Sci., Vol. 50 (1), pp. 39–53, 2018.

[19] G. Golub, C. Van Loan, 'Matrix Computation', Baltimore: John Hopkins University Press, 1989.

[20] A. Tikhonov, V. Arsenin, 'Method of Solving ill-post problems', Moscow, Nauka, 1979 (in Russian).

[21] M. Viberg, 'Subspace-based Method for the Identification of linear time-invariant systems', Automatica, vol. 31, No 12, pp. 1835-1851, 1996.

# 9

# Strategy for Simulation Complex Hierarchical Systems Based on the Methodologies of Foresight and Cognitive Modeling

**N. D.Pankratova[1], G.V.Gorelova[2] and V.A. Pankratov[1]**

[1]Institute for Applied System Analysis, Igor Sikorsky Kyiv Polytechnic Institute, Ukraine
[2]Engineering and Technology Academy of the Southern Federal University, Russia
Corresponding author N.D. Pankratova <natalidmp@gmail.com>

## Abstract

In this chapter, the strategy of complex hierarchical systems simulation based on the mathematical support of foresight and cognitive modeling methodologies is considered. It was proposed to use these methodologies together: at the first stage, apply the foresight methodology with the aim of creating alternatives of scenarios and use the obtained results at the second stage as initial data for cognitive modeling for constructing scenarios of the desired future and ways of their implementation. Using the foresight process at the first stage of modeling allows with the help of expert assessment procedures to identify critical technologies and create alternatives of scenarios with quantitative characteristics. The methods Strengths, Weaknesses, Opportunities, Threats (SWOT) analysis, Technique for Order of Preference by Similarity to Ideal Solution (TOPSIS), ViseKriterijumska Optimizacija i Kompromisno Resenje (VIKOR), Delphi, analytic hierarchy, and morphological analysis are used. The methodology of cognitive modeling of the studied systems is based on the meta-model of the study; it is a set of cognitive models and methods: specifying vertices and relationships between them, analyzing the stability of paths and cycles, numerical stability, impulse exposure, connectivity, complexity, sensitivity of the system, scenario analysis to anticipate possible development situations in the system, and decision-making. Scenarios of the education system's future are modeled taking into account the coronavirus pandemic and the unstable economy of the modern world.

**Keywords**. Foresight, qualitative analysis methods, cognitive modeling, pulse modeling, scenario analysis, education system future.

## 9.1 Introduction

As a strategy for the innovative development of complex hierarchical systems (CHSs) at the level of a megalopolis, large enterprise, region in many countries of the world with a high status, the methodology of foresight has proved itself, which allows us to answer the question "what will happen if ...?" and build alternatives to evidence-based scenarios [1]. Given the current trends of the production factor transformation, every large company, industry, or country in the world can not only but must also develop a foresight methodology as a fundamental tool for developing its own policies and strategies in the face of significant changes, new challenges, and big risks that the future carries for mankind [2–9].

The development of the innovation strategy of the CHS belongs to the class of poorly structured tasks in which the goals, structure, and conditions are only partially known and are characterized by a large volume of non-factors: inaccuracy, incompleteness, uncertainty, and vagueness of data describing the object. In contrast to decision-making problems with quantitative values of variables and relations between them, which are solved by methods and means of the operations research theory, econometrics, and other similar methods, specific methods of decision-making support are needed to solve poorly structured problems. The multifactorial, multiparametric, heterogeneous, and poorly structured information of the subject area of the object of study used at different stages of the foresight process leads to difficulties associated with the format for presenting knowledge, constructing questionnaires, processing results, and coordinated management of the foresight process as a whole. Unformalized, heterogeneous, and poorly structured data from the subject area require a single structural description language and a single presentation format [10].

The urgency of research direction is to identify new trends, strategic scientific directions, technological achievements, etc., which in the short or long term will have a significant impact on economic, social, and technical development in the future. To solve those tasks, it is expedient to create and intensively develop modern concepts, models, approaches to combining the methodologies of foresight, and cognitive modeling in order to determine the economic perspectives of scientific and technical systems and ways to implement them. In this paper, it is proposed to use these methodologies together: at the first stage, apply the methodology of foresight and use the results as input data in the second stage – for cognitive modeling of scenarios creation [11]. Such synthesis of methodologies allows to propose a scientifically grounded strategy for implementing of scenario for priority alternative of the various CHS.

## 9.2 Theoretical Foundation of Foresight and Cognitive Modeling Methodologies

### 9.2.1 Foresight Methodology of Complex System

The scenario creation for priority alternative is possible only with the use of a system approach taking into account the totality of the properties and characteristics of the studied systems, as well as the features of the methods and procedures used to create them. Based on a comparison of the characteristics of the qualitative analysis methods, the requirements for their application, the disadvantages, and advantages of each of them, researchers of foresight problems should choose the rational combination of methods and establish the correct sequence for their use, taking into account the totality of requirements for systems and the features of the tasks to be solved.

The methodological and mathematical support of a systematic approach to solving the problems of developing CHS in the form of a two-stage model based on a combination of foresight and cognitive modeling methodologies is developed and its scheme is presented in Figure 9.1. The involvement of scanning methods, sociological, technological, economical, environmental and political (STEEP) analysis, brainstorming, and SWOT analysis at the first level of the stage allows using expert assessment to identify critical technologies in economic, social, environmental, technical, technological, informational, and other directions. The basis of this level is the analysis of subsystems, which are connected by direct and feedback links to the monitoring system and field tests. The quantitative data obtained after analysis and processing are the initial ones for solving of foresight tasks. At the second level, using the qualitative methods (morphological analysis, Delphi, hierarchy analysis (MAI) and its modification, cross-analysis, etc.), the tasks of assessing the behavior of CHS and preparing for decision-making in the form of alternatives to scenarios are solved [1, 3–9, 12].

The selection of critical technologies and the construction of rational alternatives of scenarios for the development of strategically important enterprises, priority industries, and the industrial sector of the state are expedient to be performed on the basis of a collection of foresight activities. For this goal, in the process of creating alternatives of scenarios for solving foresight problems, it becomes necessary to involve expert assessment methods, among which are the most commonly used methods of SWOT analysis, analytic hierarchy, Delphi methods, and morphological analysis [1, 2–9, 12]. For the construction of scenarios that correspond to selected alternatives, cognitive modeling is involved [11], which makes it possible to obtain a valid scenario for decision-making based on the proposed mathematical apparatus with practical accuracy.

In this paper, to identify critical technologies, the SWOT analysis method is used. For the purpose of ranking, the obtained critical technologies and identifying the most topical ones, the Technique for Order Preference by Similarity to an Ideal Solution (TOPSIS) method is applied [5]. The method TOPSIS of multicriterial analysis (ranking) of alternatives in addition to estimating the distance from the considered alternative to the ideal solution allows to take into account the distance to the worst solution. The tradeoff in choosing the best alternative is based on the fact that the chosen solution must be at the same time as close to the ideal as possible and most remote from the worst solution. The obtained rating makes it possible to take into account the weight characteristics of critical technologies that are the vertices of the cognitive map when constructing a cognitive model. According to the VIKOR method, a compromise solution to the problem should be an alternative, which is closest to the ideal solution. Moreover, to assess the degree of the alternative proximity to the ideal solution, a multicriteria measure is used [2]. As soon as the critical technologies are identified, we cross to the system approach's second level, using the qualitative methods for creating alternatives of CHS scenarios [1, 4, 6, 7, 9, 11].

The proposed strategy for the development of CHS based on the synthesis of methodologies foresight and cognitive modeling allows to build a science-based strategy to implement priority alternative scenario of complex systems for different nature and offers a unique opportunity within a single

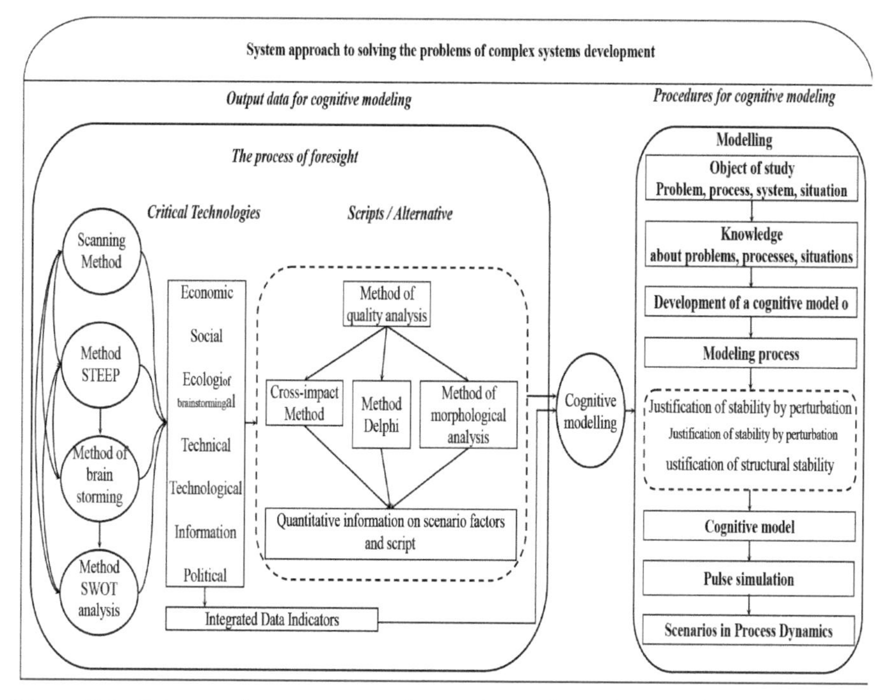


**Figure 9.1** Scheme of system approach to solving the problems of complex systems development.

software and analytical complex to solve problems of strategic planning and operational response.

Based on the expert procedure analysis of innovation activities, the following principles can be used. Instead of the potential realization principle, the *possible realization principle* is suggested [1]. For certain innovation objects (scientific ideas or technical solutions, projects of industrial products, or manufacturing technologies), the initial expert estimation results cannot guarantee their practical feasibility or prove the impossibility of being realized. This principle postulates that for the listed innovation objects, based on the expert estimation results of the presented information, a reliable estimation cannot be obtained *a priori* that would allow one, for the object under investigation being grounded and valid, to exclude the possibility of being unrealizable. The estimation for an innovation project retains the uncertainty of the conclusion about the possibility of realization until, theoretically or experimentally, the possibility of technical or technological realization is proved for the product.

Instead of the truth invariance principle, which postulates the invariability of the theoretical or technical statement, judgment, conclusion, or opinion about the object for a comparatively long time, a completely different principle is needed. Such a necessity is brought about by the previous principle and the innovation activity practice, as expert estimations under conditions of conceptual uncertainty cannot stay the same for a long time. In scientific research and experimental design processes, not only new knowledge about a product under development is accumulated, but the conception about the product's characteristics, use, and application areas may change; new inventions, technical solutions, and other know-how may emerge. Thus, the new principle must reflect the probabilistic characteristics of invariance in time of the initial estimation results of an innovation object, and that is why we shall call it the *probabilistic invariance principle*.

Initial expert estimation results of certain innovative ideas or technical solutions, industrial products, or manufacturing technologies are probabilistic and do not guarantee that they will be saved in time. This principle postulates that initial expert estimation results obtained under conditions of conceptual uncertainty as positive or negative findings, proposals, or recommendations are not invariant and may substantially change, be confirmed, or be disproved as time passes. Therefore, we do not exclude in a certain time frame both safekeeping of truth expert statements, opinions, or findings, and the possibility of disproving them.

The following definitions of latent indices of innovation products are proposed. *Practical necessity:* the presence of a relatively high market need in innovation products that are proposed in the investigated project or that have a certain demand and sale in domestic and foreign markets. *Technology possibility:* the presence or possibility to develop materials and components, equipment, and technologies for serial production of innovation products. *Economy expediency:* the presence of real conditions and the proved prospects of the demand and selling market to obtain an acceptable level of technical and economic effectiveness of innovation products.

### 9.2.2 Methodology of Cognitive Modeling of Complex Systems

The methodology of cognitive modeling for complex systems was developed while taking into account the properties of socio-economic, socio-technical, ecological, political, other complex systems, and those patterns to which these systems are a subject [11]. From these positions, the developed cognitive methodology implements an interdisciplinary approach; it is also a system that unites models, approaches, and methods for solving problems in many areas. The methodology is based on the meta-model of the study, it is a collection of cognitive models (cognitive map, weighted oriented graph, functional graphs, etc.) and methods that include an analysis of the stability of paths and cycles, connectivity, complexity, sensitivity of the system, scenario analysis to foresight the possible development of situations in the system, decision-making, etc. (it includes a software system for cognitive modeling). The toolkit of cognitive modeling has been tested in studies on many complex systems [13–17].

Cognitive modeling is carried out in the following stages.

*Pre-project stage:* collection of information, taking into account the results of the foresight process, determination of the nature and development trends of the system systematization of theoretical conceptual positions, analysis of the main properties and characteristics of the system, formulation, and clarification of the research objectives.

*First stage:* Cognitive model development. Designing of the cognitive model $G$; definition of vertices, relations between them, weights, functional dependencies, etc. Building a cognitive model of a complex system in the form of a cognitive map or building a cognitive model of a complex system in the form of a parametric functional.

*Second stage:* Analysis of the properties of the system in cognitive model $G$: determination of the properties of the stability of the model $G$ to disturbances, definition, and analysis of cycles and structural stability of the model $G$, definition, and analysis of paths on $G$ and simplex analysis of the model $G$.

*Third stage:* Scenario analysis, pulse modeling, etc.

The completion of cognitive modeling is development and evaluation of management decisions to improve the development processes of the system under study.

Consider these stages in more detail.

*First stage – Development of the cognitive model:* In the process of developing a cognitive model, there is a cognitive structuring of the subject area. This is the identification of objects (elements, concepts, essences, etc.) of the target system, desirable and undesirable states of the system, the most significant (basic) management factors, environmental factors that affect the transition of the system from state to state, and the establishment on the qualitative/quantitative level of connections (mutual influences) between objects. This is a cyclic process. It begins with the development of simpler mathematical forms of a complex system in the form of a cognitive map $G$, for which the results of the foresight phase are used as input data. This process can end with more complex forms, for example, in the form of parametric vector functional graphs. For clarity of the further presentation, we give several well-known formulas [13–17].

Parametric functional vector graph, cortege

$$G_F = < G, X, F, q >, \tag{9.1}$$

where $G = < V, E >$ is the cognitive map, where $V$ is the set of vertices (objects and concepts), vertices $V_i \in V, i = 1, 2, ..., k$ are elements of the investigated system; $E$ is the set of arcs, $5_{ij} \in E, \quad i, j = 1, 2, \ldots, n$ reflects the interconnection between vertices $V_i$ and $V_j$; $X$ is the set of vertex parameters $X = \{E\left(V_i\right)\}$, that is, each vertex is associated with a vector of independent parameters $E(V_i)$; $X : V \rightarrow \theta$, $\theta$ is the space of the parameters of vertices, the set of real numbers $F = F(X, E) = F(x_i,\ x_j,\ e_{ij})$ is the functional of the transformation of arcs, which assigns to each arc either a sign ("+", "–"), ("–") (then this is a sign digraph), or the weighting factor $\omega_{ij}$ (then it is a weighted sign digraph), or a function $F(x_i,\ x_j,\ e_{ij}) = f_{ij}$ (then this is a functional graph). In the ongoing studies of cognitive modeling for complex systems for constructing models of type (9.1), the idea of composing cognitive maps with known models of system dynamics was realized [18–20].

The cognitive map $G$ corresponds to the square matrix of relations $A_G$ in the following form:

$$A_G = \{a_{ij}\} = \begin{cases} 1, & \text{if } V_i \text{ is connected with } V_j \\ 0, & \text{otherwise} \end{cases}. \tag{9.2}$$

The ratio $a_{ij}$ can take the value "+1" or "–1." The relation between variables (interaction of factors) is a quantitative or qualitative description of the effect of changes in one variable on others at the corresponding vertexes.

Vector functional graph has the form

$$\Psi = \langle G, X, F\left(X, E\right), \theta\rangle\,, \tag{9.3}$$

where $G$ is a cognitive map; $X$ is the set of vertex parameters; $\theta$ is the space of vertex parameters; $F(X,\ E)$ is the arc transformation functional. If

$$F(X,E) = F(x_i, x_j, e_{ij}) = \begin{cases} +\omega_{ij}, \text{ if rising/falling } X_i, \\ \text{entails rising/ falling } X_j, \\ -\omega_{ij}, \text{ if rising/falling } X_i, \\ \text{entails falling/rising } X_j, \end{cases} \tag{9.4}$$

then there is a weighted sign digraph, in which $\omega_{ij}$ is the weight coefficient.

In the process of studying complex systems, various hierarchical cognitive models (9.1) can be developed, as well as models of cognitive models interaction (9.4) [13–17]

$$I_G = \langle G_k, G_{k+1}, E_k \rangle, k \geq 2, \tag{9.5}$$

where $I_G$ is hierarchical cognitive map and $G_k$ and $G_{k+1}$ are cognitive maps of $k$ and $k+1$ levels, some vertices of which are connected by arcs $e_k$. If the arc transformation functionals are defined, then a hierarchical cognitive model takes place.

Note that the hierarchical cognitive map model can represent the levels of the management hierarchy of the investigated system, and the lower levels can be a cognitive map that "unfolds" the cognitive top-level map.

*Second stage – Analysis of the system properties of the cognitive model G*: The following are investigated: the stability to disturbances, structural stability, paths, cycles, complexity, connectivity ($q$-connectedness analysis), sensitivity, etc. [18]. The results of the analysis compares with the available information on complex system.

The stability of the graph by the perturbation and value is based on the concept of the process of propagation of perturbations of a graph. Determine the value at the vertex $V_i$ at the moment of time $t$ through$V_i(t), i \in [1,n], t = 0, 1, ....$ Suppose that the value $V_i(t+1)$ depends on $V_i(t)$ and on the vertices adjacent to $V_i$. Thus, if a vertex $V_j$ is adjacent to $V_i$ and if $p_j(t)$ represents the change in $V_j$ at the moment of time $t$, it should be considered that the impact of this change on $V_i$ at the moment of time $t+1$ will be described by the function $f(V_i, V_j)P_j(t)$, where through $f(V_i, V_j)$, the weight function of connection between the vertices $V_j$ and $V_i$ is denoted [18]. Thus, we have the following rule of perturbation propagation:

$$V_i(t+1) = V_i(t) + \sum_{j=1}^{N} f(V_j, V_i) \cdot P_j(t) \; \forall i = \overline{1,n}, \tag{9.6}$$

$$P_j(t+1) = V_j(t+1) - V_j(t).$$

The vertex is called stable by perturbation if the sequence $\{|P_j(t)|\}_{t=1}^{\infty}$ is limited. The vertex is called stable by value if the sequence $\{|V_j(t)|\}_{t=1}^{\infty}$ is limited. The graph is stable by perturbation (value) if all its vertices are stable. Such a result: from the stability by value should be the stability by the perturbation. Thus, the stability by value is reduced to a limited matrix series, and the stability by the perturbation to a limited matrix sequence. Thus, the stability by value is reduced to a limited matrix series $\sum_{t=0}^{\infty} A^t$, and the stability by the perturbation to a limited matrix sequence $M_i\{A^t\}_{t=1}^{\infty}$.

Take the following stability criteria by the perturbation.

*Criterion 1:* The system in the form of signed weighted directed graph *G* with the adjacency matrix $A$ is stable by perturbation iff the spectral radius of the adjacency matrix is $\rho(A) = \max_i |\lambda_i| \leq 1$, where $\{\lambda_i\}_{i=1}^{M}$ is the eigenvalues $A$ and is the basis of eigenvectors.

Let us consider the existing possibilities of mathematical analysis for the stability description of the system. We represent the expression (9.6) in matrix form:

$$\begin{aligned} V(t+1) &= V(t) + A \cdot P(t), \\ P(t+1) &= V(t+1) - V(t), \end{aligned} \tag{9.7}$$

where $A$ is the adjacency matrix of the graph, *V*(*t*) is the vector of values at the vertices $V_1, V_2, \ldots, V_n$ at the time instant $t$, and $P(t)$ is the vector of actions at the vertices $V_1, V_2, \ldots, V_n$ at the instant $t$. Performing sequential transformations in Equation (9.7), we have

$$V(1) = V(0)+A \cdot P(0),\ V(2) = V(1)+A \cdot P(1) = V(0)+A \cdot P(0)+A \cdot P(1),$$

$$P(1) = V(1)-V(0) = A \cdot P(0),\ P(2) = V(2)-V(1) = A \cdot P(1) = A^2 P(0),$$

$$\begin{aligned} V(t+1) &= V(0) + (A + A^2 + A^3 + \ldots + A^{t+1})P(0) = \\ &= V(0) + (I + A + A^2 + A^3 + \ldots + A^t)P(1); \end{aligned} \tag{9.8}$$

$$P(t+1) = A^{t+1} \cdot P(0). \tag{9.9}$$

Thus, the stability in value was reduced to the boundedness of the matrix series $\sum_{t=0}^{\infty} A^t$, and the stability in perturbation was reduced to the boundedness of the matrix sequence $M_t = \{A^t\}_{t=1}^{\infty}$.

We give a proof of the criterion 1. In accordance with Equation (9.9), perturbation stability occurs because the matrix sequence $M_t = \{A^t\}_{t=1}^{\infty}$ is limited. Let $V_J$ be the Jordan basis $A$;then $A = V_J^{-1} A_J V_J$, where $A_j$ is the Jordan form $A$. We write the matrix $A$ in the Jordan basis [19]

$$A_J = \begin{bmatrix} J_1 & 0 & \cdots & 0 \\ 0 & J_2 & \cdots & 0 \\ \vdots & \vdots & \ddots & \vdots \\ 0 & 0 & \cdots & J_m \end{bmatrix},$$

where $J_1$, $J_2$, ..., $J_m$ and $i = 1, 2, \ldots, m$ are Jordan cells of the $m \times m$ dimension in the form

$$J_i = \begin{bmatrix} \lambda_i & 1 & 0 & \cdots & 0 \\ 0 & \lambda_i & \ddots & \ddots & \vdots \\ \vdots & \ddots & \ddots & 1 & 0 \\ 0 & \ddots & 0 & \lambda_i & 1 \\ 0 & \cdots & 0 & 0 & \lambda_i \end{bmatrix},$$

corresponding to elementary divisors $(\lambda - \lambda_1)^{p_1}, (\lambda - \lambda_2)^{p_2}, \ldots, (\lambda - \lambda_u)^{p_m}$, $(p_1 + p_2 + \cdots + p_m = n)$. Then

$$A_J{}^t = \begin{bmatrix} J_1^t & 0 & \cdots & 0 \\ 0 & J_2^t & \cdots & 0 \\ \vdots & \vdots & \ddots & \vdots \\ 0 & 0 & \cdots & J_m^t \end{bmatrix}. \tag{9.10}$$

It follows that boundedness $\{A^t\}_{t=1}^{\infty}$is equivalent to boundedness $\{J_1{}^t\}_{t=1}^{\infty}$for all Jordan cells matrix $A$.

According to Equation (9.10), for a Jordan cell we have

$$J^t = \begin{bmatrix} \lambda^t & \frac{t\lambda^{t-1}}{1!} & \frac{t(t-1)\lambda_i^{t-2}}{2!} & \cdots & \frac{t!\lambda^{t-n+1}}{(t-n+1)!(n-1)!} \\ 0 & \lambda_i^t & \ddots & \ddots & \vdots \\ \vdots & \ddots & \ddots & \frac{t\lambda^{t-1}}{1!} & \frac{t(t-1)\lambda^{t-2}}{2!} \\ 0 & \ddots & 0 & \lambda^t & \frac{t\lambda^{t-1}}{1!} \\ 0 & \cdots & 0 & 0 & \lambda^t \end{bmatrix}$$

$$= \begin{bmatrix} \lambda^t & C_t^1\lambda^{t-1} & C_t^2\lambda^{t-2} & \cdots & C_t^{n-1}\lambda^{t-n+1} \\ 0 & \lambda^t & \ddots & \ddots & \vdots \\ \vdots & \ddots & \ddots & C_t^1\lambda^{t-1} & C_t^2\lambda^{t-2} \\ 0 & \ddots & 0 & \lambda^t & C_t^1\lambda^{t-1} \\ 0 & \cdots & 0 & 0 & \lambda^t \end{bmatrix},$$

which should be limited sequence $\{C_t^i\lambda^{t-i}\}_{t=0}^{\infty}\ \forall i = 0, 1, \dots, n-1$. Then if $|\lambda| < 1$, then $|C_t^i\lambda^{t-i}| \underset{t\to\infty}{\longrightarrow} 0\ \ \forall i = 0, 1, \dots, n-1$ since the polynomial grows more slowly than the power function decreases. Therefore, sequences are not limited $\forall i = 0, 1, \dots, n-1$ ;if $|\lambda| = 1$, then

$$\begin{cases} |C_t^i\lambda^{t-i}| \underset{t\to\infty}{\longrightarrow} \infty & \forall i = 1,2,\dots,n-1, \\ |C_t^i\lambda^{t-i}| = |\lambda^t| = 1 & \text{при } i = 0. \end{cases}$$

Therefore, sequences $\{C_t^i\lambda^{t-i}\}_{t=0}^{\infty}$ will be bounded $\forall i = 0, 1, \dots, n-1$ iff $n = 1$and the Jordan cell has dimension 1. It follows that the sequence $\{A^t\}_{t=1}^{\infty}$is limited $\forall i = 0, 1, \dots, n-1$ iff all eigenvalues of the matrix $A$ modulo less than 1 or do not exceed 1 and the Jordan form of the matrix is diagonal. Thus, the graph is stable in perturbation since $\rho(A) = \max_i |\lambda_i| \leq 1$, as required to prove.

Now we formulate a criterion for stability with respect to the initial value.

*Criterion 2:* The system in the form of signed weighted directed graph *G* with the adjacency matrix *A* is stable by value iff spectral radius of the adjacency matrix is $\rho(A) = \max_i |\lambda_i| < 1$, where $\{\lambda_i\}_{i=1}^{M}$ are characteristic numbers *A*, or $\rho(A) = 1$, but the Jordan form of the matrix is diagonal and there is no eigenvalue equal to 1.

Let us give a proof of this criterion. Write the matrix equality:

$$(I + A + A^2 + \cdots + A^t)(I - A) = I - A^{t+1}. \tag{9.11}$$

Then assume that the system is stable in value, i.e. the matrix series $\sum_{t=0}^{\infty} A^t$ is bounded (in norm) by the constant C. Then from expression (9.10), we get

$$||A^{t+1}|| = ||A^{t+1} - I + I|| \leq ||A^{t+1} - I|| + ||I|| = ||(I + A + A^2 + \cdots + A^t)(I - A)|| + 1 \leq$$
$$\leq ||I + A + A^2 + \cdots + A^t|| * ||I - A|| + 1 \leq C||I - A|| + 1.$$

Therefore, the sequence $\{A^t\}_{t=1}^{\infty}$is bounded, whence from criterion 1, we obtain the first part of the necessary statement, i.e. $\rho(A) = \max_i |\lambda_i| < 1$.

It is also necessary to prove that there is no eigenvalue equal to 1. Assume the converse, that is, $\exists x \neq 0 : \quad Ax = x$. Then

$$||(I + A + A^2 + \cdots + A^t)x|| = ||(t+1)x|| \underset{t\to\infty}{\longrightarrow} \infty,$$

$$||I+A+A^2+\cdots+A^t|| = \sup_{x\neq 0} \frac{||(I + A + A^2 + \cdots + A^t)x||}{||x||} \geq t+1 \underset{t\to\infty}{\longrightarrow} \infty.$$

The amount received is not limited. Therefore, an eigenvalue equal to 1 does not exist.

We assume that the spectral radius of the matrix $A$ is 1, but the matrix has a basis from eigenvectors and 1 is not a characteristic number of the adjacency matrix. Then, by criterion 1, we obtain that the sequence $\{A^t\}_{t=1}^{\infty}$ is bounded by the constant $C$. In addition, since 1 does not belong to the spectrum of the matrix $A$, then 1 belongs to the resolvent set, that is, $\exists\ (I - A)^{-1}$. Then, after multiplying both sides of Equation (9.6) on $(I - A)^{-1}$the right, we get

$$I + A + A^2 + \cdots + A^t = (I - A^{t+1})(I - A)^{-1}.$$

From here

$$||I + A + A^2 + \cdots + A^t|| = ||(I - A^{t+1})(I - A)^{-1}|| \leq ||I - A^{t+1}|| * ||(I - A)^{-1}|| \leq$$
$$\leq (||I|| + ||A^{t+1}||) * ||(I - A)^{-1}|| \leq (1 + C) * ||(I - A)^{-1}||.$$

Therefore, the matrix series $\sum_{t=0}^{\infty} A^t$ is limited.

Thus, the graph is stable by the value of iff $\rho(A) = \max_i |\lambda_i| < 1$,or, $\rho(A) = 1$ as required to prove. Note that a graph is numerically stable if the spectral radius of the adjacency matrix is less than 1.

The study of structural stability is carried out by analyzing all cycles of graph G. Among the cycles, cycles of positive and negative feedback are distinguished. A positive feedback cycle is a cycle in which there are no or an even number of negative arcs; this is a cycle of process accelerators in the system. A negative feedback cycle is a cycle in which there is an odd number of negative arcs; this is a process stabilizer cycle. A condition for the structural stability of the system is the presence of an odd number of negative cycles in it [11].

Investigations of the connectivity and complexity of the system are necessary for solving problems about the possibility of managing the system, selecting management methods, and assessing the conditions necessary for the implementation of management.

The analysis of the connectivity of the system under cognitive modeling can be carried out both on the basis of graph theory and using the language of algebraic topology, which makes it possible to analyze the structure of a complex multidimensional geometric formation, the simplicial complex. This is possible with minimal *a priori* information regarding the objects and phenomena under study. The mathematical foundations for the polyhedral (simplicial) analysis were laid by Droucer, and further analysis was obtained in the works of the British physicists R.H. Atkin and J Casti [20–22].

To perform the simplicial analysis, it is necessary to isolate the simplexes generated by each vertex of the cognitive model to determine the chains of bonds ($q$-connections) between them and to construct simplicial complexes.

The loosely coupled system allows for offline block management. This can have both positive and negative sides. The negative effect arises from the fact that autonomous management within a real unified social and economic system can prove to be not only costly (fragmentation of funds) but also harmful because of the non-observance of the systemic nature principle. A tightly coupled system with a rigid structure along with the advantages of an "easier" centralized control is less flexible and effective in a rapidly changing environment; it is more difficult to adapt to dynamic changes. Under such conditions, systems with a tunable structure may be more effective. But in any case, it is necessary to study the structural properties of the existing complex system.

*Third stage – Scenario analysis, pulse modeling:* At the third stage of cognitive modeling, to determine the possible development of processes in a complex system and develop the scenarios development, we used the impulse process model (modeling the propagation of disturbances in cognitive models), which is a transition from a model (9.6) to the following model [23–25]:

$$x_i(n+1) = x_i(n) + \sum_{i=1}^{k-1} f(x_i, x_j, e_{ij}) P_j(n) + Q_i(n), \tag{9.12}$$

where $x\,(n)$ and $x\,(n+1)$ are the values of the indicator at the vertex $V_i$ at the modeling steps at time $t = n$ and the next $t = \ n+1$; $P_j\,(n)$ is the pulse that existed at the vertex $V_i$ at the moment $t = n$; $Q_{V_i}\,(n) = \{q_1,\ q_2,\ ...,\ q_k\}$ is the vector of external pulses (disturbing or controlling actions) introduced to the vertexes $V_i$ at time moment $n$. The scenario analysis is carried out by the means of an impulse simulation. To generate possible scenarios of the system development at the vertices of the cognitive map, hypothetical perturbing/control actions (impulses) $Q = \{q_{ij}\}$ are introduced. A set of implementations of impulse processes is a "scenario of development," which indicates possible trends in the development of situations in the system. The situation in impulse simulation is characterized by a set of all perturbing effects of $Q$ and $X$ values at each step of the simulation. The scenario answers the question: "What will happen if ...?" The development scenario is one of the hypothetical variants of the future processes in system – the foresight of the future.

The completion of cognitive modeling should be the choice of the desired scenario for the development of the system, and the development and justification of management decisions are aimed at implementing the desired scenario, preventing the consequences of unwanted scenarios.

If the decision-maker is not satisfied with the results of cognitive modeling, model adjustment is necessary. You can change the number of vertices and the relationships between them, rearranging the structure of the initial cognitive model.

Cognitive modeling of complex systems required the development of special software cognitive modeling large system (CMLS) [26], with which all stages of modeling are performed.

Consider the possibilities of cognitive modeling for complex systems using the example of the education system relationship study with the socio-economic environment in a pandemic.

### 9.2.3 Relationship of the Education System with the Socio-Economic Environment

In the context of the coronavirus pandemic and the unstable economy of the modern world, one of the essential means for stabilizing of the socio-economic and geopolitical situation in the future may be the education system. To study the possible consequences of the education system interaction with the socio-economic environment of the country in a pandemic, a corresponding cognitive study was conducted. The goal of cognitive modeling was to identify the structure (cognitive map) of causal relationships of the studied complex system and possible scenarios for the development of situations (scientific prediction) under various perturbations and effects on the system. The results of the scenario analysis are necessary for the development and decision substantiation of the system development strategy in the desired direction – improving of the quality population life and improving of the quality of human capital.

*Stage I – Cognitive model development:* To conduct cognitive modeling, information was obtained as a result of applying the foresight methodology at the initial study. The main factors and their interconnections determining the complex education system (CES) and its influence on the socio-economic environment were identified. In developing the cognitive model of the CES, the cognitive model of Shukshunova and Ovsyannikov was used [27]. It was decided to develop a hierarchical cognitive model consisting of two levels, namely, "Social" (first level) and "Education System" (second level). Table 9.1 indicates the vertices concepts for the cognitive model and their purposes are determined. The hierarchical cognitive map $I_G$ constructed using the software system CMLS [26] is shown in Figure 9.2. The image of this cognitive map contains arcs in accordance with Equation (9.4) in the form of solid ($\omega_{ij} = +1$) and dash-dotted ($\omega_{ij} = -1$) lines; $\omega_{ij} = +1$ means that amplification/attenuation of the signal at the vertex $V_i$ leads to amplification/attenuation of the signal at the vertex $V_j$; $\omega_{ij} = -1$ means that amplification/attenuation of the signal at the vertex $V_i$ leads to attenuation/amplification of the signal at the vertex $V_j$. The vertices of the first level in Figure 9.2 are highlighted in red, and the vertices of the second level are highlighted in black.

*Stage II – Analysis of the cognitive model properties:* An analysis was made of the graph $I_G$ properties, its stability (to perturbations and the initial value, structural stability), and the analysis of paths and cycles. The analysis was carried out both to verify that the $I_G$ model is consistent to a real complex system and to understand and explain its properties.

*Graph properties analysis*: An analysis was made of the number of incoming and outgoing arcs at each vertex of the cognitive map. Vertices in which the number of arcs is large can be considered the most important in the model. In this case, these vertices can be attributed as $V_1$ (Quality of the nation), $V_5$ (Indicators of anomie and meaninglessness of life), $V_{13}$ (Education status), and $V_{24}$ (Personal and physical qualities of graduates).

*Analysis of stability to disturbances and the initial value*: The calculation of the characteristic equation roots for the adjacency matrix cognitive map $I_G$ (Figure 9.3) indicates the instability of the model to disturbances and the initial value since the absolute maximum root of the equation $|M| = 1,526 > 1$.

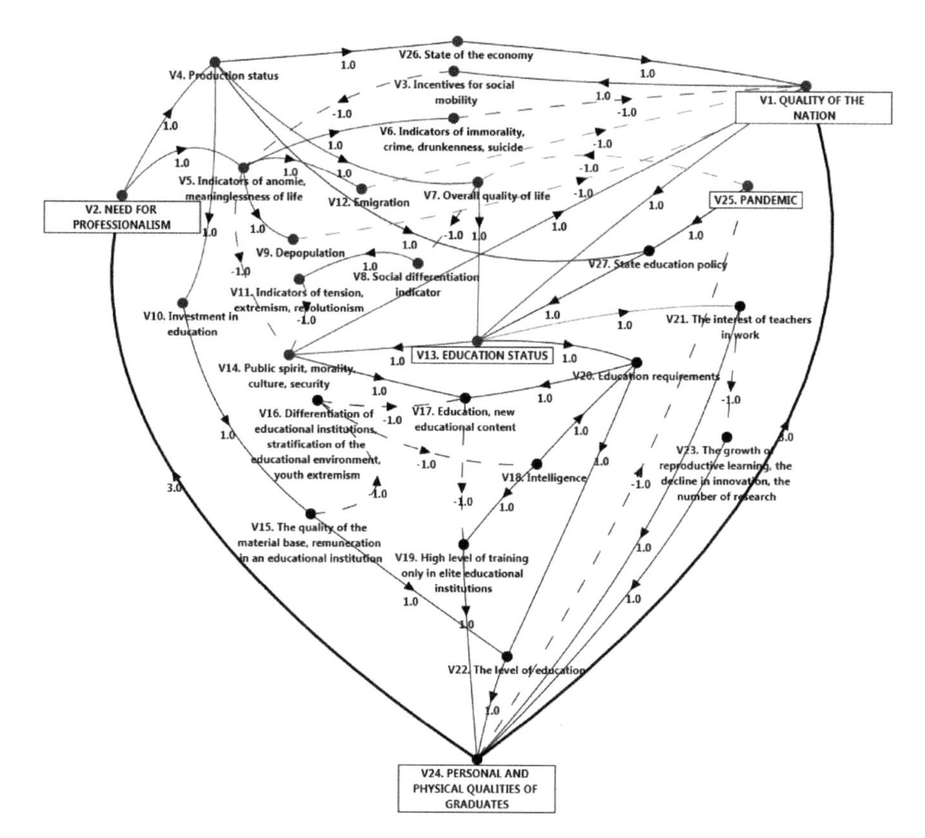


**Figure 9.2** Hierarchical cognitive model $I_G$ "The education system in the socio-economic environment."

**Eigenvalues**

| # | Real part | Imaginary part | Module (1.5266) |
|---|---|---|---|
| 0 | 0.0261 | 1.4925 | 1.4925 |
| 1 | 0.0261 | -1.4925 | 1.4925 |
| 2 | -0.6059 | 1.2378 | 1.2378 |
| 3 | -0.6059 | -1.2378 | 1.2378 |
| 4 | -1.3456 | 0.0 | 1.3456 |
| 5 | -1.1037 | 0.2448 | 1.1037 |
| 6 | -1.1037 | -0.2448 | 1.1037 |
| 7 | 1.5266 | 0.0 | 1.5266 |
| 8 | 0.5124 | 1.004 | 1.004 |

**Figure 9.3** Fragment of the characteristic equation roots calculation for the adjacency matrix cognitive map.

**Table 9.1** Vertices of the cognitive map.

| Code | Vertices | Assignment |
|---|---|---|
| $V_1$ | Quality of the nation | Target |
| $V_2$ | Need for professionalism | Manageable |
| $V_3$ | Incentives for social mobility | Manageable |
| $V_4$ | Production status | Basic |
| $V_5$ | Indicators of anomie and meaninglessness of life | Indicative |
| $V_6$ | Indicators of immorality, crime, drunkenness, and suicide | Indicative |
| $V_7$ | Overall quality of life | Target |
| $V_8$ | Social differentiation indicator | Perturbing |
| $V_9$ | Depopulation | Perturbing |
| $V_{10}$ | Investment in education | Managing |
| $V_{11}$ | Indicators of tension, extremism, and revolutionism | Perturbing |
| $V_{12}$ | Emigration | Basic |
| $V_{13}$ | Education status | Indicative |
| $V_{14}$ | Public spirit, morality, culture, and security | Indicative |
| $V_{15}$ | The quality of the material base and remuneration in an educational institution | Managing |
| $V_{16}$ | Differentiation of educational institutions, stratification of the educational environment, and youth extremism | Perturbing |
| $V_{17}$ | Education and new educational content | Managing |
| | Table 9.1 continuation | |
| $V_{18}$ | Intelligence | Indicative |
| $V_{19}$ | High level of training only in elite educational institutions | Perturbing |
| $V_{20}$ | Education requirements | Managing |
| $V_{21}$ | The interest of teachers in work | Basic |
| $V_{22}$ | The level of education | Basic |
| $V_{23}$ | The growth of reproductive learning, the decline in innovation, and the number of research | Perturbing |
| $V_{24}$ | Personal and physical qualities of graduates | Target |
| $V_{25}$ | Pandemic | Perturbing |
| $V_{26}$ | State of the economy | Basic |
| $V_{27}$ | State education policy | Managing |

*Analysis of cycles and structural stability of the cognitive map* $I_G$. A computational experiment to determine the cycles of the cognitive model made it possible to find their number and distinguish the cycles of negative and positive feedback. There are 72 cycles in total in the system. Figure 9.4 shows one of the cycles of positive feedback (a sign of positive, amplifying, feedback is an even, in this case six, number of negative arcs).

This cycle is formed by a closed chain of causal relationships $V_{13} \rightarrow V_{21} \rightarrow V_{23} \rightarrow V_{24} \rightarrow V_{25} \rightarrow V_7 \rightarrow V_8 \rightarrow V_{11} \rightarrow V_{14} \rightarrow V_1 \rightarrow V_{13}$. That is, if changes occur in any of the vertices, then, being locked to this

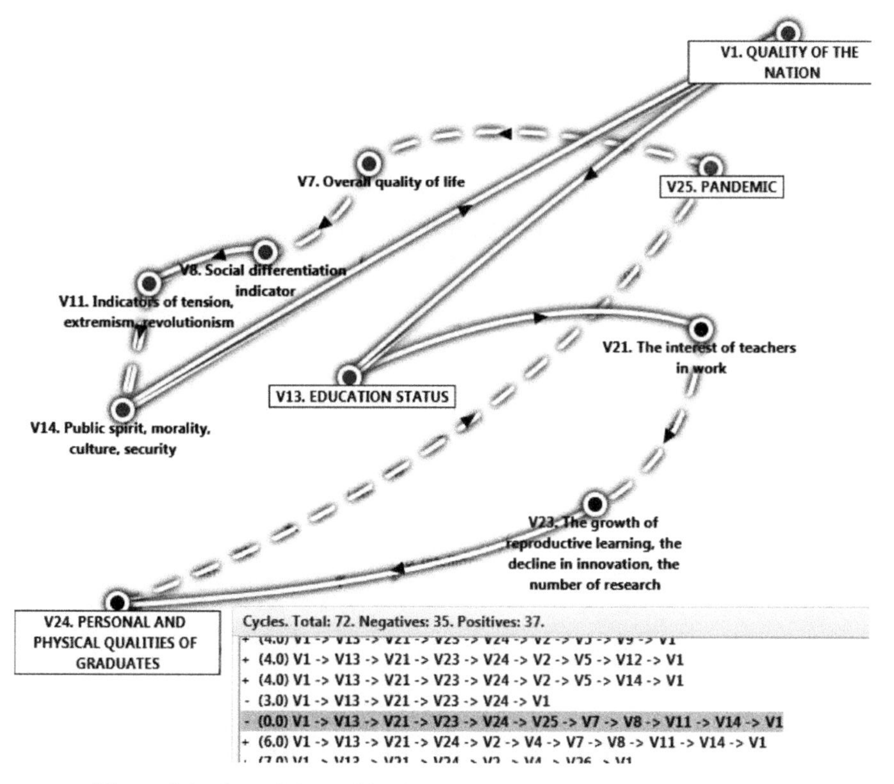


**Figure 9.4** One of the positive feedback cycles of the cognitive map $I_G$.

vertex, the changes are amplified. Changes in the designated cycle can be interpreted as follows. Let the status of the education system decrease ($V_{13}$), and this reduces the interest of teachers in labor ($V_{21}$), which leads to an increase in reproductive learning, a decrease in innovation and the number of research ($V_{23}$), which in turn causes a decrease in the personal and physical qualities of graduates ($V_{24}$), and this can lead to a weakening of the confrontation of the pandemic ($V_{25}$), it is spreading, an increase in the pandemic leads to a decrease in the general standard of living ($V_7$), which enhances social differentiation ($V_8$), indicators of tension, extremism, and revolutionism increase ($V_{11}$), decrease citizenship, morality, culture, security ($V_{14}$), all this reduces the quality of the nation ($V_1$), which in turn can lead to a decrease in the status of education ($V_{13}$).

The presence of such a cycle would be undesirable if the negative feedback cycles compensating it were not present in the system.

*Structural stability:* Among the 72 cycles in the system, 35 cycles of negative (stabilizing) feedback and 37 cycles of positive feedback (accelerator cycles) are observed. This fact indicates the structural stability of the model: if an odd number of negative feedback cycles are observed in the system, then the system is structurally stable.

*Cognitive map path analysis*: Let us analyze the possible ways of influencing the vertex $V_{25}$ "Pandemic" on the vertex $V_1$ "Quality of the nation" (Figure 9.5). There are 72 such paths in this case, and, among them, there are 35 negative (–) attenuating the signal from the initial vertex, and 37 positive (+) amplifying the signal from the initial vertex. In Figure 9.6, at the (+) path, $V_{25} \rightarrow V_7 \rightarrow V_8 \rightarrow V_{11} \rightarrow V_{14} \rightarrow V_{17} \rightarrow V_{24} \rightarrow V_2 \rightarrow \rightarrow V_4 \rightarrow V_{26} \rightarrow V_1$is shown

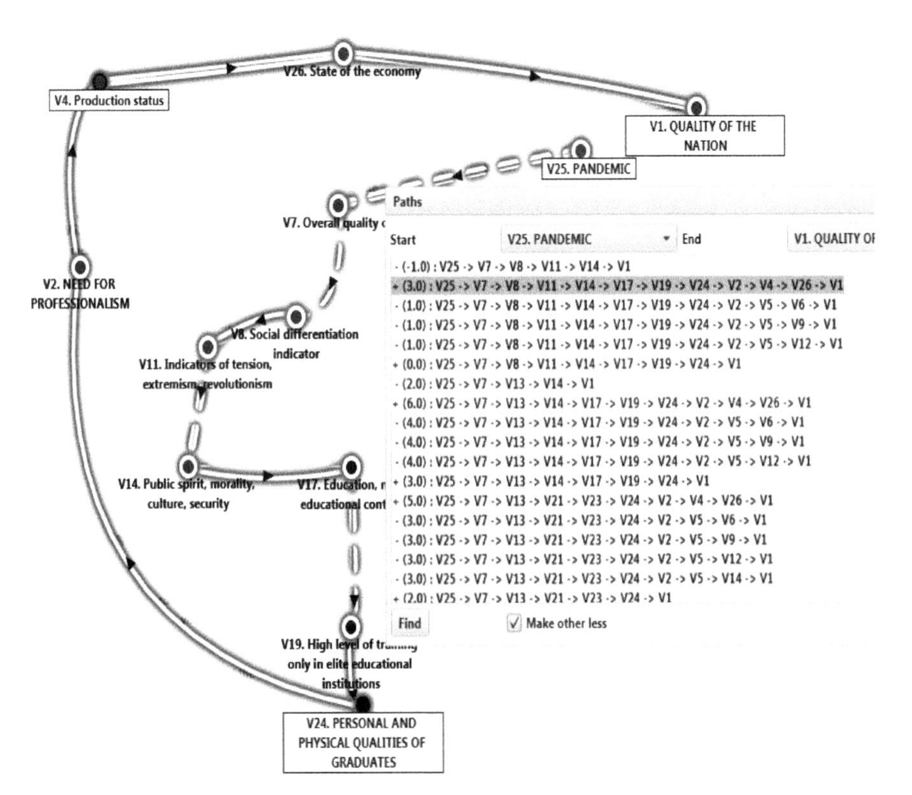


**Figure 9.5** Example of the paths from the vertex $V_{25}$ "Pandemic" to the vertex $V_1$ "Quality of the nation."

Given the signs of arcs along this path, the growth of the pandemic ($V_{25}$) reduces the general standard of living ($V_7$), the indicator of social differentiation increases ($V_8$), the indicators of tension, extremism, and revolutionism ($V_{11}$) increase, the citizenship, morality, culture, and security decrease ($V_{14}$), the need for improving education is growing, and the new content of education ($V_{17}$) is decreasing. A high level of training is only in elite educational institutions, while the personal and physical qualities of graduates deteriorate ($V_{24}$), the need for professionalism decreases ($V_2$), the need for graduates in production decreases ($V_4$), the state of the economy worsens ($V_{26}$), and the quality of the nation worsens ($V_1$). Thus, this possible "+" path is extremely undesirable if it can be realized because it "Reinforces the negative impact of the pandemic." But from the top of $V_{25}$, there are also (–) paths; for example, the path $V_{25} \rightarrow V_{27} \rightarrow V_{13} \rightarrow V_{20} \rightarrow V_{17} \rightarrow V_{19} \rightarrow V_{24} \rightarrow V_1$. We will interpret the path under the assumption that the pandemic has decreased, and not its growth, as in the previous case. Then reducing the pandemic ($V_{25}$) will reduce the efforts of the state in the field of education ($V_{27}$) and educational status ($V_{13}$) that were aimed at countering the pandemic, the education requirements ($V_{20}$) due to the pandemic can also be weakened, which can lead to a certain decrease in efforts to educate and new content of education ($V_{17}$). Note that only caused by the specifics of the pandemic can affect the concentration of efforts to increase a high level of training in elite educational institutions ($V_{19}$), which will lead to an improvement in the personal and physical qualities of graduates ($V_{24}$), contributing to an improvement in the quality of the nation ($V_1$).

In Figure 9.6, another example of the paths from the top of $V_{25}$ to the top of $V_{27}$ "State education policy" is presented. There are 10 such paths, of which the number of paths (–) is 4. One of them is shown in Figure 9.6.

Analyzing these paths, we can conclude that it is necessary to conduct a preliminary state policy in the field of education in anticipation of the pandemics possibility in the world.

The analysis of possible paths from any vertices $V_i$ to the vertices $V_j$ is necessary from at least two positions: first, this analysis allows us to analyze the inconsistency of the corresponding cause–effect chains with theoretical, practical, and expert assumptions, and, second, their analysis and comparison allows to choose and recommend for the development strategy of a complex system the most effective way to achieve the desired goal.

*Stage III – Scenario analysis, pulse modeling.* To solve the problems of predicting the possible development of situations on the cognitive map

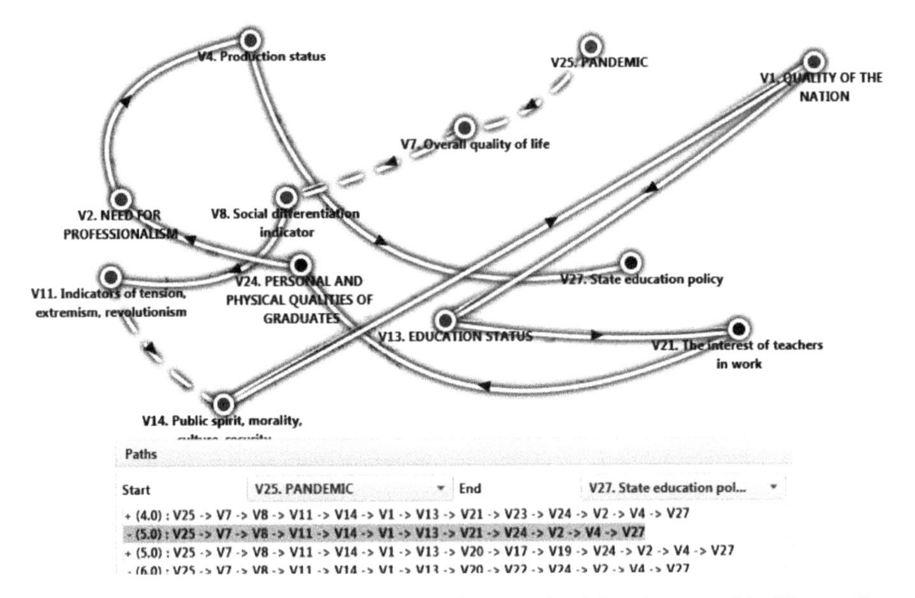


**Figure 9.6** Example of paths from the vertex $V_{25}$ "Pandemic" to the vertex $V_{27}$ "State policy in the sphere of education."

$I_G$ (development scenarios), pulse modeling was performed using CMLS according to formula (9.12). A preliminary plan for a computational experiment was developed, which also took into account the strategies proposed at the stage of using foresight methods to assess their possible consequences. The design of the experiment was built in the sequence: introduction of perturbations $q_i$ of different signs ($q_i = +1$ or $q_i = -1$) at one vertex; introduction of disturbances in two, three, and several vertices. The selection of the vertices into which the pulses triggering the system were introduced occurred in accordance with the data in Table 9.1 from the column "Vertex assignment." Each such impulse was interpreted accordingly as disturbing or controlling. Thus, scenarios were considered which answered the question: "What will happen if changes occur in the peaks ...?" The perturbation vector had the form:$Q = \{q_i\}$. Table 9.2 shows a fragment of the experimental design (test report).

To determine scenarios for the development of situations on the $I_G$ cognitive map with possible changes in disturbing and controlling influences at the vertices of the cognitive map, more than 20 scenarios were developed and studied. Here are the most interesting of them.

**Table 9.2** Test report.

| Scenario | Vector perturbation |
|---|---|
| | The impact on one vertex |
| Scenario No. 1 | $q_{25}$ = +1; $Q_1$ ={$q_1$=0;...0;...$q_{25}$ = +1;..$q_{27}$ = 0}. |
| | ... |
| | The impact on two vertices |
| Scenario No. 2 | $q_{25}$ = +1; $q_{27}$ = +1; $Q_2$ = {$q_1$ = 0; ... $q_{25}$ = +1; 0; $q_{27}$ = +1} |
| | ... |
| | The impact at several vertices |
| Scenario No. 3 | $q_4$ = +1; $q_{25}$ = +1; $q_{26}$ = +1; $q_{27}$ = +1; $Q$ = {$q_1$ = 0; ... $q_4$ = +1;... $q_{25}$ = +1; $q_{26}$ = +1; $q_{27}$ = +1} |

Since the coronavirus pandemic is currently making significant changes not only in the economies of countries but also in human society, requiring appropriate management decisions to overcome various negative consequences, we will first consider the scenario of perturbation spreading along the $I_G$ cognitive map under the assumption of a growing pandemic. The simulation process in the CMLS system is started by applying a single impulse $q_{25} = +1$ to the $V_{25}$ vertex, provided that there are no initial pulses at the other vertices.

*Scenario No. 1*: Let changes begin to occur at the peak of $V_{25}$ "Pandemic," the disturbing momentum $q_{25} = +1$; perturbation vector $Q_1 = \{q_1 = 0; \ldots q_{25} = +1; \ldots q_{27} = 0\}$.

The parts of the results of a computational experiment are presented in Figure 9.7(a) and (b), constructed according to Table 9.2. The division of graphs into parts of 6–7 vertices in each is due to the need to facilitate their visual perception and analysis. The number of modeling steps increases until the tendency of the processes development at the vertices ceases to change.

An analysis of the results of pulse modeling of Scenario No. 1 allows us to make the following observations. The growth of a pandemic in the absence of counteraction to it can lead to the following consequences. Figure 9.7(a): the quality of the nation ($V_1$) begins to decline sharply from the 5th step of modeling; the need for professionalism ($V_2$) remains the same for some time, but from step 11, the need for former specialists increases, and from step 12, it begins to fall sharply and this trend continues in the future; social mobility ($V_3$) decreases from the 6th cycle of modeling; the state of production ($V_4$) from the 12th cycle of simulation even begins to increase because, however, there is a slight decline in the pandemic, but further deterioration of the state

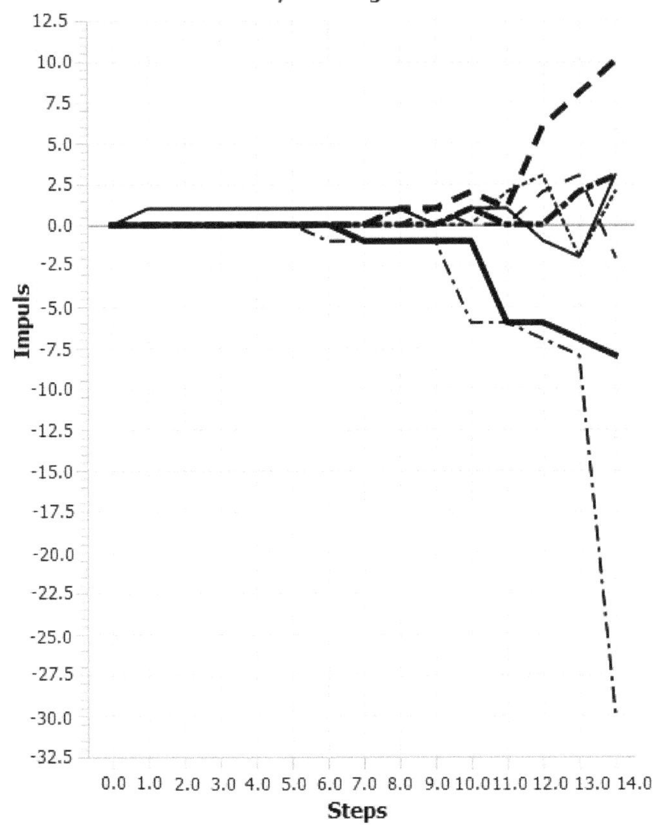


**Figure 9.7(a)** Charts of impulse processes in vertices $V_{25}$, $V_1$, $V_2$, $V_3$, $V_{24}$, $V_4$, and $V_5$; Scenario No. 1.

of production is possible; all this leads to an increase in indicators of anomie, the meaninglessness of life ($V_5$).

Similarly, each clock cycle of the simulation can be interpreted in Figure 9.7(b). In general, we can conclude that Scenario No.1 is the scenario for not responding to a pandemic, is pessimistic, and undesirable: the general

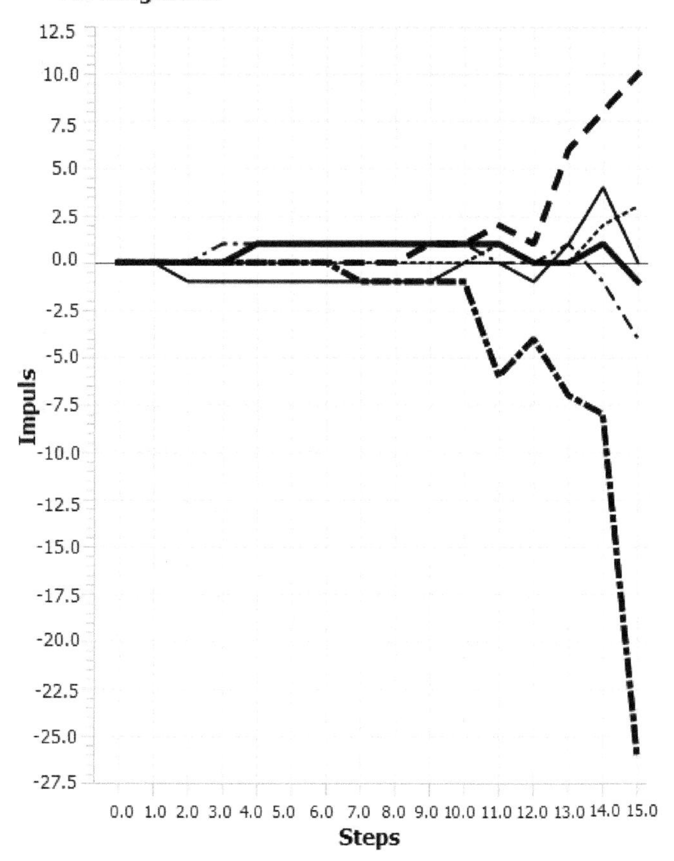


**Figure 9.7(b)** Charts of impulse processes in vertices $V_7$, $V_8$, $V_9$, $V_{10}$, $V_{11}$, $V_{13}$, and $V_{12}$; Scenario No. 1.

standard of living is declining, the state of the education system is falling, the level of education is declining, and emigration is increasing; pulse processes are subsequently oscillatory in nature, which may indicate the "desire" of a complex system to organize itself, preventing negative trends.

Consider a scenario in which a reasonable public education policy begins to counteract the effects of a pandemic since one of the goals of our study was to determine whether the educational system and qualified graduates of higher education institutions can somehow resist the catastrophic consequences of a pandemic.

*Scenario No. 2*: Let a pandemic arise, but the state begins to take reasonable measures to support the education system to counter the pandemic in the present and future. Disturbing effect $q_{25} = +1$; control action $q_{27} = +1$; impact vector $Q_2 = \{q_1 = 0; ...0; ...q_{25} = +1; q_{26} = 0; q_{27} = +1\}\}$. The simulation results are presented in Figure 9.8(a) and (b).

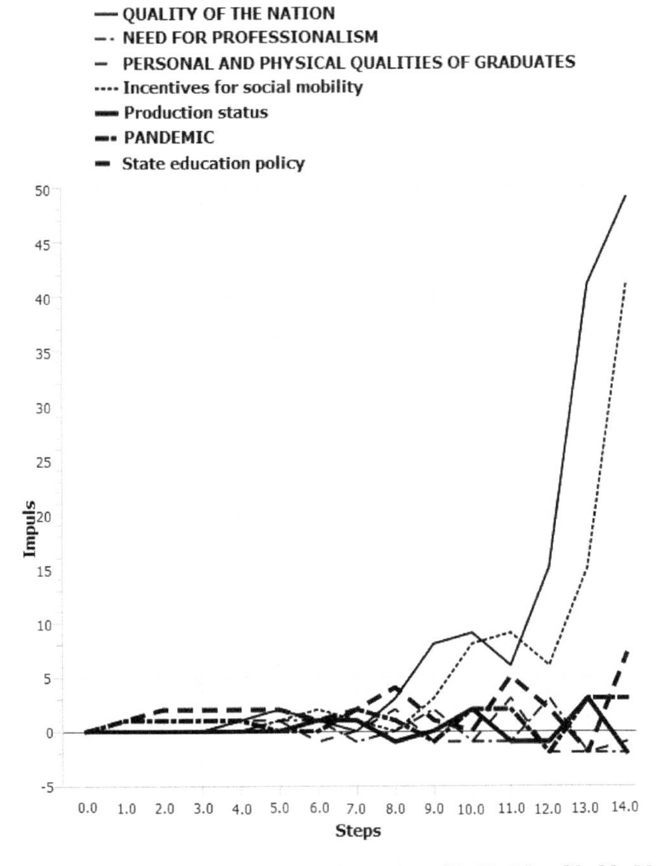


**Figure 9.8(a)** Charts of impulse processes in vertices $V_1, V_2, V_{24}$, $V_3, V_4, V_{25}$, and $V_{27}$; Scenario No. 2.

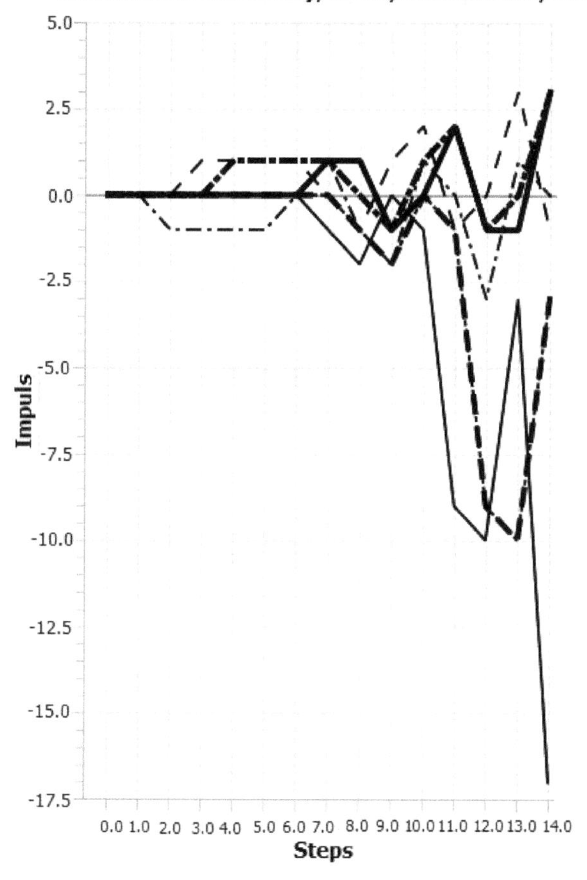


**Figure 9.8(b)** Charts of impulse processes in vertices $V_5, V_7, V_8, V_9$, $V_{10}, V_{11}$, and $V_6$; Scenario No. 2.

As can be seen from Figure 9.8(a) and (b), the state's efforts to confront the pandemic ($q_{25} = +1$) only by improving the education system ($q_{27} = +1$) does not produce long-term and stable results, although temporary improvement in situations can be expected. These improvements follow the corresponding fluctuations of the processes at the vertices $V_{25}$ and $V_{27}$ (Figure 9.8(a)). The graphs in Figure 9.8(b) also illustrate the result of the above analysis of the stability of the model to perturbations, and with respect to the initial value, such a system is not impulse stable.

Consider another scenario obtained when exposed to four vertices.

*Scenario No. 3:* Let a pandemic arise, and the state begins to apply reasonable measures not only to support the education system, but it stimulates the economy and promotes the development of production. Disturbing effect $q_{25} = +1$; control actions $q_{27} = +1$, $q_{26} = +1$, $q_4 = +1$; impact vector $Q_3 = \{q_1 = 0;\ ...\ 0;\ q_4 = +1; 0;\ ...\ q_{25} = +1;\ q_{26} = 0; q_{27} = +1\}$. The simulation results are presented in Figure 9.9(a) and (b).

The analysis of the graphs in Figure 9.9(a) and (b) allows us to make the following observation. With a good state of production ($q_4 = +1$) and the country's economy ($q_{26} = +1$), and with a well-considered state policy in the field of education ($q_{27} = +1$), the emerging of a pandemic ($q_{25} = +1$) will not immediately begin to affect socio-economic indicators, and even with a three-stroke simulation, a pandemic decline will begin to occur. Further, in the system, impulse processes, although they acquire a rather pronounced oscillatory character, show positive trends in increasing target and indicative

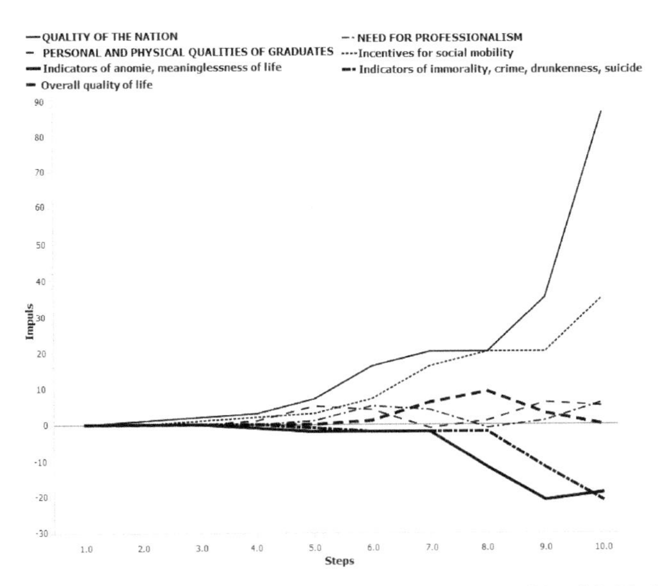


**Figure 9.9(a)** Charts of impulse processes in vertices $V_1, V_2, V_{24}, V_3, V_5, V_6$, and $V_7$; Scenario No. 3.

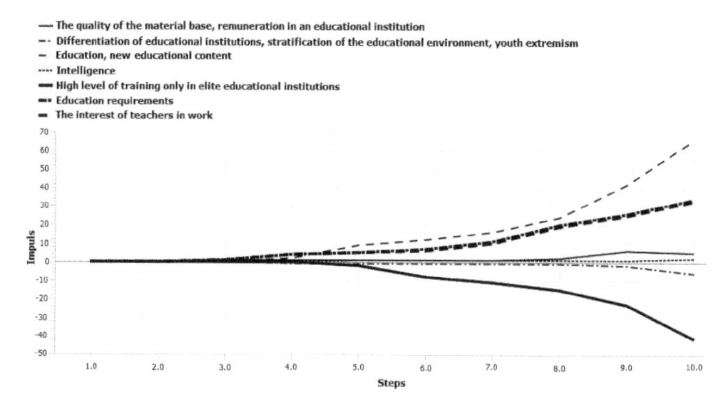


**Figure 9.9(b)** Charts of impulse processes in vertices $V_{15}, V_{16}, V_{17}, V_{12}, V_{19}, V_{20}$, and $V_{21}$; Scenario No. 3.

indicators, for example, quality of the life, education status, state of economy, and a decrease in negative ones, such as emigration, indicators of anomie, meaninglessness of life, and others.

## 9.3 Conclusion

To develop a strategy for CHS, it was first proposed to use jointly the methodologies of foresight and cognitive modeling: at the first stage, apply the foresight methodology with the aim of creating alternatives of scenarios and use the obtained results at the second stage as initial data for cognitive modeling for constructing scenarios of the desired future and ways of their implementation.

The strategy of combining foresight methodologies and cognitive modeling of CHSs that complement each other is an effective tool for studying complex systems, their structure, and behavior. Methods of foresight and cognitive modeling allow to understand, describe, and explain a complex system in its integrity, to represent the system as a whole without losing details. Modeling a complex system and its properties, predicting its future development is a prerequisite for the development and adoption of managerial decisions aimed at the sustainable development of any socio-economic, environmental, and political system. The given example of a cognitive study of the socio-economic system considered in the context of a pandemic illustrates the possibilities of the cognitive modeling methodology and the corresponding software tool. Based on the study, the fundamental possibility of confronting the pandemic in the future was determined by improving the education system, improving the quality of university graduates, the impact of sound public policies on the education system, and the quality of life of the population. These studies may continue in the direction of a deeper detailed study and cognitive description of society and the impact of higher education on the quality of a nation.

## 9.4 Acknowledgment

The authors of this chapter thank the editors for the opportunity to publish their investigations in the River Publishers book "Advanced Control Systems: Theory and Applications" which will provide an opportunity to bring the scientific community to the discussion of our research.

## References

[1] M.Z Zgurovsky, N.D. Pankratova, 'System analysis: Theory and Applications', Springer, 2007.
[2] Abbas Mardani, Edmundas Zavadskas, Kannan Govindan, Aslan Senin, Ahmad Jusoh, 'VIKOR Technique: A Systematic Review of the State of the Art Literature on Methodologies and Applications. Sustainability', 8, p1-38, 2016.
[3] P.A. Mikhnenko, 'Dynamic modification of SWOT analysis', Economic analysis: theory and practice, 18 (417), pp. 60–68, 2015.
[4] K. Gopalakrishnan, V. Vijayalakshmi, 'Using Morphological Analysis for Innovation and Resource and Development: An Invaluable Tool for Entrepreneurship', Annual Research Journal of Symbiosis Centre for Management Studies, Pune Vol. 2, Issue 1, March, pp. 28–36, 2014.
[5] M. Socorro García-Cascale, M.Teresa Lamata, 'On rank reversal and TOPSIS method' Mathematical and Computer Modelling, vol. 56, 5–6, pp. 123–132, 2012.
[6] T.Ritchey, 'Futures Studies using Morphological Analysis', Adapted from an article for the UN University Millennium Project: Futures Research Methodology Series, 2005.
[7] Wolfgang Weimer-Jehle, 'Cross-impact balances: A system-theoretical approach to cross-impact', Technological Forecasting & Social Change 73, pp. 334–361, 2006.
[8] Nesrin Alptekin, 'Integration of SWOT Analysis and TOPSIS Method In Strategic Decision Making Process', The Macrotheme Review, 2(7), Winter, 2013.

[9] [M.Hallowell, J.A. Gambatese, 'Qualitative Research: Application of the Delphi Method to CEM Research', Journal of Construction Engineering and Management, 136(1), pp.99-107, January 2010. DOI: 10.1061/(ASCE) CO.1943-7862.0000137

[10] N.D. Pankratova, V.V. Savastiyanov, 'Foresight Process Based on Text Analytics', Int. J. "Information Content and Processing", 1(1), ITHEA, Sofia, 2014.

[11] Innovative development of socio-economic systems based on foresight and cognitive modelling methodologies. In editors Gorelova G.V., Pankratova N.D. Kiev, Nauk. Dumka, 2015. (In Russian)

[12] N.D.Pankratova, I.O.Savchenko, 'Morphological analysis. Problems, theory, application', Kiev, Nauk. Dumka, 2015 (In Ukraine).

[13] N.D.Pankratova, G.V.Gorelova, V.A.Pankratov, 'Strategy for the Study of Interregional Economic and Social Exchange Based on Foresight and Cognitive Modelling Methodologies'. In Proceedings of the 8th Int. Conf. on "Mathematics. Information Technologies. Education, Shatsk, Ukraine, June 2–4, pp.136–141, 2019.

[14] N.A Abramova, Z.K.Avdeeva, 'Cognitive analysis and management of the development of situations: problems of methodology, theory and practice', Problems of control, 3, pp.85–87, 2008.

[15] Z.K., Avdeeva, S.V.Kovriga, 'On Governance Decision Support in the Area of Political Stability Using Cognitive Maps', In 18-th IFAC Conference on Technology, Culture and International Stability (TECIS2018), 51(30), pp.498–503, 2018. doi:10.1016/j.ifacol.2018.11.264

[16] P Langley. J.E Laird, S Rogers. 'Cognitive architectures: Research issues and challenges. Cognitive Systems Research', vol.10(2),141-160, 2009. DOI: 10.1016/j.cogsys.2006.07.004

[17] S.V.Kovriga, V.I.Maksimov, 'Cognitive technology of strategic management of the development of complex socio-economic objects in an unstable external environment', 1st issue of Cognitive Analysis and Situational Management (CASC'2001), 2001.

[18] M.Z. Zgurovsky, V.A Pankratov. 'Strategy of innovative development of the region based on the synthesis of foresight methodology and cognitive modelling', System research and information technology, No.2, pp.7–17, 2014 (In Russian).

[19] F.R. Gantmakher, 'Matrix theory', M ,Nauka, 1967 (In Russian).

[20] R.H.Atkin, 'Combinatorial Connectivies in Social Systems. An Application of Simplicial Complex Structures to the Study of Large Organisations', Interdisciplinary Systems Research, 1997.
[21] R.H. Atkin, J Casti, 'Polyhedral Dynamics and the Geometry of Systems', RR-77-International Institute for Applied Systems Analysis, Laxenburg, Austria, 1977.
[22] J Casti. 'Connectivity, Complexity, and Catastrophe in Large-scale Systems'. A Wiley – Interscience Publication International Institute for Applied Systems Analysis. JOHN WILEY and SONS. Chichester – New York – Brisbane – Toronto, 1979.
[23] F.Roberts, 'Graph Theory and its Applications to Problems of Society, Society for Industrial and Applied Mathematics', Philadelphia, 1978. doi:10.1137/1.9781611970401
[24] V.Kulba, D.A.Kononov, S.S.Kovalevsky, S.A.Kosyachenko, R.M.Nizhegorodtsev, I.V Chernov, 'Scenario analysis of the dynamics of behavior of socio-economic systems'. M.: IPU RAS, 2002.
[25] V.I.Maksimov, 'Cognitive technology – from ignorance to understanding'. 1st work "Cognitive analysis and management of the development of situations", (CASC'2001).1. pp.4–18, 2001.
[26] Program for cognitive modeling and analysis of socio-economic systems at the regional level. Certificate of state registration of computer programs N 2018661506, 2018.
[27] V.E.Shukshunov, A.A. Ovsyannikov, 'A system model of the organizational and economic reform of education in Russia', M., MANVSh, 1998.

# 10

# Special Cases in Determining the Spacecraft Position and Attitude Using Computer Vision System

**V. Gubarev, N. Salnikov, S. Melnychuk, V. Shevchenko, L. Maksymyuk**

Space Research Institute, National Academy of Sciences of Ukraine and State Space Agency of Ukraine, Glushkov Av. 40, 4/1, Kyiv, 03680, Ukraine
E-mail: v.f.gubarev@gmail.com, salnikov.nikolai@gmail.com, melnychuk89s@gmail.com, vovan@gmail.com, hatahatky@gmail.com

## Abstract

This paper investigates the accuracy of the relative pose estimation using monocular computer vision systems. The orientation and position of the target solid are determined by a set of control points located on its surface, which are recognized on the captured digital image. Their measured positions contain a sampling error, the propagation of which leads to inaccurate solution. The magnitude of the final error depends on the spatial location of the used control points and therefore varies with the movement and rotation of the target. In such an application as measuring the relative pose of the target spacecraft during the approach and docking in space, it is necessary to control the magnitude of the solution errors and ensure that the specified accuracy of the solution is obtained. For guaranteed estimation, a problem is considered under conditions of bounded uncertainty. We start with the assumption that an image shift of less than 1 pixel cannot be detected. This condition applied to control points determines the maximum possible guaranteed accuracy of the solution. Typical cases of the spatial location of control points are considered and the guaranteed accuracy of measuring position and orientation is determined.

**Keywords:** Pose estimation, perspective-n-point problem, computer vision, guaranteed estimation.

## 10.1 Introduction

Determining the position and orientation of the camera from the observed reference points with known coordinates is one of the main tasks of photogrammetry and machine vision. In photogrammetry, this problem is known as "location determination problem," "space resection problem," or "camera resection problem." In machine vision, it is called "perspective-n-point (PnP) problem."

The first methods for solving the problem were proposed more than 170 years ago. Historically, the problem has emerged from practical tasks in the mapping and aircraft navigation. Currently, this problem arises in many applications related to the monitoring and control of moving objects in medicine, industrial and autonomous robotics, augmented reality, etc.

The need for a reliable and high-precision solution to this problem exists in one general scheme for determining the relative position and attitude of spacecraft during autonomous approach and docking in orbit [1, 2]. This work is the development of a research program to create such a computer vision system [3]. Due to the critical fulfillment of accuracy requirements, the problem of estimating the solution error arises. When designing and developing vision-based docking systems, an analysis of potential measurement capabilities is required.

The initial data of the problem are given known 3D control points and corresponding 2D image observations, obtained by perspective projection. The transformation of the projection is determined by the internal camera parameters. In photogrammetry, the case when the camera parameters are unknown is often considered [4, 5]. Then they distinguish the task of identifying camera parameters – "internal calibration," and the task of determining its position and orientation – "external calibration." In machine vision, camera parameters are generally considered known. Problems are distinguished by the number of used control points – P3P, P4P, P5P, P6P, ..., PnP.

The minimum number of control points for a problem with a calibrated camera is 3. Depending on the spatial location of control points and the center of perspective (CP), the number of solutions varies. In non-degenerate cases, the P3P problem has up to four solutions [6]. The P4P problem has a unique solution, when no more than two control points lie on a single line and all control points lie in a common plane, not containing CP [7]. In other non-degenerate cases, P4P and P5P problems may have non-unique solutions. When $n \geq 6$, the problem can be reduced to linear using direct linear transform (DLT) method [4, 8]. Hence, it will have multiple solutions only in degenerate cases.

Two main definitions of a problem exist [9]. The first – distance-based definition – refers to distances from CP to control points. Images of any pair of points define an angle between directions on them from the CP. On the other hand, the distance between the reference points in 3D space is determined from the source data. The cosine theorem yields a quadratic equation for distances. Enumerating all pairs gives a system of such equations. Solution for P3P is considered in [10, 6, 11], for P4P in [7], [12–15], and [9], and for P5P and PnP in [14]. Pose of the camera is determined by these distances [16–19].

The second definition – transform-based definition – consists in directly determining the conversion between 3D coordinates of control points and 2D image coordinates through space rotation, translation, and projection. Unlike distance-based definition, this approach does not lead to the possibility of obtaining additional physically impossible solutions when $3 < n$ [9]. This approach is commonly used for the general PnP problem. Many methods for solving the problem have been developed. There are iterative methods [20–22] such as POSIT [23], LHM [24], [25], NPL [26], SP [27], PPnP [28], ASPnP [29], OPnP [30], UPnP [31], and MLPnP [32], and non-iterative methods such as EPnP [33], REPPnP [34], CEPPnP [35], DLS [36], and RPnP [37].

Spatial configuration of control points and CP determines not only the number of solutions in noise-free case but also a solution sensitivity in the presence of noise. In [38–40], we considered failure cases for P3P problem, when solution becomes unstable.

In this article, we consider an achievable accuracy of a PnP problem solution in the presence of bounded noise in initial data. We will examine the correctness of the problem of high-precision relative pose determination within the guaranteed approach to uncertainty. We assume that 2D position on image can be measured with an error limited by some constant value. The motivation for this definition is the discreteness of digital photo sensor. Known specific characteristics of camera allow numerically assessing the maximum possible positioning error in each particular case.

## 10.2 PnP Problem Statement

Position and orientation are determined by the relative pose of the corresponding reference frames. We denote the world reference frame (WRF) $O_W x_W y_W z_W$ with unit vectors $\vec{x}_W, \vec{y}_W, \vec{z}_W$ and origin in $O_W$ and the camera reference frame (CRF) $O_C x_C y_C z_C$ with unit vectors $\vec{x}_C, \vec{y}_C, \vec{z}_C$ and origin in projective center $O_C$. Pose of WRF in CRF is defined by $3 \times 4$ matrix

$$E = [R, t], \tag{10.1}$$

where $R = U^T \cdot [\vec{x}_W, \vec{y}_W, \vec{z}_W]$ is an orthogonal rotation matrix, $t = U^T \cdot (O_W - O_C)$ is a translation vector, and $U = [\vec{x}_C, \vec{y}_C, \vec{z}_C]$.

The position in WRF is denoted by vector $P = (x, y, z)^T$ or extended vector $\tilde{P} = (x, y, z, 1)^T$ in homogeneous coordinates. The position on the image is expressed in pixel units and is denoted by the vector $p = (u, v)^T$ or extended vector $\tilde{p} = (u, v, 1)^T$ in homogeneous coordinates. According to the ideal pinhole camera model, the relationship between coordinates of a control point is given by

$$A\tilde{p} = KE\tilde{P}, \tag{10.2}$$

where $A$ is a scalar factor. $K$ is a camera intrinsic calibration matrix

$$K = \begin{pmatrix} f_u & s & u_0 \\ 0 & f_v & v_0 \\ 0 & 0 & 1 \end{pmatrix}, \tag{10.3}$$

where $f_u, f_v$ are scale factors in image $u$ and $v$ axes, $s$ is a skewness parameter, and $u_0, v_0$ are coordinates of the principal point. The set of control point is given as

$$\tilde{P}_i = (x_i, y_i, z_i,\ 1)^T, \qquad \tilde{p}_i = (u_i, v_i,\ 1)^T, \qquad i = \overrightarrow{1, n}, \tag{10.4}$$

where $n$ is a number of control points.

Equations (10.1)–(10.4) present the PnP problem statement in a transform-based definition. We should derive a distance-based definition. With given $K$, each $\tilde{p}_i$ defines a ray, on which the $i$th control point lies (Figure 10.1).

Ray direction in CRF is defined by a unit vector $e_i$

$$e_i = \frac{K^{-1} \cdot \tilde{p}_i}{\sqrt{(K^{-1} \cdot \tilde{p}_i)^T \cdot (K^{-1} \cdot \tilde{p}_i)}}. \tag{10.5}$$

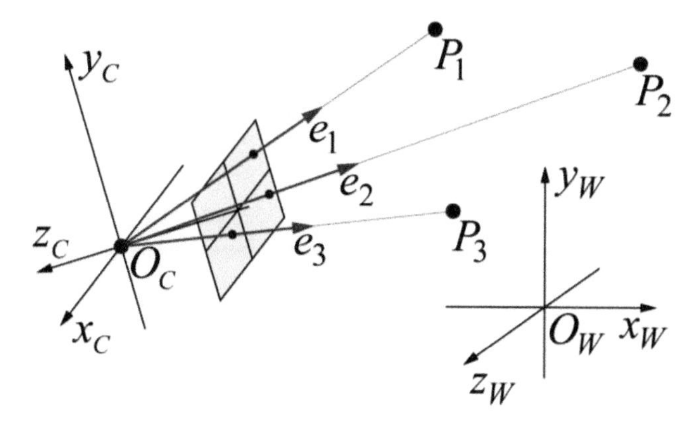


**Figure 10.1** P3P problem.

Considering two points $\tilde{p}_i$ and $\tilde{p}_j$, we can apply law of cosines to the triangle $O_C P_i P_j$. We denote Euclidian distance from $O_C$ to $P_i$ as $h_i$ and from $O_C$ to $P_j$ as $h_j$. Hence,

$$h_i^2 + h_j^2 - 2h_i h_j \cos\theta_{ij} = d_{ij}^2, \tag{10.6}$$

where $\theta_{ij}$ is an angle between rays $e_i$ and $e_j$. So $\cos\theta_{ij} = {e_i}^T e_j$, and $d_{ij}$ is Euclidian distance between $i$th and $j$th control points; so $d_{ij}^2 = \left(\tilde{P}_i - \tilde{P}_j\right)^T \cdot \left(\tilde{P}_i - \tilde{P}_j\right)$. Every pair of control points provides one quadratic equation (10.6) to the system

$$F_{ij}\left(h_i, h_j\right) = 0, \qquad i = \overrightarrow{1, n-1}, \;\; j = \overrightarrow{i+1, n}. \tag{10.7}$$

After finding positive distances $h_i$, $i = \overrightarrow{1, n}$, we have coordinates of control points in CRF and in WRF; so Equation (10.2) transforms into the 3D–3D absolute pose problem [19]

$$P_i^C = E\tilde{P}_i, \quad i = \overrightarrow{1, N}, \tag{10.8}$$

where $P_i^C = h_i e_i$.

## 10.3 PnP Problem Under Uncertainty

Solution to the PnP problem is determined by the spatial configuration of the control points and the center of the projection. As the initial data, we use only those points whose images $\tilde{p}_i$ are distinct. If their number is at least

3, we consider this case non-degenerate. Then the PnP problem can have from one to four different solutions, and for each solution, there is a certain neighborhood in which there is no other one.

In practice, the initial data (10.3) and (10.4) of the problem are determined with some error. From here arises a statement of the problem containing uncertainty. The final solution of PnP will depend on the method of representing the uncertainty of the source data. The simple application of a specific solution method to inaccurate source data can lead to more or fewer solutions compared to a noiseless case. When minimizing the criteria taking into account a large number of control points with inaccurate coordinates, the number of solutions can be significantly greater than 4 [30].

In some practical cases, such as tracking objects, an initial approximation is known which simplifies the selection of the right solution. However, the question of the accuracy of the solution, which can be guaranteed, remains unclear.

We assume that the parameters of the camera $K$ and the position of the control points $\tilde{P}_i,\ i = \overrightarrow{1,n}$ are set without errors. Position of the control points on the image is measured with an error

$$p_i = p_i^* + \xi_i, \qquad i = \overrightarrow{1,n}, \tag{10.9}$$

where $p_i^* = (u_i^*, v_i^*)^T$ is a true position of $i$th control point on image plane, and $\xi_i = (\xi_i^u, \xi_i^v)^T$ is a measurement error. Uncertainty will be considered in terms of a guaranteed approach. We assume that positioning on the image is carried with 1 pixel precision; so we have bounded error

$$|\xi_i^u| < 1,\ |\xi_i^v| < 1, \qquad i = \overrightarrow{1,n}, \tag{10.10}$$

where $\hat{p}_i$ is the measured position of the $i$th reference point and $p_i$ is true position.

Since the initial data contain uncertainty, a solution can be found only with some limited accuracy. Two different relative poses $E$ and $E'$ will be considered indistinguishable if the images of all control points differ by less than one pixel. Hence, for each possible $E$, there exists a set of indistinguishable poses

$$G(E) = \left\{ E' : \ \forall i = \overrightarrow{1,n} : \ \left\| p_i - p_i' \right\|_\infty < 1 \right\}, \tag{10.11}$$

where $p_i'$ is a position of $i$th reference point on image plane for relative pose $E'$. The size of $G(E)$ defines the potential guaranteed precision of vision-based relative pose estimation. To measure this quantity, it is necessary to parameterize space of matrices $E$ and determine the limits of variation of these parameters.

## 10.4 Rotation Parameterization

A rigid body is 3D space that has six degrees of freedom. Position and orientation can be represented in a more compact form than the $3 \times 4$ matrix $E = [R, t]$. Rotation matrix $R$ is orthonormal, and there are various methods for parameterizing it. Since our goal is to estimate size of the set $G(E)$, we choose the parameterization of the minimum dimension.

Any rotation in 3D space can be represented by the axis of rotation $\vec{r}$ and the magnitude of rotation $\gamma$. In such form, we can evaluate the size of set $G(E) \subset R^6$ in rotational subspace along direction determined by the axis $\vec{r}$.

We denote matrices of basic rotation in space around the coordinate axes by an angle $\varphi$ as

$$R_X(\varphi) = \begin{pmatrix} 1 & 0 & 0 \\ 0 & c\varphi & -s\varphi \\ 0 & s\varphi & c\varphi \end{pmatrix}, \; R_Y(\varphi) = \begin{pmatrix} c\varphi & 0 & s\varphi \\ 0 & 1 & 0 \\ -s\varphi & 0 & c\varphi \end{pmatrix},$$
$$R_Z(\varphi) = \begin{pmatrix} c\varphi & -s\varphi & 0 \\ s\varphi & c\varphi & 0 \\ 0 & 0 & 1 \end{pmatrix}, \tag{10.12}$$

where $s$ denotes sine function of the next specified argument and $c$ denotes cosine.

Unit vector $\vec{r} \in R^3$ can be expressed in spherical coordinates by two angles $\alpha$ and $\beta$. We define the first two turns on angle $\alpha$ and $\beta$; so the last rotation on angle $\gamma$ will be done around the resulting axis Z. We choose

$$R(\alpha, \beta, \gamma) = [R_Y(\alpha) R_X(\beta)] \cdot R_Z(\gamma) \cdot [R_Y(\alpha) R_X(\beta)]^T. \tag{10.13}$$

Hence, axis $\vec{r}$ is the third column of the matrix $R_Y(\alpha) R_X(\beta)$:

$$\vec{r}(\alpha, \beta) = \begin{pmatrix} \sin\alpha\cos\beta & -\sin\beta & \cos\alpha\cos\beta \end{pmatrix}^T, \tag{10.14}$$

where $\alpha \in [-\pi, \pi)$, $\beta \in \left[-\frac{\pi}{2}, \frac{\pi}{2}\right]$. Rotation matrix (10.13) for axis $\vec{r}$ and angle $\gamma$ is

$$R(\alpha,\beta,\gamma) = I^{3\times3} + \begin{pmatrix} 0 & -c\alpha c\beta & -s\beta \\ c\alpha c\beta & 0 & -s\alpha c\beta \\ s\beta & s\alpha c\beta & 0 \end{pmatrix} s\gamma +$$
$$+ \begin{pmatrix} 1 - s^2\alpha c^2\beta & s\alpha\frac{s2\beta}{2} & -\frac{s2\alpha}{2}c^2\beta \\ s\alpha\frac{s2\beta}{2} & c^2\beta & c\alpha\frac{s2\beta}{2} \\ -\frac{s2\alpha}{2}c^2\beta & c\alpha\frac{s2\beta}{2} & 1 - c^2\alpha c^2\beta \end{pmatrix}(c\gamma - 1), \tag{10.15}$$

where $I^{3\times3}$ is an identity matrix. When analyzing the effect of rotation, we will fix variables $\alpha$ and $\beta$ and consider the rotation $R_{\alpha\beta}(\gamma)$ along the shortest arc.

## 10.5 Sensitivity of Image

We fix the camera position and assume the world moving. We consider how much the image position of a control point changes with a small translation and a small rotation of the world. To simplify the analysis, we will take zero-skew camera ($s = 0$) and $u_0 = v_0 = 0,\ f_u = f_v = f$, where $f$ is a focus length in pixels. Instead of Equation (10.3), we gain

$$K = \begin{pmatrix} f & 0 & 0 \\ 0 & f & 0 \\ 0 & 0 & 1 \end{pmatrix}. \tag{10.16}$$

Consider a control point, given by the vector $P = (x, y, z)^T$ in non-homogeneous coordinates in WRF. We choose WRF orientation that coincides with the orientation of CRF; so $R = I^{3\times3}$. The position of WRF relative to CRF is denoted by $t^0 = \left(t_x^0, t_y^0, t_z^0\right)^T$. Image of this control point is determined by the vector $p = (u, v)^T$

$$u = f \cdot \frac{x + t_x^0}{z + t_z^0}, \qquad v = f \cdot \frac{y + t_y^0}{z + t_z^0}. \tag{10.17}$$

Applying rotation (10.15) by an angle $\gamma$ around a certain fixed axis and translating by the vector $t = (t_x, t_y, t_z)^T$ will result in image shift from $p$ to $p' = (u', v')^T$

$$u' = f \cdot \frac{A}{C}, \qquad v' = f \cdot \frac{B}{C}, \tag{10.18}$$

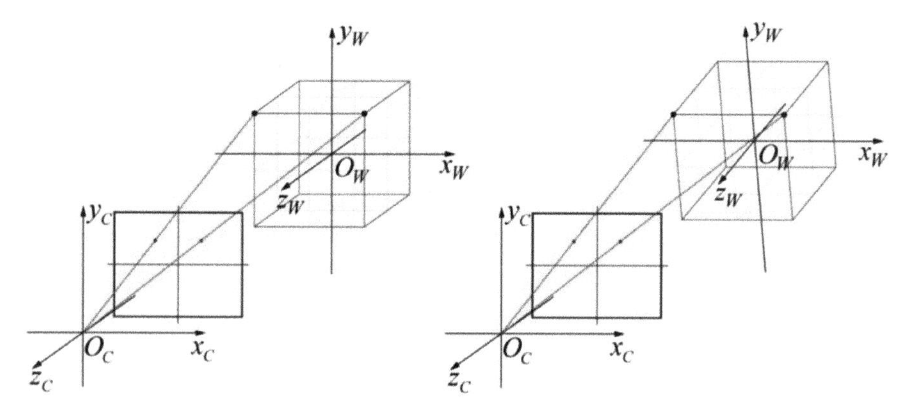


**Figure 10.2** Rotation and translation of WRF with two stationary points.

where we denote

$$\begin{aligned}
A &= \left[1 + \left(1 - s^2\alpha c^2\beta\right)(c\gamma - 1)\right]x + \left[-c\alpha c\beta s\gamma + s\alpha\frac{s2\beta}{2}(c\gamma - 1)\right]y+ \\
&+ \left[-s\beta s\gamma - \frac{s2\alpha}{2}c^2\beta\,(c\gamma - 1)\right]z + t_x^0 + t_x, \\
B &= \left[c\alpha c\beta s\gamma + s\alpha\frac{s2\beta}{2}(c\gamma - 1)\right]x + \left[1 + c^2\beta\,(c\gamma - 1)\right]y+ \\
&+ \left[-s\alpha c\beta s\gamma + c\alpha\frac{s2\beta}{2}(c\gamma - 1)\right]z + t_y^0 + t_y, \\
C &= \left[s\beta s\gamma - \frac{s2\alpha}{2}c^2\beta\,(c\gamma - 1)\right]x + \left[s\alpha c\beta s\gamma + c\alpha\frac{s2\beta}{2}(c\gamma - 1)\right]y+ \\
&+ \left[1 + \left(1 - c^2\alpha c^2\beta\right)(c\gamma - 1)\right]z + t_z^0 + t_z.
\end{aligned} \tag{10.19}$$

We are interested in the case when the simultaneous rotation $R$ and translation $t$ will retain the image position (10.18) of the control point. Obviously, such combinations of $R$ and $t$ exist. An example of such a transformation would be the rotation of space around this control point. Similarly, we can construct a transformation that preserves the image position of a pair of control points (Figure 10.2).

It is also obvious that in non-degenerate case with three or more control points, there are no such $R$ and $t$ corresponding to arbitrarily small values of rotation and translation that do not change images. Otherwise, the number of solutions would have to be infinite. However, it is possible that there are such small $R$ and $t$ that lead to an immeasurably small displacement of images.

We consider the position of the certain control point in the image (10.18) as a function

$$u = u\left(q\right), \qquad v = v\left(q\right), \tag{10.20}$$

where $q = (\gamma, t_x, t_y, t_z)^T$. The Taylor expansion in the point $q_0 = (0, 0, 0, 0)^T$ gives a linear approximation

$$\begin{pmatrix} u \\ v \end{pmatrix} \approx \begin{pmatrix} u_0 \\ v_0 \end{pmatrix} + J_{\alpha,\beta}\left(q_0\right) \cdot \Delta q, \tag{10.21}$$

where $u = u\left(\Delta q\right), \ v = v\left(\Delta q\right), \Delta q = \left(\Delta\gamma, \Delta t_x, \Delta t_y, \Delta t_z\right)^T$. Matrix $J_{\alpha,\beta}$ is the Jacobian

$$J_{\alpha,\beta} = \begin{pmatrix} \frac{\partial u}{\partial \gamma} & \frac{\partial u}{\partial t_x} & \frac{\partial u}{\partial t_y} & \frac{\partial u}{\partial t_z} \\ \frac{\partial v}{\partial \gamma} & \frac{\partial v}{\partial t_x} & \frac{\partial v}{\partial t_y} & \frac{\partial v}{\partial t_z} \end{pmatrix}, \tag{10.22}$$

where the first column represents partial derivatives by rotation, and columns 2–4 represent partial derivatives by translation. Indices $\alpha, \beta$ indicate the fixed axis of rotation. In terms of Equation (10.19), it is

$$J_{\alpha,\beta} = \frac{f}{C^2} \begin{pmatrix} \frac{\partial A}{\partial \gamma}C - \frac{\partial C}{\partial \gamma}A & C & 0 & -A \\ \frac{\partial B}{\partial \gamma}C - \frac{\partial C}{\partial \gamma}B & 0 & C & -B \end{pmatrix}, \tag{10.23}$$

where

$$\begin{aligned}
\tfrac{\partial A}{\partial \gamma} &= \left[-\left(1 - s^2\alpha c^2\beta\right)s\gamma\right]x + \left[-c\alpha c\beta c\gamma - s\alpha\tfrac{s2\beta}{2}s\gamma\right]y + \left[\tfrac{s2\alpha}{2}c^2\beta s\gamma - s\beta c\gamma\right]z, \\
\tfrac{\partial B}{\partial \gamma} &= \left[c\alpha c\beta A\gamma - s\alpha\tfrac{s2\beta}{2}s\gamma\right]x + \left[-c^2\beta s\gamma\right]y + \left[-s\alpha c\beta c\gamma - c\alpha\tfrac{s2\beta}{2}s\gamma\right]z, \\
\tfrac{\partial C}{\partial \gamma} &= \left[\tfrac{s2\alpha}{2}c^2\beta s\gamma + s\beta c\gamma\right]x + \left[s\alpha c\beta c\gamma - c\alpha\tfrac{s2\beta}{2}s\gamma\right]y + \left[-\left(1 - c^2\alpha c^2\beta\right)s\gamma\right]z.
\end{aligned}$$

Substitution of $\gamma = t_x = t_y = t_z = 0$ gives expressions $A = x + t_x^0$, $B = y + t_y^0$, $C = z + t_z^0$, $\frac{\partial A}{\partial \gamma} = \left[-c\alpha c\beta\right]y + \left[-s\beta\right]z$, $\frac{\partial B}{\partial \gamma} = \left[c\alpha c\beta\right]x + \left[-s\alpha c\beta\right]z$, $\frac{\partial C}{\partial \gamma} = \left[s\beta\right]x + \left[s\alpha c\beta\right]y$; so Equation (10.23) takes on the form

$$J_{\alpha,\beta}\left(\vec{0}\right) = \frac{f}{\tilde{z}^2}\begin{pmatrix} \left(\left[-c\alpha c\beta\right]y + \left[-s\beta\right]z\right)\tilde{z} - \left(\left[s\beta\right]x + \left[s\alpha c\beta\right]y\right)\tilde{x} & \tilde{z} & 0 & -\tilde{x} \\ \left(\left[c\alpha c\beta\right]x + \left[-s\alpha c\beta\right]z\right)\tilde{z} - \left(\left[s\beta\right]x + \left[s\alpha c\beta\right]y\right)\tilde{y} & 0 & \tilde{z} & -\tilde{y} \end{pmatrix}, \tag{10.24}$$

where $\tilde{x} = x + t_x^0, \ \tilde{y} = y + t_y^0, \ \tilde{z} = z + t_z^0$. With respect to Equation (10.14), it converts into

$$J_{\alpha,\beta}\left(\vec{0}\right) = \frac{f}{\tilde{z}^2}\begin{pmatrix} \left(r_y z - r_z y\right)\tilde{z} - \left(r_x y - r_y x\right)\tilde{x} & \tilde{z} & 0 & -\tilde{x} \\ \left(r_z x - r_x z\right)\tilde{z} - \left(r_x y - r_y x\right)\tilde{y} & 0 & \tilde{z} & -\tilde{y} \end{pmatrix}. \tag{10.25}$$

Consider linearization of the image alteration. We denote the shift of the $i$th control point on the image plane by the vector $\Delta p^i = \left(u^i - u_0^i, v^i - v_0^i\right)^T$; so

$$\Delta p^i = J^i_{\alpha,\beta} \cdot \Delta q, \tag{10.26}$$

where $J^i_{\alpha,\beta}$ is a $J_{\alpha,\beta}\left(\vec{0}\right)$ in a corresponding control point $P^i = \left(x^i, y^i, z^i\right)^T$. Every control point provides two rows (10.26) to a general compound sensitivity matrix

$$M_{\alpha,\beta} = \begin{pmatrix} J^1_{\alpha,\beta} \\ J^2_{\alpha,\beta} \\ \vdots \\ J^n_{\alpha,\beta} \end{pmatrix} \tag{10.27}$$

of a linear system

$$M_{\alpha,\beta} \cdot \Delta q = \Delta p^{\text{total}}, \tag{10.28}$$

where $\Delta p^{\text{total}} = \left(\Delta u^1, \Delta v^1, \Delta u^2, \Delta v^2, \ldots, \Delta u^n, \Delta v^n\right)^T$. If the matrix $M_{\alpha,\beta}$ is not overdetermined, such a non-zero vector $\Delta q^* = \left(\Delta\gamma_1^*, \Delta t_x^*, \Delta t_y^*, \Delta t_z^*\right)^T$ exists, which satisfies

$$M_{\alpha,\beta} \cdot \Delta q = \vec{0}. \tag{10.29}$$

Hence, rotating by an angle $\varepsilon \cdot \Delta\gamma_1^*$ and translating by a vector $\varepsilon \cdot \left(\Delta t_x^*, \Delta t_y^*, \Delta t_z^*\right)^T$ will shift images of control points by a magnitude of the second order of smallest $\varepsilon$. Conversely, if system (10.29) is inconsistent, then there is no such rotation and translation that would nullify the linear components of the image shift magnitudes.

We consider general non-degenerate case of $n \geq 3$, which allows up to four different solutions. We denote by $B_0$ a neighborhood of the solution $E$ in which there are no other solutions. Also, we denote by $B_1$ a neighborhood of the solution $E$, where the Taylor series of a function (10.20) for all control points is convergent. The change of pose from $E$ to $E' \in B_0 \bigcap B_1$ shifts image points by non-zero values, which can be represented as the sum of linear and non-linear components. Linear component is determined by Equation (10.28). There are two alternatives.

1) If the linear component is non-zero, there is some neighborhood $B_2$ in which it is dominant. If the indistinguishable set of poses $G(E)$ is contained in $B_0 \bigcap B_1 \bigcap B_2$, then it can be estimated by the linear

component. In this case, system (10.29) is overdetermined and has only a trivial solution. For any non-zero $\Delta q$ solution of Equation (10.29) gives a residual vector, which determines the shift of control points in the image. If it does not exceed 1 pixel, such $\Delta q$ defines an unmeasured transformation. We want to determine bounds for value of $\Delta q$, which correspond to such a situation.

2) Otherwise, when the linear component is zero or the indistinguishable set of poses $G(E)$ is big enough so that it is not contained in $B_0 \bigcap B_1 \bigcap B_2$, $G(E)$ should be evaluated by a non-linear component.

In this paper, we consider only the first alternative. It corresponds to a situation of sufficiently high-precision measurements, which arises in practice when determining the position and orientation of a target spacecraft at extremely small distances.

## 10.6 Estimating an Indistinguishable Set

We consider an overdetermined homogeneous linear system

$$Hw = \vec{0}, \tag{10.30}$$

where $H$ is a $m \times k$ matrix, $w$ is a $k \times 1$ vector, $\vec{0}$ is a $m \times 1$ null vector, and $k < m$. We divide the system (10.30) into two parts:

$$H = [-b, H_1], \qquad w = \begin{pmatrix} \varepsilon \\ w_1 \end{pmatrix}, \tag{10.31}$$

where $b$ denotes the first column of $H$ with the opposite sign, $H_1$ is the matrix of the remaining columns, scalar $\varepsilon$ is the first component of $w$, and $w_1$ is a vector of remaining components. Then we have overdetermined non-homogeneous linear system

$$H_1 w_1 = \varepsilon \cdot b. \tag{10.32}$$

Least-squares method gives a solution

$$w_1 = \varepsilon \cdot \left(H_1{}^T H_1\right)^{-1} H_1{}^T b \tag{10.33}$$

that minimizes $l_2$-norm of the residual vector

$$\xi = H_1 w_1 - \varepsilon \cdot b. \tag{10.34}$$

Substituting Equation (10.33) gives

$$\xi = \varepsilon \cdot \left( H_1 \left( {H_1}^T H_1 \right)^{-1} {H_1}^T - I \right) b. \tag{10.35}$$

The norm is expressed as

$$\|\xi\|_2 = \sqrt{\xi^T \xi} = \varepsilon \cdot \sqrt{b^T \left( I - H_1 \left( {H_1}^T H_1 \right)^{-1} {H_1}^T \right) b}. \tag{10.36}$$

For a given value of the $l_2$-norm of the residual, we can obtain bounds for the first component of $w$

$$\|\xi\|_2 \leq \Xi \quad \Rightarrow \quad |\varepsilon| \leq \frac{\Xi}{\sqrt{b^T \left( I - H_1 \left( {H_1}^T H_1 \right)^{-1} {H_1}^T \right) b}}. \tag{10.37}$$

Similarly, we can obtain bounds for any component of the vector $w$.

The above approach is applicable to Equation (10.29). To determine the bounds of indistinguishable parameters $\Delta q$, it is necessary to set the residual constraint in the $l_2$-norm. Since the image indistinguishability is driven by the condition $\left|\Delta u^i\right| < 1, \quad \left|\Delta v^i\right| < 1, \quad i = \overrightarrow{1, n}$, i.e. in terms of $l_\infty$- norm, we can get overestimation

$$\left\|\Delta p^{\text{total}}\right\|_2 = \sqrt{\sum_{i=1}^{n} (\Delta u^i)^2 + \sum_{i=1}^{n} (\Delta v^i)^2} < \sqrt{\sum_{i=1}^{n} |\Delta u^i| + \sum_{i=1}^{n} |\Delta v^i|} < \sqrt{2n}. \tag{10.38}$$

Hence, value $\Xi = \sqrt{2n}$ in Equation (10.37) defines estimates for the magnitude of rotation and translation, which are guaranteed to be detected by a digital camera sensor, when rotation is performed around fixed axis $\vec{r}\,(\alpha, \beta)$. We denote these magnitudes by

$$\varepsilon_{rot}^{\alpha,\beta}, \quad \varepsilon_x^{\alpha,\beta}, \quad \varepsilon_y^{\alpha,\beta}, \quad \varepsilon_z^{\alpha,\beta}. \tag{10.39}$$

For each direction of the rotation axis, the values (10.39) will be different. The size of the set of indistinguishable poses will be determined by the worst cases. We denote

$$\begin{gathered} \varepsilon_{\text{rot}} = \max_{\alpha,\beta} \left( \varepsilon_{\text{rot}}^{\alpha,\beta} \right), \quad \varepsilon_x = \max_{\alpha,\beta} \left( \varepsilon_x^{\alpha,\beta} \right), \\ \varepsilon_y = \max_{\alpha,\beta} \left( \varepsilon_y^{\alpha,\beta} \right), \quad \varepsilon_z = \max_{\alpha,\beta} \left( \varepsilon_z^{\alpha,\beta} \right); \end{gathered} \tag{10.40}$$

so any separate or joint rotation and translation, which exceeds $\varepsilon_{\text{rot}}$ and $\varepsilon_x$, $\varepsilon_y$, $\varepsilon_z$ respectively, will lead to a measurable image displacement of at least one control point

$$\begin{bmatrix} \varepsilon_{\text{rot}} \leq |\Delta\gamma| \\ \varepsilon_x \leq |\Delta t_x| \\ \varepsilon_y \leq |\Delta t_y| \\ \varepsilon_z \leq |\Delta t_z| \end{bmatrix} \quad \Rightarrow \quad \left\| \Delta p^{\text{total}} \right\|_\infty > 1. \tag{10.41}$$

## 10.7 Design of Experiment

We will use the spatial configuration of several control points (Figure 10.3) for evaluating the size of the indistinguishable set (10.40). Control points $A$, $B$, $C$, $D$, $E$, and $F$ are placed in the corners of the union cube, which is rigidly fixed to WRF

$$P_A = \frac{1}{2}(-1,1,1)^T, P_B = \frac{1}{2}(1,1,1)^T, P_C = \frac{1}{2}(1,-1,1)^T,$$

$$P_D = \frac{1}{2}(-1,-1,1)^T, P_E = \frac{1}{2}(1,1,-1)^T, P_F = \frac{1}{2}(1,-1,-1)^T.$$

Such configuration imitates a typical case in the relative pose determination problem of approaching spacecraft. Suppose that control points lie on the surface of the target spacecraft when camera is mounted on the Chaser spacecraft. The task is to measure the position and orientation of target relative to the calibrated camera on Chaser.

Position of the target body relative to the camera is given by the transform $E = [R, t]$ without rotation

$$R = \begin{pmatrix} 1 & 0 & 0 \\ 0 & 1 & 0 \\ 0 & 0 & 1 \end{pmatrix}, \qquad t = \begin{pmatrix} t_x \\ t_y \\ t_z \end{pmatrix}.$$

Images of control points are obtained by a zero-skew camera (10.3) with parameters

$$s = 0, \quad u_0 = 0, \quad v_0 = 0, \quad f_u = f_v = f = 10000.$$

Such a large value of the parameter $f$ was chosen for modeling high-precision measurements.

We will determine the values (10.40) of rotation and translation of the target body, which guarantee a change in the position of at least one control

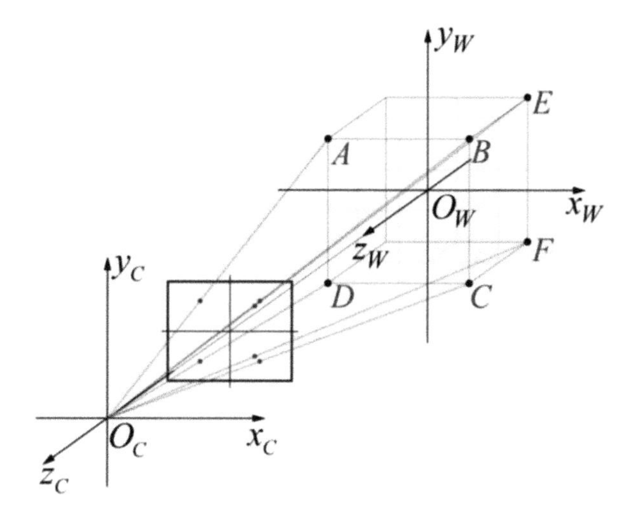


**Figure 10.3** Spatial configuration of control points.

point in the image by more than 1 pixel. We will use different sets of control points and vary the distance from the camera.

Since deriving Equation (10.21), we fixed the direction of the axis, around which the body rotated. Defining the worst-case axis, selection in Equation (10.40) is performed numerically among all the possible values

$$\alpha \in [0, \pi], \quad \beta \in \left[-\frac{\pi}{2}, \frac{\pi}{2}\right].$$

In each experiment, we choose:

- set of control points.
- vector $t$

and determine:

- Values $\varepsilon_{\text{rot}}, \ \varepsilon_x, \ \varepsilon_y, \ \varepsilon_z.$
- Corresponding worst-case rotation axes

$$\{\alpha_{\text{rot}}, \beta_{\text{rot}}\} = \arg\max\left(\varepsilon_{\text{rot}}^{\alpha,\beta}\right), \quad \{\alpha_x, \beta_x\} = \arg\max\left(\varepsilon_x^{\alpha,\beta}\right),$$
$$\{\alpha_y, \beta_y\} = \arg\max\left(\varepsilon_y^{\alpha,\beta}\right), \quad \{\alpha_z, \beta_z\} = \arg\max\left(\varepsilon_z^{\alpha,\beta}\right).$$

- Conditional number

$$\Theta = \frac{s_1\left(M_{\alpha,\beta}\right)}{s_4\left(M_{\alpha,\beta}\right)},$$

where $s_1$ and $s_4$ denote the largest and smallest singular values derived from singular value decomposition (SVD), for worst cases $\{\alpha_{\text{rot}}, \beta_{\text{rot}}\}$, $\{\alpha_x, \beta_x\}$, $\{\alpha_y, \beta_y\}$, $\{\alpha_z, \beta_z\}$.

- Minimal by $l_2$-norm reachable residual vector of image shift

$$\Delta p^{\text{total}} = \left(\Delta u^1, \Delta v^1, \Delta u^2, \Delta v^2, \ldots, \Delta u^n, \Delta v^n\right)^T,$$

received when rotation performs around the axis $\{\alpha_{\text{rot}}, \beta_{\text{rot}}\}$ by the angle $\varepsilon_{\text{rot}}$. This value is determined by Equation (10.29)

$$M_{\alpha_{\text{rot}}, \beta_{\text{rot}}} \cdot (\varepsilon_{\text{rot}}, \Delta t_x, \Delta t_y, \Delta t_z) = \vec{0}.$$

This residual vector will serve as a test of whether the linear approximation is correct at a given point. The linear component of image shift will dominate when

$$\left\|\Delta p^{\text{total}}\right\|_2 \approx \sqrt{2n}.$$

Otherwise, image shift is determined by a non-linear dependency, and received values $\varepsilon_{\text{rot}}$, $\varepsilon_x$, $\varepsilon_y$, $\varepsilon_z$ are not consistent estimates of the size of set $G(E)$.

- Overestimation constant, generated by using $l_2$-norm minimization, is determined by the value $\left\|\Delta p^{\text{total}}\right\|_\infty$.

We will build graphs for:

- $\varepsilon_{\text{rot}}(t)$; and $\varepsilon_x(t)$, $\varepsilon_y(t)$, $\varepsilon_z(t)$;
- $\Theta(t)$;
- $\left\|\Delta p^{\text{total}}\right\|_2$ and $\left\|\Delta p^{\text{total}}\right\|_\infty$ from $t$.

## 10.8 Numerical Simulations

First, we consider minimal configurations with $n = 3$. The distance from the camera varies from 0.8 to 10 m. Accordingly, the position parameters changed from $t = (0, 0, -0.8)^T$ to $t = (0, 0, -10)^T$.

**Case 1.** Control points *A*, *B*, and *C*. The points lie on a plane parallel to the image plane. Results are shown below (Figures 10.4 and 10.5).

Dependency of linear estimates of guaranteed detectable pose changes (Figure 10.4) coincides with the number of conditionality (Figure 10.5, left). Residual vector confirms the suitability of linear approximation. Using $l_2$-norm, we got overestimation; so instead of the 1 pixel, we have minimum shift of control points from 1.5 to 2.5 pixels.

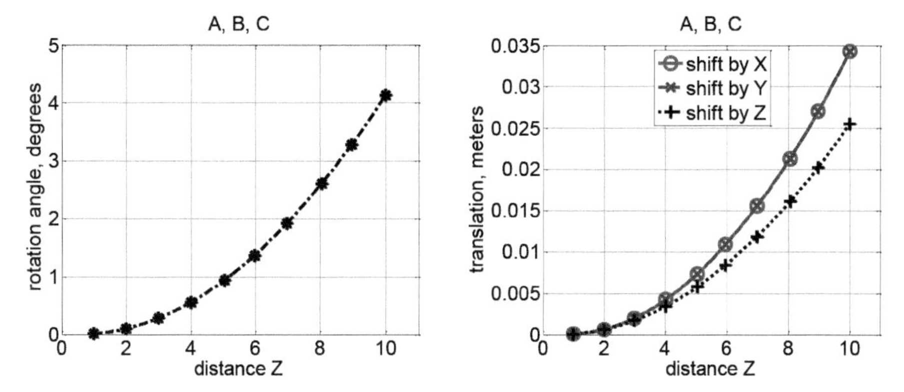


**Figure 10.4** Distinguishable rotation angle (left) and translation distance (right).

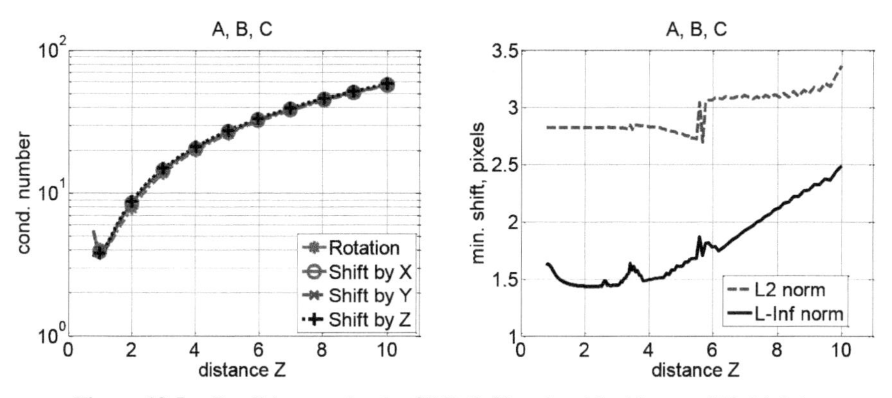


**Figure 10.5** Condition number by SVD (left) and residual image shift (right).

With increasing distance, the accuracy of guaranteed detection decreases quadratically both in rotation and in all displacement directions.

No bad cases were found across the entire range of $[0.8, \quad 10]$ meters.

**Case 2.** Control points $A$, $B$, and $E$. Plane $A,B,E$ is parallel to camera vision line.

Figures 10.6 and 10.7 show that there is a bad case near the lower bound of range. We consider the interval $[0.8, \quad 1.5]$ to be bad. Hence, there is some axis such that the "visible" effect of rotation around it can be greatly compensated by translation. In this situation pose, observability is reduced. It is confirmed by the number of conditionality.

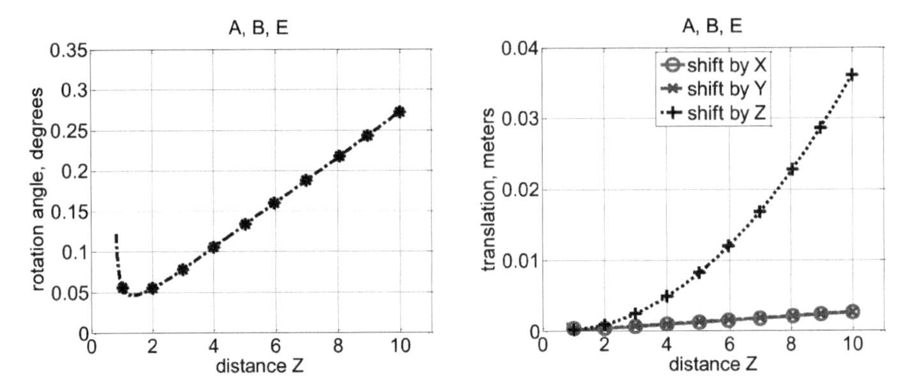


**Figure 10.6** Distinguishable rotation angle (left) and translation distance (right).

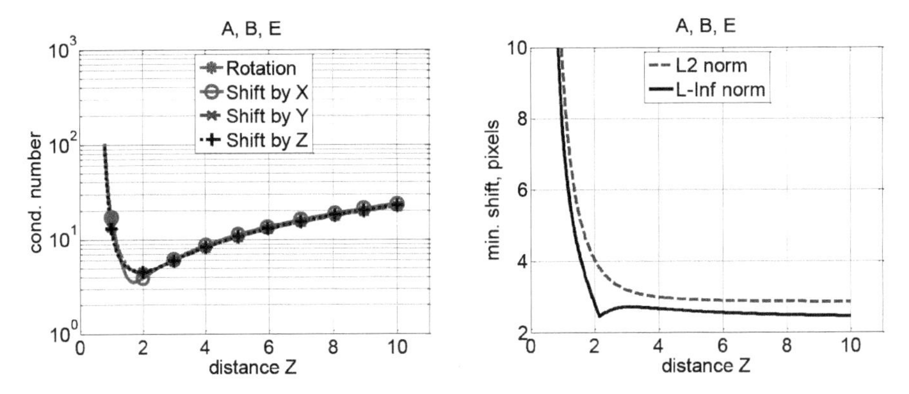


**Figure 10.7** Condition number by SVD (left) and residual image shift (right).

Translations by $X$ and $Y$ directions are distinguishable much better than distance $Z$. Sensitivity of longitudinal displacements decreases linearly, when sensitivity to distance changes decreases quadratically.

Image sensitivity to any small rotation also decreases linearly with the growth of distance.

Compared to the case ($A$,$B$,$C$), combination ($A$,$B$,$E$) gives much less condition number. Accordingly, the accuracy of the worst case is increased (about 16 times).

**Case 3.** Control points $A$, $B$, and $F$. This configuration is of an intermediate type between case 1 (plane $A$,$B$,$C$ is parallel to the image plane) and case 2 (plane $A$,$B$,$E$ is perpendicular to the image plane). The results are shown in Figures 10.8 and 10.9.

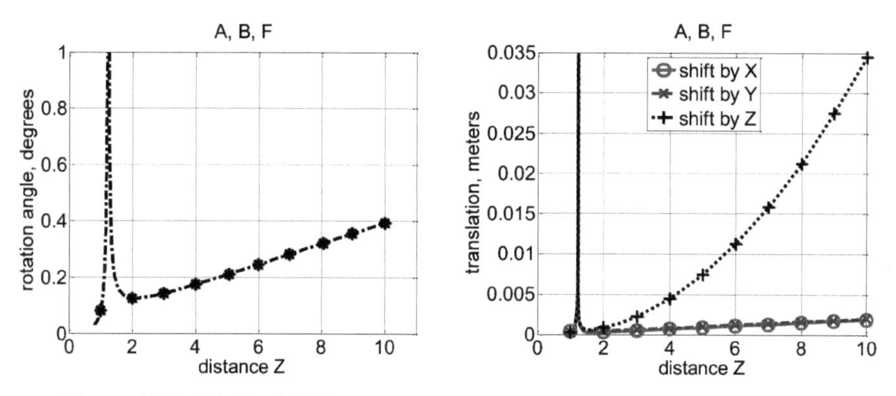


**Figure 10.8** Distinguishable rotation angle (left) and translation distance (right).

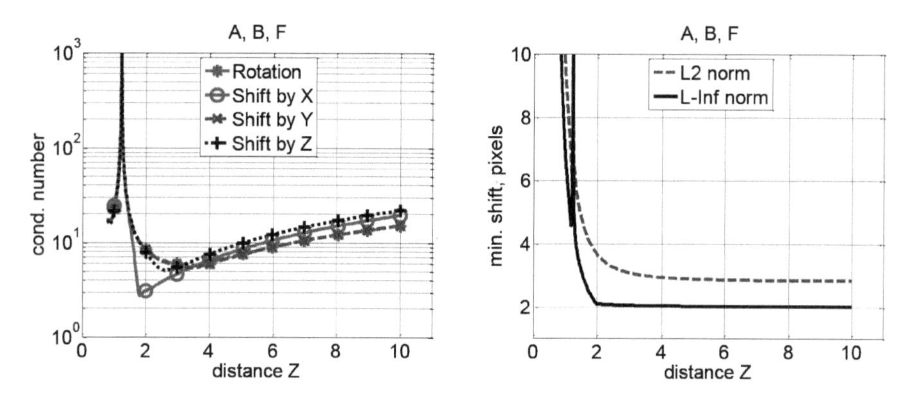


**Figure 10.9** Condition number by SVD (left) and residual image shift (right).

Bad case was found inside the range. In the point $t_z \approx -1.224756$, the matrix $M_{\alpha,\beta}$ is rank-deficient for such $\alpha, \beta$ when the rotation axis is

$$r = (0.298374065963878, \; -0.785628427171974, \; -0.541997132076797)^T.$$

Outside a bad interval $[0.8, \;\; 1.5]$ of distance, the dependencies are similar to the previous combination (*A*,*B*,*E*).

At the point $t_z \approx -1.224756$, a peak is observed in all directions of translation and in rotation. This means that optimal image shift compensation caused by change of one parameter performs by fitting of all the others.

**Case 4.** Control points *A*, *C*, *E*. The results are shown in Figures 10.10 and 10.11.

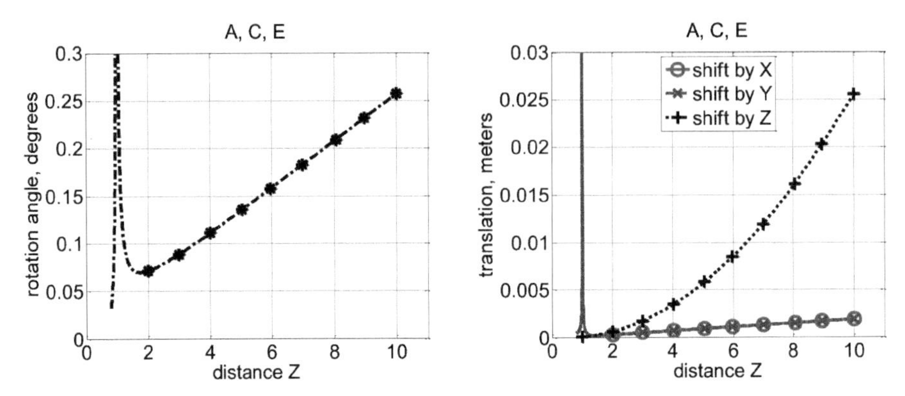


**Figure 10.10** Distinguishable rotation angle (left) and translation distance (right).

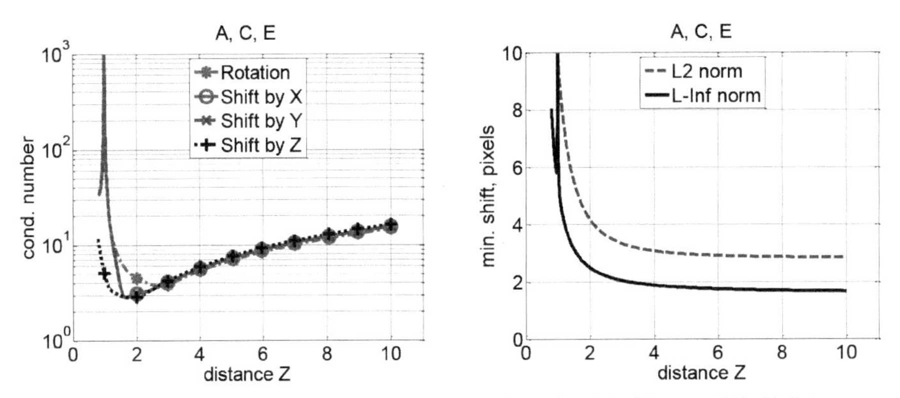


**Figure 10.11** Condition number by SVD (left) and residual image shift (right).

Bad case was found inside the range. In the point $t_z = -1$, the matrix $M_{\alpha,\beta}$ has rank 2, for such $\alpha, \beta$ when rotation axis is

$$r = \left( -\frac{\sqrt{2}}{2}, \ -\frac{\sqrt{2}}{2}, \ 0 \right)^T.$$

As shown in Figure 10.10 (left), there is no peak for the *Z* component. It means that the image shift caused by rotation around axis $r$ and translation along *X* and *Y* are linear dependent. They can be compensated by each other, leading to poor observability. At the same time, the component *Z* is detected with great accuracy.

**Case 5.** Control points *C*, *E*, and *F*. This case is very similar to the previous ones. The results will be given in the table below.

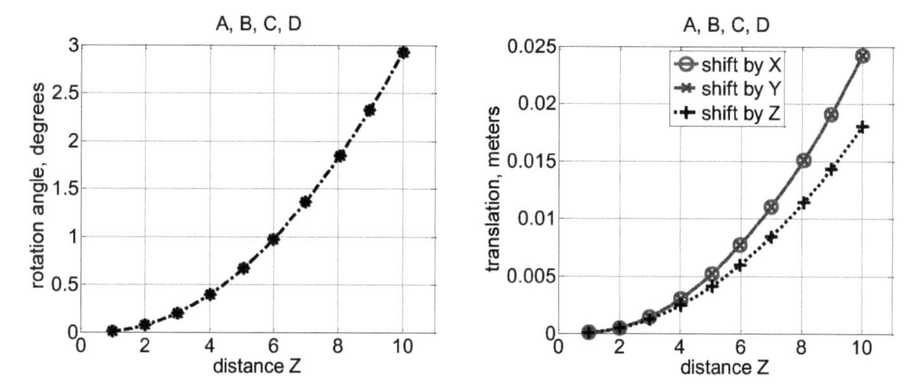


**Figure 10.12** Distinguishable rotation angle (left) and translation distance (right).

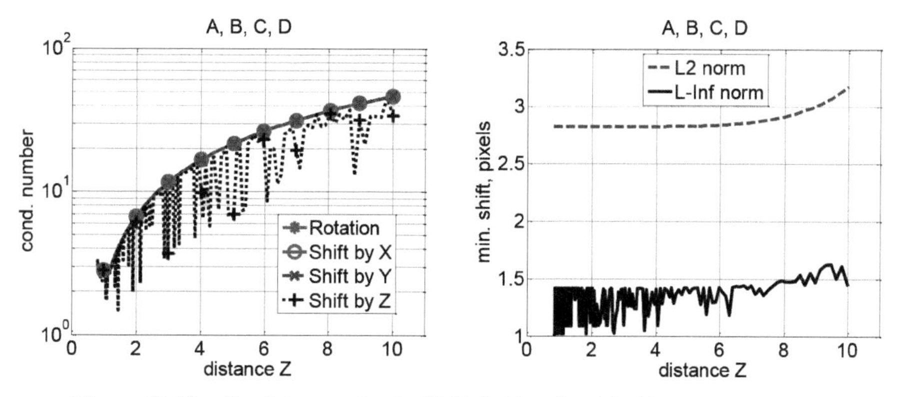


**Figure 10.13** Condition number by SVD (left) and residual image shift (right).

Next we consider cases with $n = 4$.

**Case 6.** Control points *A*, *B*, *C*, *D*, a planar configuration. Compared with the combination (*A*,*B*,*C*), a slight increase in accuracy is obtained, as shown in Figures 10.12 and 10.13.

**Case 7.** Control points *A*, *B*, *E*, and *F*. This case combines cases 2 (*A*,*B*,*E*) and 3 (*A*,*B*,*F*). We give here only the second graphs. Peak from case 3 disappeared (Figure 10.14).

**Case 8.** Control points *A*, *B*, *C*, and *E*. This case combines cases 1 (*A*,*B*,*C*) and 2 (*A*,*B*,*E*) or 2 (*A*,*B*,*E*) and 4 (*A*,*C*,*E*). The results will be given in the table below.

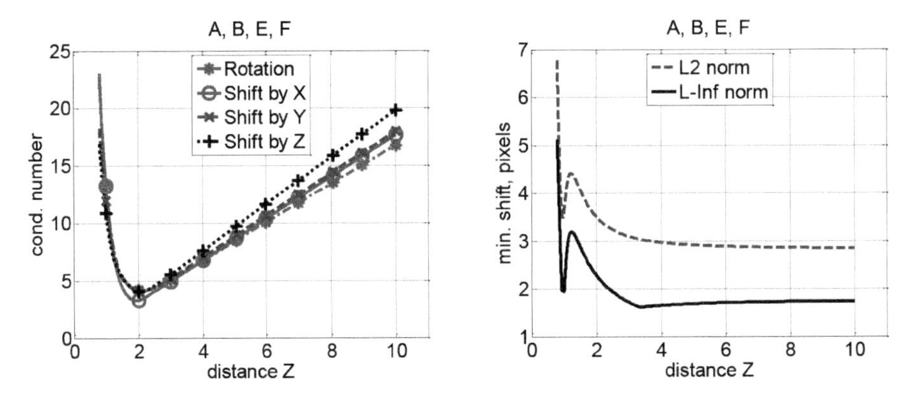


**Figure 10.14** Condition number by SVD (left) and residual image shift (right).

**Case 9.** Control points *B*, *C*, *E*, and *F*. It is another planar case, an extension of case 5 (*C*,*E*,*F*) or case 2 (*A*,*B*,*E*). The results will be given in the table below.

**Case 10.** Control points *A*, *B*, *C*, and *D* and one additional point (0.1,0.15,0.5) that lies on the same plane inside the square *ABCD*. It is an extension of case 6. The results will be given in the table below.

**Case 11.** Control points *A*, *B*, *C*, and *D* and two additional points (0.1,0.15,0.5) and (0.1,0.4,0.5). The results will be given in the table below.

**Case 12.** Control points *A*, *B*, *C*, *D*, *E*, and *F*. A larger number of points makes it possible to avoid degeneracy. The results are shown in Figures 10.15 and 10.16.

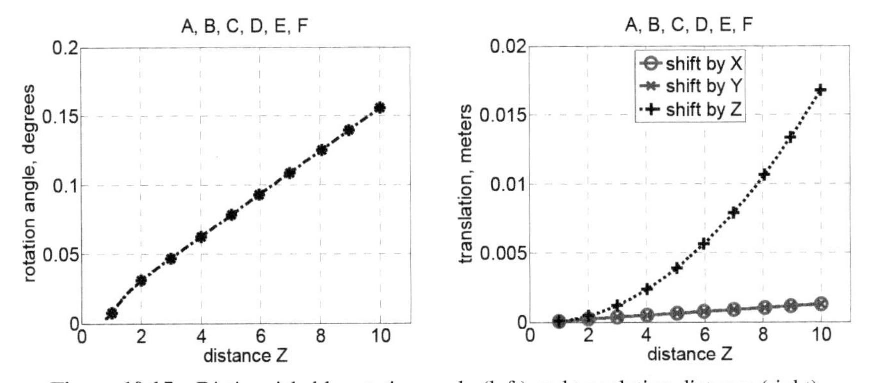


**Figure 10.15** Distinguishable rotation angle (left) and translation distance (right).

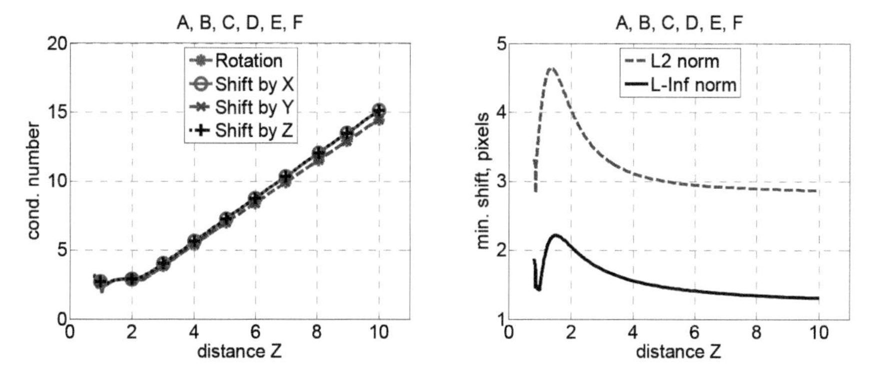


**Figure 10.16** Condition number by SVD (left) and residual image shift (right).

**Table 10.1** Planar perpendicular case.

| Control points | Cond. value | $\varepsilon_{\text{rot}}$, deg. | $\varepsilon_x$, cm | $\varepsilon_y$, cm | $\varepsilon_z$, cm | Bad case |
|---|---|---|---|---|---|---|
| | (at the distance 10 m) | | | | | |
| *A,B,C* | 56–58 | 4.13 | 3.43 | 3.43 | 2.55 | No |
| *A,B,C,D* | 34–46 | 2.92 | 2.42 | 2.42 | 1.80 | No |
| *A,B,C,D*, (0.1,0.15,0.5) | 48–50 | 2.73 | 2.25 | 2.27 | 1.80 | No |
| *A,B,C,D*, (0.1,0.15,0.5), (0.1,0.4,0.5) | 50-53 | 2.72 | 2.16 | 2.26 | 1.74 | No |

The experimental results are collected in two tables. Table 10.1 relates to the case when control points lie on a plane parallel to the picture plane. This case corresponds to the situation when Chaser spacecraft approaches the almost flat docking surface of the target spacecraft. In this case, it is usually required that the mooring angles be zero.

It can be seen that the condition number is quite large and the guaranteed measurement accuracy is relatively low, except for distance. No cases of linear degeneration were found.

Table 10.2 relates to other cases, when the distances from CP to control points are different. This situation arises when fragments of not only the docking surface are in the field of view.

Guaranteed measurement accuracy is significantly higher, except for distance. With a small number of points, degeneration of the linear component is possible. Additional control points eliminate these cases.

**Table 10.2** Cases with the different depth.

| Points | Cond. value | $\varepsilon_{\text{rot}}$, deg. | $\varepsilon_x$, cm | $\varepsilon_y$, cm | $\varepsilon_z$, cm | Bad case |
|---|---|---|---|---|---|---|
| | | | (at the distance 10 m) | | | |
| *A,B,E* | 22–24 | 0.27 | 0.26 | 0.27 | 3.61 | <1.5 |
| *A,B,F* | 15–22 | 0.39 | 0.18 | 0.21 | 3.45 | <1.5 |
| *A,C,E* | 14–16 | 0.26 | 0.19 | 0.19 | 2.55 | <1.5 |
| *C,E,F* | 24–28 | 0.29 | 0.30 | 0.30 | 4.44 | <2 |
| *A,B,E,F* | 17–20 | 0.23 | 0.17 | 0.18 | 2.78 | <0.9 |
| *A,B,C,E* | 16–17 | 0.21 | 0.19 | 0.19 | 2.23 | No |
| *B,C,E,F* | 20–23 | 0.16 | 0.20 | 0.20 | 2.79 | <1.2 |
| *A,B,C,D,E,F* | 14–15 | 0.15 | 0.13 | 0.13 | 1.68 | No |

To obtain the highest possible resolution and avoid cases of degeneration, the set of control points must contain:

- subset of points that lie on a plane perpendicular to the image plane;
- subset of points with maximum possible depth difference.

## 10.9 Conclusion

Evaluation of guaranteed detectable changes in relative attitude and position has been done through an analysis of linearized dependencies. The proposed method estimates the value of PnP solution errors, caused by limited accuracy of positioning in the image. The numerical simulation shows that in most cases of spatial configurations of control points, this method can be successfully applied.

The estimates obtained are consistent when linear approximation of the resulting image shift is correct. The correctness condition can be easily checked. It may not be performed only in certain cases when a small number of control points are used. When using a sufficient number of non-coplanar reference points, the method gives lower bounds for rotation and translation magnitudes, which induce measurable change of image in the worst-case combination. These bounds are overestimated due to difference of norms used.

Simulations show that optimal configuration of control points must include two subsets. The first should consist of points lying on a plane parallel or almost parallel to the picture plane. The second should consist of points with different distances from the camera. The first subset provides the accuracy of range estimation and eliminates the occurrence of degeneracy.

The second one provides high-precision detection of rotation and lateral displacements.

The proposed method can be used in the development of a computer vision system for spacecraft relative pose estimation using monocular camera.

## References


[1] R. Opromolla, et. al., 'A review of cooperative and uncooperative spacecraft pose determination techniques for close proximity operations', Progress in Aerospace Sciences, vol. 93, pp. 53–72, 2017.

[2] J.F. Shi, placeS. Ulrich, S. Ruel, 'Spacecraft pose estimation using a monocular camera', AIAA Space 2015 Conference and Exposition. Issue AIAA 2015–4429.

[3] V. Gubarev, N. Salnikov, S. Melnychuk, 'Ellipsoidal Pose Estimation of an Uncooperative Spacecraft from Video Image Data', in Control Systems: Theory and Applications. River Publishers Series in Automation, Control and Robotics, pp. 169-195, 2018.

[4] Y.I. Abdel-Aziz, H.M. Karara 'Direct linear transformation from Comparator coordinates into object space coordinates in close-range photogrammetry', American society for photogrammetry and remote sensing, no. 2, pp.103-107, 2015.

[5] C. Wu, 'P3.5P: Pose estimation with unknown focal length', 2015 IEEE Conference on Computer Vision and Pattern Recognition (CVPR), Boston, MA, pp. 2440-2448, 2015.

[6] X.-S. Gao, X.-R. Hou, J. Tang and H.-F. Cheng, 'Complete solution classification for the perspective-three-point problem', IEEE Transactions on Pattern Analysis and Machine Intelligence, vol. 25, no. 8, pp. 930-943, 2003.

[7] M.A. Fischler, R.C. Bolles, 'Random sample consensus: a paradigm for model fitting with applications to image analysis and automated cartography', Communications of the ACM, vol. 24, no. 6, pp. 381-395, 1981.

[8] P.D. Fiore, 'Efficient linear solution of exterior orientation', IEEE Transactions on Pattern Analysis and Machine Intelligence, vol. 23, no. 2, pp. 140-148, 2001.

[9] Z.Y. Hu, F.C. Wu, 'A note on the number of solutions of the noncoplanar P4P problem', IEEE Transactions on Pattern Analysis and Machine Intelligence, vol. 24, no. 4, pp. 550-555, 2002.

[10] R.M. Haralick, D. Lee, K. Ottenburg, M. Nolle, 'Review and analysis of solutions of the three point perspective pose estimation problem', International Journal of Computer Vision, vol. 13, no. 3, pp. 331-356, 1994.

[11] L. Kneip, D. Scaramuzza, R. Siegwart, 'A novel parametrization of the perspective-three-point problem for a direct computation of absolute camera position and orientation', CVPR 2011, pp. 2969-2976, 2011.

[12] R. Horaud, B. Conio, O. Leboulleux, B. Lacolle, 'An analytic solution for the perspective 4-point problem', Computer Vision Graphics and Image Processing, Elsevier, vol. 47, no. 1, pp. 33–44, 1989.

[13] M.A. Abidi, T. Chandra, 'A new efficient and direct solution for pose estimation using quadrangular tergets: algorithm and evaluation', IEEE Transactions on Pattern Analysis and Machine Intelligence, vol. 17, no. 5, pp. 534-538, 1995.

[14] L. Quan, Z. Lan, 'Linear n-point camera pose determination', IEEE Transactions on Pattern Analysis and Machine Intelligence, vol. 21, no. 8, pp. 774-780, 1999.

[15] M.-A. Ameller, B. Triggs, L. Quan, 'Camera Pose Revisited – New Linear Algorithms', Proc. European Conf. Computer Vision, 2000.

[16] O.D. Faugeras, M. Hebert, 'The representation, recognition, and locating of 3-D objects', The International Journal of Robotics Research, vol. 5, no. 3, pp. 27-52, 1986.

[17] B. Horn, 'Closed-form solution of absolute orientation using union quaternion', Journal of the Optical Society of America A, vol. 4, issue 4, pp. 629-642, 1987.

[18] B. Horn, H. Hugh, N. Shahriar, 'Closed-Form Solution of Absolute Orientation using Orthonormal Matrices', Journal of the Optical Society of America A, vol. 5, issue 7, pp. 1127-1135, 1988.

[19] S. Umeyama, 'Least-squares estimation of transformation parameters between two point patterns', IEEE Transactions on Pattern Analysis and Machine Intelligence, vol. 13, no. 4, pp. 376-380, 1991.

[20] D. Lowe, 'Solving for the parameters of object models from image descriptions', Proc. ARPA Image Understanding Workshop, pp. 121–127, 1980.

[21] D. Lowe, 'Fitting Parameterized Three-Dimensional Models to Images', IEEE Transactions on Pattern Analysis and Machine Intelligence, vol. 13, no. 5, pp. 441–450, 1991.

[22] J. Yuan, 'A general photogrammetric method for determining object position and orientation', IEEE Transactions on Robotics and Automation, vol. 5, no. 2, pp.129-142, 1989.

[23] D. Dementhon, L. Davis, 'Model-based object pose in 25 lines of code', International Journal Computer Vision, vol. 15, p. 123-141, 1995.

[24] C.-P. Lu, G.D. Hager, E. Mjolsness, 'Fast and Globally Convergent Pose Estimation from Video Images', IEEE Transactions on Pattern Analysis and Machine Intelligence, vol. 22, no. 6, pp. 610–622, 2000.

[25] S. Malik, G. Roth, C. McDonald, 'Robust 2D Tracking for Real-Time Augmented Reality', Proc. Conf. Vision Interface, vol. 1, no. 2, pp. 12, 2002

[26] A. Ansar, K. Daniilidis, 'Linear pose estimation from points or lines', IEEE Transactions on Pattern Analysis and Machine Intelligence, vol. 25, no. 5, pp. 578–589, 2003.

[27] G. Schweighofer, A. Pinz, 'Robust Pose Estimation from a Planar Target', IEEE Transactions on Pattern Analysis and Machine Intelligence, vol. 28, no. 12, pp. 2024–2030, 2006.

[28] V. Garro, F. Crosilla, A. Fusiello, 'Solving the PnP Problem with Anisotropic Orthogonal Procrustes Analysis', 2012 Second International Conference on 3D Imaging, Modeling, Processing, Visualization & Transmission, pp. 262-269, 2012.

[29] Y. Zheng, S. Sugimoto, M. Okutomi, 'ASPnP: An accurate and scalable solution to the Perspective-n-Point problem', IEICE Transactions on Information and Systems, E96.D(7), pp. 1525-1535, 2013.

[30] Y. Zheng, et. al., 'Revisiting the PnP Problem: A Fast, General and Optimal Solution', 2013 IEEE International Conference on Computer Vision, Sydney, pp. 2344-2351, 2013.

[31] Kneip, H. Li, Y. Seo, 'UPnP: An Optimal O(n) Solution to the Absolute Pose Problem with Universal Applicability', in: D. Fleet, T. Pajdla, B. Schiele, T. Tuytelaars (eds) Computer Vision – ECCV 2014, Lecture Notes in Computer Science, vol 8689, pp. 127-142, 2014.

[32] S. Urban, J. Leitloff, placeS. Hinz, 'MLPnP – A real-time maximum likelyhood solution to the perspective-n-point problem', ISPRS Ann. Photogramm. Remote Sens. Spatial Inf. Sci., III-3, pp. 131–138, 2016.

[33] V. Lepetit, F. Moreno-Noguer, P. Fua, 'EPnP: An Accurate O(n) Solution to the PnP Problem', International Journal of Computer Vision, 81, 155, 2009.

[34] L. Ferraz, X. Binefa, F. Moreno-Noguer, 'Very Fast Solution to the PnP Problem with Algebraic Outlier Rejection', 2014 IEEE Conference on

Computer Vision and Pattern Recognition, Columbus, OH, 2014, pp. 501-508, 2014.
[35] L. Ferraz, X. Binefa, F. Moreno-Noguer, 'Leveraging feature uncertainty in the PnP problem', Proceedings of the Brittish Machine Vision Conference, BMVA Press, 2014.
[36] J.A. Hesch, S.I. Roumeliotis, 'A Direct Least-Squares (DLS) Method for PnP', 2011 International Conference on Computer Vision, pp. 383-390, 2011.
[37] S. Li, C. Xu, M. Xie, 'A Robust O(n) Solution to the Perspective-n-Point Problem', IEEE Transactions on Pattern Analysis and Machine Intelligence, vol. 34, no. 7, pp. 1444-1450, 2012.
[38] A.D.N. Smith, 'The Explicit Solution of Single Picture Resection Problem with a Least Squares Adjustment to Redundant Control', Photogrammetric Record, vol. 5, no. 26, pp. 113-122, 1965.
[39] E.H. Thompson, 'Space Resection: Failure Cases', Photogrammetric Record, vol. 5, No. 27, pp. 201-204, 1966.
[40] B.P. Wrobel, 'Minimum Solutions for Orientation', in: Gruen A., Huang T.S. (eds) Calibration and Orientation of Cameras in Computer Vision. Springer Series in Information Sciences, vol 34, pp. 7-62, 2001.

# 11

# On Determining the Spacecraft Orientation by Information from a System of Stellar Sensors

**Dmitriy V. Lebedev**

International Research and Training Center for Information Technologies and Systems, Kiev, Ukraine
Corresponding Author: ldv1491@gmail.com

## Abstract

Increasing requirements on the accuracy of information on the spacecraft orientation parameters when solving problems both onboard the spacecraft and on the ground processing information coming from orbit stimulate the search for such algorithmic solutions for processing information from onboard orientation sources that would satisfy these requirements. Given this trend, the problem of minimizing the influence of residual uncertainty in the angular position of stellar sensors in the coordinate system associated with the spacecraft on the accuracy of the information systems under consideration is considered. Accuracy characteristics of satellite orientation, which are realized by using the proposed algorithm, are studied. Simulation confirms the effectiveness of the proposed algorithmic solutions.

**Keywords:** Spacecraft, stellar sensor, orientation, Rodrigues–Hamilton parameters, modified Rodrigues parameters, quaternion of rotation, inertial space, coordinate system.

## 11.1 Introduction

Increasing the requirements for accuracy in controlling the orientation of spacecraft (SC) requires continuous improvement and perfection of the accuracy characteristics of orientation sensors. Improved accuracy in determining the orientation, for example, of the Earth remote sensing satellite is also required for the implementation of high-precision coordinate reference images of objects of interest on the Earth's surface according to orbital data [1]. The high-precision orientation studied, in particular, in [2] is also required to implement some control algorithms for moving objects.

The highest accuracy in determining the orientation of the SC is currently provided by stellar sensors (SSs) [3]. Nevertheless, a significant increase in the accuracy of determining the orientation parameters of the SC is achieved via processing of measurement information going from a system, which consists of several SSs [4–6]. Usually, some basic stellar sensor (BSS), whose orientation in the inertial coordinate system (CS) is identified with the orientation of the SC in inertial space, is selected from the system of stellar sensors (SSs). Given the known mutual orientation of the CS of the BSS with the CSs of each of the other SS, results of measurements of directions to the stars in the field of view of each sensor are projected into the CS of the BSS. The overdetermined system of linear algebraic equations formed in this way is solved with respect to the required parameters of the SC orientation using one of the known methods.

It should be noted that even high-precision pre-flight calibration of the mutual angular position of the SSs and BSS is unable to eliminate errors that occur during the operation of the SC. Their sources are vibration and shock loads at the moments of launch and separation of the SC, cyclic temperature changes during the flight, deformations of the bases, and details of sensors caused by aging and influence to the space environment [7]. The BSS is also exposed to the influence of the perturbing factors mentioned above, which naturally affects the accuracy of determining the orientation of the SC. An effective method for compensating the influence of these perturbations on the accuracy of the SC orientation is in-flight calibration of the mutual orientation of the sensor CSs [4, 7].

One of the requirements for a new generation of SSs is a high accuracy of determining the orientation (from 0.1 to 0.01 arc sec.) when the rate of updating information is high enough [3]. In this work, it was also noted that in the near future, the use of lasers as transmitters of information will be possible to transfer the one both from orbit to earth surface and between SCs. Again, this requires pointing the transmitter to the receiver and its stabilization with accuracy better than 0.5 arc sec. Naturally, that the possibility to achieve the accuracy characteristics pointed out needs to analyze how the various disturbing factors affect a quality of the operation of the system for determining the SC orientation parameters.

It should be noted that when using a single SS or the SSs system, the orientation of the own (instrument) CS of the SS or BSS relative to the inertial geocentric CS is determined. However, the installation of an SS on an SC is naturally accompanied by errors that, during the long-term functioning of an object, can evolve due to the influence of the various kinds of disturbances.

In this chapter, as applied to an astro-measuring system containing two SSs fixed rigidly to the SC body, we propose the procedure to minimize an influence of the residual uncertainty of the angular position of the SS in the CS connected with the SC on the accuracy of the functioning of this information system without using the flight procedure calibration of the relative orientation of the SSs. An algorithmic solution of this problem is presented, the use of which makes it possible to counterbalance the influence of the disturbing factor on the accuracy of calculating the SC orientation parameters. The implementation of this algorithm does not provide for the use of information with respect to the mutual orientation of SS1 and SS2. Based on computational experiment, characteristics of accuracy of the SC orientation in inertial space, which is realized by using the proposed algorithm, are investigated.

## 11.2 Systems of Coordinates: Formulation of the Problem

Introduce the right orthogonal CSs and bases composed of orts of axes of these CSs to be needed further:

1) inertial CS $I$ (bases $\mathbf{I}$) with the vertex in the Earth center, the axis $X$ directed to the vernal equinox, and the axis $Z$ oriented along the axis of the world in the direction of Polar star;
2) CS $E$ ($\mathbf{E}$) rigidly connected with the body of the moving object;
3) CSs $E_1^*$, $E_2^*$ and their bases $\mathbf{E}_1^*$ and $\mathbf{E}_2^*$ characterizing the design orientation of SS1 and SS2 in the CS $E$;
4) CSs $E_1$, $E_2$ and their bases $\mathbf{E}_1$, $\mathbf{E}_2$ characterizing the current orientation of SSs in the CS $E$.

If the mutual orientation of the CSs introduced above is characterized by the corresponding quaternions of rotation (QRs), then the connection between the indicated bases is illustrated by the scheme depicted in Figure 11.1.

It is required to minimize the influence of the residual uncertainty in the orientation of SSs, characterized by quaternions $\mathbf{\Lambda}_{\mathrm{E}_1^*\mathrm{E}_1}$ and $\mathbf{\Lambda}_{\mathrm{E}_2^*\mathrm{E}_2}$ in the

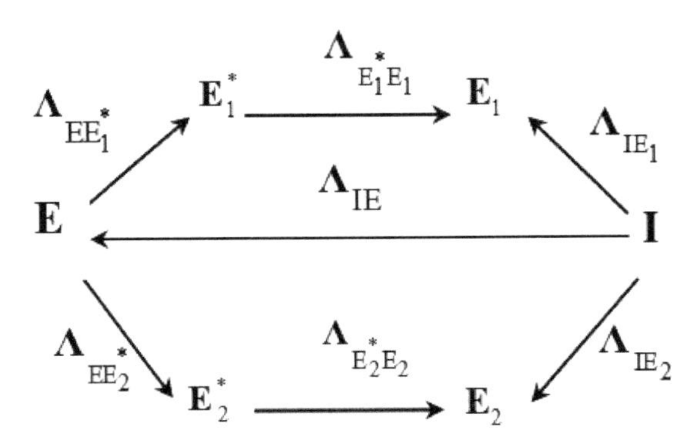


**Figure 11.1** Connection between bases of coordinate systems.

CS **E**, on the accuracy of calculating the SC orientation in inertial space. The solution should not contain in-flight calibration of the mutual orientation of SSs.

## 11.3 Correspondence of Three-Dimensional and Four-Dimensional Parameters of a Group of Three-Dimensional Rotations

When describing the motion around the center of mass of dynamic objects schematized by a solid model, the four-dimensional kinematic Rodrigues–Hamilton parameters (RHP) $\lambda_i$ $(i = 0, 1, 2, 3)$ that are components of the unit QR $\mathbf{\Lambda}$ (see, for example, [8]) are widely used. Nevertheless, in some cases, it may be effective to use three-dimensional kinematic parameters for mathematical description and solving problems associated with the orientation of a rigid body.

Based on the fundamental Euler's theorem on the motion of a rigid body relatively to fixed point [9], a synthesis of three-dimensional and four-dimensional kinematic parameters of the orientation of objects in three-dimensional space is possible [10]. An example of using this type of parameters is given below.

For further presentation of the material, the so-called modified Rodrigues parameter (MRP) and formulas that establish their connection with the RHP will be required. Recall that the mutual orientation of orthogonal trihedrons can be characterized by the Rodrigues vector

$$\boldsymbol{\rho} = (\rho_1, \rho_2, \rho_3) = tg(\varphi/2)\mathbf{e},$$

in which **e** is the unit vector of the Euler's axis of rotation and $\varphi$ represents the angle of rotation around the axis **e**. The three coordinates of the vector $\boldsymbol{\rho}$ are Rodrigues parameters [11]. Formulas

$$\boldsymbol{\rho} = \boldsymbol{\lambda}/\lambda_0\,, \quad \lambda_0 = \frac{1}{\sqrt{1+\|\boldsymbol{\rho}\|^2}}, \quad \boldsymbol{\lambda} = \frac{\boldsymbol{\rho}}{\sqrt{1+\|\boldsymbol{\rho}\|^2}}$$

illustrate the one-to-one correspondence between of the Rodrigues vector and the QR $\boldsymbol{\Lambda} = (\lambda_0,\, \boldsymbol{\lambda}),\ \ \boldsymbol{\lambda} = (\lambda_1,\, \lambda_2,\, \lambda_3)$. As for the MRP, we note that they exist in two forms [11]: the positive form defined in terms of the RHP by the formula

$$\boldsymbol{\theta} = \frac{\boldsymbol{\lambda}}{1+\lambda_0} = \mathbf{e}\ tg(\varphi/4), \ \ \boldsymbol{\lambda} = (\lambda_1,\, \lambda_2,\, \lambda_3) \tag{11.1}$$

and the negative form as a ratio

$$\boldsymbol{\Psi} = \frac{\boldsymbol{\lambda}}{1-\lambda_0} = \mathbf{e}\ ctg(\varphi/4). \tag{11.2}$$

It is important to note that the formulas (11.1) and (11.2) represent three-dimensional stereographic projections of a three-dimensional sphere $\lambda_0^2 + \lambda_1^2 + \lambda_2^2 + \lambda_3^2 = 1$, having infinitely distant points at $\lambda_0 = -1$ and $\lambda_0 = 1$, respectively. Taking into account the correspondence of the three-dimensional and four-dimensional parameters of the group of three-dimensional rotations [10], the mutual orientation of the bases $\mathbf{E}_i$ and $\mathbf{E}_j$ will be characterized by the normalized quaternion whose components are the RHP or the three-dimensional vector of the MRP [11]

$$\boldsymbol{\Theta} = (\vartheta_1,\, \vartheta_2,\, \vartheta_3) = \mathbf{e}\ tg(\varphi/4). \tag{11.3}$$

The quaternion $\boldsymbol{\Lambda} = (\lambda_0,\, \boldsymbol{\lambda}),\ \ \boldsymbol{\lambda} = (\lambda_1,\, \lambda_2,\, \lambda_3)$ is one-to-one connected with the three-dimensional vector $\boldsymbol{\Theta}$ relations [12]

$$\begin{gathered} \boldsymbol{\Theta} = \boldsymbol{\lambda}/(1+\lambda_0), \\ \lambda_0 = (1-\boldsymbol{\Theta}^T\boldsymbol{\Theta})/(1+\boldsymbol{\Theta}^T\boldsymbol{\Theta}),\ \boldsymbol{\lambda} = 2\boldsymbol{\Theta}/(1+\boldsymbol{\Theta}^T\boldsymbol{\Theta}). \end{gathered} \tag{11.4}$$

Direct and inverse quaternion kinematic equations correspond to direct and inverse kinematic equations for the vector $\boldsymbol{\Theta} = (\vartheta_1,\, \vartheta_2,\, \vartheta_3)$ [12]:

$$\begin{aligned} \dot{\boldsymbol{\Theta}} &= \frac{1}{4}(1-\boldsymbol{\Theta}^T\boldsymbol{\Theta})\boldsymbol{\omega} + \frac{1}{2}\boldsymbol{\Theta}\times\boldsymbol{\omega} + \frac{1}{2}(\boldsymbol{\Theta}^T\boldsymbol{\omega})\boldsymbol{\Theta}, \\ \boldsymbol{\omega} &= \frac{4[(1-\boldsymbol{\Theta}^T\boldsymbol{\Theta})\dot{\boldsymbol{\Theta}} - 2(\boldsymbol{\Theta}\times\dot{\boldsymbol{\Theta}}) + 2\boldsymbol{\Theta}(\dot{\boldsymbol{\Theta}}^T\boldsymbol{\Theta})}{(1+\boldsymbol{\Theta}^T\boldsymbol{\Theta})^2} \end{aligned} \tag{11.5}$$

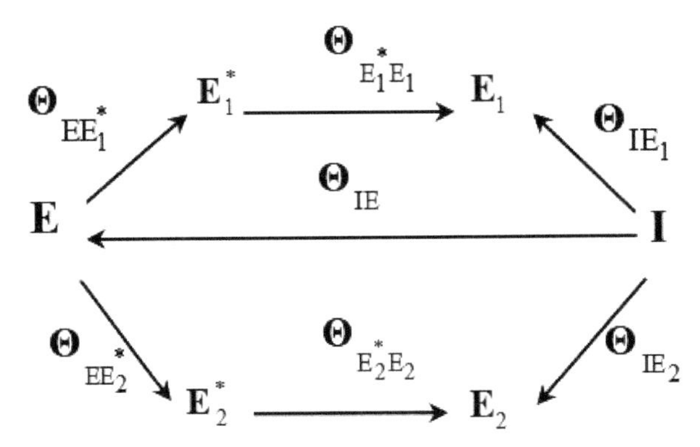


**Figure 11.2** Connection between vectors of modified Rodrigues parameters.

In [12], it was noted that the relations (11.4) allow us to reduce the problems of smoothing quaternion data to the usual problem of smoothing vector measurements.

The scheme shown in Figure 11.2, visually similar to the scheme in Figure 11.1, illustrates the relationship between the three-dimensional vectors of the MRP. As following from Figures 11.1 and 11.2, when forming readings, for example, of the SS1, information is needed on the mutual orientation of the bases $\mathbf{I}$ and $\mathbf{E}_1$. In terms of quaternions and vectors of the MRP, the carriers of this information are $\mathbf{\Lambda}_{\mathrm{IE}_1}$ and $\mathbf{\Theta}_{\mathrm{IE}_1}$, determined by the following system of equalities:

$$\begin{aligned} \mathbf{\Lambda}_{\mathrm{IE}_1} &= \mathbf{\Lambda}_{\mathrm{IE}} \circ \mathbf{\Lambda}_{\mathrm{EE}_1^*} \circ \mathbf{\Lambda}_{\mathrm{E}_1^*\mathrm{E}_1}, \\ \mathbf{\Theta}_{\mathrm{IE}_1} &= \mathbf{\Theta}_{\mathrm{IE}} + \mathbf{\Theta}_{\mathrm{EE}_1^*} + \mathbf{\Theta}_{\mathrm{E}_1^*\mathrm{E}_1}. \end{aligned} \tag{11.6}$$

The relationship between the quaternion and the vector of the MRP is established by formulas (11.4).

## 11.4 Algorithms for Determining the Orientation Parameters of the Spacecraft

From Figure 11.2 and the second formula in equalities (11.6), it follows that the orientation of SS1 and SS2 in inertial space are characterized by the vectors $\mathbf{\Theta}_{\mathrm{IE}_1}$ and $\mathbf{\Theta}_{\mathrm{IE}_2}$ of the MRP determined by formulas

$$\mathbf{\Theta}_{\mathrm{IE}_1} = \mathbf{\Theta}_{\mathrm{IE}} + \mathbf{\Theta}_{\mathrm{EE}_1^*} + \mathbf{\Theta}_{\mathrm{E}_1^*\mathrm{E}_1} \tag{11.7}$$

$$\mathbf{\Theta}_{\mathrm{IE}_2} = \mathbf{\Theta}_{\mathrm{IE}} + \mathbf{\Theta}_{\mathrm{EE}_2^*} + \mathbf{\Theta}_{\mathrm{E}_2^*\mathrm{E}_2}. \tag{11.8}$$

Introduce the notation

$$\mathbf{\Psi}_1 = \mathbf{\Theta}_{\mathrm{IE}} + \mathbf{\Theta}_{\mathrm{E}_1^*\mathrm{E}_1},\ \mathbf{\Psi}_2 = \mathbf{\Theta}_{\mathrm{IE}} + \mathbf{\Theta}_{\mathrm{E}_2^*\mathrm{E}_2}. \qquad (11.9)$$

And note that the vectors $\mathbf{\Psi}_1$ and $\mathbf{\Psi}_2$ are calculated also by the formulas

$$\mathbf{\Psi}_1 = \mathbf{\Theta}_{\mathrm{IE}_1} - \mathbf{\Theta}_{\mathrm{EE}_1^*},\ \mathbf{\Psi}_2 = \mathbf{\Theta}_{\mathrm{IE}_2} - \mathbf{\Theta}_{\mathrm{EE}_2^*}. \qquad (11.10)$$

After summing the expressions (11.9), the orientation of the SC in inertial space **I** is determined by the following relationships:

$$\mathbf{\Theta}_{\mathrm{IE}} = {}^1\!/_2(\mathbf{\Psi}_1 + \mathbf{\Psi}_2) - \boldsymbol{\delta},\ \ \boldsymbol{\delta} = {}^1\!/_2(\mathbf{\Theta}_{\mathrm{E}_1^*\mathrm{E}_1} + \mathbf{\Theta}_{\mathrm{E}_2^*\mathrm{E}_2}). \qquad (11.11)$$

In the steps $n$ and $n + 1$ of measuring, from two values $\mathbf{\Theta}_{\mathrm{IE}}(n)$ and $\mathbf{\Theta}_{\mathrm{IE}}(n{+}1)$ of the SC orientation parameters calculated by the formulas

$$\mathbf{\Theta}_{\mathrm{IE}}(n) = {}^1\!/_2(\mathbf{\Psi}_1(n) + \mathbf{\Psi}_2(n)) - \boldsymbol{\delta}, \qquad (11.12)$$

$$\mathbf{\Theta}_{\mathrm{IE}}(n+1) = {}^1\!/_2(\mathbf{\Psi}_1(n+1) + \mathbf{\Psi}_2(n+1)) - \boldsymbol{\delta}, \qquad (11.13)$$

we form recurrence relations for determining the current orientation of the SC according to the reading of SS1 and SS2. Subtracting the expression (11.12) from the equality (11.13) and taking into account the formulas (11.9), we represent the desired relation in the form

$$\begin{gathered}\mathbf{\Theta}_{\mathrm{IE}}(n+1) = \mathbf{\Theta}_{\mathrm{IE}}(n) + {}^1\!/_2(\Delta\mathbf{\Theta}_{\mathrm{IE}_1}(n) + \Delta\mathbf{\Theta}_{\mathrm{IE}_2}(n)),\\ \Delta\mathbf{\Theta}_{\mathrm{IE}_i}(n) = \mathbf{\Theta}_{\mathrm{IE}_i}(n+1) - \mathbf{\Theta}_{\mathrm{IE}_i}(n), \quad i = 1,\ 2.\end{gathered} \qquad (11.14)$$

After calculations performed before, the vector $\mathbf{\Theta}_{\mathrm{IE}}(n+1)$ is transformed (via the equalities (11.4)) in the QR $\mathbf{\Lambda}_{\mathrm{IE}}(n+1)$ characterizing the current orientation of the SC. Note that the algorithm (11.14) exploited for processing measurement information going from SS1 and SS2 has the following features:

1) To calculate $\mathbf{\Theta}_{\mathrm{IE}}$ via the formulas (11.14), information about the initial value of $\mathbf{\Theta}_{\mathrm{IE}}$ is needed (some analog of the inertial navigation system).
2) In the case when the vector $\boldsymbol{\delta}$ remains constant throughout the calculation cycle, the current value of the SC orientation vector $\mathbf{\Theta}_{\mathrm{IE}}(n{+}1)$formed by results of increments in readings of SSs on the next calculation cycle does not depend on the errors $\mathbf{\Theta}_{\mathrm{E}_1^*\mathrm{E}_1}$ and $\mathbf{\Theta}_{\mathrm{E}_2^*\mathrm{E}_2}$ in the installing of SSs in the bases **E**.

3) The error in setting the initial value of the SC orientation parameter vector $\boldsymbol{\Theta}_{\mathrm{IE}}$,which refers to the unrecoverable error (according to the terminology of [13]), is preserved in consequent calculations performed by the algorithm (11.14).

Consider the situation when, during the operation of the measuring system, one from SSs, for example, SS2, failed. Let us evaluate the accuracy of determining the SC orientation for this case, utilizing the reading of the SS1. With the known vector $\boldsymbol{\Theta}_{\mathrm{EE}_1^*}$ and current reading $\boldsymbol{\Theta}_{\mathrm{IE}_1}$ of the SS1, it follows from equality (11.6) that the SC orientation in the situation under consideration is estimated by relation

$$\boldsymbol{\Theta}_{\mathrm{IE}} = \boldsymbol{\Theta}_{\mathrm{IE}_1} - \boldsymbol{\Theta}_{\mathrm{EE}_1^*} - \boldsymbol{\Theta}_{\mathrm{E}_1^*\mathrm{E}_1}. \tag{11.15}$$

## 11.5 Accuracy Analysis of Determining the Parameters of the SC Orientation

Now, let us evaluate the accuracy of determining parameters of the SC orientation in inertial space for the case when the one moves in a circular orbit around the Earth in the mode of triaxial orbit orientation.

Determine the following characteristics of the orbit:

1) radius (distance from the center of the Earth to the center of mass of the SC) equal to 7061 km;
2) inclination equal to 98°;
3) the longitude of the ascending node equal to 142°.

The angular velocity of the satellite rotation is equal to 0.06097°/sec; the period of revolution is equal to 98.41 minutes.

The six types of SSs will be exploited to analyze the accuracy of determining the parameters of the SC orientation. Their accuracy characteristics are given in Table 11.1. The data for sensor variants I–IV have been taken from [14]. The accuracy characteristics of SSs in option V and VI are hypothetical. Each type of sensors will be used in the three versions of processing measurement information (for example, SS of type I will be used in options I.1, I.2, I.3, etc.).

In option I.1, the information going from SS1 will be processed according to the formula

$$\boldsymbol{\Theta}_{\mathrm{IE}}(n) = \boldsymbol{\Theta}_{\mathrm{IE}_1}(n) - \boldsymbol{\Theta}_{\mathrm{EE}_1^*}(n) \tag{11.16}$$

with an error $\boldsymbol{\delta} = -\boldsymbol{\Theta}_{\mathrm{E}_1^*\mathrm{E}_1}$ to determine the orientation of the SC in the case of the failure of the SS2.

**Table 11.1** Characteristics of stellar sensors.

| SS variant | SS type | $\sigma_x$ | $\sigma_y$<br>Ang. sec | $\sigma_z$ |
|---|---|---|---|---|
| I | BOKZ – M | 2 | 2 | 20 |
| II | BOKZ – M60 | 1.5 | 1.5 | 15 |
| III | BOKZ – M60/1100<br>ASRO15 | 1 | 1 | 11 |
| IV | SED 36 | 0.3 | 0.3 | 2 |
| V | Perspective | 0.1 | 0.1 | 1 |
| VI | Perspective | 0.01 | 0.01 | 0.1 |

In option I.2, the information from two SSs will be processed according to the formula

$$\boldsymbol{\Theta}_{\mathrm{IE}}(n) = {}^{1}\!/_{2}(\boldsymbol{\Psi}_1(n) + \boldsymbol{\Psi}_2(n)), \tag{11.17}$$

whose error is estimated by the ratio $\delta = {}^{1}\!/_{2}(\boldsymbol{\Theta}_{\mathrm{E}_1^*\mathrm{E}_1} + \boldsymbol{\Theta}_{\mathrm{E}_2^*\mathrm{E}_2})$.

In option I.3, the proposed algorithm (11.14) will be tested. If the vectors $\boldsymbol{\Theta}_{\mathrm{E}_1^*\mathrm{E}_1}$ and $\boldsymbol{\Theta}_{\mathrm{E}_2^*\mathrm{E}_2}$ will be constant at each step of processing information, using this algorithm, then calculation errors will not be observed.

For all variants of measurement information processing algorithms, the SC orientation will be calculated after each second of flight on the time interval equal to 11,000 seconds.

Let us evaluate the effect of the residual uncertainty in the orientation of SS1 and SS2 in CSs $\mathrm{E}_1$ and$\mathrm{E}_2$, accordingly, due to errors in their installation in CSs $\mathrm{E}_1^*$ and $\mathrm{E}_2^*$ and (or) their drift during the SC operation. Considering the mentioned errors to be small, the mutual orientation of the bases $\mathbf{E}_i$ and $\mathbf{E}_i^*$ $(i = 1,\ 2)$ are characterized by the quaternions

$$\boldsymbol{\Lambda}_{\mathrm{E}_1^*\mathrm{E}_1} = 1 + \boldsymbol{\alpha}/2, \boldsymbol{\Lambda}_{\mathrm{E}_2^*\mathrm{E}_2} = 1 + \boldsymbol{\beta}/2,$$

where $\boldsymbol{\alpha} = \{\alpha_i\},\ \boldsymbol{\beta} = \{\beta_i\}\ \ (i = 1,\ 2,\ 3)$.

The design orientations of SS1 and SS2 in CS $E$ are characterized by quaternions

$$\boldsymbol{\Lambda}_{\mathrm{EE}_1^*} = [\cos(\theta/2),\, \sin(\theta/2),\, 0,\, 0]\,,\ \boldsymbol{\Lambda}_{\mathrm{EE}_2^*} = [\cos(\theta/2),\, -\sin(\theta/2),\, 0,\, 0]\,.$$

In computational experiment, the following values of coordinates of the vectors $\boldsymbol{\alpha}$, $\boldsymbol{\beta}$ (angular sec) and the angle $\theta$ (degrees) were used:

$$\boldsymbol{\alpha} = [-40,\ 60,\ 5]^T,\ \boldsymbol{\beta} = [-30,\ -50,\ 1]^T, \theta = 5.$$

In using the algorithm (11.14), the exact initial values of the vector $\boldsymbol{\Theta}_{\mathrm{IE}}$ (the quaternion $\boldsymbol{\Lambda}_{\mathrm{IE}}$ ) of the SC orientation parameters were given. The

**Table 11.2** Results of simulation.

| SS type | | $\varepsilon$, ang. sec | SS type | | $\varepsilon$, ang. sec |
|---|---|---|---|---|---|
| I | I.1 | $42.251 \pm 0.732$ | IV | IV.1 | $33.513 \pm 0.504$ |
| | I.2 | $24.285 \pm 0.468$ | | IV.2 | $16.511 \pm 0.249$ |
| | I.3 | $16.804 \pm 0.462$ | | IV.3 | $1.743 \pm 0.057$ |
| II | II.1 | $38.823 \pm 0.630$ | V | V.1 | $33.428 \pm 0.501$ |
| | II.2 | $21.387 \pm 0.378$ | | V.2 | $16.424 \pm 0.246$ |
| | II.3 | $12.603 \pm 0.348$ | | V.3 | $0.840 \pm 0.024$ |
| III | III.1 | $36.015 \pm 0.555$ | VI | VI.1 | $33.401 \pm 0.501$ |
| | III.2 | $18.902 \pm 0.303$ | | VI.2 | $16.396 \pm 0.246$ |
| | III.3 | $8.402 \pm 0.231$ | | VI.3 | $0.084 \pm 0.003$ |

results of computer simulation are presented in Table 11.2. They estimate the calculation accuracy of the SC orientation parameters (angular sec) for various types of SSs. The data presented in Table 11.2 have such structure of the error $\varepsilon$:

$$\varepsilon = m \pm 3\sigma,$$

where $m$ denotes the expectation and $\sigma$ represents the standard deviation of error $\varepsilon$.

Analyzing the results of computer simulation of the calculation of the SC orientation parameters, we note the following:

1) Information processing from two SSs according to the algorithm (11.17) approximately doubles the accuracy of determining the orientation of the SC in inertial space as compared with the data received from one SS of the same type.
2) The accuracy of determining the orientation of the SC increases essentially when measurement information coming from two SSs is processed according to the algorithm (11.14), and there is a high accuracy of information about the initial orientation of SC.
3) If the accuracy characteristics $\sigma_x$, $\sigma_y$, and $\sigma_z$ of SSs are related to each other by the relation of the form $\sigma_x : \sigma_y : \sigma_z = 1 : 1 : 10$, then the errors in calculating the SC orientation parameters according to the algorithm (11.14) using information from the *i*th and *j*th types of SSs are specified by the ratio

$$\varepsilon(i) = \varepsilon(j) \frac{\sigma_z(i)}{\sigma_z(j)}, \tag{11.18}$$

where $\sigma_z(i)$ and $\sigma_z(j)$ are the standard deviations of the SSs errors of types *i* and *j*, respectively.

## 11.6 Effect of Satellite Initial Orientation Error on the Accuracy of Determining Its Current Orientation

To analyze the influence of the error in measurement information on the SC orientation at the moment, when functioning the SSs start, we form an inaccurate initial value of the vector $\mathbf{\Theta}_{\text{IE}}$ of the MRP. Based on the information about the value of the argument of latitude, inclination of the orbit, and the current value of longitude of the ascending node, we calculate exact initial value of the orientation quaternion $\mathbf{\Lambda}^*_{\text{IE}}$ (and the corresponding vector $\mathbf{\Theta}^*_{\text{IE}}$). Next, we take three angles $\vartheta$, $\phi$, and $\varphi$ and form the corresponding quaternion of orientation $\mathbf{\Lambda}_0$ and the vector $\mathbf{\Theta}_0$ of the MRP characterizing the error of initial orientation of the SC. Then the initial orientation of the satellite (the vector $\mathbf{\Theta}_{\text{IE}}$) in inertial space is determined by the sum

$$\mathbf{\Theta}_{\text{IE}} = \mathbf{\Theta}^*_{\text{IE}} + \mathbf{\Theta}_0.$$

In computer simulation, two options of setting of angles $\vartheta$, $\phi$, and $\varphi$, namely

$$\begin{aligned} \vartheta = 0,5\sigma_x,\ \phi = 0,5\sigma_y,\ \varphi = 0,5\sigma_z, \\ \vartheta = -0,5\sigma_x,\ \phi = -0,5\sigma_y,\ \varphi = -0,5\sigma_z, \end{aligned} \tag{11.19}$$

have been used. In expression (11.19), the mean square errors $\sigma_x$, $\sigma_y$, and $\sigma_z$ correspond to the accuracy characteristics of the used SSs from Table 11.1.

Having to do with these two options, the option for which the error is greater is selected.

The results of computer simulation obtained in this way are shown in Table 11.3. Column "a" of Table 11.3 contains the values of the errors in the calculation of the SC orientation parameters in the presence of accurate information about its orientation at the moment when SSs system comes to play. Column "b" contains data obtained in the absence of such information.

On the one hand, an analysis of the data given in Table 11.3 indicates slight effect of the system of parametric disturbances on the accuracy of

**Table 11.3** Spacecraft orientation errors.

| SS type | $\varepsilon$, ang. sec | |
|---|---|---|
| | a | b |
| I | $16.804 \pm 0.462$ | $18.017 \pm 0.489$ |
| II | $12.603 \pm 0.348$ | $13.513 \pm 0.369$ |
| III | $8.402 \pm 0.231$ | $9.009 \pm 0.245$ |
| IV | $1.743 \pm 0.045$ | $1.871 \pm 0.048$ |
| V | $0.840 \pm 0.024$ | $0.901 \pm 0.025$ |
| VI | $0.084 \pm 0.003$ | $0.090 \pm 0.002$ |

calculating the orientation parameters of the SC and, on the other hand, makes it possible to conclude that the proposed algorithmic solution to the problem of increasing the accuracy of determining the orientation parameters of the SC in inertial space is efficient enough. Note that the data contained in columns "a" and "b" of Table 11.3 are interconnected by relations similar to equality (11.18) (except for data related to the SS of type IV).

## 11.7 Conclusion

Increasing demands on the accuracy of the SC orientation system require a continuous improvement and perfection of the accuracy characteristics of orientation sensors. Among them, a special place is occupied by the SSs, which provide the highest accuracy up to now. A significant increase in the accuracy of determining the SC orientation parameters is achieved via processing of measurement information going from a system of several SSs. As a result, the orientation of the BSS is determined. Its orientation is identified with the orientation of the SC in inertial space. To implement this approach, current information on the mutual orientation of the BSS CS with the CSs of each of the other SSs is required.

Since even a high-precision pre-flight calibration of the mutual angular positions of the SSs and BSS is not capable of eliminating the errors arising during the long-term operation of the SC, the flight mission of the SC must have a procedure for periodic calibration of the mutual orientation of SSs. As applied to the astro-measuring system containing two sensors (SS1 and SS2) fixed rigidly in the SC body, a solution is proposed to minimize the influence of the residual uncertainty with respect to the angular positions of SSs in the CS associated with the SC on the accuracy of the operation of this information system.

An algorithm for processing measurement information coming from two SSs was obtained in terms of vectors of the MRP. It turned out that in this case, these parameters are effective means for describing the orientation of the SC. The implementation of the algorithm does not provide the use of information about the mutual orientation of SS1 and SS2. The three-dimensional vector of the MRP obtained during data processing is, if necessary, recalculated into the SC orientation quaternion.

The results of computer modeling show the effectiveness of the proposed algorithm for processing information coming from two SSs. This algorithm essentially increases the accuracy of calculating the SC orientation parameters in inertial space in the presence of parametric disturbances caused by the residual uncertainty in the angular positions of SSs in the CS $E$ associated with the SC.

## References

[1] V. Kuntsevich, V. Gubarev, Y. Kondratenko, D. Lebedev, V. Lysenko, (Eds), 'Control Systems: Theory and Applications. Series in Automation, Control and Robotics', River Publishers, Gistrup, Delft, 2018.

[2] Y. Kondratenko, A. Chikrii, F., Gubarev, J. Kacprzyk, (Eds), 'Advanced Control Techniques in Complex Engineering Systems: Theory and Applications. Dedicated to Professor Vsevolod M. Kuntsevich.Studies in Systems, Decision and Control', Vol. 203. Cham: Springer Nature Switzeland AG, 2019.

[3] M. Prohorov, A. Zaharov, A. Mironov, F. Nikolaev, M. Tuchin, 'Modern star sensors', Proc. In the 38th Int. Student Scientific Conf., pp. 170-186, Russia, Feb., 2009, [in Russian].

[4] N. Efimenko, 'Determining spacecraft orientation using information from two jointly processed star trackers'. Kosmichna nauka i tehnologia, v.20, N̲o 3, pp. 22-27, 2014 , [in Russian].

[5] G. Avanesov, R. Bessonov, A. Kurkin, A. Nikitin, V. Sazonov. 'Estimate of Accuracy of Determining the Orientation of the Star Sensor System according to the Experimental Data'. Cosmic Research. 2018. Vol. 56. No.1. pp. 38-53.

[6] G. Avanesov, R. Bessonov, A. Kurkin, A. Nikitin, V. Sazonov. 'Determining Spacecraft Motion ftom Four Star System Measurement'. Cosmic Research. 2018. Vol. 56. No.3. pp. 232-250

[7] B. Suhovilov, 'Method of pair distances in problem of star-trackers in-flight calibration of spacecrafts attitude systems'. Vestnik YuUrGU, N̲o 23, pp. 35–41, 2007, [in Russian].

[8] V. Branets, I. Shmyglevskiy, 'Application of quaternions in problems of rigid body orientation' ,[in Russian], Moskow: Nauka Publ., 1973.

[9] V. Zhurablev, 'Fundamentals of theoretical mechanics', [in Russian], Moskow: Nauka. Fizmatlit, 1997.

[10] S. Perelyaev, 'On the correspondence of three-dimensional and four-dimensional parameters of a group of three-dimensional rotations', Solid mechanics, 2009, no. 2, pp. 47-58.

[11] M. Shuster, 'A survey of attitude representations', J. Astronaut. Sci. 1993. V. 41. N̲o 4. P. 439- 517.

[12] E. Somov, S. Butyrin, 'Technology for processing accompanying measurement information for high-precision coordinate reference of satellite images', Bulletin of the Samara Scientific Center of the Russian Academy of Sciences. 2009.Vol. 11. No. 5. pp. 156-163, [in Russian].
[13] N. Bakhvalov,'Numerical methods (analysis, algebra, ordinary differential equations)', [in Russian], The main edition of the physical and mathematical literature of the Nauka publishing house, M., 1975, 632 p.
[14] S. Dyatlov, R. Bessonov, 'A survey of stellar sensors of spacecrafts orientation', Proc. In The 1st All-Russian Scientific and Technological Conf. 'Contemporary Problems of Spacecraft Attitude Determination and Control' pp. 11-31, Russia, Tarusa, 22-25 Sept., 2008, [in Russian].

# 12

# Control Synthesis of Rotational and Spatial Spacecraft Motion at Approaching Stage of Docking

**V. Volosov, S. Melnychuk, N. Salnikov, V. Shevchenko**

Space Research Institute, National Academy of Sciences of Ukraine and State Space Agency of Ukraine, Glushkov Av. 40, 4/1, Kyiv, 03680, Ukraine
E-mail: wwolosov@gmail.com, melnychuk89s@gmail.com, salnikov.nikolai@gmail.com, vovan_16@ukr.net

### Abstract

In this chapter, the control of rotational and translational motion for shock-free docking with a non-cooperative rotating spacecraft is considered. It is assumed that the relative position and orientation of target spacecraft are measured by a computer vision system installed on an active spacecraft. The angular motion parameters of the target spacecraft, including ratios of the body inertia moments are estimated by the proposed ellipsoidal estimation algorithm. The spacecraft orbital motion parameters are assumed to be known. The control problem is solved by using the Lyapunov function method separately for kinematic and dynamic equations of the relative spacecraft motion.

**Keywords:** Spacecraft docking, angular motion ellipsoidal estimation, translational and angular motion control, Lyapunov function.

## 12.1 Introduction

The number of space objects in near-Earth space is steadily growing. Some of these objects are idle spacecraft (SC), which can be repaired by refueling.

Development of technologies for servicing SC capable of docking with a non-cooperative space object is carried out in many scientific centers. Such servicing SCs should be equipped with a manipulator that provides reliable capture of the target when docking.

The SC docking is the most difficult maneuver. Note that the first automatic docking was carried out by the Soviet Union with unmanned vehicles Cosmos-186 and Cosmos-188 in October 1967. The USA Northrop Grumman's servicing spacecraft Mission Extension Vehicle-1 (MEV-1) successfully docked with a non-cooperative Intelsat communications satellite in February 2020 to extend the operation of the 19-year-old satellite for another five years. The docking process is getting greatly complicated when the space object does not have an attitude control system or, for some reason, it does not work. It is known [1, 2] that such a space object eventually begins to rotate around the center of mass. This occurs due to the action of the gravitational moment [3], the Sun radiation, or a stream of solar wind particles. In such situation, docking in automatic mode is the only possible way.

Successful rapprochement and docking is impossible without a high-precision determination of the relative position and orientation of a non-cooperative or passive spacecraft (PSC). In this paper, we use an ellipsoidal estimation algorithm which allows us to increase an accuracy of direct relative pose measurements, which are performed by computer vision system (CVS) mounted on the active servicing spacecraft (ASC). Design and algorithms of such CVS has been considered, for example, in [4]. In the case of a tumbling satellite, special requirements are imposed on the estimation accuracy of rotational motion. In this paper, we consider an ellipsoidal filter for estimating the attitude quaternion, angular velocity, and the unknown ratio of the inertia moments of an SC.

The following rendezvous and docking stages [5–7] are generally accepted: launch of ASC into vicinity of PSC, long-range guidance, short-range guidance, hovering, and actual docking. We will consider the last two stages, where features associated with the rotation of PSC are the most important. At the hovering stage, the relative position and rotational motion of the PSC are determined with high accuracy using a CVS device and an ellipsoidal filter. This information is necessary at the final stage in order to impose restrictions on the final attitude and position as well as on relative velocity and angular velocity. The control problems at docking stage and their solutions can be found in [6–11]. In this paper, we consider the use of Lyapunov method for closed-loop control synthesis. The peculiarity of our approach is in construction of Lyapunov functions separately for kinematic and dynamic equations of translational and angular motions. It allowed to obtain simple and robust control algorithms for translational and rotational SC motion. The efficiency of the obtained solutions is illustrated by numerical simulation.

## 12.2 Equation of the Spacecraft Relative Motion in the Docking Stage

### 12.2.1 Equation of the Relative Motion of the Spacecraft Center of Mass

Let us consider the motion of an active and non-cooperative (passive) SC during their rapprochement. The motion parameters of the ASC and the PSC will be denoted by subscripts $a$ and $p$, respectively. We will neglect disturbances caused by the non-sphericity of the Earth's gravitational field (EGF). Motion of the SC centers of mass in vector form relative to the Earth-centered inertial reference frame (IRF) [5, 12–14] is described by the following equations:

$$\frac{d^2 r_p}{dt^2} + \mu \frac{r_p}{||r_p||^3} = 0, \tag{12.1}$$

$$\frac{d^2 r_a}{dt^2} + \mu \frac{r_a}{||r_a||^3} = a. \tag{12.2}$$

Here, $r_p$and $r_a$ are the current radius vectors of the PSC and ASC in the IRF, $||r_p||$ and $||r_a||$ are the lengths (Euclidean norms) of the corresponding vectors (distance from the SC to the Earth's center), $a$ is the controlling acceleration vector of the ASC, $\mu$ =398 600.4 km$^3$/s$^2$ is the Earth's gravitational constant (see Figure 12.1). We introduce the vector $\rho = r_a - r_p$ of relative position of the ASC as shown in Figure 12.1.

Point $O_E$ in Figure 12.1 is the center of mass of the Earth. Subtracting Equation (12.1) from Equation (12.2), we obtain the following equation:

$$\frac{d^2 \rho}{dt^2} + \mu \left( \frac{r_a}{||r_a||^3} - \frac{r_p}{||r_p||^3} \right) = u, \rho = r_a - r_p, u = a_a. \tag{12.3}$$

Substituting the expression $r_a = r_p + \rho$ in Equation (12.3), we obtain

$$\frac{d^2 \rho}{dt^2} + \mu \frac{r_p + \rho}{||r_p + \rho||^3} - \mu \frac{r_p}{||r_p||^3} = u. \tag{12.4}$$

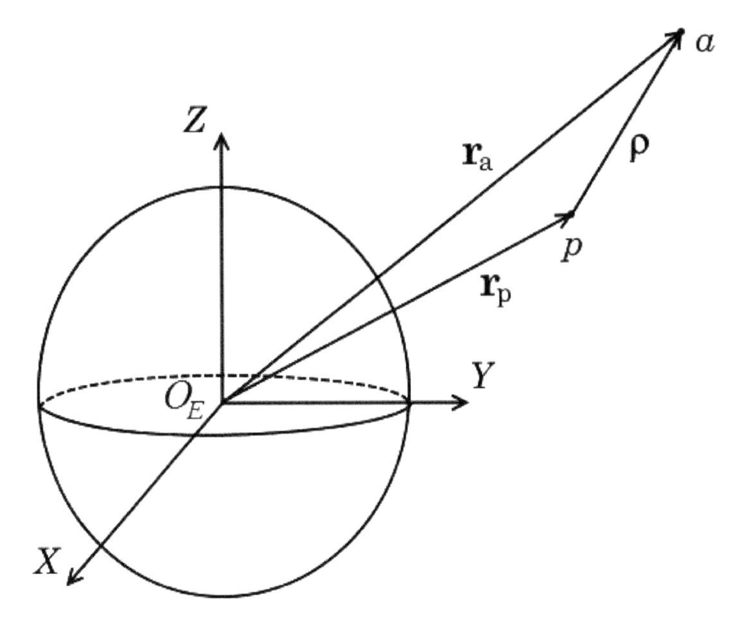


**Figure 12.1** Position of the spacecraft in IRF.

Equation (12.4) describes the motion of an ASC relative to a passive one in the IRF.

For convenience of description, we present the equation of the relative motion of an ASC in the right-handed local orbital frame (LOF) $O_P x_0 y_0 z_0$ with the origin at the center of mass (point $O_P$) of the PSC. We assume that the PSC moves over a circular orbit of radius$R = ||r_p||$. By analogy with [6, 15, 16], we direct the axis $O_p y_0$ along the current geocentric vertical, and the axis $O_P x_0$ lies in the orbit plane with a positive direction along the SC velocity. LOF $O_P x_0 y_0 z_0$ rotates relative to IRF with the PSC orbital angular velocity. In this case, the angular velocity vector $\Omega(\Omega = (0, 0, -\omega_0)^T$, where $\omega_0 = \sqrt{\mu/R^3}$ [6, 15, 17] is directed opposite to the positive direction of the axis $O_P z_0$ (see Figure 12.2).

Using the well-known rules for the differentiation of vectors in a rotating coordinate system [18, 19] and Equation (12.4), we obtain the equation for the relative motion of the ASC in the LOF

$$\ddot{\rho} + 2\breve{\Omega}\dot{\rho} + \breve{\Omega}\breve{\Omega}\rho + \mu\frac{r_p + \rho}{||r_p + \rho||^3} - \mu\frac{r_p}{||r_p||^3} = u, \tag{12.5}$$

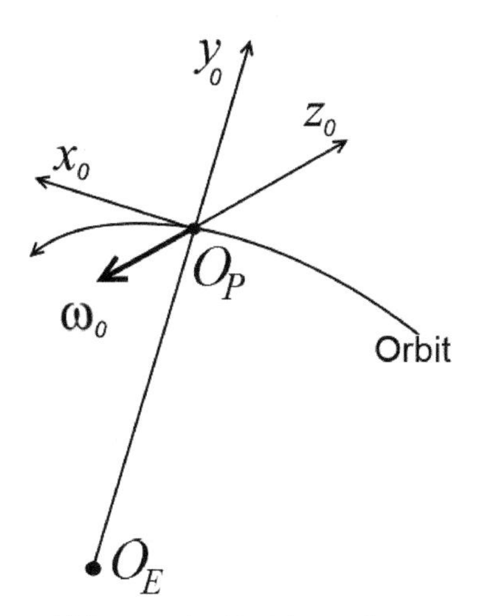


**Figure 12.2** Local orbital frame $O_P x_0 y_0 z_0$.

where

$$\rho = \begin{pmatrix} x \\ y \\ z \end{pmatrix}, \dot{\rho} = \begin{pmatrix} \dot{x} \\ \dot{y} \\ \dot{z} \end{pmatrix}, \ddot{\rho} = \begin{pmatrix} \ddot{x} \\ \ddot{y} \\ \ddot{z} \end{pmatrix}, \Omega = \begin{pmatrix} 0 \\ 0 \\ -\omega_0 \end{pmatrix},$$

$$\breve{\Omega} = \begin{pmatrix} 0 & \omega_0 & 0 \\ -\omega_0 & 0 & 0 \\ 0 & 0 & 0 \end{pmatrix}, \breve{\Omega}\breve{\Omega} = \begin{pmatrix} -\omega_0^2 & 0 & 0 \\ 0 & -\omega_0^2 & 0 \\ 0 & 0 & 0 \end{pmatrix},$$

$$r_p = \begin{pmatrix} 0 \\ R \\ 0 \end{pmatrix}, r_p + \rho = \begin{pmatrix} x \\ R+y \\ z \end{pmatrix}, ||r_p + \rho|| = \sqrt{x^2 + (y+R)^2 + z^2}.$$

The vector–matrix equation (12.5) has the following scalar representation:

$$\left.\begin{aligned} &\ddot{x} + 2\omega_0\dot{y} - \omega_0^2 x + \frac{\omega_0^2 R^3 x}{(\sqrt{x^2+(y+R)^2+z^2})^3} = u_x \\ &\ddot{y} - 2\omega_0\dot{x} - \omega_0^2(y+R) + \frac{\omega_0^2 R^3}{(\sqrt{x^2+(y+R)^2+z^2})^3}(y+R) = u_y, \\ &\ddot{z} + \frac{\omega_0^2 R^3}{(\sqrt{x^2+(y+R)^2+z^2})^3} z = u_z. \end{aligned}\right\} \quad (12.6)$$

Since at the docking phase $||\rho|| << R$, we use the linear representation of Equation (12.6) obtained by expanding non-linear functions in the left-hand sides of Equation (12.6) at a point $x = y = z = 0$. After that, the following system of linear equations can be written:

$$\left.\begin{aligned} \ddot{x} + 2\omega_0\dot{y} = u_x, \\ \ddot{y} - 2\omega_0\dot{x} - 3\omega_0^2 y = u_y, \\ \ddot{z} + \omega_0^2 z = u_z. \end{aligned}\right\} \tag{12.7}$$

which is known as Hill's or Clohessy–Wiltshire equations [6, 14, 16].

To formulate the control problem, we introduce here the notation $d = d(t)$ for the current distance between the SC during the docking process:

$$d(t) = \|\rho(t)\| = \sqrt{x^2(t) + y^2(t) + z^2(t)}.$$

### 12.2.2 Equation of the Spacecraft Relative Angular Motion

Let us obtain the equations of relative angular motion using some of the assumptions accepted in [20]. We will use two body frames (BFs) $O_a x_a y_a z_a$(with the basis $E_a$) and $O_p x_p y_p z_p$ (with the basis $E_p$) with their origins $O_a$ and $O_p$ placed at the centers of mass of the active and the PSC, respectively.

Let the relative attitude of the active and PSC (that is the relative attitude of the bases $E_a$ and $E_p$) be determined by the direction cosine matrix $S$, $S^{-1} = S^{\mathrm{T}}$. Then

$$X_p = SX_a, \tag{12.8}$$

where $X_a = (x_a, y_a, z_a)^{\mathrm{T}}$ and $X_p = (x_p, y_p, z_p)^{\mathrm{T}}$ are the column vectors composed of projections of an arbitrary vector $X = (x, y, z)$ on the axis of the bases $E_a$ and $E_p$. The coincidence of the BF axes $O_a x_a y_a z_a$ and $O_p x_p y_p z_p$ corresponds to the case $S = I_3$.

The angular motion of an ASC is described by the dynamic Euler equation

$$J\dot{\omega} + \breve{\omega} J\omega = M, \tag{12.9}$$

where $J = J^{\mathrm{T}} > 0$ is a positive definite symmetric matrix of the inertia tensor of the ASC in its BF, satisfying the conditions of its physical realizability [21]; $\omega = (\omega_1, \omega_2, \omega_3)^{\mathrm{T}}$ is the vector of the absolute angular velocity of the ASC and $M = (M_1, M_2, M_3)^{\mathrm{T}}$ is the vector of control moments, both given by their projections on the coordinate axes of the BF $O_a x_a y_a z_a$. Matrix form of cross-product was used in Equation (12.9) with the matrix

$$\breve{\omega} = \begin{pmatrix} 0 & -\omega_3 & \omega_2 \\ \omega_3 & 0 & -\omega_1 \\ -\omega_2 & \omega_1 & 0 \end{pmatrix}.$$

As it directly follows from Equation (12.9), the control moments change the angular velocity of the SC and, thereby, the relative orientation matrix $S$ in Equation (12.8) according to the generalized Poisson matrix equation (see, for example, [22])

$$\dot{S} = S\breve{\omega} - \breve{\omega}_* S, \tag{12.10}$$

where $\omega_* = (\omega_{*1}, \omega_{*2}, \omega_{*3})^T$ is an angular velocity vector of PSC, given by its projections on coordinate axis of BF $O_p x_p y_p z_p$.

Elements $s_{ij}$ of the attitude matrix $S$ can be parameterized by the vector of Euler–Rodrigue–Hamilton parameters (quaternion components) $\Lambda^{\mathrm{T}} = (\lambda_0, \lambda^{\mathrm{T}})$, where $\lambda_0$ is a scalar part of the quaternion, and vector part $\lambda = (\lambda_1, \lambda_2, \lambda_3)^{\mathrm{T}}$, $||\Lambda|| = 1$ [23–25]. The direction cosine matrix $S(\Lambda)$ corresponding to $\Lambda$ is determined by the following expression [26, 27]:

$$S(\Lambda) = I_3 - 2\lambda_0\breve{\lambda} + 2\breve{\lambda}\breve{\lambda}. \tag{12.11}$$

In some cases, it is preferable to use, instead of Equation (12.10), containing nine matrix elements, the following equations:

$$\dot{\Lambda} = 0,5Q(\Lambda)\omega_r, \ \omega_r = \omega - S(\Lambda)\omega_*, \tag{12.12}$$

with a four-dimensional state vector $\Lambda \in R^4$ [27], where a $4 \times 3$ matrix

$$Q(\Lambda) = \begin{pmatrix} -\lambda^T \\ \lambda_0 I_3 + \breve{\lambda} \end{pmatrix}, \operatorname{rank} Q(\Lambda) = 3 \,\forall ||\Lambda|| \neq 0.$$

Let us note that as it can be seen from Equation (12.11), two values of the vector $\Lambda$ correspond to the same relative attitude of the SC since $S(\Lambda) = S(-\Lambda)$.

### 12.2.3 Control Problem Statement at the Docking Stage

By analogy with [20], we formulate the control problem as a common synthesis problem of control accelerations $U = (u_x, u_y, u_z)$ in Equation (12.7) and control moments $M = M_c$ in the dynamic Euler equation (12.9) of the motion of an ASC when the following relations hold:

$$d(t_E) = ||\rho(t_E)|| = 0, ||\dot{\rho}(t_E)|| = 0, S(t_E) = I_3, ||\omega_r(t_E)|| = 0. \tag{12.13}$$

Here, finite time moment $t = t_E$ can be arbitrary or fixed (terminal control).

Taking into account [20], the heterogeneity of the EGF is also neglected due to the relative small size of the SC spatial coordinates change during time interval of the final docking stage.

We note that conditions (12.13), generally speaking, are idealized. Due to the inevitable measurement errors, they can be fulfilled only approximately. Therefore, the urgent task is to obtain guaranteed estimates of the accuracy of the conditions (12.13) implementation. To solve this problem, methods for constructing reachable sets of dynamical systems can be applied, as it was done, for example, in [28]. It is also necessary to take into account that, in practice, the initial conditions for spatial and angular coordinates (see Equations (12.6), (12.7), (12.9), and (12.12)) can be known only in the form of their belonging to closed bounded sets. However, reachable sets of even linear systems perturbed structurally or/and parametrically can have a complex structure. Therefore, the actual future problem-motivated practice of controlling the SC docking is a problem of developing methods for ellipsoidal approximation of the trajectories of dynamical systems or their integral funnels [29–31].

## 12.3 Parameter Estimation of the PSC Rotational Motion

An increase in the accuracy of CVS used in SC relative navigation is associated with the need to store a larger amount of information in the computer memory and, as a result, it reduces the performance due to an increase in the processing time of this information. In addition, accuracy is limited by the resolution of the photosensitive camera matrix and the resolution of the optical system. Therefore, the accuracy of determining the position and attitude of the non-cooperative SC (PSC) cannot be arbitrarily increased, and the values of the position and orientation parameters of the PSC determined by the CVS will inevitably contain some limited errors.

The accuracy of the pose estimates can be improved by filtering using a dynamic PSC model, which will also determine the approach speed and the angular velocity of the SC rotation. These data are necessary for constructing controlled motion at the stage of SC rapprochement and docking.

The dynamics of SC orbital motion and rotation around the center of mass can be considered separately. Numerous observations indicate that SC with an idle control system acquires rotational motion while in orbit [1,2]. This subsection discusses the use of a dynamic filter to estimate angular motion parameters.

The most common filtering approach uses the assumption of the stochastic nature of measurement errors. This allows to apply the Kalman filter and its numerous modifications [32, 33]. However, the justified use of such filters requires sufficiently accurate knowledge of the distribution laws of random variables or their main stochastic moments: mathematical expectation and dispersion. Obtaining such information is associated with a large number of experiments requiring large expenditures of time and resources. In addition, the properties of random variables, as practice shows, are often not stationary [34].

These drawbacks are absent in the ellipsoidal estimation algorithms proposed by the authors of this article [35–39]. They are insensitive in the sense of preserving convergence to the difference between *a priori* assumptions about the properties of uncertainty used in the algorithms from their real properties. Uncertainty here is understood as ambiguous estimates of phase coordinates, parameters, intensity of possible realizations of measurement noise, and, generally speaking, structural disturbances of the mathematical models of dynamical systems. In [35, 36], algorithms for ellipsoidal estimation are proposed, which have the property of robustness to these differences. The efficiency of using these algorithms for control of a space object is illustrated in [37].

Here we use one of the modifications [38, 39] of the algorithms for guaranteed estimation using ellipsoids [29, 40–45], which are based on the use of minimal *a priori* information on uncertain values. It is assumed that only the sets of their possible values are known. The main advantages of the modification used are the high convergence rate, ease of implementation, applicability to non-linear systems, and resistance to possible violations of *a priori* hypotheses about the properties of uncertain values.

### 12.3.1 Problem Statement of the Angular Motion Parameters Estimation

The dynamical equations of the PSC angular motion are considered in the BF and have the following form:

$$J\dot{\omega} + \omega \times J\omega = M, \tag{12.14}$$

where $\omega = (\omega_1, \omega_2, \omega_3)^{\mathrm{T}}$ is an angular velocity vector and $M = (M_1, M_2, M_3)^{\mathrm{T}}$ is torque vector acting at the PSC, which are given by their coordinates in BF. Diagonal matrix of inertia moments $J = \mathrm{diag}(J_1, J_2, J_3)$ is given by the main inertia moments $J_i$, $i = \overline{1, 3}$, with respect to $i$-th coordinate axis. Note that the subscript "*p*" denoting the variables related to the PSC is omitted in all variables for notation simplification.

Overall moment $M$ acting at the PSC can be caused by the action of the gravitational moment, the Sun radiation, and particle flows [3, 46].

However, all these moments are rather weak, and their effect is manifested in a large time interval. In addition, they are not measured. An account of their action with the use of analytical expressions needs knowledge of many additional parameters including the PSC orbital parameters that could make the estimation problem solution to be more difficult. Therefore, we will neglect the action of these moments and assume that $M = 0$ in Equation (12.14).

Under space flight conditions, the inertia moments of the SC can change, for example, due to solar panels deployment, a change in the SC configuration, fuel consumption, and due to other reasons. Obtaining accurate information about the value of the PSC inertia moments in space flight is generally impossible. Therefore, the diagonal matrix $J$ in Equation (12.14) is also not assumed to be known.

It is easy to verify that in the absence of a torque of external forces, solutions of Equation (12.14) depend only on the ratios of the inertia moments. We introduce the following notations:

$$p_1 = J_1 J_3^{-1}, \; p_2 = J_2 J_3^{-1}.$$

Then Equation (12.14) can be written in the following component-wise form:

$$\begin{cases} \dot{\omega}_1 = p_1^{-1}(p_2 - 1)\omega_2\omega_3; \\ \dot{\omega}_2 = p_2^{-1}(1 - p_1)\omega_1\omega_3; \\ \dot{\omega}_3 = (p_1 - p_2)\omega_1\omega_2. \end{cases} \tag{12.15}$$

Assume that in the considered time interval, the moments of inertia are constant. Therefore, for the vector $p = (p_1, p_2)^{\mathrm{T}}$, we have the equation

$$\dot{p} = 0. \tag{12.16}$$

Let us consider that the angular position of the PSC and the ASC is determined relative to the IRF and is given by the quaternion. The PSC attitude change relative to the IRF is described by the following equation:

$$\dot{\Lambda} = \frac{1}{2} Q(\Lambda) \cdot \omega, \tag{12.17}$$

where $\Lambda = (\lambda_0, \lambda_1, \lambda_2, \lambda_3)^{\mathrm{T}}$ is the PSC attitude quaternion with respect to IRF [23, 47, 48], $\lambda_0$ is its scalar part, and $(\lambda_1, \lambda_2, \lambda_3)^{\mathrm{T}}$ is its vector part. The matrix

$$Q(\Lambda) = Q(\lambda_0, \lambda_1, \lambda_2, \lambda_3) = \begin{pmatrix} -\lambda_1 & -\lambda_2 & -\lambda_3 \\ \lambda_0 & -\lambda_3 & \lambda_2 \\ \lambda_3 & \lambda_0 & -\lambda_1 \\ -\lambda_2 & \lambda_1 & \lambda_0 \end{pmatrix}.$$

Vector $\Lambda$ in Equation (12.17) and in all the following equations is normalized

$$||\Lambda|| = \lambda_0^2 + \lambda_1^2 + \lambda_2^2 + \lambda_3^2 = 1. \tag{12.18}$$

Initial conditions for Equations (12.14) and (12.17), vectors $\Lambda_0 = \Lambda(t_0)$, and $\omega_0 = \omega(t_0)$ are unknown.

CVS installed at the ASC measures attitude quaternion $\Lambda_k^{p|a}$ of the PSC relative to the ASC BF in discrete time moments $t_k = t_0 + k \cdot \Delta t,\ k = 1,\ 2, \ldots$, where $\Delta t$ is a measurement period. We note that if the PSC is not observed by the ASC camera, then there are no measurements available. At these time moments, it is important to have estimates of the angular motion parameters.

It is assumed that ASC navigation devices determine its orientation relative to the IRF with high accuracy, i.e. the corresponding quaternion $\Lambda_k^a$ is known for each moment of time. For the quaternion in Equation (12.17), at the time $t_k$, we can write the following exact relation:

$$\Lambda_k = \Lambda_k^{p|a} \circ \Lambda_k^a,$$

where $\circ$ denotes the quaternion multiplication [23, 47, 48]. Multiplying the last equation by conjugated quaternion $\bar{\Lambda}_k^a$ from the right, we obtain

$$\Lambda_k^{p|a} = \Lambda_k \circ \bar{\Lambda}_k^a.$$

Using expressions for quaternion multiplication, the last equation can be written in the following vector–matrix equality:

$$\Lambda_k^{p|a} = C(\bar{\Lambda}_k^a)\Lambda_k, \tag{12.19}$$

where matrix

$$C(\bar{\Lambda}) = \begin{pmatrix} \lambda_0 & \lambda_1 & \lambda_2 & \lambda_3 \\ -\lambda_1 & \lambda_0 & -\lambda_3 & \lambda_2 \\ -\lambda_2 & \lambda_3 & \lambda_0 & -\lambda_1 \\ -\lambda_3 & -\lambda_2 & \lambda_1 & \lambda_0 \end{pmatrix}.$$

As a result of CVS measurement of the PSC attitude relative to the ASC, we obtain the quaternion

$$\tilde{\Lambda}_k^{p|a} = \Lambda_k^{p|a} + \xi_k, \tag{12.20}$$

where $\xi_k = (\xi_{0,k}, \ldots, \xi_{3,k})^{\mathrm{T}}$ is a vector of bounded measurement errors, satisfying the inequalities

$$|\xi_{i,k}| \leq c_i \quad \forall i. \tag{12.21}$$

Here, positive numbers $c_i$, $i = \overline{0,\ 3}$, are assumed to be known.

Substituting Equation (12.19) into Equation (12.20), we obtain

$$\tilde{\Lambda}_k^{p|a} = C(\bar{\Lambda}_k^a)\Lambda_k + \xi_k. \tag{12.22}$$

From here and Equation (12.21), it follows that the unknown attitude quaternion $\Lambda_k$ at each moment of time $k = 1, 2, \ldots$ must satisfy the following inequalities:

$$|C_{i,k}\Lambda - \tilde{\lambda}_{i,k}^{p|a}| \leq c_i, \quad i = \overline{0,\ 3}. \tag{12.23}$$

Here, $C_{i,k}$ is the $i$-th row of the matrix $C(\bar{\Lambda}_k^a)$.

Differential equations (12.15)–(12.17) represent the equations of a non-linear dynamic system in continuous time, the state vector of which has the form

$$x = (\omega^{\mathrm{T}}, \Lambda^{\mathrm{T}}, p^{\mathrm{T}})^{\mathrm{T}}.$$

At discrete moments of time, it must satisfy the inequalities (12.23) associated with the measurements. In addition, the normalization condition (12.18) must be satisfied for the component $\Lambda$ of the vector $x$.

In the ideal case, it is required to construct the method of calculation of estimates $\hat{x}_k$ of the state vector $x_k = x(t_k)$, for which the following limit holds:

$$\lim_{k\to\infty} ||\hat{x}_k - x_k|| = 0.$$

Its existence is associated with a realization of certain properties of measurement errors, which are difficult to verify and provide in practice. That is why we consider the solution of a weakened problem. It is required that the estimates $\hat{x}_k$ obtained by virtue of Equations (12.15)–(12.17) satisfy inequalities (12.23) associated with the measurements starting from some finite time moment $K$.

Taking into account *a priori* restrictions on the measurements errors of the form (12.21), we apply the guaranteed approach to the state vector estimation [29, 40–45]. We will use the modified ellipsoidal estimation algorithm [38, 39], which is applicable for non-linear systems and is not sensitive to violence of *a priori* information about the initial state vector.

### 12.3.2 Non-Linear Ellipsoidal Estimation Method

Let us briefly describe the guaranteed approach to the state vector estimation of dynamical systems. Equations (12.15)–(12.17) can be written as

$$\dot{x} = f(x(t), \zeta(t)), \quad t \geq t_0, \tag{12.24}$$

where $x(t)$ is the state vector at the instant of continuous time $t$, $\zeta(t)$ is the vector of uncontrolled disturbances, in this case, the vector of moments acting at the SC,

$$\zeta(t) \in \mathrm{Z} \quad \forall t, \tag{12.25}$$

where Z is some bounded closed set, $\mathrm{Z} \subset R^3$. We assume that the functions in Equation (12.24) are such that the classical conditions for the existence and uniqueness of a solution of Equation (12.24) are satisfied.

The restrictions on the state vector $x_k$ given by inequalities (12.23) can be written in the following form:

$$|\tilde{C}_{i,k}x_k - \tilde{\lambda}_{i,k}^{p|a}| \leq c_i, \quad i = \overline{0, 3}. \tag{12.26}$$

Here, $\tilde{C}_{i,k} = (\Theta_{1\times 3}, C_{i,k}, \Theta_{1\times 2})$ is row-vector $\tilde{C}_{i,k} = (\Theta_{1\times 3}, C_{i,k}, \Theta_{1\times 2})$, and $\Theta_{n\times m}$ is an $n \times m$–zero matrix. The inequalities are considered for discrete time moments $t_k$.

The procedure for constructing state vector estimates according to the ellipsoidal estimation method [29, 40–45] is as follows. Suppose that at time moment $t_k$, it is known that the state vector

$$x_k = x(t_k) \in E_k, \tag{12.27}$$

where, for definiteness, we will assume that the set $E_k$ is an ellipsoid $E_k = \{x : (x - \hat{x}_k)^T H_k^{-1}(x - \hat{x}_k) \leq 1\}$, characterized by a center vector $\hat{x}_k$ and a positive definite symmetric matrix $H_k = H_k^T > 0$. An ellipsoid $E_k$ satisfying Equation (12.27) is usually called a set or ellipsoidal estimate of the vector $x_k$. The center of the ellipsoid vector $\hat{x}_k$ is taken as a point estimate. Considering solutions of Equation (12.24) on the time interval $[t_k, t_{k+1}]$ for all possible initial conditions satisfying Equation (12.27) and for all possible realizations of perturbations $\zeta(\cdot)$ satisfying condition (12.25) on this interval, we can obtain the set

$$X_{k+1|k} = \{x = x(t_{k+1}, x_k, \xi(\cdot)), \forall x_k \in E_k \; \forall \zeta(\tau) \in \mathrm{Z} \; \forall \tau \in [t_k, t_{k+1}]\}$$

of possible values of the vector $x_{k+1} = x(t_{k+1})$ at a discrete time instant $k + 1$. This set is generally not an ellipsoid. The construction of this set can be accomplished, for example, by integrating Equation (12.24) on the interval $[t_k, t_{k+1}]$ under various initial conditions satisfying Equation (12.25) and for all possible realizations of perturbations $\zeta(\tau) \in Z \; \forall \tau \in [t_k, t_{k+1}]$. In general case, this is obviously a very time-consuming process, requiring a large amount of calculations.

On the other hand, the state vector $x_{k+1}$ at the time moment $t_{k+1}$ must satisfy the conditions (12.26), which can be written as

$$x_{k+1} \in \bar{X}_{k+1} = \{x : |\tilde{C}_{i,k+1}x - \tilde{\lambda}^{p|a}_{i,k+1}| \leq c_i, \quad i = \overline{0, 3}\}.$$

The set $\bar{X}_{k+1}$ contains state vectors that are compatible with the measurements under given *a priori* restrictions on the values of the measurement errors. As a result, we can conclude that

$$x_{k+1} \in X_{k+1} = \bar{X}_{k+1} \bigcap X_{k+1|k}.$$

Despite the obvious simplicity and logical rigor of the approach to refining the set of possible values of the state vector, its practical implementation encounters in the general case insurmountable computational difficulties associated with the construction of sets $X_{k+1|k}$, $\bar{X}_{k+1}$, $X_{k+1}$, and the implementation of set-theoretic operations on the sets, in this case, set intersection operation.

One approach to reduce the computational complexity of solving set estimation problems is to use the ellipsoid method. In accordance with this method, the ellipsoidal approximations for sets $X_{k+1|k}$ and $X_{k+1}$ are constructed in the form of ellipsoids $E_{k+1|k}$ and $E_{k+1}$ containing them, and the solution of the estimation problem reduces to constructing a sequence $\{E_k\}_{k=0}^{\infty}$ of ellipsoidal estimates, $x_k \in E_k$, for the vector $x_k$ in accordance with the following recursive procedure:

$$x_{k+1} \in E_{k+1} = \left[\bar{X}_{k+1} \bigcap \left[X_{k+1|k}\right]_E\right]_E = \left[\bar{X}_{k+1} \bigcap E_{k+1|k}\right]_E. \quad (12.28)$$

Here, $[X]_E$ denotes some operation of covering bounded set $X \subset R^n$ by an ellipsoid ($R^n$ is $n$-dimensional real Euclidean space). The described method for constructing ellipsoidal estimates is called ellipsoidal estimation. For linear systems, many operations $[X]_E$ have been developed, which are optimal and suboptimal in some sense. In the case of non-linear systems, these operations $[X]_E$ can generally be realized only numerically and requires a large amount of computation. In [38, 39], by analogy with the extended Kalman filter, an approach was proposed that allows to reduce the amount of computation when constructing a sequence of ellipsoidal estimates. In accordance with this approach, at a time $t_{k+1}$, the following ellipsoid is constructed:

$$E_{k+1|k} = \{x : (x - \hat{x}_{k+1|k})^T H_{k+1|k}^{-1} (x - \hat{x}_{k+1|k}) \leq 1\}.$$

Its center vector $\hat{x}_{k+1|k}$ is found by integrating the following equation:

$$d\tilde{x}/dt = f(\tilde{x}(t), \hat{\zeta}(t)), \quad \tilde{x}(t_k) = \hat{x}_k,\ t \in [t_k, t_{k+1}], \tag{12.29}$$

setting $\hat{x}_{k+1|k} = \tilde{x}(t_{k+1})$. Here, $\hat{\zeta}(t)$ is an estimate of the unknown vector $\zeta(t)$ which is chosen from some considerations. It is proposed to take the ellipsoid matrix as follows:

$$H_{k+1|k} = \mathrm{A}_k H_k \mathrm{A}_k^{\mathrm{T}}, \tag{12.30}$$

where $n \times n$–matrix ($n$ is a dimension of the vector $x$)

$$\mathrm{A}_k = \exp(\partial_x f(0,5 \cdot (\hat{x}_{k+1|k} + \hat{x}_k), \hat{\zeta}_{k+1/2})\Delta t). \tag{12.31}$$

Here, $\partial_x f(\cdot\ ,\cdot)$ is the Jacobi matrix of the function $f(x\ ,\zeta)$ with respect to a variable $x$,

$$\hat{\zeta}_{k+1/2} = \frac{1}{\Delta t} \int_{t_k}^{t_{k+1}} \hat{\zeta}(\tau) d\tau. \tag{12.32}$$

Thus, we used the linearization and averaging procedures of Equation (12.24) to construct the ellipsoid matrix. The ellipsoid $E_{k+1|k}$ is some approximation of the set $X_{k+1|k}$, for which $E_{k+1|k} \bigcap X_{k+1|k} \neq \emptyset$. However, the ellipsoid $E_{k+1|k}$ does not necessarily contain the set $X_{k+1|k}$. Therefore, in this case, the situation could take place when $\bar{X}_{k+1} \bigcap E_{k+1|k} = \emptyset$, and the algorithm (12.28) stops. To avoid this situation, an additional ellipsoid expansion operation is introduced into the algorithm.

Let us consider the ellipsoid

$$\tilde{E}_{k+1|k} = \{x : (x - \hat{x}_{k+1|k})^T \tilde{H}_{k+1|k}^{-1} (x - \hat{x}_{k+1|k}) \leq 1\},$$

in which the center coincides with the center of the ellipsoids $E_{k+1|k}$, and the matrix

$$\tilde{H}_{k+1|k} = \alpha^2 H_{k+1|k}, \tag{12.33}$$

where the number $\alpha \geq 1$. The ellipsoid $\tilde{E}_{k+1|k}$ is obtained at uniform expansion of the ellipsoid $E_{k+1|k}$ in $\alpha$ times with respect to its center.

If the function $f(\cdot)$ in Equation (12.24) is such that the matrix $\mathrm{A}_k$ defined by Equation (12.30) is non-singular and bounded, the perturbations are bounded, then obviously there exists a finite number $\alpha \geq 1$ such that

$$X_{k+1|k} \subseteq \tilde{E}_{k+1|k}.$$

The described procedures (12.29)–(12.32) for constructing an ellipsoid $E_{k+1|k}$ and its extension (12.33) are used in the ellipsoidal estimation algorithm for the state vector $x_k$ of the system (12.24) to construct a sequence of ellipsoids $\{E_k\}_{k=0}^{\infty}$, which centers $\hat{x}_k \in \bar{X}_k$ for granted. This sequence is constructed in the following way. Let the ellipsoid $E_k$ be already built at the discrete time moment $k$. Then, parameters of the ellipsoid $E_{k+1|k}$ are determined using the formulas (12.29)–(12.32). If

$$E_{k+1|k} \bigcap \bar{X}_{k+1} \neq \emptyset, \tag{12.34}$$

then an ellipsoid, containing a set $E_{k+1|k} \bigcap \bar{X}_{k+1}$ and which multidimensional volume is less than the volume of $E_{k+1|k}$, i.e. $V(E_{k+1}) < V(E_{k|k+1})$, is taken as ellipsoid $E_{k+1}$. The set $\bar{X}_{k+1}$ is the intersection of multidimensional layers. The construction of $E_{k+1}$ can be carried out using successive iterative optimal procedures [41] for covering by an ellipsoid the intersection of a layer with the ellipsoid obtained at the previous iteration. This iterative procedure stops after finite number of steps and the resulting ellipsoid is taken as $E_{k+1}$, and according to construction procedure, its center $\hat{x}_{k+1} \in \bar{X}_{k+1}$. Ellipsoid $E_{k+1|k}$ is taken as initial ellipsoid for the procedure.

If the condition (12.34) is not fulfilled, then an ellipsoid $\tilde{E}_{k+1|k}$ is taken instead of the ellipsoid $E_{k+1|k}$, and parameter $\alpha \geq 1$ is chosen so that the condition (12.34) is true. After that, the construction of the ellipsoid is carried out as described above. The center of the obtained ellipsoid $E_{k+1}$ satisfies $\hat{x}_{k+1} \in \bar{X}_{k+1} \quad \forall k$.

### 12.3.3 Estimation of the Quaternion, Angular Velocity, and Ratios of Inertia Moments

Consider the implementation of the above approach to Equations (12.15)–(12.17). Let it be known at the time moment $t_k$ that the state vector $x_k \in E_k$, where the ellipsoid $E_k = \{x : (x - \hat{x}_k)^T H_k^{-1}(x - \hat{x}_k) \leq 1\}$. It can be written that

$$x_k = \hat{x}_k + \Delta x_k,$$

where vector $\Delta x_k \in E_k(0) = \{x : x^{\mathrm{T}} H_k^{-1} x \leq 1\}$ and $E_k(0)$ is ellipsoid with the center at the origin. Let us denote $\tilde{x}(t) = (\tilde{\omega}^{\mathrm{T}}, \tilde{\Lambda}^{\mathrm{T}}, \tilde{p}^{\mathrm{T}})^{\mathrm{T}}$ the solution of Equations (12.15)–(12.17) on the interval $[t_k, t_{k+1}]$ with the initial condition $\tilde{x}(t_k) = \hat{x}_k$.

Substitute the expression

$$x(t) = \tilde{x}(t) + \Delta x(t)$$

in Equations (12.15)–(12.17). After discarding terms of order greater than the first, we obtain linear differential equations for $\Delta x(t) = (\Delta\omega^{\mathrm{T}}, \Delta q^{\mathrm{T}}, \Delta p^{\mathrm{T}})^{\mathrm{T}}$. In particular, from Equation (12.15), we have the following equation:

$$\begin{cases} \Delta\dot{\omega}_1 = \tilde{p}_1^{-1}(\tilde{p}_2 - 1)[\Delta\omega_2\tilde{\omega}_3 + \tilde{\omega}_2\Delta\omega_3] - [\tilde{p}_1^{-2}(\tilde{p}_2 - 1)\Delta p_1 - \tilde{p}_1^{-1}\Delta p_2]\tilde{\omega}_2\tilde{\omega}_3; \\ \Delta\dot{\omega}_2 = \tilde{p}_2^{-1}(1 - \tilde{p}_1)[\Delta\omega_1\tilde{\omega}_3 + \tilde{\omega}_1\Delta\omega_3] - [\tilde{p}_2^{-2}(1 - \tilde{p}_1)\Delta p_2 + \tilde{p}_2^{-1}\Delta p_1]\tilde{\omega}_1\tilde{\omega}_3; \\ \Delta\dot{\omega}_3 = (\tilde{p}_1 - \tilde{p}_2)[\Delta\omega_1\tilde{\omega}_2 + \tilde{\omega}_1\Delta\omega_2] + [\Delta p_1 - \Delta p_2]\tilde{\omega}_1\tilde{\omega}_2; \end{cases}$$

which can be written in matrix notations as

$$\Delta\dot{\omega} = A_\omega(\tilde{p}, \tilde{\omega})\Delta\omega + A_p(\tilde{p}, \tilde{\omega})\Delta p.$$

Here matrices

$$A_\omega(p, \omega) = \begin{pmatrix} 0 & p_1^{-1}(p_2 - 1)\omega_3 & p_1^{-1}(p_2 - 1)\omega_2 \\ p_2^{-1}(1 - p_1)\omega_3 & 0 & p_2^{-1}(1 - p_1)\omega_1 \\ (p_1 - p_2)\omega_2 & (p_1 - p_2)\omega_1 & 0 \end{pmatrix},$$

$$A_p(p, \omega) = \begin{pmatrix} -p_1^{-2}(p_2 - 1)\omega_2\omega_3 & p_1^{-1}\omega_2\omega_3 \\ -p_2^{-1}\omega_1\omega_3 & -p_2^{-2}(1 - p_1)\omega_1\omega_3 \\ \omega_1\omega_2 & -\omega_1\omega_2 \end{pmatrix}.$$

Linearization of Equation (12.17) results in the following equation:

$$\Delta\dot{\Lambda} = \frac{1}{2}Q(\tilde{\Lambda}) \cdot \Delta\omega + \frac{1}{2}Q(\Delta\Lambda) \cdot \tilde{\omega} = \frac{1}{2}Q(\tilde{\Lambda}) \cdot \Delta\omega + \frac{1}{2}A_\Omega(\tilde{\omega})\Delta\Lambda,$$

where matrix

$$A_\Omega(\omega) = \begin{pmatrix} 0 & -\omega_1 & -\omega_2 & -\omega_3 \\ \omega_1 & 0 & \omega_3 & -\omega_2 \\ \omega_2 & -\omega_3 & 0 & \omega_1 \\ \omega_3 & \omega_2 & -\omega_1 & 0 \end{pmatrix}.$$

The resulting equations describing the time variation of the increments in the linear approximation can be written as

$$\begin{pmatrix} \Delta\dot{\omega} \\ \Delta\dot{\Lambda} \\ \Delta\dot{p} \end{pmatrix} = \begin{pmatrix} A_\omega(\tilde{p},\tilde{\omega}) & \Theta_{3\times4} & A_p(\tilde{p},\tilde{\omega}) \\ 0.5\Omega(\tilde{\Lambda}) & 0.5A_\Omega(\tilde{\omega}) & \Theta_{4\times2} \\ & \Theta_{2\times9} & \end{pmatrix} \begin{pmatrix} \Delta\omega \\ \Delta\Lambda \\ \Delta p \end{pmatrix}. \quad (12.35)$$

Here, the variables with the wave are the solution of the system of equations (12.15)–(12.17) on the interval $[t_k, t_{k+1}]$. Accordingly, the matrix $\mathrm{A}_{k+1}$ used in calculation of the ellipsoid matrix $H_{k+1|k}$, due to Equation (12.28) has the following form:

$$\mathrm{A}_{k+1} = \exp\left[\begin{pmatrix} A_\omega(p_k,\omega_{k+1/2}) & \Theta_{3\times4} & A_p(p_k,\omega_{k+1/2}) \\ 0.5\Omega(\Lambda_{k+1/2}) & 0.5A_\Omega(\omega_{k+1/2}) & \Theta_{4\times2} \\ & \Theta_{2\times9} & \end{pmatrix}\Delta t\right]. \quad (12.36)$$

Here, $\omega_{k+1/2} = 0.5(\tilde{\omega}_{k+1} + \hat{\omega}_k)$, and quaternion $\Lambda_{k+1/2}$ is equal to normalized quaternion $0.5(\tilde{\Lambda}_{k+1} + \hat{\Lambda}_k)$. Provided that $\Delta x_k$ takes all possible values from the ellipsoid $E_k(0)$, $\Delta x_k \in E_k(0)$, the vector $\Delta x_{k+1}$ could also belong to the ellipsoid

$$\Delta x_{k+1} \in E_{k+1|k}(0) = \{x : x^{\mathrm{T}} H^{-1}_{k+1|k} x \le 1\},$$

where matrix

$$H_{k+1|k} = \mathrm{A}_{k+1} H_k \mathrm{A}^{\mathrm{T}}_{k+1}.$$

Finally, we assume

$$E_{k+1|k} = \{\tilde{x}_{k+1}\} + E_{k+1|k}(0) = \{x : (x - \hat{x}_{k+1|k})^{\mathrm{T}} H^{-1}_{k+1|k}(x - \hat{x}_{k+1|k}) \le 1\}.$$

Here $\hat{x}_{k+1|k} = \tilde{x}_{k+1}$, and addition in the formula above is understood as the vector sum of sets.

Consider a method of constructing the ellipsoid $E_{k+1}$. The set $\bar{X}_{k+1}$ associated with Equation (12.22) of the PSC quaternion measurement can be represented as follows:

$$\bar{X}_{k+1} = \bigcap_{j=1}^{4} \bar{X}_{j,k+1},$$

where each of the sets

$$\bar{X}_{j,k+1} = \{x \in R^n : |C_{i,k+1}x - \tilde{\lambda}^{p|a}_{j-1,k+1}| \le c_{j-1}\}, \quad j = \overline{1,\ 4} \quad (12.37)$$

is connected with one of the inequalities (12.23). In Equation (12.37), $C_{i,k+1}$ is the $i$th row of the matrix $C_{k+1} = (\Theta_{4\times3} : C(\bar{\Lambda}^a_{k+1}) : \Theta_{4\times2})$.

The set $\bar{X}_{j,k+1}$ is a multidimensional layer in $R^n$, as it follows from Equation (12.37). The construction of an ellipsoid $E_{k+1}$ containing the intersection of the sets $\bar{X}_{j,k+1}$, $j = \overline{1, 4}$, with the ellipsoid $E_{k+1|k}$ is carried out iteratively using the algorithm [39]. We assume that $\bar{E}_{s=0} = E_{k+1|k}$ for the initial ellipsoid. An ellipsoid of minimal volume containing an intersection $\bar{E}_0 \bigcap \bar{X}_{1,k+1}$ is taken as $\bar{E}_1$. In general, according to the construction procedure, an ellipsoid $\bar{E}_{s+1}$ contains an intersection $\bar{E}_s \bigcap \bar{X}_{[1+(s+4)mod4],k+1}$. Its center vector $x_{s+1} = (\omega^{\mathrm{T}}_{s+1}, \Lambda^{\mathrm{T}}_{s+1}, p^{\mathrm{T}}_{s+1})^{\mathrm{T}} \in \bar{X}_{[1+(s+4)mod4],k+1}$.

In the process of constructing a sequence $\{\bar{E}_s\}^N_{s=0}$ of ellipsoids $\bar{E}_s$, it may turn out that the normalization condition (12.18) can be violated for the quaternion estimate $\Lambda_s$. The normalization of the quaternion estimate $\Lambda_s$ is carried out using its projection onto the unit sphere $||\Lambda|| = 1$, and it is made for the entire ellipsoid $\bar{E}_s$ using the space dilation–compression operator for the part of variables. The ellipsoid $n \times n$-matrix $H_s$ can be uniquely represented in the form $H_s = G^{\mathrm{T}}_s G_s$, where $G_s$ is some non-singular matrix. For any vector $x \in \bar{E}_s$, the following representation is valid:

$$x = x_s + G_s\varepsilon, \quad ||\varepsilon|| \leq 1.$$

Then for part of the state vector, for the vector $\Lambda$, we obtain

$$\Lambda = \Lambda_s + G_s(4:7, 1:n)\varepsilon, \quad ||\varepsilon|| \leq 1. \tag{12.38}$$

Here, the $4 \times n$-matrix $G_s(4 : 7, 1 : n)$ is composed of the rows of the matrix $G_s$ from the fourth to the seventh. Let the number $\nu_s = |\Lambda_s|^{-1} = (\lambda^2_{0,s} + \lambda^2_{1,s} + \lambda^2_{2,s} + \lambda^2_{3,s})^{-1/2} \neq 1$. We multiply Equation (12.38) by the number $\nu_s$ and form the following ellipsoid:

$$E_s(\nu_s) = \{x = x_s(\nu_s) + G_s(\nu_s)\varepsilon, \quad ||\varepsilon|| \leq 1\},$$

where

$$x_s(\nu_s) = (\omega^{\mathrm{T}}_s, \nu_s\Lambda^{\mathrm{T}}_s, p^{\mathrm{T}}_s)^{\mathrm{T}}, G_s(\nu_s) = \begin{pmatrix} G_s(1:3, 1:n) \\ \nu_s G_s(4:7, 1:n) \\ G_s(8:9, 1:n) \end{pmatrix}.$$

The center of the ellipsoid $E_s(\nu_s)$ satisfies the normalization condition (12.18). The area of possible values of the remaining coordinates of the vectors $x$ from the ellipsoid $E_s(\nu_s)$will be the same as that for the ellipsoid $\bar{E}_s$ at this transformation. We assume that $\bar{E}_s = E_s(\nu_s)$ and continue the process of ellipsoids construction. The process will stop [39] for some finite $s = N$. The center vector of the resulting ellipsoid $\bar{E}_N$ will satisfy the inequalities (12.37) and the normalization condition (12.18). Finally, we assume $E_{k+1} = \bar{E}_N$. Numerical simulation of the proposed estimation algorithm was performed to test its efficiency and properties.

### 12.3.4 Numerical Simulation of the Estimation Algorithm

The characteristics of the satellite TOPEX/Poseidon were taken as a model example for the PSC. This satellite has a mass of about 2200 kg, the dimensions of the main body of the satellite are 5.5 m× 6.6 m× 2.8 m, and the dimensions of solar panels are 8.9 m×3.3 m. Moments of inertia of the satellite are $J_1 = 3616\text{kg}^2$, $J_2 = 7618\text{kg}^2$, and$J_3 = 8098\text{kg}^2$. Simulation of the estimation algorithm was performed at various initial values of the angular velocity $\omega(t_0)$ and attitude quaternion $\Lambda(t_0)$ of the PSC with respect to the IRF. These values are considered to be unknown and were used at integration of Equations (12.15) and (12.16) to calculate the so-called true values of these parameters in time.

Plots of the time variation of coordinates of the angular velocity vector $\omega(t)$ and the attitude quaternion $\Lambda(t)$ calculated at initial values of $\omega(t_0) = (0.005, 0.010, 0.020)^{\mathrm{T}}$ and $\Lambda(t_0) = (0.1005, 0.5025, 0.3015, 0.8040)^{\mathrm{T}}$ are shown in Figures 12.3 and 12.4, respectively.

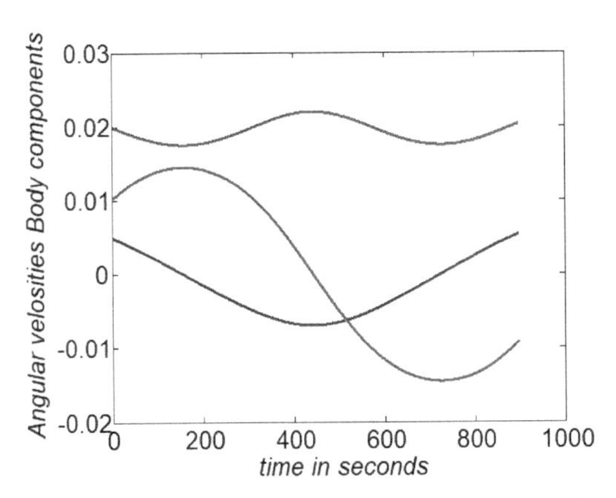


**Figure 12.3** Time variation of the coordinates of the angular velocity vector in the BF.

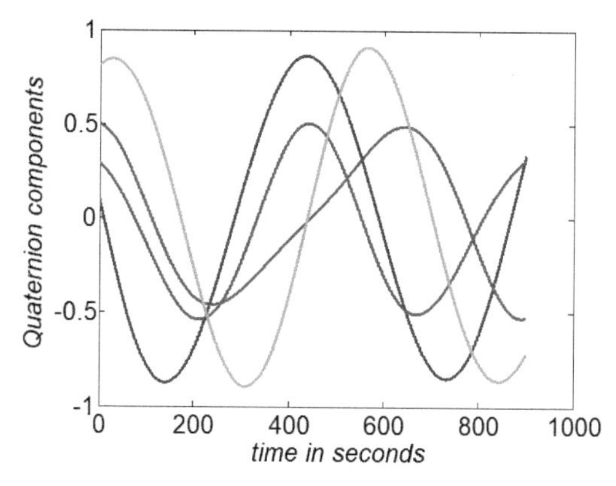


**Figure 12.4** Time variation of the components of the quaternion.

During the simulation, it was assumed that the attitude of the ASC BF relative to the IRF did not change, i.e. $\omega^a = 0$. Without loss of generality, it was assumed that the axes of the ASC BF and IRF coincided. In this case, the attitude quaternion of the ASC will remain constant, $\Lambda_k^a \equiv (1,\ 0,\ 0,\ 0)^{\mathrm{T}}$, and the matrix $C(\bar{\Lambda}_k^a) \equiv I$.

The measured value $\tilde{\Lambda}_k^{p|a}$ of the attitude quaternion of the PSC BF relative to the ASC BF was obtained during simulation in accordance with Equation (12.22). Measurement errors $\xi_{i,k}$ was modeled as random independent processes such as white noise, and the error value is uniformly distributed in the interval $[-c_i, c_i]$, $c_i = 0.02$. Such a value and the distribution of the measurement errors of the quaternion corresponds to the CVS operation algorithm considered in [4].

The duration of the estimation process was 900 s, which corresponds to 15 min. Quaternion measurement period $\Delta t = 0.48$ s. Information on the relative value of the inertia moments was not used in the estimation algorithm. The initial estimate of the parameter vector $p$ was $\hat{p}_0 = (\hat{p}_{1,0}, \hat{p}_{2,0})^{\mathrm{T}} = (0.5, 0.5)^{\mathrm{T}}$. The true values of the ratios of the inertia moments were as follows: $p_1^* = J_1 J_3^{-1} = 3616/8098 = 0.4465$, $p_2^* = J_2 J_3^{-1} = 7618/8098 = 0.9407$, i.e. the searched for parameter vector $p^* = (p_1^*, p_2^*)^{\mathrm{T}} = (0.4465, 0.9407)^{\mathrm{T}}$.

The estimation errors of the PSC angular velocity vector $e_k^{\omega} = ||\omega(t_k) - \hat{\omega}_k||_{\infty}$ and the PSC attitude quaternion relative to the IRF $e_k^{\Lambda} = ||\Lambda(t_k) - \hat{\Lambda}_k||_{\infty}$ at discrete time $t_k = k \cdot \Delta t$ at simulation of the estimation algorithm are shown in Figures 12.5 and 12.6, respectively. As it can be seen from

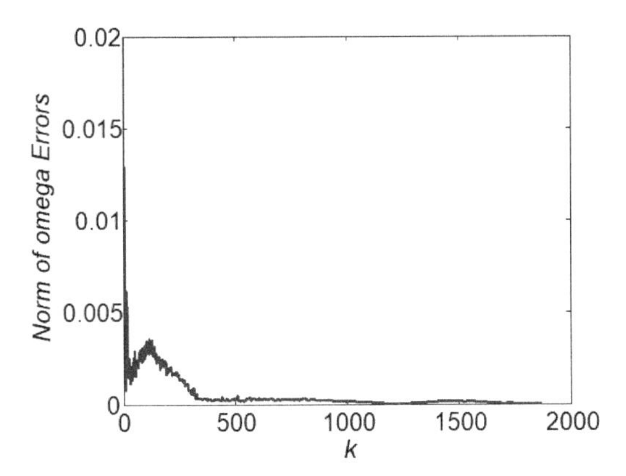


**Figure 12.5** Time evolution of the estimation error of angular velocity.

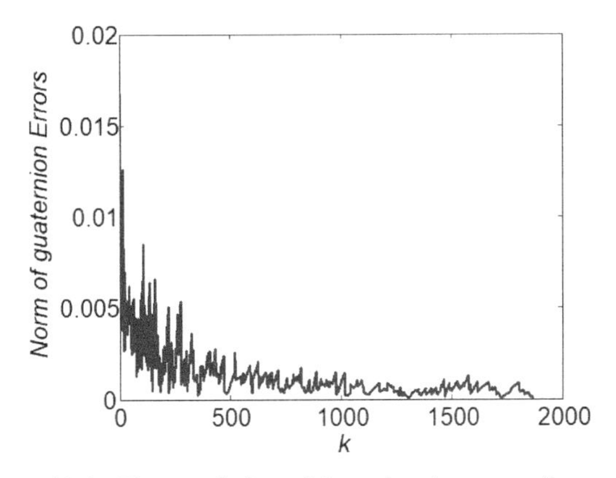


**Figure 12.6** Time evolution of the estimation error of quaternion.

these figures, the estimation accuracy of the angular velocity at the end of the interval of algorithm operation reached 0.000025 rad/s, which is less than $0.5\%$ of the nominal value. The quaternion estimation accuracy for the same time reached the value of 0.0015, which is 10 times less than the maximum value of $\xi_{i,k}$, i.e. $c_i$.

The change over time of the estimation error $e_k^p = ||p^* - \hat{p}_k||_\infty$ of the vector of ratios of the PSC inertia moments is shown in Figure 12.7.

The value of the estimate of the vector of inertia moment ratios $\hat{p}_{1874} = (0.4543, 0.9401)^{\mathrm{T}}$.The discrepancy with the true value is in the third digit.

The results of numerical simulation showed the efficiency of the proposed ellipsoidal filter in solving the estimation problem of the parameters of the

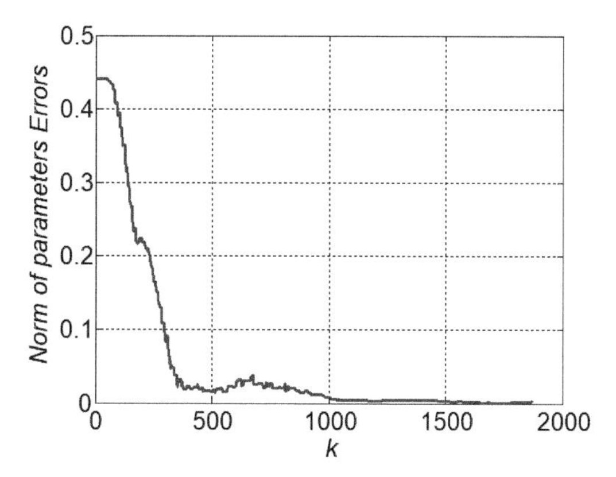


**Figure 12.7** Time evolution of the estimation error of ratios of the PSC inertia moments.

relative rotational motion of two SCs, when the CVS is considered as a sensor of attitude quaternion. The PSC attitude quaternion with respect to the IRF and the PSC angular velocity vector are restored too. The accuracy of the estimated signals is significantly increased compared to direct measurements due to the effective filtering properties of the ellipsoidal estimation.

The problem was solved under some simplifying assumptions, which should not significantly affect the research results. In particular, small but constantly acting disturbing moments applied to the PSC were not taken into account. These and other small details will inevitably require attention when implementing the proposed algorithms using real devices and prototype models.

The extremely important application of the proposed algorithms is worth mentioning. According to the authors' knowledge, these algorithms are absolutely indispensable in the implementation of the ASC automatic docking with a PSC being in rotational motion. Only knowledge of high-precision values of attitude parameters and angular velocity will allow to perform a maneuver for approaching and docking with a tumbling satellite.

## 12.4 Synthesis of Spacecraft Motion Control at Docking

### 12.4.1 Synthesis of Motion Control of the Center of Mass of Active Spacecraft

We consider the control synthesis problem (see conditions (12.13)) with not fixed time, namely the control synthesis that ensures $||\rho(t)|| \to 0$ and

$||\dot{\rho}(t)|| \to 0$ at $t \to \infty$. To solve the control synthesis problem, we use an analog of the decomposition method of control problem of the SC angular motion into a kinematic and dynamic problem. The method, apparently, was first given in [23] and partly in previous works of the authors.

The equations of the controlled motion of the center of mass (12.7) can be written in the following normal Cauchy form:

$$\left.\begin{array}{l} \dot{X} = V, \\ \dot{V} = A_{12}X + A_{22}V + U \end{array}\right\} \tag{12.39}$$

where $X = (x, y, z)^{\mathrm{T}}$, $V = (\dot{x}, \dot{y}, \dot{z})^{\mathrm{T}}$, $U = (u_x, u_y, u_z)^{\mathrm{T}}$, and matrices

$$A_{12} = \begin{pmatrix} 0 & 0 & 0 \\ 0 & 3\omega_0 & 0 \\ 0 & 0 & -\omega_0^2 \end{pmatrix}, \; A_{22} = \begin{pmatrix} 0 & -2\omega_0 & 0 \\ 2\omega_0 & 0 & 0 \\ 0 & 0 & 0 \end{pmatrix} = -A_{22}{}^T. \tag{12.40}$$

Equation (12.39) can be represented in the following form:

$$\dot{Z} = AZ + BU, \tag{12.41}$$

where $Z = (z_1, z_2, z_3, z_4, z_5, z_6)^{\mathrm{T}} = (X^{\mathrm{T}}, V^{\mathrm{T}})^{\mathrm{T}}$, and

$$A = \begin{pmatrix} O_3 & I_3 \\ A_{21} & A_{22} \end{pmatrix}, B = \begin{pmatrix} O_3 \\ I_3 \end{pmatrix}.$$

It is easy to verify that $\det(AB \vdots B) = 1$, and it means that the pair $(A, B)$ is controllable.

Taking into account the representations (12.39) and (12.41) of Equation (12.7), the control synthesis problem of the center of mass motion can be interpreted as a synthesis problem of the control $U = U(Z)$ that ensures the asymptotic stability condition $||Z(t)|| \to 0$ of the solution $Z(t) \equiv 0$ of Equation (12.41). By analogy with [23], we consider two stages of solving the latter problem.

Stage 1. Choose the "kinematic" Lyapunov function

$$W_C = \frac{1}{2} X^T X,$$

whose derivative according to Equation (12.39) is

$$\dot{W}_C = X^T \dot{X} = X^T V. \tag{12.42}$$

Setting in Equation (12.42)

$$V = V_C = -QX, \; Q = \mathrm{diag}\{q_{11}, q_{22}, q_{33}\}, \; Q > 0,$$

we get $\dot{W}_C = -X^T QX < 0$ at $X \neq 0$.

Stage 2. We choose the "dynamic" Lyapunov function

$$W_d = \frac{1}{2}(V - V_C)^T (V - V_C),$$

whose derivative according to Equation (12.40) is

$$\dot{W}_d = (V - V_C)^T (A_{12}X + A_{22}V + U - \dot{V}_C). \tag{12.43}$$

Assuming in Equation (12.43) the control

$$U = -A_{12}X - A_{22}V + \dot{V}_C - \frac{1}{2\tau}(V - V_C), \dot{V}_C = -Q\dot{X} = -QV, \tag{12.44}$$

we get

$$\dot{W}_d = -\frac{1}{2\tau}(V - V_C)^T (V - V_C) = -\frac{1}{\tau} W_d, \dot{W}_d < 0 \; at \; V \neq V_C. \tag{12.45}$$

Here the matrix $Q$ and the positive number $\tau$ are chosen by the control system developer. This implies the asymptotic stability of system (12.41). Let us note that relation (12.45) is the equation of a stable aperiodic link $\tau\dot{W}_d + W_d = 0$ whose solution $W_d(t) \to 0$ ($||V(t) - V_C(t)|| \to 0$) at $t \to \infty$.

As a result, for the desired control and the corresponding closed-loop equation, we obtain

$$U = -C^{\mathrm{T}} Z, \dot{Z} = A_C Z, \; A_C = (A - BC^{\mathrm{T}}), \tag{12.46}$$

where $C^{\mathrm{T}} = (A_{12} + 0,5\tau^{-1}Q)C_1^{\mathrm{T}} + (A_{22} + Q + 0,5\tau^{-1}I_3)C_2^{\mathrm{T}}$,

$$C_1^{\mathrm{T}} = (I_3, \Theta_3) = \begin{pmatrix} 1 & 0 & 0 & 0 & 0 & 0 \\ 0 & 1 & 0 & 0 & 0 & 0 \\ 0 & 0 & 1 & 0 & 0 & 0 \end{pmatrix},$$

$$C_2^{\mathrm{T}} = (\Theta_3, I_3) = \begin{pmatrix} 0 & 0 & 0 & 1 & 0 & 0 \\ 0 & 0 & 0 & 0 & 1 & 0 \\ 0 & 0 & 0 & 0 & 0 & 1 \end{pmatrix}.$$

Let us note that choosing control $U$ in Equation (12.39) in the form (12.44) and neglecting the term $\dot{V}_C$, we obtain the equation of a stable aperiodic link

$$T\dot{V} + V = V_C, T = 2\tau. \tag{12.47}$$

Equation of the form (12.47) was used in the algorithm for the angular motion control of the Soviet-Russian manned orbital complex "Mir," and an angular velocity vector in it was taken instead of the linear velocity vector $V$ [49].

### 12.4.2 Synthesis of Spacecraft Angular Motion Control

The control vector $U$, obtained at control synthesis solution in Section 12.4.1, is given by its projections on the LOF axis $O_P x_0 y_0 z_0$ with the origin at the center of mass of the passive SC. This control acceleration vector must be created by the active SC engines. By analogy with [50], we can write

$$U = S(\Lambda)R,$$

where $S(\Lambda)$ is direction cosine matrix (12.11) and $R$ is the control acceleration vector of the ASC, given by the projections in the BF. According to the formula (12.46), $U = -C^{\mathrm{T}}Z$. Therefore, for arbitrary initial condition $\Lambda(t_0)$, the demand $S(t_E) = I_3$ of the control problem formulation (12.13) can be fulfilled only as a result of the angular motion control of the ASC. Along the way, we note that the direction cosine matrix $S(\Lambda)$(see formula (12.11)) will be the same for quaternion $\Lambda$ and $-\Lambda$. The matrix $S(t_E) = I_3$ corresponds to quaternions $\Lambda_E = (\pm 1, 0, 0, 0)$.

The control moment synthesis is based on Equations (12.9) and (12.12) of the relative attitude evolution. Recall that in those equations, vector $\omega_*$ of the absolute angular velocity of the PSC is given by its projection on the axis of the PSC BF $O_p x_p y_p z_p$. Moreover, the true value of this vector is not known; instead, its ellipsoidal estimates are supposed to be used in control algorithms (see Section 12.3). Here, for simplicity, we will assume that the PSC is ideally oriented in the LOF, i.e. the axes of its BF $O_p x_p y_p z_p$ and the LOF $O_p x_0 y_0 z_0$ coincide. With this assumption, the PSC angular velocity $\omega_* = (0, 0, -\omega_0)^T, \omega_0 = \sqrt{\mu/R^3}$.

The synthesis of the control moments in Equation (12.9) under the assumptions made are carried out with the use of the results of [51]. The problem of stabilization of the SC attitude to an arbitrary quaternion $\Lambda_E$ with respect to the LOF was solved in [51]. Then it was a more general problem than the one considered here, where we have to ensure the fulfillment of the condition $S(\Lambda_E) = I_3$, $\Lambda_E = (\pm 1, 0, 0, 0)$. Solution of the problem [51] was obtained with the use of the decomposition method considered above for the kinematic and dynamic problems and Lyapunov functions.

Formulas for the control moment $M = M_C$ in Equations (12.10) and (12.13) are obtained by adaptation of the corresponding formulas [51] for $\Lambda_S = \Lambda_E = (\pm 1, 0, 0, 0)$ and have the following form:

$$M_C = \breve{\omega} J \omega + J \dot{\omega}_C - \tau^{-1} J(\omega - \omega_C), \tag{12.48}$$

where $\omega_C = S(\Lambda)\omega_* - \lambda_0 \lambda$, $\dot{\omega}_C = -0,5(\lambda_0^2 I_3 + \lambda_0 \breve{\lambda} - \lambda\lambda^T)\omega_r$.

### 12.4.3 Computer Simulation of Control Algorithm

The following computational experiments were performed to evaluate the principal operability and efficiency of the obtained algorithms.

The simulation of the control algorithm (12.46) of the motion of the center of mass of the ASC relative to the PSC was performed at the following parameters $Q = \text{diag}\{q_{11}, q_{22}, q_{33}\}, q_{jj} = 0,01, j = 1,2,3,$ and $\tau = 0,5$ s. It was assumed that a PSC moves in a circular orbit of radius $R = 7070$ km (orbit altitude is 700 km; Earth radius is $R_E = 6370$ km), orbital angular velocity $\omega_0 = \sqrt{\mu/R^3}$, $\mu = 398600, 4$ km$^3$/s$^2$ is the Earth's gravitational constant. Time evolution of the distance between satellites $d(t) = (x^2(t) + y^2(t) + z^2(t))^{1/2}$ and the speed of approach $||\dot{d}(t)|| = (\dot{x}^2(t) + \dot{y}^2(t) + \dot{z}^2(t))^{1/2}$ are shown in Figures 12.8 and 12.9.

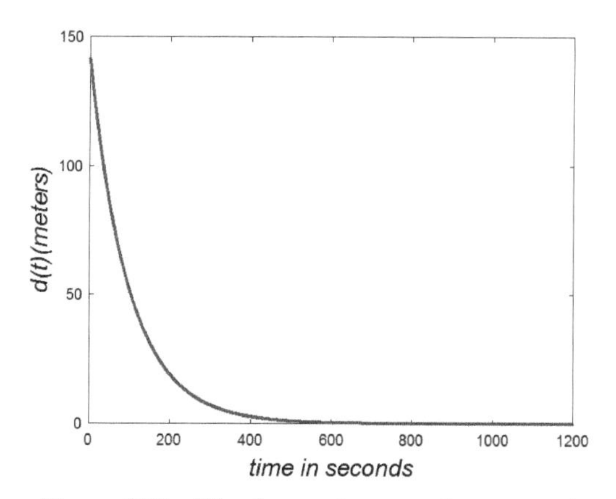


**Figure 12.8** The distance between the spacecraft .

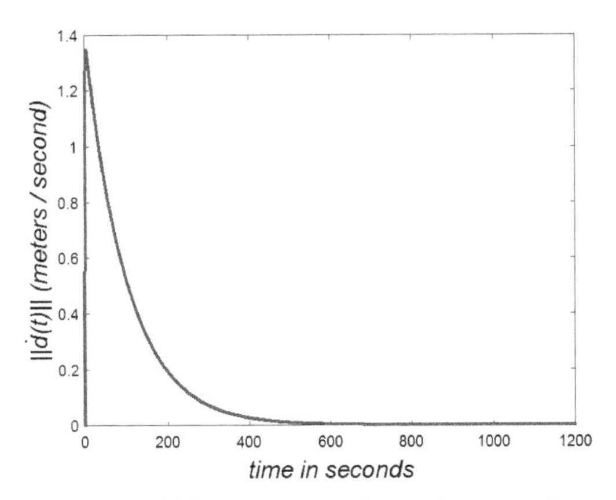


**Figure 12.9** The approach speed spacecraft.

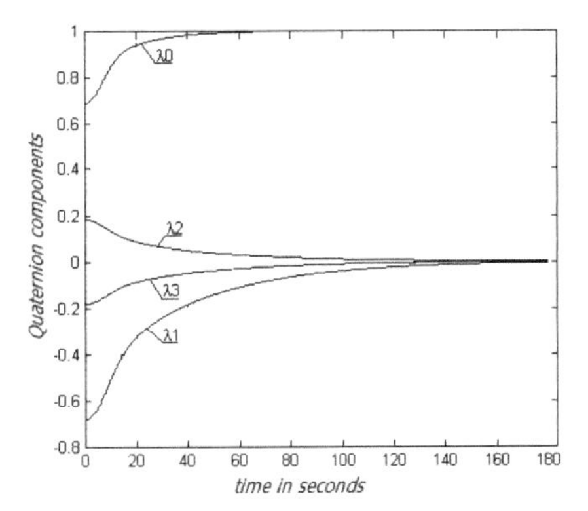


**Figure 12.10** Quaternion components $\lambda_j(t), j = \overline{0,3}$.

At the initial moment of time, the distance between the satellites and the approach velocity (see Equations (12.7) and formulas (12.39)) were assumed to be as follows: $x = 100$ m, $y = 0$, $z = 100$ m, and $\dot{x} = \dot{y} = \dot{z} = 0$.

These plots illustrate the possibility of use of the proposed control algorithm for the implementation of soft docking. The duration of the docking process and the required magnitude of the control accelerations created by the thrusters can be adjusted by changing the parameters of the algorithm – matrix $Q$ and the value of $\tau$.

The control algorithm (12.48) of the relative attitude of the satellites was simulated with the parameter $\tau = 40$ s and the inertia moments matrix of the active satellite $J = \text{diag}\{4000, 2000, 4000\}$ kgm$^2$. The initial values of the

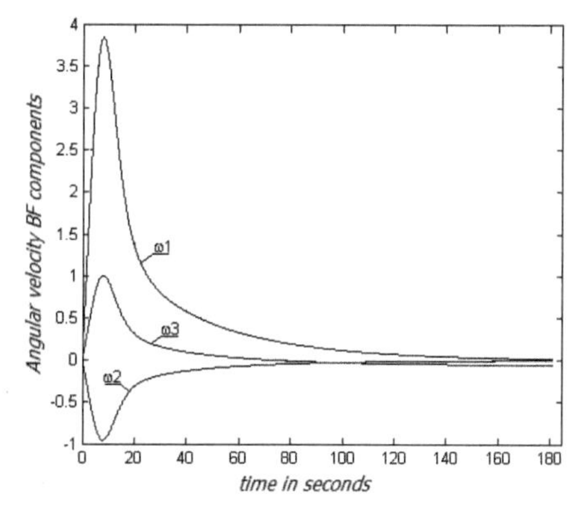


**Figure 12.11** Angular velocity BF components $\omega_i(t), i = \overline{1,3}$.

parameters of the relative attitude (see Equations (12.10) and (12.13)) were assumed as $\omega^{\mathrm{T}}(t_0) = (0,0,0)$ and $\Lambda^{\mathrm{T}}(t_0) = (0.7, -0.68, 0.18, -0.18)$, which corresponds to the values $\vartheta(t_0) = 90°$, $\gamma(t_0) = -90°$, and $\psi(t_0) = 120°$, where $\vartheta, \gamma,$ and $\psi$ are the pitch, roll, and yaw angles, respectively [12].

Transient processes of establishing relative attitude $\lambda_J(t), j = \overline{0,3}$, and $\omega_i(t), i = \overline{1,3}$ are shown in Figures 12.10 and 12.11.

The plots illustrate the process of sustainable establishment of relative attitude of the BFs associated with the satellites: $\lambda_0(t) \to 1$, $\lambda_J(t) \to 0, j = \overline{1,3}$; $\omega_1(t) \to 0$, $\omega_2(t) \to 0$, $\omega_3(t) \to \omega_{*3} = -\sqrt{\mu / R^3} \approx -0,061 \deg/\mathrm{s}$. The results of the computational experiments illustrate their performance and the fundamental possibility of implementing a soft dock with a non-cooperative rotating SC.

## 12.5 Conclusion

The problems of controlling an ASC at the stage of close proximity and soft docking with a non-cooperative rotating SC are considered. Mathematical models of relative spatial and angular motions of the SC interconnected through control channels are proposed. In control algorithms, it is supposed to use information on the relative pose from the cameras of an ASC pre-processed with a regularized ellipsoidal filter in order to increase its accuracy. Ellipsoidal filtering algorithms have the property of insensitivity to violation of *a priori* information about uncertainty (the property of robustness in the terminology of Andronov–Pontryagin [52]) and are the original development of the authors of this paper. Their effectiveness is confirmed by application to a number of technical objects at solving control problems. Using a unified methodological approach based on the use of generalizations of the direct Lyapunov method, new algorithms for controlling motion of the center of mass and angular motion of the ASC at the docking stage are obtained. The performed computer simulation has illustrated the operability of the proposed algorithms and the fundamental possibility of their use in real systems of soft docking after performing certain studies and improvements. In particular, development of terminal control algorithms for an ASC with bounded controls should be further considered.

## References

[1] N. Koshkin et al. 'Monitoring of Space Debris Rotation Based on Photometry'. Odessa Astronomical Publications, vol. 31, pp. 179-185, 2018.

[2] S. Efimov, D. Pritykin, V. Sidorenko, 'Long-term attitude dynamics of space debris in Sun-synchronous orbits: Cassini cycles and chaotic stabilization', Celestial Mechanics and Dynamical Astronomy, 130:62, 2018.

[3] A.P. Markeev, 'Teoreticheskaya Mekhanika'. (Theoretical Mechanics). (in Russian) Moscow, CheRo, 1999.

[4] V. Gubarev, N. Salnikov, S. Melnychuk, 'Ellipsoidal Pose Estimation of an Uncooperative Spacecraft from Video Image Data', in Control Systems: Theory and Applications. River Publishers Series in Automation, Control and Robotics, pp. 169-195, 2018.

[5] Yu.A.Ermilov, E.E.Ivanova, S.V.Pantyushin. 'Upravleniye sblizheniem cosmicheskikh apparatov'.(Spacecraft Proximity Control). (in Russian) Moscow, Nauka, 1977.

[6] W. Fehse 'Automated Rendezvous and Docking of Spacecraft', New York: Cambridge University Press, 2003.

[7] G. Rouleau et al., 'Autonomous Capture of a Tumbling Satellite', Proceedings of the IEEE International Conference on Robotics and Automation, pp. 3855 – 3860. June 2006.

[8] N. Philip, M. Ananthasayanam, 'Relative Position and Attitude Estimation and Control Schemes for the Final Phase of an Autonomous Docking Mission of Spacecraft', Acta Astronautica, vol. 52, No. 7, pp. 511–522, 2003.

[9] L. Sun, W. Huo, '6-DOF Integrated Adaptive Backstepping Control for Spacecraft Proximity Operations', IEEE Transactions on Aerospace and Electronic Systems, Vol. 51, No. 3, pp. 2433–2443, 2015.
[10] H. Dong, Q. Hu, M.R. Akella, 'Dual-Quaternion-Based Spacecraft Autonomous Rendezvous and Docking Under Six-Degree-of-Freedom Motion Constraints', Journal of Guidance, Control, and Dynamics, **41**(5), pp. 1150-1162, 2018.
[11] B-Z. Zhou, X-F. Liu, G-P. Cai, 'Motion-Planning and Pose-Tracking Based Rendezvous and Docking with a Tumbling Target', Advances in Space Research, vol. 65, Issue 4, pp. 1139-1157, 2020.
[12] V.P. Legostayev (ed.) 'Raketno-cosmicheskaya tekhnika', (Rocket and space technology) (in Russian), Book 1. Moscow, Mashinostroyenie, 2012.
[13] N.M.Ivanov, L.N.Lysenko. 'Balistika i navigatsia cosmicheskikh apparatov', (Ballistics and navigation of spacecraft) (in Russian), Moscow, Drofa, 2004.
[14] Yu.M. Ovchinnikov. 'Vvedenie v dinamiku cosmicheskogo poliota',(Introduction to the space flight dynamics) (in Russian), Moscow: Moscow Institute of Physics and Technology, 2016.
[15] B.V. Raushenbah, E.N. Tokar 'Upravlenie orientatsiey cosmicheskikh apparatov',(Spacecraft attitude control) (in Russian), Moscow, Nauka, 1974.
[16] H. Schaub, and J.L. Junkins. 'Analytical Mechanics of Space Systems'. Reston, VA, American Institute of Aeronautics and Astronautics, 2009.
[17] V.K. Abalakin et al. 'Spravochnoe rukovodstvo po nebesnoy mekhanike i astrodinamike'(Reference Manual on Celestial Mechanics and Astrodynamics) (in Russian), Moscow, Nauka, 1976.
[18] N.N. Bucholz. 'Osnovnoy kurs teoreticheskoy mekhaniki'(Basic course of theoretical mechanics) (in Russian), Part II. Moscow, Nauka,1969.
[19] I.V. Ostoslavskiy, I.V.Strazheva. 'Dinamika poleta.Traektorii letatelnykh apparatov',(Flight dynamics. Aircraft trajectories) (in Russian), Moscow, Mashinostroyenie, 1969.
[20] N.E. Zubov. 'Algorithm for automatic control of the rotational and translational motion of the spacecraft in the rapprochement process', Soviet Journal of Computer and Systems Sciences, âĎŰ 3, pp.166–178, 1990.
[21] N.V. Butenin, Yu.Ya.Luntz, D.R.Merkin. 'Kurs teoreticheskoy mekhaniki',(Theoretical Mechanics Course) (in Russian), Vol.2., Moscow, Nauka, 1979.

[22] N.T. Kuzovkov, O.S.Salychev. 'Inertzialnaya navigatsia i optimalnaya filtratsia'. (Inertial navigation and optimal filtering) (in Russian), Moscow, Mashinostroyenie, 1982.

[23] V.N. Branets, I.P. Shmyglevskiy. 'Primenenie kvaternionov v zadachakh orientatsii tverdogo tela',(The use of quaternion in rigid body attitude problems) (in Russian), Moscow, Nauka, 1973.

[24] Yu.F. Golubev. 'Algebra kvaternionov v zadachakh kinematiki tverdogo tela',(Quaternion algebra in rigid body kinematics) (in Russian), M.V.Keldysh. Institute of Applied Mathematics Press, N 39. 2013.

[25] Yu.N. Chelnokov 'Kvaternionnye modeli i metody dinamiki, navigatsii i upravlenia dvizheniem'(Quaternion models and methods in dynamics, navigation, and motion control) (in Russian), Moscow, Fizmatlit, 2011.

[26] J. Wittenburg. 'Dynamics of Systems of Rigid Bodies'. Stuttgart, B.G.Teubner. 1977.

[27] V.V. Volosov, L.I.Tyutyunnik. 'Synthesis of spacecraft attitude control algorithms using quaternions'. Space Science and Technology, 5(4), pp. 61—69, 1999.

[28] K.N. Kurskaya, E.N. Ostapenko, N.A. Repyakh. 'Reachability areas for soft meeting points with radial control'(in Russian), Problems of mechanics and control, Interuniversity collection of scientific papers, Issue 42, pp. 31-35, 2010.

[29] A.B. Kurzhanski, I. Valyi. 'Ellipsoidal Calculus for Estimation and Control', Birkhauser, Boston, 1997.

[30] V.N. Ushakov, A.R. Matviychuk, A.V. Ushakov. 'Approximation of reachability sets and integral funnels of differential inclusions'. Tomsk State University Journal. Udmurt University, Issue 4, pp. 23–39, 2011.

[31] N.A. Babiy, V.V. Volosov, V.N. Shevchenko. 'External ellipsoidal approximations of reachable sets of dynamical systems'. Cybernetics and Computer Engineering, Issue 177, pp. 16 –27, 2014.

[32] Kalman Filtering and Neural Networks/edited by S. Haykin. – New York, Toronto, John Wiley&Sons Inc., 2001.

[33] M.S. Arulampalam, S. Maskell, N. Gordon, T. Clapp. 'A tutorial on particle filters for online nonlinear/non-Gaussian bayesian tracking', IEEE Trans. on Signal processing, vol. 50, No.2, pp. 174-188, 2002.

[34] I.I. Gorban. 'The Statistical Stability Phenomenon', Springer, 2016.

[35] V.V. Volosov, L.I. Tyutyunnik. 'Development and Analysis of Robust Algorithms for Guaranteed Ellipsoidal Estimation of the State of Multidimensional Linear Discrete Dynamic Systems. Part 1'. Journal of Automation and Information Sciences, vol.32, No. 3, pp. 37-46, 2000.

[36] V.V. Volosov, L.I. Tyutyunnik. 'Development and Analysis of Robust Algorithms for Guaranteed Ellipsoidal Estimation of the State of Multidimensional Linear Discrete Dynamic Systems. Part 2'. Journal of Automation and Information Sciences, vol.32, No. 11, pp. 13-23, 2000.

[37] V.V. Volosov, I.A. Kutsenko, Yu.A. Selivanov. 'Development and Investigation of the Robust lgorithms of Ellipsoidal Estimation of the Inertia Characteristics of a Spacecraft Controlled by Powered Gyroscopes'. Journal of Automation and Information Sciences, vol.37, No. 8, pp. 44-57, 2005.

[38] N.N. Salnikov, 'On One Modification of Linear Regression Estimation Algorithm Using Ellipsoids', Journal of Automation and Information Sciences, vol. 44, No. 3, pp. 15–32, 2012.

[39] N.N. Salnikov, 'Estimation of State and Parameters of Dynamic System with the Use of Ellipsoids at the Lack of a Priori Information on Estimated Quantities', Journal of Automation and Information Sciences, vol. 46, No. 4, pp. 60–75, 2014.

[40] F.C. Schweppe, 'Uncertain dynamic systems', Englewood Cliffs, N.J., Prentice-Hall, 1973.

[41] V.V.Volosov, 'Method of constructing ellipsoidal estimates in problems of non-stochastic filtering and identification of the parameters of control systems', Journal of Automation and Information Sciences, vol. 24, No. 3, pp. 22–30, 1991.

[42] F.L. Chernousko, 'State estimation for dynamic systems', Boca Raton, CRC Press, 1994.

[43] S.B. Chabane, et. al., 'A New Approach for Guaranteed Ellipsoidal State Estimation', Preprints of the 19th World Congress, The International Federation of Automatic Control, Cape Town, South Africa, August 24-29, pp. 6533–6538, 2014.

[44] F. Blanchini, S. Miani, 'Set-Theoretic Methods in Control', Springer International Publishing, Switzerland, 2015.

[45] A. Poznyak, A. Polyakov, V. Azhmyakov, 'Attractive Ellipsoids in Robust Control', Springer International Publishing, Switzerland, 2014.

[46] V.A. Sarychev, P. Paglione, A.D. Guerman. 'Stability of Equilibria for a Satellite Subject to Gravitational and Constant Torques', Journal of Guidance Control and Dynamics, vol. 31, No. 2, pp. 386–394, 2008.

[47] V.F. Zhuravlev, 'Osnovy teoreticheskoy mekhaniki' (Basis of theoretical mechanics) (in Russian), Moscow, Fizmatlit Publ., 2001.

[48] C. Jekeli, 'Inertial navigation systems with geodetic applications', Berlin, New York, Walter de Gruyter, 2001.

[49] V.A.Sarychev, M.Yu.Belyaev, S.G.Zaykov, V.V.Sazonov, B.P.Teslenko, 'Mathematical modeling of Eulerian turns of the Mir orbital complex by using gyrodines'. Cosmic Research, 29(4), pp. 458–468, 1991.

[50] N.E. Zubov, E.A. Mikrin, S.S. Negodyaev, I.N. Lavrentyev, 'Synthesis of the approach of the spacecraft with the polar scheme of the propulsion system using the free trajectory method based on the optimal control algorithm with a predictive model'(in Russian), Proceedings of Moscow Institute of Physics and Technology(State University), Volume 2, No. 3, pp. 168–173, 2010.

[51] V.V. Volosov, V.M. Shevchenko, 'Synthesis of control of the angular motion of a spacecraft on the basis of generalization of the direct Lapunov method'(in Russian), Space science and technology, N. 4, pp. 3–13, 2018.

[52] Yu.I. Neymark. 'Dynamic system as the main model of modern science', Automation and Remote Control, **60**:3, pp. 458–461, 1999.

# 13

# Intelligent Algorithms for the Automation of Complex Biotechnical Objects

**V. Lysenko, N. Zaiets, A. Dudnyk, T. Lendiel, K. Nakonechna**

Heroiv Oborony Str.15, building 11, of.311, Kyiv, National University of Life and Environmental Sciences of Ukraine, Ukraine, 03041
E-mail: lysenko@nubip.edu.ua

## Abstract

Traditional automation systems implement the simplest stabilization algorithms that do not provide high energy and resource efficiency. Since information about the state of the biological component of the control object and the forecast of natural disturbances are not taken into account, the solutions to the problem of energy-efficient control of biotechnological objects is presented in this chapter. The modern intelligent algorithms for processing control object information and applying the results are considered. Appropriate control strategies allow maximizing the production profit of the poultry houses and greenhouses.

The intelligent control system that takes into account the forecast of temperature changes is described and its use is shown (for example, poultry farms). The problem of an adequate analysis of the beginning one temperature image change to another is solved; use an intelligent pattern recognition system based on probabilistic neural networks is proposed. The functional diagram of the intelligent temperature control system in the poultry farms is given.

The option of developing an intelligent control system for an industrial greenhouse is shown. It allows taking into account the forecast results of the external disturbances (temperature and solar radiation intensity). Such a control system is used to formulate control strategies for greenhouse complexes, which significantly reduce energy costs in the vegetable production.

**Keywords:** Biotechnical objects, natural disturbances, forecasting, pattern recognition, neural network, intelligent system, temperature images.

## 13.1 Introduction

The agricultural production sector is filled with modern high-tech enterprises and the presence of a biological component part is their distinctive feature. These enterprises comprise poultry farms and industrial greenhouse, in the first place. In Ukraine, the energy share in the cost of production for such enterprises sometimes reaches up to 80% (for greenhouse) and 20% (for poultry farms). Given the high energy cost and its shortage, the measures aimed at reducing the energy consumption are of current interest.

Provided that the costs of arrangement for the production manufacture, material and operational maintenance, wages and general expenses of production can be precisely calculated taking into consideration their long-term use, the energy costs being dependent on the natural disturbances affecting the production facility can be as much predictable as possible to forecast the natural disturbances. The other constituents of the expenditures are utterly predictable for each separate company [26–28].

Thus, the task of selecting the control actions in the process of maintaining the biological facilities of agricultural production is highly sensitive to the capability of the control system to forecast the natural disturbances.

Despite the available modern technological equipment, the simplest control algorithms for electro-technical systems which accompany the manufacturing process of the corresponding products are implemented in the poultry houses and greenhouses. As a rule, these are stabilization algorithms which are proposed to maintain the manufacturing parameters to maximize the productivity of poultry and plants and are determined by the biologists on the basis of research in zootrons and phytotrons. These algorithms are not energy-efficient since they do not take into consideration the following factors: the condition of the biological component part which determines its productivity and the character of the natural disturbances. In addition, as shown using the example of the poultry house, the capacity of the typical operating mechanisms is often insufficient to maintain a biotechnical object (such as poultry) at a temperature that ensures its maximum productivity

(Figure 13.1). That being said, the specific energy consumption for the temperature ranges of 15°C–18°C, 18°C–21°C, and 21°C–24°C is lower than that for 12°C–15°C which provides the maximum egg-laying capacity (the stay period of the fowl in this zone is about 56%).

The condition of plants which are grown inside the greenhouse is affected by solar radiation as well. Thus, the analysis and forecasting of the factors like temperature and solar radiation are necessary in order to use the results of the forecast in the formation of the management strategies for the electrotechnical systems with the aim of reducing the energy costs in agricultural production [2, 8].

Taking into consideration the fact that the natural disturbances such as temperature of the environment, solar radiation, and wind direction and strength can be described in terms of uncertainty, it is advisable to use intelligent control algorithms for managing complex biotechnical facilities when the formation of the control strategies is carried out with the use of a special knowledge base which is formed on the basis of the information analysis on the natural disturbances, the conditions of the biological component part, the pricing policy on the product market, and the rules for using the results of such an analysis.

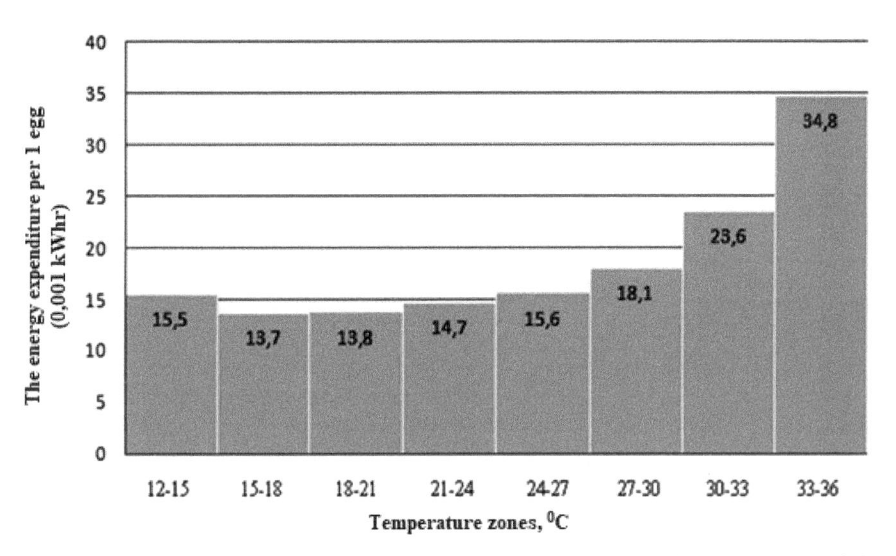


**Figure 13.1** Specific energy costs for different temperature zones through the use of the stabilization system.

A generalized analysis of the technological, zootechnical, cultivation, and organizational processes which occur in greenhouses and poultry houses has been carried out in order to define the peculiar features of biotechnical facilities. As a result of the analysis of these processes, it has been established that greenhouses and poultry houses are complex systems and their study and research requires a systematic analytical approach [3, 5].

Thus, poultry which are used for industrial production are homothermic organisms maintaining body temperature at a constant level as their characteristic feature. The thermoregulatory effect which is built on the cybernetic principle of feedback control includes the activity of a number of body systems which are aimed at increasing the heat production and limiting the heat loss when cooling the body as well as limiting the heat production and increasing the heat loss when the body is heated. This is well illustrated in Figure 13.2. By the critical point, we mean the ambient temperature at which the lowest level of metabolism in the body takes place while the mechanism of heat loss is largely considered discontinued. However, experiments prove that there is a seasonal change of the critical point [1] as well as its change depending on the behavior of the biotechnical object.

Thermoregulations provide us with a vivid picture of the dependence of metabolism on the surrounding temperature. The quantitative characteristics of this dependence for different species of animals are still unexplored to a certain extent. The reference [2] contains the data which enables us to provide an approximate characterization of the zones for certain farm animals.

The range of zones largely depends on the age of the animals, and, therefore, they are significantly narrowed for the young-stock. Furthermore,

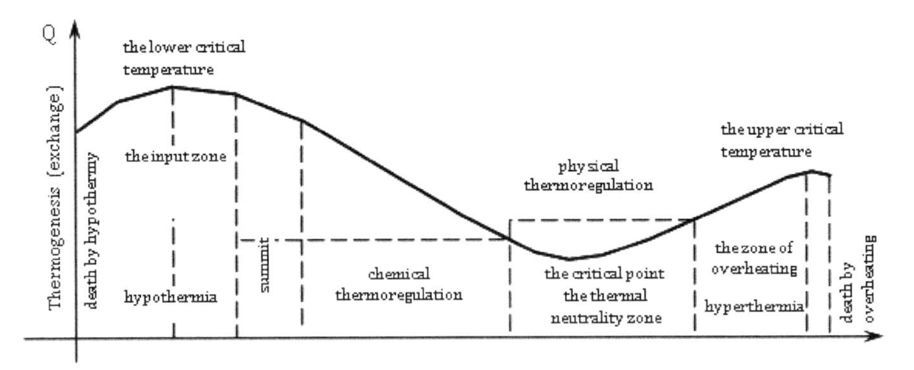


**Figure 13.2** Thermoregulation of the homothermic organisms.

a significant narrowing of the margins for all the zones occurs in case of the indoor air pollution (e.g. $CO_2$ [4]).

Thus, the characteristics of a biotechnical object can be presented in the form of a number of conditions. Here, each condition will be characterized not only by such indicators of animal's health as gas exchange intensity, body temperature, heart rate, etc., but also by the technological indicators, including productivity, weight loss, and feeding costs. At the outer margins of the overheating and cooling zones, there are absorption margins, on reaching which the object ceases to exist.

The crops which are grown in a greenhouse are living organisms. The plant in the greenhouse is affected by a complex set of factors which can be conveniently classified into four groups [7]: biotic (depends on the plant only), abiotic (do not depend on the plant), geographic, and anthropogenic (driven by the human activity).

The formation of chlorophyll occurs in the plant cells when influenced by the sunlight. During the process of photosynthesis (assimilation), chlorophyll absorbs carbon which, in turn, enters a biochemical reaction under the action of light which results in the formation of organic compounds such as starch, sugar, proteins, fats, organic acids, etc. The plants simultaneously emit oxygen.

Respiration (dissimilation) which involves the constant absorption of oxygen and emission of carbon is another process peculiar to plants. The relationship between the photosynthesis and respiration changes during the day. During the daytime, the absorption of carbon is about 10 times faster than the decomposition of the organic substances during respiration. Photosynthesis consumes approximately 2%–5% of all the incoming solar energy. This energy is mainly spent on moisture evaporation and heat exchange between the air and the soil. The temperature which affects the plant in the greenhouse is considered to be an abiotic factor. We distinguish between the soil and the air temperatures. The thermal regime of the plants is formed under the influence of the radiation balance, the heat exchange with the environment, and the evaporation of moisture by the plants. The temperature, as well as light, affects the biochemical processes in the cells. The temperature influences the processes of photosynthesis and respiration in a different way. However, at temperatures below 10°C, both processes halt. The temperature which provides maximum photosynthesis can vary depending on the species of the plant. It constitutes from 26°C to 30°C for tomatoes. However, for most plants, the process of photosynthesis halts at the temperatures below 10°C and above 50°C [6, 8].

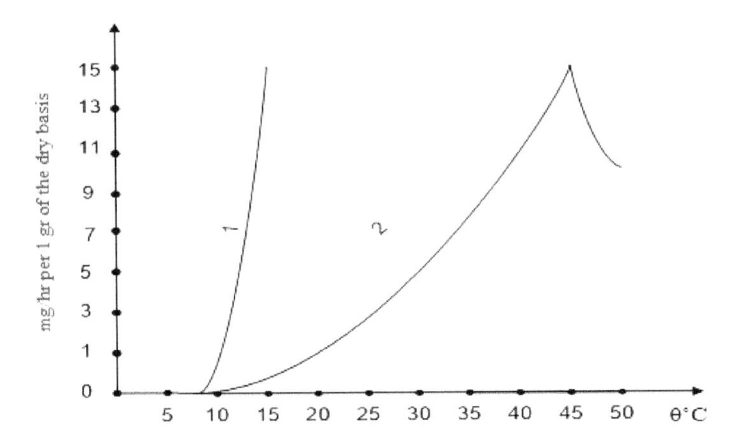


**Figure 13.3** The dependence of (1) assimilation and (2) dissimilation on the temperature.

Respiratory processes inside the plants also depend on the temperature. The temperature which maximizes this process constitutes from 36°C to 40°C for most plants. The increase in the organic mass depends on the interaction of photosynthesis and respiration. Since both processes depend on the external factors, the accumulation of the organic substance can be considered to be the difference between the amount of the substance which was formed from the photosynthesis and the substance which was decomposed during respiration (Figure 13.3).

To ensure maximum productivity in the plant growth, it is necessary to carry out the control actions which make it possible to maintain the microclimate, moisture, and nutrient properties of the soil at the levels which would maximize the organic substance growth at any time during their development while taking into consideration the physiological characteristics of the plants [10, 11].

Ensuring the above-mentioned conditions of the plant development requires significant investment into the construction of the facilities and the technological equipment as well as significant energy expenditures which are necessary to maintain the specified parameters. Such control systems do not use an efficiency indicator which would maximize the difference between the profits from the products grown and the costs of growing them.

The assessment criteria for the production of the crop products in the greenhouse can be as follows: the accuracy of the growing conditions' maintenance which preclude the biotechnical object from dying regardless of the changes in the external disturbances or provide the lowest percentage of losses within the individual least durable representatives; the quality indicators of the received products; the highest productivity of the biotechnical object during its management; the minimum costs spent on physical, energy, and labor resources for the management of a biotechnical object.

## 13.2 Intelligent Automation Systems for Biotechnical Facilities

### 13.2.1 Traditional Automation Systems for Biotechnical Facilities and their Drawbacks

Industrial production in the agricultural sector is secured by the automated control systems. However, while designing and implementing these control systems, it is necessary to take into consideration the peculiar features of the biological component part which constitutes a characteristic difference of the automation systems for the biotechnical facilities such as poultry farms and greenhouse in comparison to the systems which secure the manufacturing processes in other production sectors.

The industrial production of poultry products is connected with the poultry operation and control in the closed facilities with controlled heating systems, potent ventilation, the regime of lighting, mechanized feeding, and egg collection [9, 13].

The major energy consumption is due to the system of the microclimate maintenance in the poultry house (ventilation and air heating) which, unlike other systems, responds to the natural disturbances. Special attention should be devoted to Viper and Amacs, the world's most common microclimate support systems, which are produced by the Big Dutchman Company in Germany [12, 14]. These systems control the supply and exhaust ventilation units, heating and humidification depending on the air temperature inside and outside the poultry house and the age of the poultry while maintaining the inside air temperature at a certain level. Furthermore, they control the manufacturing processes and register all the necessary data on the production, growth, food and water consumption, and so on.

However, all the above-mentioned control systems of the electro-technical complexes in an industrial poultry house do not take into consideration the dynamic changes within the control facility, including poultry as a biological component part of this facility, the possible natural disturbances, including

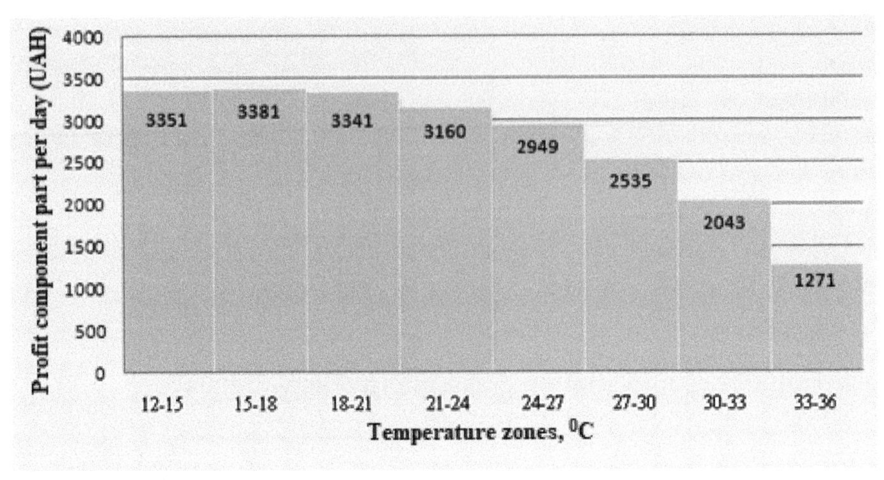


**Figure 13.4** The energy component part of the profit for different temperature zones under the use of a stabilization system.

temperature disturbances, on the technological facility during the whole period of managing the poultry and implement only the modes of stabilization for the certain technological parameters, which does not provide for the maximization of the enterprise profits (this is the main criterion of effective performance under the conditions of intense rivalry) (Figure 13.4 – the maximum energy component part of the profit does not correspond to the comfortable temperature zone of 12°C–15°C which is a necessity for the maximum egg-laying capacity).

In greenhouse, the technology efficiency and productivity of a biotechnical facility is ensured through the operation of a number of control systems each of which is specified by its own energy characteristics, cost of maintenance, and impact on the manufacturing costs. The analysis conducted has made it possible to identify the following automated control systems: ventilation, curtaining, heating, air recirculation, water supply, and sewerage, establishing the microclimate and electric supplementary lighting, and electric lighting and nutrition of the plants.

The main energy consumers in greenhouses are the pumping groups with electric drives for the heating units, electro-technical plant irrigation units, evaporative cooling and air humidification units, electric motors of the curtaining and ventilation systems, the technological equipment of the preparation unit for mineral fertilizers, and the electric supplementary lighting system of the plants. The energy consumption is also maintained due to the use of natural gas for heating.

However, the algorithms realized by the modern systems in greenhouse the technology efficiency and productivity of a biotechnical facilities is ensured through the operation of a number of control systems each of which is specified by its own energy characteristics, cost of maintenance, and impact on the manufacturing costs. Remain the simplest and the same is valid for the poultry production: it is the stabilization of the microclimate parameters at a level which corresponds to a certain productivity of the biotechnical facility.

As a result of the analysis of the data [11] on the functioning of greenhouse control systems, it has been determined that the vast majority of greenhouses in Ukraine spend almost 70% of their income from the product sales on energy resources (electricity, gas, diesel fuel, etc.), with their flows being regulated by means of using the electro-technical systems which are based on stabilization algorithms without forecasting the natural disturbances or considering the condition of the plants, and the same is valid for the poultry farms. It is clear that these control algorithms are not energy-efficient, and the same is valid for the poultry farms.

When creating an energy-efficient control system for biotechnical facilities, intelligent energy flow control algorithms are worthy of note. The results of the previous studies have made it possible to conclude that additional information on the forecast values of the ambient temperature, solar radiation, information on the state of the biological burden, and the rules of using such information allows us to create a knowledge base and use it for the formation of the control influences on the biotechnical facilities being directed at minimizing the energy consumption while ensuring the product manufacturing of the required quality and volume [9].

The analysis of the already existing designs has shown that the current systems, being a component part of the technology, implement the simplest stabilization algorithms which do not provide for high energy and resource efficiency due to the fact that the information on the condition of the biological component part of the control facility and forecasting of the natural disturbances are not taken into consideration. The solution to the task of energy-efficient control of the biotechnical facilities is possible through using modern intelligent algorithms for processing the information which comes from the control facility and applying the results obtained to form the appropriate control strategies which are aimed at maximizing the profits following from the production results.

Thus, *the objective of the research* is to develop the intelligent control systems for biotechnical facilities taking into consideration the forecasting of the natural disturbances in order to increase their energy efficiency (in the context of poultry farms and greenhouse).

It is obvious that the value and utilization efficiency of the information on the condition of the biological component part of the control facility is determined in the first place by the ability to find it and be provided with a quality access to it. The operations aimed at processing the currently available information are also becoming more advanced: they comprise not only the simplest arithmetic operations but also database operations such as search, sampling, selection, filtering and sorting, statistical computations, numerical methods, simulation modeling, methods of image recognition and decision-making, game theory, expert systems, neural networks (NNs), genetic algorithms, cognitive modeling, and others.

The creation of intelligent systems consists of the following three stages:

1) creation of a material support system (this task has been solved by and large since intelligent systems can be created even on the basis of modern personal computers);
2) creation of a potential artificial intelligence system, that is to say a software shell, a development system (there are very few currently existing systems of this kind);
3) training and self-training of a potential artificial intelligence system and its transformation into an actual intellectual system.

When controlling the biotechnical facilities, intelligent systems are used for the purpose of solving the tasks of forecasting, classification, and control. From a mathematical standpoint, the required forecasted values of the temperature and solar radiation are a time-series. Until recently, statistical methods were the main methods of forecasting the time-series. However, they cannot forecast the complicated influence processes of environmental factors due to the nonlinearity character of the data model. In these cases, the apparatus of NNs comes to rescue. NNs which are used to forecast the external natural disturbances and radiation constitute a mathematical model of parallel computing which contains the simple processor elements, that is to say, artificial neurons which are interacting with each other. Almost any task can be traced to a problem solved by an NN [21]. Thus, let us build several NNs for the ambient temperature and radiation forecasting and compare their performance.

In comparison with the classical methods of data analysis, NNs possess the following advantages [15–20]:

1) constant optimization of their own structure for the purpose of minimizing the forecast error in real time;
2) higher potential opportunities while analyzing the complicated dynamic systems and patterns;
3) the ability to solve the tasks successfully on the back of incomplete, misleading, and internally contradictory input information.

The task of selecting the control actions while managing the biotechnical facilities of agricultural production largely depends on the ability of the control system to forecast the natural disturbances. The analysis of the natural disturbances on the territory of Ukraine has made it possible to establish that for the biotechnological facilities where animals are kept, this is, in the first place, the temperature (poultry houses, dairy byres, etc.), while for the constructions of the greenhouse where the plants vegetate, these are the temperature and solar radiation. Here are a few examples of implementing and using the intelligent systems for biotechnical facilities of the agricultural sector.

### 13.2.2 Synthesis of an Intelligent Control System Taking into Account the Forecasting of the Changes in Temperature Images in the Context of a Poultry House

The characteristics of the change in the natural disturbances in the poultry house depend on the climatic zone where the production is located. The annual time-series of fluctuations $\Theta 1, \Theta 2, \ldots, \Theta n$ of the air temperature, which were observed in the specified area, have been built on the basis of the data obtained from the Hydrometeorological Center of Ukraine (HMCU). These samples represent a nonstationary random process, and determining its statistical characteristics is a very complicated stochastic problem which is almost impossible to solve. However, the analysis of change in the individual sections of annual samples (of the time-series) has proved that they can be forecasted due to the fact that they are the samples of either stationary random processes or stationary processes with additive deterministic components, that is to say, quasi-stationary processes. Each of the annual samples can be presented in the form of 45–70 stationary or quasi-stationary sections as mentioned above. Five hundred and sixty-nine sections have been allocated during the period of 10 years. A cluster analysis of these time-series segments has been performed in order to classify these facilities by their characteristic features and to forecast their further appearance.

The task of the algorithm for pattern recognition of areas with temperature disturbances is to sequentially redefine the deterministic and statistical characteristics of temperature changes outside the technological facility where the biological component part is being managed.

The initial characteristics of these changes are determined on the basis of the HMCU forecast which forecasts the possible temperature fluctuations during the night hours ($\Theta_{n.h.}$ and $\Theta_{n.l}$) and during the day hours ($\Theta_{d.h.}$ and $\Theta_{d.l.}$). Due to this information, the approximate (forecasted) values of the mathematical expectation $\mathrm{m}_{\Theta_\mathrm{n}}$ and the amplitude of fluctuations $\mathrm{A}_{\Theta_\mathrm{n}}$in the temperature changes of the following day are calculated

$$\mathrm{m}_\Theta = \frac{\Theta_{n.h} + \Theta_{d.h.}}{4} + \frac{\Theta_{d.l.} + \Theta_{n.l.}}{4} \tag{13.1}$$

$$\mathrm{A}_\Theta = \frac{\Theta_{d.l.} - \Theta_{n.l.}}{4} + \frac{\Theta_{d.h.} - \Theta_{n.h.}}{4} \tag{13.2}$$

Thereupon, a comparison of the forecasted mathematical expectation $\mathrm{m}_\Theta$to the mathematical expectation of the current day temperature changes $\mathrm{m}_\Theta$is carried out. If $(\mathrm{m}_\Theta - 1) \ \le \mathrm{m}_\Theta \le (\mathrm{m}_\Theta + 1)$, then the temperature section still coincides with the image defined for the previous day. If the forecasted values $\mathrm{m}_\Theta$ deviate by more than 1°C, it is assumed that the temperature image is expected to change.

The image recognition algorithm is based on comparing the degree of proximity *L* of the object (the sections of the air temperature change) $\omega$ to any of the classes $\Omega_q$, where $q = \overline{1,m}$. Five classes have been previously identified in our system, [9]; therefore, *m* = 5. The root mean square distance (expressed in °C) between the object $\omega$ and the set of objects $\omega_{q1}, \omega_{q2}, \ldots,$ $\omega_{qk}$, which belong to the $\Omega_q$ class (*k* stands for the number of images (objects) of the $\Omega_q$ class), has been imputed as the degree of proximity:

$$L\left(\omega, \Omega_q\right) = \sqrt{k_q^{-1} \sum_{k=1}^{k_q} d^2\left(\omega, \omega_{qk}\right)}. \tag{13.3}$$

The decision to assign object $\omega$ to the $\Omega_q$ class is made if

$$L\left(\omega, \Omega_q\right) = \min L\left(\omega, \Omega_i\right), \tag{13.4}$$

where $i = \overline{1,m}$.

Thus, when a new image appears, it is determined to which class it belongs in the first place. It was proposed to use an intelligent pattern recognition system based on probabilistic NNs to solve this task, proceeding from the necessity for an adequate analysis of the start of the change of one image to another and taking into consideration the functional peculiarities of the systems based on the methods of mathematical statistics.

The Bayesian formula is one of the main in the Elementary Probability Theory, which allows us to determine the probability of a certain event (hypothesis) in the presence of circumstantial confirmations (data) only [22–25] which may contain inaccuracies, which is exceptionally important for solving our task. The Bayesian formula record is as follows:

$$P(A|B) = \frac{P(A|B)P(A)}{P(B)}, \tag{13.5}$$

where $P(A)$ stands for the *a priori* probability of the hypothesis $A$; $P(A|B)$ stands for the probability of the hypothesis A in case of occurrence of the event B (the *a posteriori* probability); $P(B|A)$ stands for the probability of occurrence of the event B in case of trueness of the hypothesis A; $P(B)$ stands for the probability of occurrence of the event B.

An important corollary of the Bayesian formula is the formula for the total probability of an event, which depends on several incompatible hypotheses (and on them only):

$$P(B) = \sum_{i=1}^{N} P(A_i)P(B|A_i), \tag{13.6}$$

where $N$ stands for the number of hypotheses.

Taking into consideration Equation (13.5), we conclude that the probability of occurrence of the event B depends on a number of hypotheses $A_i$, if the degrees of reliability of these hypotheses are known (for example, experimental data on the external air temperature).

The Bayesian network built on these theoretical principles is a probabilistic model which represents a set of variables and their probabilistic dependence relationships. Formally, a Bayesian network is a directed acyclic graph with variables being its nodes while the ribs encode the conditional dependences between the variables. The nodes can represent variables of any type and can be weighted parameters, hidden variables, or hypotheses. If the rib connects the node A to node B, then A is named the father of B, and B is called the descendant of A. The set of ancestor nodes of the node $Xi$ is denoted as parents ($X_i$). Then, the common distribution of values in the nodes can be conveniently recorded as a result of local distributions:

$$P(X_1, ..., X_n) = \prod_{i=1}^{n} P(X_i \,|\text{parents}(X_i)) \tag{13.7}$$

where $n$ stands for the number of local distributions.

Probabilistic neural networks (PNNs) are a special case of Bayesian networks, a type of NNs which are effectively used to solve the classification tasks in which the probability density of belonging to classes is estimated with the help of the kernel approximation. When solving the classification tasks, the network outputs can be advantageously interpreted as probability estimates of whether an element belongs to a certain class. In fact, the network trains itself to estimate the probability density function.

A total of 132 temperature images with the corresponding numerical values of the input parameters have been generated [8].

During the synthesis of the PNN classifier of the temperature images, the following were used as input values (Table 13.1):

1) the mathematical expectation ($m_0$);
2) the amplitude of temperature fluctuations ($A$);
3) the minimum standard deviation ($\sigma_{\text{min}}$);
4) the maximum standard deviation ($\sigma_{\text{max}}$).

The uncertainty losses are never significant enough according to the HMCU, given that the parameters of the images of temperature disturbances become more reliable after 20 hours of observations.

The network output is the number of the class (image) to which the resulting set of the input values belongs.

The NN layer of addition is to have one element for each element from the training data set – 132. All the elements of this layer are connected only to the elements of the layer of samples which belong to the corresponding image.

**Table 13.1** The variation ranges of the input values.

| $m_0$, °C | A, °C | $\sigma_{\text{min}}$, °C | $\sigma_{\text{max}}$, °C |
|---|---|---|---|
| −24 to +18 | 0–10 | 0.5–2.5 | 3–5 |

The activity of the sample layer element is equal to

$$O_j = \exp\left(\frac{-\sum(w_{ij} - x_i)^2}{\sigma^2}\right), \tag{13.8}$$

where *w* stands for the value of the weighting numbers.

The weighting values of the relationships leading from the elements of the sample layer to the elements of the addition layer are equal to 1.

The addition layer element simply summarizes the output values of the sample layer elements. This sum estimates the value of the probability distribution density function for the set of instances of the corresponding image. The output elements are threshold discriminators which indicate the element of the addition layer with the maximum value of activation (that is to say, they indicate one of the 132 temperature images).

A network of this kind does not require the training which is required for the perceptron-type networks, radial-basis function, etc., since all the parameters of the PNN network, such as the number of elements and values of weights, are determined directly by the training data.

The procedure of using the PNN network is relatively simple: once a network is built, an unknown instance can be led to the network input and the output layer can indicate the image to which the sample most likely belongs resulting from a direct passage through the network. Twenty possible sets of input parameters belonging to different classes have been created to examine the quality of classification.

The probabilistic NN has correctly classified all the sets with a clear advantage at the output of the addition layer of the probability distribution density of the corresponding winning images (Figure 13.5).

Within the scope of the task, we are interested not so much in the discrete classification of images but in the value of the output of the addition layer which calculates the probability distribution density for the set of instances of the corresponding image. That is to say, we can monitor the dynamics of changes in the temperature images at the output of this layer.

An experimental sample of the intelligent control system was installed in the poultry house No 4 at the State Enterprise "Training and Research Poultry Breeding Plant" named after Frunze; its functional diagram is presented in Figure 13.6.

In the figure, *V* stands for the reconstruction of the control device action; *U* stands for the decision of the control strategy; *f* stands for the image type and its parameters; $Q_n$ stands for the temperature forecast; $\theta_z$ stands for the

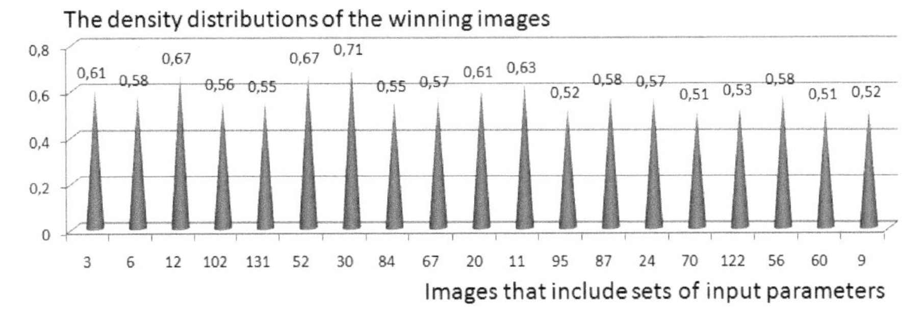


**Figure 13.5** The probability density distribution of the winning images during the research conducted.

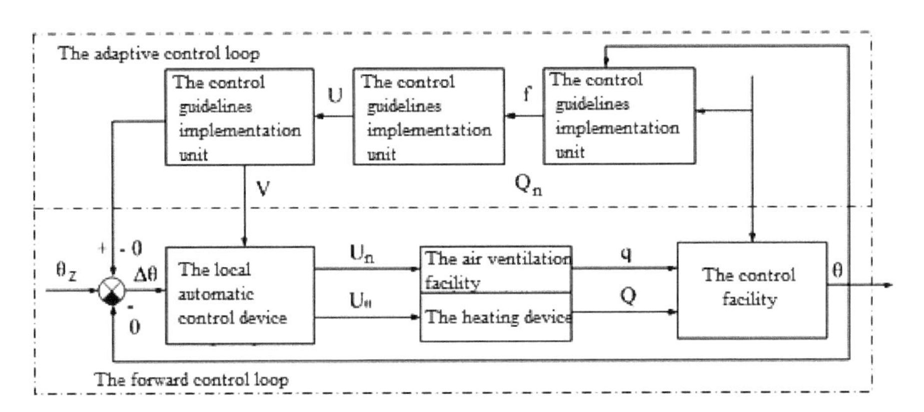


**Figure 13.6** The functional diagram of the intelligent control system of the poultry house temperature.

ambient temperature; $U_v$ stands for the ventilation control; $U_n$ stands for the heating control; q stands for the air flow; $Q$ stands for the flow of the heated air; $\theta_z$ stands for the given temperature; stands for the temperature which was measured in the poultry house; $\Delta$ stands for the temperature deviation (Patent No. 44637 UA, IPC G05B 13/00 (2009).

The architecture of the computer-integrated control system is built in such a way so as to ensure the maximum reliability of its operation. The industrial controller possesses adequate resources (modules, an expansion base) to connect a large number of sensors and uses open protocols, that is to say, enables control by means of actuators in accordance with different control principles.

### 13.2.3 Synthesis of the Intelligent Control System Taking into Account the Forecast of the External Natural Disturbances and Radiation in the Context of a Greenhouse

Appropriate technological conditions for plant growth and development are created in the greenhouse. The ambient temperature and solar radiation are the main natural disturbances which affect the vegetation of plants in the greenhouse. Therefore, the development of an intelligent system which can make it possible to predict these natural disturbances is an urgent task.

We use the software package Statistica Neural Networks and the multilayer perceptron which is traditional for solving the tasks on time-series forecasting [9] for the synthesis and investigation of the corresponding NNs, the criterion being the minimization of the NN error [25]. In the context of our task, its advantage over the similar items under development lies in the implementation of a functional block for optimizing the architecture of the neuromodels which uses linear approaches and a simulated annealing method on the basis of the Gibbs probability distribution:

$$P(\overline{x^*} \to \overline{x_{i+1}}\,|\overline{x_i}) = \left\{ \begin{array}{ll} 1, F(\overline{x^*}) - F(\overline{x_i}) < 0 & \\ \exp(-\frac{F(\overline{x^*})-F(\overline{x_i})}{Q_i}), & F(\overline{x^*}) - F(\overline{x_i}) \geq 0 \end{array} \right\}, \tag{13.9}$$

where $Q_i > 0$ are the elements of an arbitrarily descending to zero sequence.

In the context of the proposed sequence (Figure 13.7), the input data is automatically divided into three blocks: training, control, and test for the effective modeling in the Statistica Neural Networks package. The presence of the three blocks is not obligatory, whereas the test block improves the quality of the further work since it ensures that there is no overfitting of the network.

#### 13.2.3.1 The Neural Network Forecasting of the External Natural Disturbances

The values of the day and time of the forecasting are used as the input data; that is to say, there are two input variables (Var1 stands for the day and Var2 stands for the time) and one output variable (Var4 stands for the ambient temperature).

The following NNs have been selected as the best as a result of solving the optimization task: the multilayer perceptron MLP 2-8-1 (with the training effectiveness of 89.3%), the multilayer perceptron MLP 2-8-1 (with the

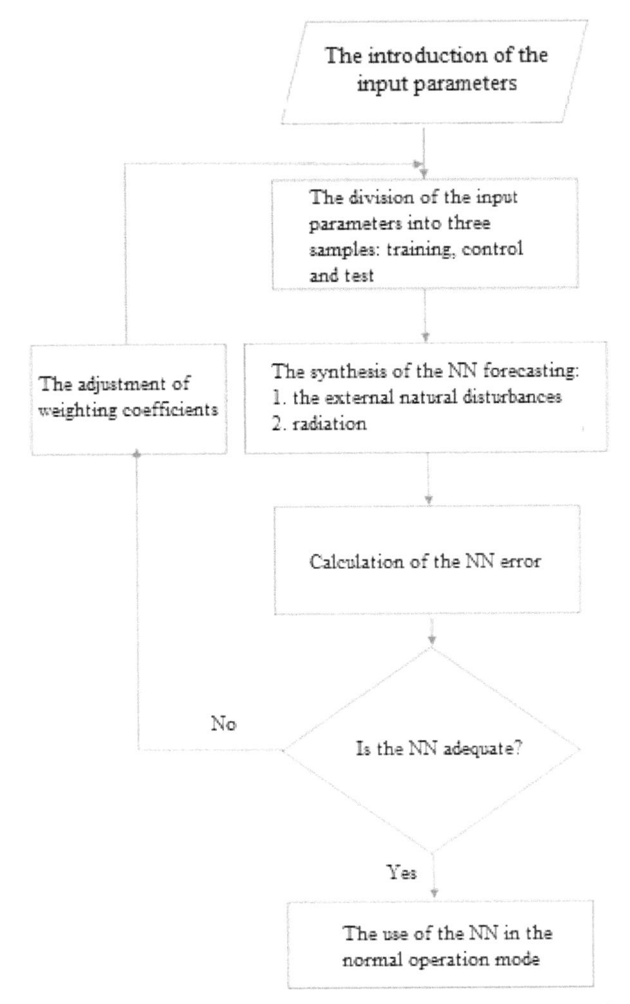


**Figure 13.7** The NN synthesis algorithm for the forecasting of the external natural disturbances and radiation.

training effectiveness of 99.7%), the multilayer perceptron MLP 2-7-1 (with the training effectiveness of 88.9%), the multilayer perceptron MLP 2-3-1 (with the training effectiveness of 87.7%), and the multilayer perceptron MLP 2-3-1 (with the training effectiveness of 87.7%).

The best result was demonstrated by the MLP 2-8-1 network, which provided the effectiveness of 99.7% in all the samples; that is to say, it presented the forecast with the maximum accuracy.

**Table 13.2** The results of solving the optimization task of the neural networks synthesis.

| Total models | | | | | | | | |
|---|---|---|---|---|---|---|---|---|
| No. | Architecture | Training effectiveness | Performance control | Performance test | Training algorithm | Error | Hidden neurons activity function | Output neurons activity function |
| 1 | MLP 2-8-1 | 0.893260 | 0.892000 | 0.894734 | BFGS 118 | Quadratic sum | Hyperbolic | Exponent |
| 2 | MLP 2-8-1 | 0.997834 | 0.994275 | 0.987809 | BFGS 259 | Quadratic sum | Hyperbolic | Hyperbolic |
| 3 | MLP 2-7-1 | 0.889667 | 0.888883 | 0.891928 | BFGS 113 | Quadratic sum | Hyperbolic | Identity |
| 4 | MLP 2-3-1 | 0.877182 | 0.864570 | 0.872726 | BFGS 112 | Quadratic sum | Logistic | Identity |
| 5 | MLP 2-3-1 | 0.877309 | 0.865464 | 0.874213 | BFGS 93 | Quadratic sum | Logistic | Hyperbolic |

To verify the correctness of the time-series construction for the forecasting of the ambient temperature and radiation, it is necessary to determine the adequacy of choice for the input variables: the day and time in increments of 1 hour. In order to find these mathematical, functional, or structural dependence relationships between two or more variables (in accordance with the accumulated experimental data) the methods of correlation analysis are of great use, which indicates the absence or presence of a connection between the variables with a certain predetermined confidence probability.

The linear correlation coefficient is widely used for the quantity evaluation of the density of the network connections. If the values of the variables $X$ and $Y$ are given, then it is calculated by the following formula:

$$r_{XY} = r_{YX} = \frac{\overline{XY} - \bar{X} \cdot \bar{Y}}{\sigma_X \cdot \sigma_Y}. \quad (13.10)$$

If $|r| < 0.30$, then the connection between the features is weak; $0.30 \leq |r| \leq 0.70$ stands for the moderate connection; $|r| > 0.70$ stands for the strong or dense connection. When $|r| = 1$, then the connection is functional. If $|r| \approx 0$, then there is no linear connection between $X$ and $Y$. However, the nonlinear interaction is possible and it requires additional verification.

All the correlation coefficients are greater than 0.7, which indicates the adequacy of the input parameters choice.

Despite the relatively low training quality of some NNs, we are going to use them for further research since during the forecasting (validation), they can potentially demonstrate the required quality while taking into consideration the internal functional peculiarities. The next step in the

**Table 13.3** The values of correlation coefficients between the input and target variables.

| | Correlation coefficients | | |
|---|---|---|---|
| | −10.500000 | −10.500000 | −10.500000 |
| | Training | Test | Test |
| 1) MLP 2-8-1 | 0.893260 | 0.892000 | 0.894734 |
| 2) MLP 2-8-1 | 0.907834 | 0.904275 | 0.907809 |
| 3) MLP 2-7-1 | 0.889667 | 0.888883 | 0.891928 |
| 4) MLP 2-3-1 | 0.877182 | 0.864570 | 0.872726 |
| 5) MLP 2-3-1 | 0.877309 | 0.865464 | 0.874213 |

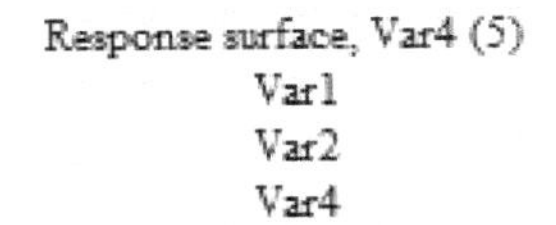


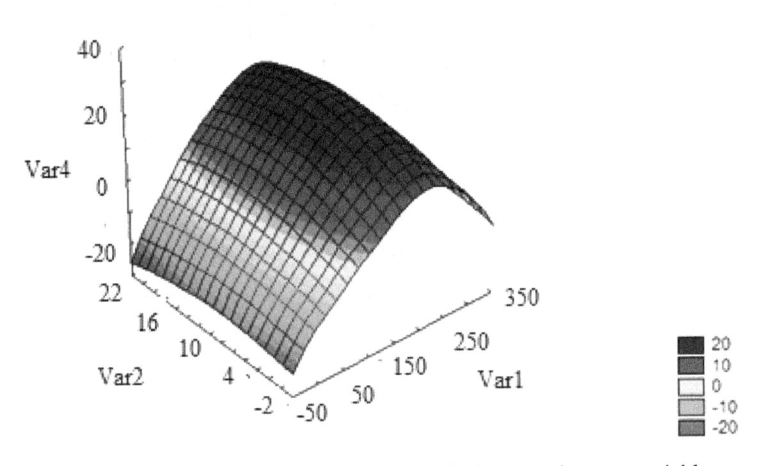


**Figure 13.8** The response surface between the input and output variables.

analysis of the time-series of the ambient temperature is obtaining the appropriate predictions which are to go beyond the training sample.

To accomplish this, we implement a projection of the time-series for each of the networks, setting the depth of the forecast at eight steps forward (Figure 13.9).

During the forecasting at 8 hours forward, the generally sufficient accuracy of the forecast is observed (the mean square error being 1.19°C–3.43°C).

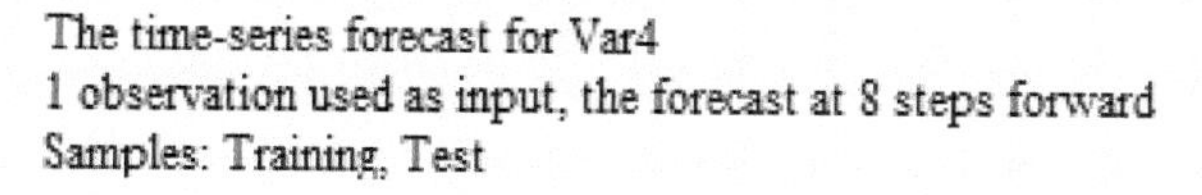


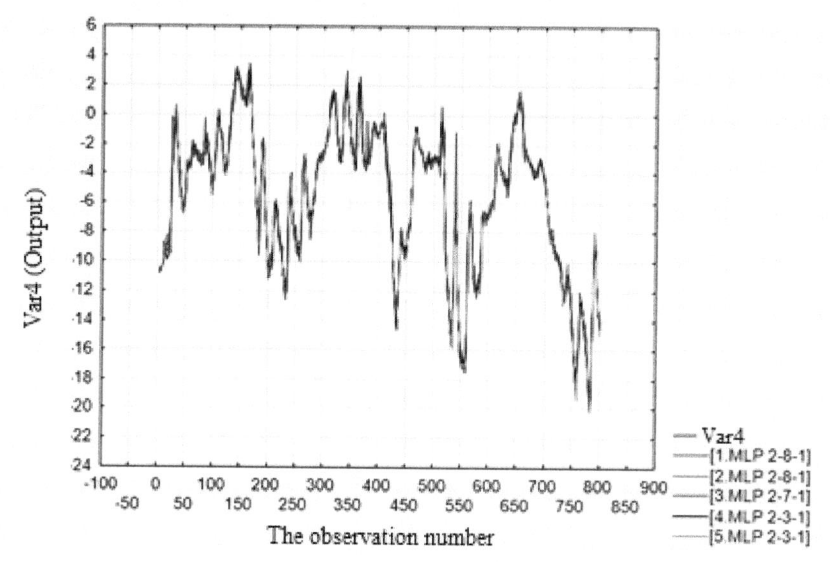


**Figure 13.9** Neural network projections of the time-series of the ambient temperature.

#### 13.2.3.2 The Intelligent Solar Radiation Forecasting System

The values of the day and time forecasting are used as input data, that is to say, there are two input variables (Var1 stands for day and Var2 stands for the time) and one output variable (Var3 stands for radiation).

The following NNs have been selected as the best as a result of solving the optimization task: the multilayer perceptron MLP 2-7-1 (with the training effectiveness of 97.64%), the multilayer perceptron MLP 2-9-1 (with the training effectiveness of 97.58%), the multilayer perceptron MLP 2-9-1 (with the training effectiveness of 97.64%), the multilayer perceptron MLP 2-9-1 (with the training effectiveness of 99.74%), and the multilayer perceptron MLP 2-4-1 (with the training effectiveness of 97.61%).

The training parameters of the synthesized NNs based on the results of solving the optimization task are given in the table below.

Although all the synthesized networks possess a fairly high training effectiveness, the best result has been demonstrated by the MLP 2-9-1

**Table 13.4** The results of solving the optimization task of the neural networks synthesis.

| Total models | | | | | | | | |
|---|---|---|---|---|---|---|---|---|
| No. | Architecture | Training effectiveness | Performance control | Performance test | Training algorithm | Error function | Hidden neurons activity function | Output neurons activity function |
| 1 | MLP 2-7-1 | 0.976400 | 0.975040 | 0.976665 | BFGS 224 | Quadratic sum | Hyperbolic | Logistic |
| 2 | MLP 2-9-1 | 0.975863 | 0.974947 | 0.977057 | BFGS 252 | Quadratic sum | Hyperbolic | Identity |
| 3 | MLP 2-9-1 | 0.976452 | 0.974613 | 0.977243 | BFGS 494 | Quadratic sum | Hyperbolic | Exponent |
| 4 | MLP 2-9-1 | 0.997452 | 0.996349 | 0.997915 | BFGS 133 | Quadratic sum | Hyperbolic | Exponent |
| 5 | MLP 2-4-1 | 0.976123 | 0.975051 | 0.976370 | BFGS 177 | Quadratic sum | Hyperbolic | Logistic |

**Table 13.5** The values of the correlation coefficients between the input and target variables.

| | Correlation coefficients | | |
|---|---|---|---|
| | Value Training | Value Test | Value Test |
| 1) MLP 2-7-1 | 0.976400 | 0.975040 | 0.976665 |
| 2) MLP 2-9-1 | 0.975863 | 0.974947 | 0.977057 |
| 3) MLP 2-9-1 | 0.976452 | 0.974613 | 0.977243 |
| 4) MLP 2-9-1 | 0.977452 | 0.976349 | 0.977915 |
| 5) MLP 2-4-1 | 0.976123 | 0.975051 | 0.976370 |

network, which demonstrated the effectiveness in all the samples of 99.74%, and secured the maximum forecast accuracy.

In order to verify the correctness of the construction of the time-series for radiation, it is also necessary to determine the adequacy of the input variables choice: the day and time in increments of 1 hour. A linear correlation coefficient has been used for the quantity evaluation of the density of the network connections (Table 13.5).

All the correlation coefficients are greater than 0.7, which indicates the adequacy of the input parameters choice.

The next step in the analysis of radiation time-series is to obtain the appropriate predictions which are to go beyond the training sample. In order to do this, we implement the projection of the time-series for each of the networks, setting the depth of the forecast at eight steps forward (Figure 13.11).

During the forecasting at 8 hours forward, generally sufficient accuracy of the forecast is observed (the mean square error being 1.76%–5.35%). Thus,

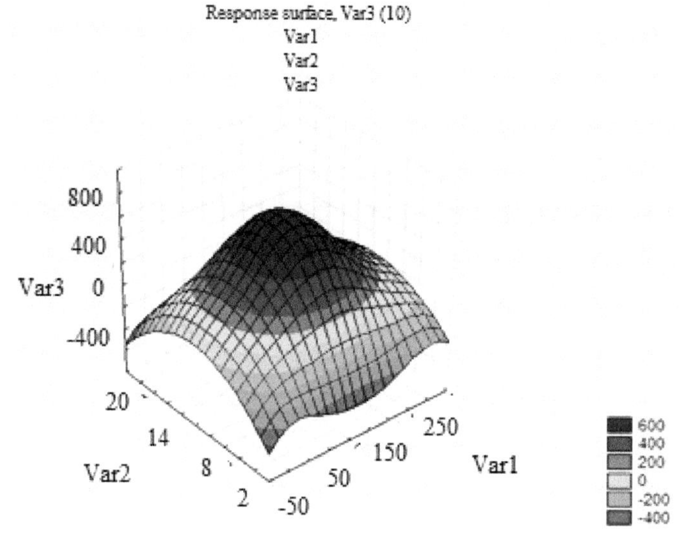


**Figure 13.10** The response surface between the input and output variables.

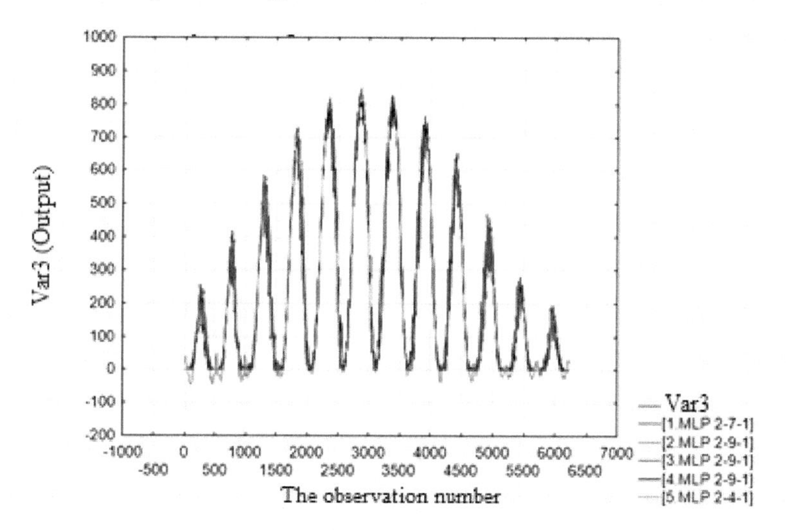


**Figure 13.11** The neural network projections of the radiation time-series.

the use of the intelligent systems based on NNs to forecast radiation by means of analyzing and processing the time-series data is accurate enough to be used in the control system.

In order to build an energy-efficient intelligent control system of the greenhouse complex, a block of fuzzy control system has been developed following from the experimental data with the aim of determining the need for switching on the existing heating circuits while taking into consideration the ambient temperature and solar radiation.

The obtained dependence of the internal temperature and intensity of solar radiation during the entire period of growing tomatoes in the greenhouse has been used to determine the influence of the forecasted value of solar radiation on the internal temperature of the greenhouse (Figure 13.12). It can be approximated by means of a second-order polynomial.

The values of the ambient temperature (tz – Inp1), the time of the day (T – Inp2), and the temperature inside the greenhouse (tv – Inp3) are applied to the input of the fuzzy control system. At the output, the heating circuits are set as follows: I stands for the under-tent heating (Out1), II stands for the heating in the plant growth zone (Out2), and III stands for the above-ground heating (Out3).

The tasks of the function parameters belonging to the internal temperature are carried out, proceeding from the technological parameters during the cultivation of tomatoes in the greenhouse [13]. That is to say, the value of

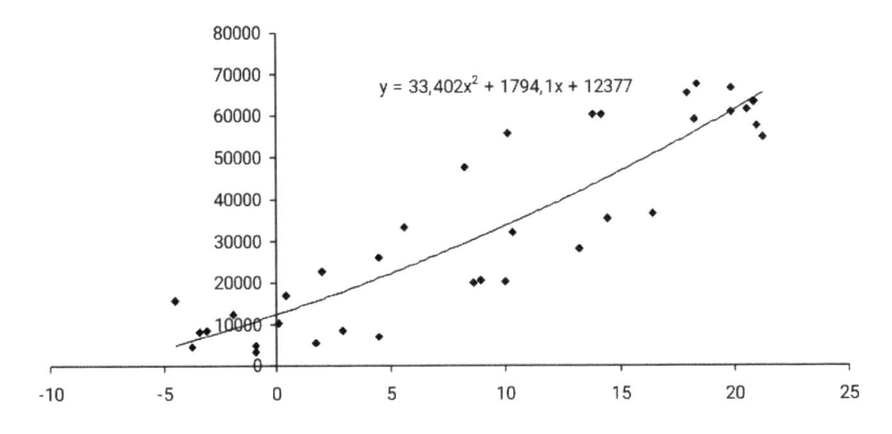


**Figure 13.12** The dependence of the solar radiation intensity and the internal air temperature.

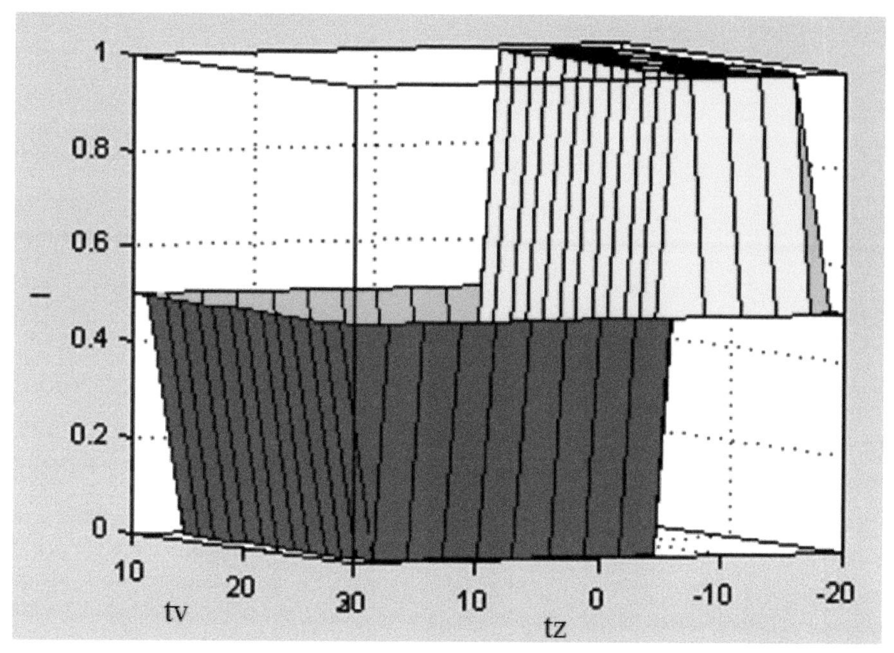


**Figure 13.13** The response surface between the input and output variables.

+18°C to + 25°C is taken as the norm (tv2), tv1 is in the range of +10°C to +18°C, and tv3 is in the range of values + 25°C to + 30°C.

The parameters of the ambient temperature membership function are taken in the range from −20°C to +30°C. The parameters of the membership function of the time variable are determined on the basis of the division into hourly loadings on the electrical power network in accordance with the Kyivenergo tariffs.

The following base of rules has been determined in order to develop a fuzzy control system expertly, resulting from the passive experiment data, as to the energy efficiency of the greenhouse production (Table 13.6).

The result of the system operation is displayed in the rules overview window, where it is possible to investigate the activation of the corresponding heating circuits, in accordance with the change of the input parameters with the purpose of ensuring the energy-efficient control.

In the figure, $E$ stands for the control error, $S_n$ stands for the forecasted value of solar radiation, $f$ stands for the disturbance vector, and $U$ stands for the control actions vector.

**Table 13.6** The base of rules for the fuzzy expert control system.

| No. | Inp1 | Inp2 | Inp3 | Out1 | Out2 | Out3 |
|---|---|---|---|---|---|---|
| 1) | tz1 | T1 | tv1 | 1 | 1 | 1 |
| 2) | tz1 | T2 | tv1 | 1 | 1 | 1 |
| 3) | tz1 | T3 | tv1 | 1 | 1 | 1 |
| 4) | tz1 | T1 | tv2 | 1 | 0 | 1 |
| 5) | tz1 | T2 | tv2 | 1 | 0 | 1 |
| 6) | tz1 | T3 | tv2 | 1 | 0 | 1 |
| 7) | tz1 | T3 | tv3 | 1 | 0 | 0 |
| 8) | tz1 | T2 | tv3 | 1 | 0 | 0 |
| 9) | tz1 | T1 | tv3 | 1 | 0 | 0 |
| 10) | tz2 | T1 | tv1 | 0 | 1 | 1 |
| 11) | tz2 | T2 | tv1 | 0 | 1 | 1 |
| 12) | tz2 | T3 | tv1 | 1 | 1 | 1 |
| 13) | tz2 | T1 | tv2 | 0 | 0 | 1 |
| 14) | tz2 | T2 | tv2 | 0 | 0 | 1 |
| 15) | tz2 | T3 | tv2 | 0 | 1 | 1 |
| 16) | tz2 | T1 | tv3 | 0 | 0 | 1 |
| 17) | tz2 | T2 | tv3 | 0 | 0 | 1 |
| 18) | tz2 | T3 | tv3 | 0 | 0 | 1 |
| 19) | tz3 | T1 | tv1 | 0 | 0 | 1 |
| 20) | tz3 | T2 | tv1 | 0 | 0 | 1 |
| 21) | tz3 | T3 | tv1 | 0 | 1 | 1 |
| 22) | tz3 | T1 | tv2 | 0 | 0 | 1 |
| 23) | tz3 | T2 | tv2 | 0 | 0 | 1 |
| 24) | tz3 | T3 | tv2 | 0 | 0 | 1 |
| 25) | tz3 | T1 | tv3 | 0 | 0 | 0 |
| 26) | tz3 | T2 | tv3 | 0 | 0 | 0 |
| 27) | tz3 | T3 | tv3 | 0 | 0 | 0 |

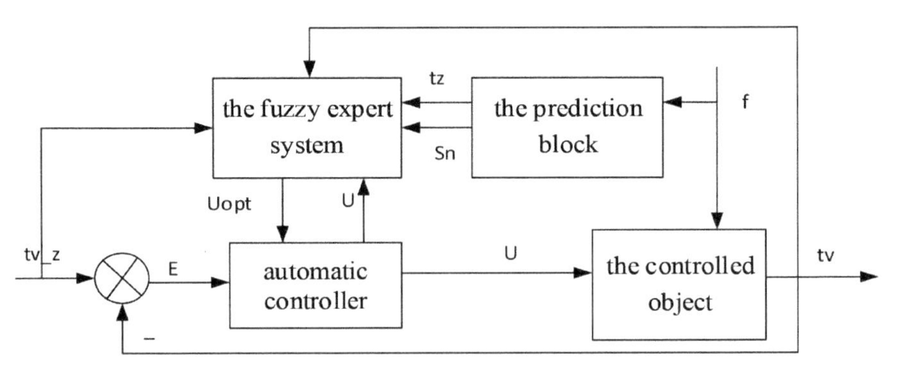


**Figure 13.14** The structure of the intelligent temperature control system in the greenhouse.

**Table 13.7** The engines loading during the day without the introduction of the forecasting system.

| Tariff | Night | Semi-peak | Peak | Semi-peak | Peak | Semi-peak | |
|---|---|---|---|---|---|---|---|
| The cost of electricity in accordance with the tariff, UAH/kW | 0.25 | 1.02 | 1.8 | 1.02 | 1.8 | 1.02 | |
| Greenhouse heating | + | + | | + | + | + | |
| The main (in the plant growth zone) | + | + | + | + | + | + | |
| Rail heating | + | + | + | + | + | | Total |
| Total consumption, kW | 157.2 | 72.2 | 72.2 | 288.8 | 144.4 | 72.2 | 807 |
| Consumed for the total sum of UAH | 39.3 | 73.644 | 129.96 | 294.576 | 259.92 | 73.644 | 871.04 |

**Table 13.8** The engine loading during the day using the forecasting system.

| Tariff | Night | Semi-peak | Peak | Semi-peak | Peak | Semi-peak | |
|---|---|---|---|---|---|---|---|
| The cost of electricity in accordance with the tariff, UAH/kW | 0.25 | 1.02 | 1.8 | 1.02 | 1.8 | 1.02 | |
| Greenhouse heating | + | + | | + | | + | |
| The main (in the plant growth zone | + | + | + | + | + | + | |
| Rail heating | + | | | | | | Total |
| Total consumption, kW | 216.6 | 52.4 | 36 | 209.6 | 72 | 52.4 | 639 |
| Consumed for the total sum of UAH | 54.15 | 53.44 | 64.8 | 213.8 | 129.6 | 53.4 | 569.2 |

The results of the intelligent temperature control system performance in the greenhouse are shown in Tables 13.7 and 13.8 with the initial conditions being as follows: the ambient temperature is below −7°C throughout the day and the required temperature inside the greenhouse is 20°C.

In general, the simulation results show that if compared to traditional approaches, the intelligent control system has better quality indicators of transients. Having compared the given data of Tables 13.7 and 13.8, it can be concluded that the introduction of the control system of switching on the engines helps reduce the energy costs by 25% to 30%. Thus, the urgent task lies in the analysis and forecasting of the factors like the temperature and solar radiation in order to use the results of the forecast in the formation of the control strategies for greenhouse complexes with the purpose of reducing the energy costs in agricultural production.

## 13.3 Conclusion

The intelligent temperature control system of the poultry house has been developed and implemented and it provides an opportunity to forecast, on the basis of the NN classifier of temperature images, the change in the ambient temperature outside the technological facility where the biological component part is managed.

The intelligent system based on NNs has been created to forecast the external forecasting and radiation. The best result of forecasting the ambient temperature has been demonstrated by the MLP 2-8-1 network, which ensured the effectiveness of 99.7% in all the samples. The best result in forecasting solar radiation has been demonstrated by the MLP 2-9-1 network, which provided the effectiveness of 99.74% in all the samples. The developed intelligent system provides means for the maximum accuracy of forecasting the external natural disturbances and radiation which is taken into consideration while molding the control strategy for natural disturbances.

The intelligent temperature control system in the greenhouse based on the analysis and forecasting of the temperature and solar radiation has been developed in order to use the forecast results in the formation of control strategies for the greenhouse complexes for the purpose of reducing the energy costs in agricultural production.

The intelligent control algorithms for biotechnical facilities are to be created and scientifically substantiated with regard to the findings of the study and being determined with consideration of the forecasted changes in the external natural disturbances and the information on the quality of the biological component part. This would enable minimizing the cost of resources and ensure the maximum profit of the enterprise.

## References

[1] Besulin, V. I. 2003. Ptakhivnytstvo i tekhnolohiya vyrobnytstva yayets' ta m'yasa ptytsi [Poultry farming and production technology of eggs and poultry meat]. Bila Tserkva State Agrarian University. 447 p. (in Ukrainian).

[2] Magdelaine Pascale.2010. Future prospects for the European egg industry. *World Poultry*. Vol. 25 No 11:14-16.

[3] Ivanov, K.P. 1990. The bioenergetic mechanisms of homeothermy [O bioénergeticheskikh mekhanizmakh gomoĭotermii.] *Zhurnal obshchei biologii*. Vol. 51, Issue 1:36-53.

[4] Yaroshenko, F.O. 2004. Ptakhivnytstvo Ukrayiny: problemy i perspektyvy rozvytku [Poultry farming of Ukraine: state, problems and prospects of development]. Agrarna nauka. 504 p. (in Ukrainian).

[5] Jan Hulzebosch. 2006. Wide range of housing options for layers. *World Poultry*. Vol. 22 No 6:20-22.

[6] Gorobets, V.G., V.I. Trokhaniak, I.L. Rogovskii, T.I. Lendiel, A.O. Dudnyk, and M.Y. Masiuk. 2018. The numerical simulation of hydrodynamics and mass transfer processes for ventilating system effective location. *INMATEH - Agricultural Engineering*. Vol. 56, Issue 3:185-192.

[7] Saleeva, I.P., A.V. Sklyar, T.E. Marinchenko, M.V. Postnova, A.V. Ivanov, and A.I. Tikhomirov. 2019. *Proceedings of International Scientific and Technical Conference Smart Energy Systems,* Article number 05070.

[8] Gorobets, V.G., V.I. Trokhaniak, I.O. Antypov, and Y.O. Bohdan. 2018. The numerical simulation of heat and mass transfer processes in tunneling air ventilation system in poultry houses. *INMATEH - Agricultural Engineering*. Vol.55, Issue 2:87-96

[9] Dudnyk, A., M. Hachkovska, N. Zaiets, T. Lendiel, and I. Yakymenko. 2019. Managing a greenhouse complex using the synergetic approach and neural networks. *Eastern-European Journal of Enterprise Technologies*. Vol. 4, Issue 2-100:72-78.

[10] Lysenko, V. P. 2014. Artificial intelligence systems: fuzzy logic, neural networks, fuzzy neural networks, genetic algorithm. NULES of Ukraine. 341 p. (in Ukrainian).

[11] Lysenko, V. and Dudnyk, A. 2016. Automation of biotechnological objects, *Proceedings of International Conference on Modern Problems of Radio Engineering, Telecommunications, and Computer Science*, pp. 44-46.

[12] Platz, S., E. Heyn, F. Hergt, B. Weigl, and M. Erhard. 2009. *Berliner und Munchener Tierarztliche Wochenschrift*. Vol. 122, Issue 7-8:235-240.

[13] Arbib M. 2003. The handbook of brain theory and neural networks. London: MIT Press. 1309 p.

[14] Sajmon Hajkin. 2006. Neyronnyye seti: polnyy kurs. [Neural networks: full course]. Moscow, Vilyams, 1104 p. (in Russian).

[15] Hawkins, Jeff. 2004. On Intelligence (1st ed.). Times Books. p. 272. ISBN 978-0805074567.

[16] Kibzun A. 2002. Teorija Veroiatnostei I Matematicheskaia Statistika. [Theory of Probability and Mathematical Statistics]. Moskva: FIZMATLIT. p. 224. (in Russian).

[17] Kruglov, V.V. and V.V. Borisov. Iskusstvennyye neyronnyye seti. Teoriya i praktika. [Artificial Neural Networks. Theory and practice]. 2001. Moscow, Goryachaya liniya - Telekom, 382 p. (in Russian).

[18] Russell S. and Norvig P. 2010. Artificial Intelligence. Upper Saddle River: Prentice-Hall.

[19] Ripley Brian D. 2007. Pattern Recognition and Neural Networks. Cambridge University Press. ISBN 978-0-521-71770-0.

[20] Luger, George & Stubblefield, William. 2004. Artificial Intelligence: Structures and Strategies for Complex Problem Solving (5th ed.), The Benjamin/Cummings Publishing Company, Inc., c. 720, ISBN 0-8053-4780-1

[21] Osowski, S. and Rudinskij, I. 2004. Nejronnye Seti Dlâ Obrabotki Informacii. Moskva: Finansy i Statistika. (in Russian).

[22] Schmidhuber, J. 2015. Deep Learning in Neural Networks: An Overview. Neural Networks. 61: 85–117.

[23] Aksenov, S.V and Novoseltsev, V.B. 2006. Organizatsiya i ispol'zovaniye neyronnykh setey (metody i tekhnologii) [Organization and use of neural networks (methods and technologies)]. Tomsk, NTL. 128 p. (in Russian).

[24] Simon H. 2008. Neural Networks. New Delhi: Prentice-Hall of India.

[25] Neural networks. StatSoft. Electronic textbook on statistics. Available at: http://www.statsoft.ru/home/textbook/modules/stneunet.html

[26] Kuntsevich, V.M., Gubarev, V.F., Kondratenko, Y.P., Lebedev, D.V., Lysenko, V.P. (Eds). 2018. Control Systems: Theory and Applications.

Series in Automation, Control and Robotics, River Publishers, Gistrup, Delft.

[27] Kondratenko, Y. P., Chikrii, A. A., Gubarev, V. F., Kacprzyk, J. (Eds). 2019. Advanced Control Techniques in Complex Engineering Systems: Theory and Applications. Dedicated to Professor Vsevolod M. Kuntsevich. Studies in Systems, Decision and Control, Vol. 203. Cham: Springer Nature Switzerland AG.

[28] J. Kacprzyk, et al. A Status Quo Biased Multistage Decision Model for Regional Agricultural Socioeconomic Planning Under Fuzzy Information. In: Kondratenko, Y.P., Chikrii, A.A., Gubarev, V.F., Kacprzyk, J. (Eds). 2019. Advanced Control Techniques in Complex Engineering Systems: Theory and Applications. Dedicated to Prof. V.M.Kuntsevich. Studies in Systems, Decision and Control, Vol. 203. Cham: Springer Nature Switzerland AG, pp. 201-226.

# 14

# Automatic Control for the Slow Pyrolysis of Organic Materials with Variable Composition

**M.V. Maksymov[1], O.I. Brunetkin[1], K.V. Beglov[1], S.V. Alyokhina[2], O.V. Butenko[1]**

[1]Shevchenko ave. 1, Odessa National Polytechnic University, Odessa, Ukraine, 65044
[2]A. Podgorny Institute of Mechanical Engineering Problems of the National Academy of Sciences of Ukraine, Kharkiv, Ukraine. V.N. Karazin Kharkiv National University, Ukraine
E-mail: prof.maksimov@gmail.com, a.i.brunetkin@gmail.com, beglov.kv@opu.ua

## Abstract

Chemical and energy equipment is calculated on the use of raw materials of a certain composition and requires fine-tuning. A slight change in the composition of the same type of substance leads to failure or even a stoppage of technological processes.

The proposed method of organic substances processing allows optimally transforming various organic substances into the gaseous state. Optimality is determined by the possibility of obtaining the maximum volume of the combustible gas mixture with maximum calorific value. Optimality is conditioned by the organization of the pyrolysis and gasification process in superadiabatic mode with the known composition of initial raw materials. The organization of such a thermal processing process allows using organic substances of high humidity. Such, in the most cases, are households and other wastes. Moreover, increasing humidity increases the share of combustible substances in the product-gas.

An imitation mathematical model was developed, which describes the dynamic properties of the pyrolysis unit.

Modeling of automatic control system has shown that at this stage, a set of single-circuit regulators is enough. Shutting off any of the regulators leads to the deterioration of regulation quality, but the system as a whole continues to function steadily.

In addition, in the process of modeling automatic control system, it was found out that the control system copes with perturbations quite well. Moreover, in case of multi-directional perturbations (reduction of product-gas consumption and decrease of calorific value or increase of product-gas consumption and increase of the calorific value of initial raw materials), the maximum calorific value of product-gas is reached.

**Keywords:** Organic material, variable composition, controlled pyrolysis, automatic control system, extreme controller, imitation model.

## 14.1 Introduction

Environmental issues play a major role in the life of modern society, affecting all aspects of its vital activity. The paradox is that measures are taken to address environmental issues, often leading to increased resource consumption and exacerbating the man-made impact on nature. Whereas, their main goal must be the effective use of resources and the reduction of man-made impact on the environment.

The development of solar and wind power industry is considered as one of the directions for solving environmental problems. Currently, in some countries, their share in the energy balance has reached a significant value (e.g. ~47% in Germany). However this is in one country and it is difficult to achieve in the group of several neighboring countries. There are circumstances that happened on a windless night or a very windy sunny day. In this case, it is necessary to compensate for underproduction at such generating capacities (at night) or to be able to sell excess electricity (during the day). Thermal, hydroelectric power plants of neighboring countries solve these tasks by starting or stopping the reserve capacity. In other words, the usage of solar and wind energy does not allow to abandon traditional generation methods in a significant part. The corresponding capacities are simply squeezed out to neighboring countries, where additional funds are required for their maintenance and, as a result, additional reserves of organic fuel. Sometimes this can lead to a paradoxical situation. For example, on February 17, 2020, because of wind turbines running at full capacity, the cost of electricity in Finland was negative: the producers paid extra so that consumers could take away the electricity produced from them. As a result, the producers suffered a loss. Thus, the lack of alternative energy in the form of poor dispatching (regulation) was manifested.

Another side of environmental problems is the task of recycling household, agricultural, and other types of organic waste. Their growing volume has led to a crisis in the possibility of storing them in different types of landfills. In Europe, as the most affected region by the problem of landfill allocation, a European program for waste management has been developed. Under this program, by 2035, EU countries will be allowed to dispose of no more than 10% of municipal waste in landfills. With the purpose to achieve these targets, waste sorting, followed by recycling (reuse) and composting, is being widely implemented. Nevertheless, currently (and in the future) for about 50% of the waste must require additional resources for its utilization or must be utilized with an additional energy sources. Currently, $\sim$100 million tons of waste per year is incinerated in the EU countries. By 2035, it is planned to bring this figure to $\sim$140 million tons. Thermal utilization of waste refers to dirty methods of waste processing but has no alternative now. One of the main reasons for the pollution from waste incineration plants is the variable and unknown composition of raw materials (garbage) to be incinerated. As a result, it is not possible to organize a controlled air supply process to optimize the combustion process.

Obtaining energy from the combustion of such significant amounts of organic waste is attractive. However, the relevance of this process can increase if it is considered with alternative sources (solar and wind turbines) in terms of joint dispatching of processes with the exclusion or reduction of fossil fuels. Implementation of such a process is complicated, as well as in case of pollution from waste incineration plants emissions, raw material composition changes, and poor controllability of the combustion process. Regarding this, it is relevant to develop a method of controlled thermal processing of organic substances with variable unknown composition.

## 14.2 Controlled Pyrolysis Model and Method

### 14.2.1 Problem Definition

One of the ways to solve the problem related to the variable composition of processed organic substances is the sorting of waste to homogenize the composition of fractions. However, even in the countries with the best-developed system of waste collection and recycling (e.g. Sweden), there is a problem of its sorting [1]. The composition of the sorted waste may vary depending on the collection region or time of year. The lack of an established sorting system for organic waste in most places is a constraint to recycling. The difficulty in recycling sorted waste, even in developed countries, is also confirmed by the independent study of The Guardian [2]. It cites a federal report in the U.S. on the amount of sorted plastic waste in that country that is sent for recycling as only 5%. The rest is sent to landfills. This situation occurred after China's refusal to import for recycling waste from the U.S. since 2017 due to poor sorting. According to The Washington Post [3], under such conditions, by 2030, the U.S. will face a shortage of processing capacity for 37 million tons of waste. The data shows that, as in previous years, the sorted waste was mostly not recycled but was taken out of the country. The situation is similar in European countries. Germany, for example, is rightly the European leader in sorting and recycling waste. The official recycling requirement for plastic waste is 36%. However, they include plastic sorted and transported to China. According to estimates of the BDE (Bundesverband der Deutschen Entsorgungs – Federal Association of German Waste, Water and Raw Materials Industries) [4], from 2017 to 2018, the volume of plastic waste imported into China fell by 95%: from 340,000 to 16,000 tons. The European Union road map for municipal waste management foresees that by 2035, no more than 10% of municipal waste will be disposed of in landfills. The rest is to be recycled or incinerated. Spiegel [5] notes that, currently, only 16% of plastic waste is recycled.

In addition to plastics, household waste also includes biodegradable components. Composting can be one way of utilizing them. Even then, the main problem is still sorting into biodegradable and non-degradable components. In [6], apart from the composting, anaerobic digestion is the most attractive solution for processing the organic part of municipal waste. This conclusion is based on the ecological purity and cost-effectiveness of this method. It should be considered that the analysis was carried out with many more parameters in mind than the two ones mentioned above: air pollution, cost, byproducts, capacity, commercial maturity, energy efficiency, and type of waste treated. It is noted, however, that the main limitations of anaerobic digestion are its low energy production efficiency and limited possibilities for waste treatment. Thus, up to 70% of the flammable gases produced are used to maintain the digestion parameters. Besides, only part of the organic waste can be biodegradable. The decomposition rate varies from one substance to another. Not all waste from the biodegradable group is economically viable in anaerobic digestion. However, the latter two parameters act as constraints that are most important in the development of alternative fuels. In addition, the anaerobic digestion process is poorly managed in terms of the number of gases produced per unit of installed equipment.

Recycling of non-degradable or poorly biodegradable parts of organic wastes at incineration plants is the dirtiest way of their utilization, not counting direct disposal at solid household waste dumps [7]. As a replacement for pyrolytic incineration, one of the most promising methods of treatment is the two-stage pyrolysis-based method with further possibility to use the obtained chemically active (combustible) substances in the managed processes. Besides, this process is more energy-efficient in comparison with anaerobic digestion. To maintain the pyrolysis process parameters, only 10% of the produced combustible substances are enough. However, this method has disadvantages. There are two types of pyrolysis: rapid and slow. Rapid pyrolysis is characterized by a large yield of the liquid fraction at some amount of coal residue and gas phase. The accumulation and storage of liquid and solid reaction products for a considerable period can be organized. This period can be used to determine the composition of the obtained products. Consequently, it is possible to organize the controlled utilization processing. The limiting factor is the absence at present of commercially grounded technology of rapid pyrolysis in the industrial scale.

The slow pyrolysis process has been in use for a long time and is more technologically mature. However, at it, most of the reaction products are in the gas phase. Therefore, they must be used immediately after formation. Besides, in most cases, the operation of the equipment for slow pyrolysis is cyclical with a change in the composition of the working substance during the cycle. This leads to a change in the composition of the reaction products. Another factor influencing the quantitative ratio of substances in the mixture of reaction products is the humidity of the initial products. Currently, the mentioned features for organization of the controlled process of slow pyrolysis products use require the fixed and known composition of initial raw materials. The slow pyrolysis process can be considered as preferable at the organic wastes processing to obtain secondary energy resources. Low- and high-temperature slow pyrolysis is distinguished. As theoretical [8] and experimental [9] studies show, with increasing process temperature, the yield of combustible gaseous phase increases. At high temperature ($\sim$1500°C) and the in presence of the necessary quantity of water vapors, the mass of carbonaceous residue and condensing phase can be minimized. This should be considered as a positive factor if the problem of immediate usage of gaseous fuel with unknown and changing composition is solved. However, the high humidity of the feedstock or the targeted introduction of water into the reaction zone requires an increase of energy consumption to maintain the required process temperature.

In many cases, pyrolysis is considered as a process of thermal decomposition of organic matter without air access. In this case, various fuels are used as an energy carrier for external heating of the reaction zone. However, more correctly, pyrolysis should be considered as a process of decomposition with lack of air [10], as a special case of the process of gasification of organic substances and, in even wider limits, as combustion with lack of air [11]. In such an approach, there is a possibility of transition from external heating to internal one due to the energy usage, emitted in the process of pyrolysis (gasification, combustion at low oxidizer excess factor). As a result, potentially in addition to fuel-saving, there can be provided uniform heating of the reaction zone in the absence of heat transfer surfaces, organization of continuous and regulated process under the condition of the controlled supply of components.

The organization of this method of pyrolysis is constrained by the uncertainty and variability in the composition of the raw materials and, in many cases, its high humidity. In the case of high moisture content, the gasification process can be organized based on the method that has been recently developed. It is based on the implementation of a process corresponding to filtration combustion in a superadiabatic mode [12]. Such a process allows us to burn organic substances of low caloric value or to carry it out in such a way that up to 95% of thermal energy of combustion is converted into the caloric value of formed products in the gas phase. Materials with low carbon content (from 10%) and humidity up to 60% can be processed. At the given composition of the recycled substance, its gasification in the superadiabatic mode allows reducing the fuel component consumption while increasing the temperature in the reaction zone. As a result, the yield of combustible gases increases with the simultaneous reduction of solid (coal) residue. The peculiarity of the combustion process in the superadiabatic mode is the oncoming movement of the raw material (fuel) and the combustion front together with its products. The combustion products ensure the drying of the raw material and its initial decomposition before entering the zone of exothermic reactions. As a result, the reaction zone does not receive remote moisture and low-temperature decomposition products, which reduces the yield of combustible gases. Another disadvantage is the impossibility to control the flow of reaction products at a given process temperature. The flow rate depends on the reactor size and the composition of the feedstock.

In the work [13], the method of determination in real-time mode of gross formula and calorific value of gaseous fuel during its combustion in a special device is offered. The values of measured gas and air consumption, as well as the temperature of combustion products, are used as input data. Such measurements are simple and allow for implementation in automatic mode. The combination of the pyrolysis technology, based on the method of gasification in the superadiabatic mode [12], with the method of determination in real time of product-gas composition [13] creates the possibility of organic substances processing of arbitrary, variable composition. Organization of controlled pyrolysis process will allow to exclude from the technological chain the stage of liquid combustible substances synthesis and their storage. The raw material base is also expanding. Unsorted household, agricultural, and other wastes of high humidity can be subjected to pyrolysis.

### 14.2.2 Purpose and Objectives of the Research

The purpose is to develop a model and method of controlled pyrolysis and gasification processes of organic substances with variable composition for the gases production with maximum calorific value. This will allow rational use of hydrocarbon substances of different origin from technological, energy, and environmental positions.

The following objectives were set to achieve the goal:

1) to develop an imitation model of pyrolysis and gasification processes allowing to obtain a product-gas composition with maximum calorific value at the minimum amount of carbon residue depending on the composition of the initial raw materials;
2) to find the area of existence of the model solutions for energy balance;
3) to develop a control method of pyrolysis of organic raw materials to provide the given flow of product-gas at the infliction of perturbations by the flow of synthesis gas and composition of initial raw materials;
4) to synthesize the system of product-gas plant control and to study its work.

### 14.2.3 Method of Problem Solving

#### 14.2.3.1 Facility Scheme Selection

In the known sources at the consideration of processes of filtration combustion in a superadiabatic mode are based on the macrokinetic model construction [12]. At the same time, it is noted [11, 12, 14] that at low velocities of components, their transformation processes close to equilibrium are realized. Calculations based on equilibrium models are constructed considering all enthalpy flows. Among other things, heat losses through the reactor walls to the external environment must be considered. Calculations of thermal flows from the central zone of the reactor to its periphery should be made considering changes in thermophysical characteristics of the charge during its thermochemical transformation. Methods of calculation of such characteristics exist [15], but they are complicated in application. Due to the small losses in comparison with heat fluxes arising in the process of considered reactions, they refuse to be considered when forming models of equilibrium processes. Moreover, losses through reactor walls can be reduced due to the application of thermal insulation or even excluded at the application of measures described in the patent [16].

The definition of the study area as "filtration combustion" in most of the known sources involves treating the occurring reactions as a process of the charge oxidation with air oxygen. Besides, these exothermic reactions are considered as a source of thermal energy to sustain the combustion process. In [14], the presence of water vapor and carbon dioxide as endothermic oxidizers is considered, but air oxygen remains the main oxidizer. In this case, air oxygen as an oxidizer is necessary as a reagent supporting exothermic reaction – a source of heat energy.

If organic substances are used as raw material, there may be another source of heat energy. For the sake of certainty, let us consider pine wood. This choice is determined by the availability of its physical and chemical properties. The possibility of using combustible substances of variable and unknown composition will be considered below. In the known sources when considering wood pyrolysis in the process of its heating, several stages with the designation of their temperature limits and direction of the corresponding energy flows are distinguished [17]:

1) to temperature $\sim$150°C with energy consumption – drying of raw materials;
2) up to temperature $\sim$300°C (in some sources $\sim$400°C) with energy consumption – initial decomposition with gaseous products release and charring beginning;
3) at further heating – rapid decomposition of raw materials with the release of a large amount of energy.

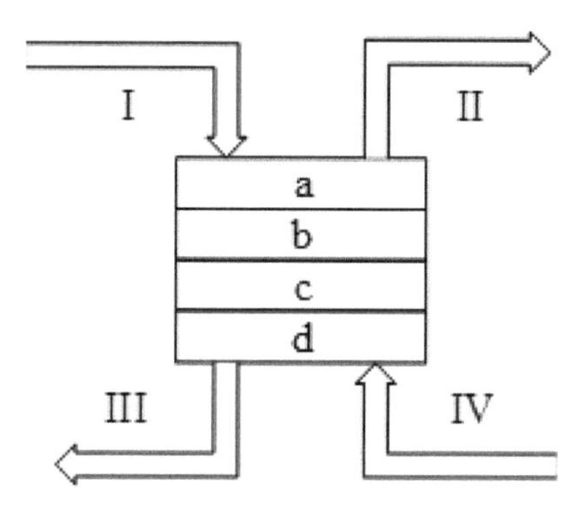


**Figure 14.1** Scheme of the process of filtration combustion of organic substances. I – charge; II – product-gas; III – solid residue; IV – oxidizer (air). a – drying zone; b – pyrolysis zone; c – oxidation zone; d – quenching zone.

Thermal energy emitted at the last stage can be used to heat the charge in a superadiabatic mode.

et's Let us consider the scheme of the filtration combustion process of organic substances in superadiabatic mode (Figure 14.1), corresponding [12]. Its distinctive feature is the oncoming motion of the charge and oxidizer. For all the advantages of such a process organization, there are some disadvantages also. Source material (charge) is dried in the upper part of the reactor. Below is the direction of raw material movement as it warms up pyrolytic processes with the extraction of part of reaction water beginning. Thus, the most of the water is removed by product-gas flow before the charge enters the oxidation zone. This increases the water content of the product-gas and thus reduces its quality. It is necessary to dehumidify the resulting gas mixture with the organization of its cooling and loss of a significant part of energy. In the oxidation zone at temperatures, there is a reaction of coke residue with oxygen, water vapor, and carbon dioxide with the formation of hydrogen and carbon oxide [12]. In the quenching zone, cooling of the solid residue by the supplied oxidizer with simultaneous heating of the latter takes place. According to such a scheme, water vapors together with oxygen and carbon dioxide act as an oxidizer, which acts in the direction of increasing the energy value and volume of product-gas with simultaneous reduction of carbon residue. However, the input of dehydrated raw materials into the oxidation zone reduces the amount of formed hydrogen. This is one of the disadvantages of the considered gasification scheme. In some cases, to increase the share of $CO+H_2$ components (synthesis gas) in the

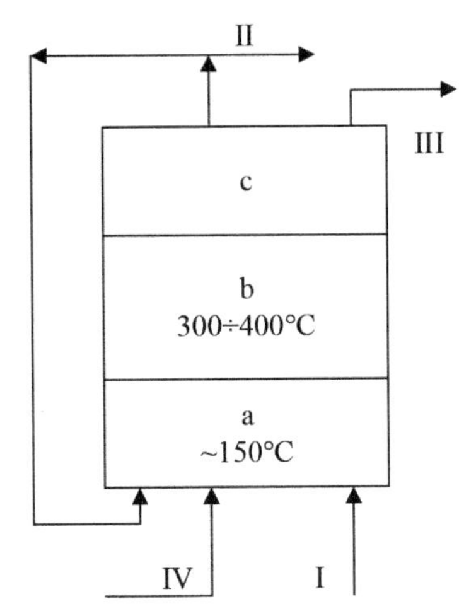


**Figure 14.2** Suggested scheme of the process of filtration combustion of organic substances. I – charge; II – product-gas; III – solid residue; IV – oxidizer (air). a – drying zone; b – pyrolysis zone; c – oxidation zone.

product-gas composition, water vapors are added to the blowing (oxidizer), which is accompanied by an increase in energy consumption for the process. Considering the described features of the process, a device operating on this principle can be positioned as a reactor for burning various, including low-energy raw materials with the associated formation of some combustible gases.

To increase the calorific value and volume of combustible gases generated during pyrolysis (gasification), a different scheme of charge and gas flows is proposed. According to the process scheme shown in Figure 14.1, thermal energy is released in the oxidation zone "c." In this scheme, the emitted energy is used for drying and is removed from the reaction zone. The idea of the proposed new process scheme (Figure 14.2) consists of drying and heating of the initial raw materials by the reaction products also after the oxidation zone but in the satellite stream. For this purpose, a part of reaction products (product-gas) at their high temperature is supplied to the chamber input together with the initial raw materials.

Then they move in the same direction. As a result, the drying and initial decomposition products enter the oxidation zone. Together with them, the energy used to form them is introduced. If there is enough energy for drying and heating of the initial raw material to the final temperature of the pre-pyrolysis zone (300–400°C) in the part of the product-gas which is fed to the reactor inlet along the recirculation line, the zone of energy release and self-decomposition of wood will start. It coincides with the zone of oxidizing processes, which take place due to oxygen from $CO_2$ and $H_2O$ and their reduction to CO and H2. In case of the energy lack and if it is necessary to increase the temperature of product-gas, the necessary amount of oxidizer (air) can be supplied at the reactor outlet.

#### 14.2.3.2 Control Object Model

The process of thermochemical decomposition of organic matter without using air organized according to the proposed scheme (Figure 14.2) can be attributed to pyrolysis. The process, organized according to the scheme of Figure 14.1, refers to combustion. As a rule, different models describe these processes. However, according to the scheme of Figure 14.2, besides pyrolysis at oxidizer feeding burning can be also organized. And these processes can smoothly pass from one to another. For this reason, it is necessary to use the uniform model, allowing to describe pyrolysis and combustion processes with identical reliability. The model suggested in [8] is used as its basis. The difference is the possibility to account for liquid water in the composition of the source substance (wood).

Let us consider the processes taking place under the assumption that the reaction products are in gaseous or condensed (carbon residue) state. This is true if the processes are in equilibrium. The law of substance conservation allows determining the list of elements and their quantity in reaction products based on the gross formula of the initial substance. It may include both water (wet wood) and air in case of oxidizer usage. Based on a limited number of basic chemical elements that can be included in the composition of the initial substance (C, H, O, N), a list of basic substances that are included in the reaction products can be determined. This list is identified based on preliminary calculations, known experimental data and includes

$$H_2O,\ CO_2,\ CO,\ CH_4,\ H_2,\ N_2,\ C_U. \tag{14.1}$$

Here, $C_U$ is the carbon residue. This condition is also fulfilled when feeding reaction products to the reactor input (Figure 14.2). The temperature in the equilibrium zone of reactions is determined based on the energy saving law. It is expressed in the equality of enthalpy in initial products and reaction products.

Using the accepted assumptions, the generalized model of processes is based on the method proposed in [18]. In its framework, we will consider the types of equations included in the model and their features.

*The law of mass action.* For each of the substances (14.1) included in the reaction products, let us write down the equation of thermochemical equilibrium. In the example of $CO_2$, it looks like

$$K_{CO_2}(T) = \frac{P_C \cdot P_O^2}{P_{CO_2}}. \tag{14.2}$$

Here $K_{CO_2}(T)$ is a constant of chemical equilibrium at temperature $T$ for equilibrium reaction of $CO_2$ formation from elements C and O (tabular value); $P_{CO_2}$, $P_C$, and $P_O$ are partial pressures of corresponding components in the mixture of combustion products. The use of partial pressures instead of concentrations is lawful in case of consideration of a certain number of moles $M_T$ of the source substance instead of one mole. The value of $M_T$ is an additional unknown value and is to be determined.

One of the features of the model when using partial pressures instead of concentrations is to record the corresponding equation for the carbon residue. More precisely, when solving a system of equations using it, the equation has the form:

$$K_U(T) = \frac{P_C}{P_U}. \tag{14.3}$$

In this equation, the $P_U$ value at a given temperature, as opposed to other partial pressure values, remains constant during the solution process.

*The substance conservation law.* The equations are written down for each of the chemical elements that determine the composition of the initial substance in the form of a balance of its quantity in the initial substance and reaction products. The equations are recorded considering the value of $M_T$ of the initial substance. As an example, for element *C*, it has the form:

$$M_T \cdot b \cdot C = P_{CO} + P_{CO_2} + P_{CH_4} + P_U + P_C. \tag{14.4}$$

The left side of the equation shows the number of carbon atoms in the source material. In this case, *b* is the number of atoms in one molecule of the source substance.

*Dalton's law.* The appearance of a variable $M_T$ to close the system requires the introduction of an additional equation. It is a sum of partial pressures of reaction products equal to the pressure in the reactor

$$P_\Sigma = \sum_i P_i. \tag{14.5}$$

Here, $P_\Sigma$ is the gas phase pressure in the reactor and $P_i$ is partial pressures of reaction product components. In this case, the peculiarity related to the presence of carbon residue is also evident. It is assumed that its volume is small in comparison with the volume of the gas phase. Therefore, the $P_U$ value is not considered in Equation (14.5).

In the stationary mode, the mixture of reaction products has the same gross formula as the initial substance. Using reaction products as a heating agent blown through the initial substance does not change the composition of the resulting mixture. The feeding of the feedstock only increases its quantity. This determines the peculiarity of the calculation.

#### 14.2.3.3 Analysis of the Control Object Model to Solve the Control Task

Chemical and thermal processes can be characterized by limit (equilibrium) parameters, within which they can be implemented. These parameters play a role in a kind of the ideal process efficiency. Such a role is played by the following calculation results. Within the process of slow pyrolysis, especially high-temperature pyrolysis, the composition of reaction products is determined not only by the chemical composition of the initial organic substance but also by its moisture. In such processes, the amount of chemically unbound water in the initial products affects the composition of the reaction products. The usage of a single model for pyrolysis and gasification modes (when air is added as an oxidizer) requires a single approach to describing the composition of the raw materials in all cases under consideration. Based on these features, the feedstock is considered as a homogeneous substance, the gross formula of which depends on the amount of water in its composition and air consumption.

To demonstrate the performance of the suggested model, absolutely dry pine wood with the gross formula $CH_{1.44}O_{0.64}N_{0.009}$ was adopted as the initial organic matter. Water is considered as a molar fraction of $\varphi$ in the wood composition. For example, for $\varphi$=0.1 is taken 0.9 wood moles and 0.1 water moles. In this case, the resulting one mole of the source substance has a gross formula of $C_{0.885}H_{1,504}O_{0,681}N_{0,008}$. The air addition affects the change in the gross formula of the raw material in a similar way and is accounted for as a mole fraction $\alpha$ of the wood composition with already added water. To simplify the calculations, air composition is taken as a mixture of oxygen (21%) and nitrogen (79%) with the gross formula $O_{0.42}N_{1.58}$. For example, adding $\alpha$ = 0.1 air to $C_{0.885}H_{1,504}O_{0,681}N_{0,008}$ (wood + $\varphi$ = 0.1 water) will make the gross formula of the raw material look like $C_{0.592}H_{1.006}O_{0.595}N_{0.528}$. In the calculation, the temperature of the equilibrium mixture of reaction products is determined, at which its enthalpy is equal to the enthalpy of the initial substance.

In [12] was noted that in the oxidation zone (Figure 14.1), the temperature is 1000–1200°C. Therefore, in the performed calculations, the temperature in the equilibrium reaction zone was considered up to the value of 1500 K. Thus, the applied model does not impose such restrictions. The results of calculations of equilibrium temperatures at various combinations of the degree of moisture of the initial raw material ($\varphi$) and the amount of supplied air ($\alpha$) are given in Table 14.1. The received values show that at the usage of the offered scheme (Figure 14.2), temperatures in an oxidation zone can reach the values corresponding to values in work [12] without air supply or at its small quantity. So, when $\alpha = 0$ (without air) and $\varphi = 0.4$, we got $T = 1380$ K (~1100°C), and for $\alpha = 0.1$ and $\varphi = 0.4$, $T = 1420$ K (~1150°C).

The feature of the proposed scheme (Figure 14.2) is the need for a significant amount of energy for the preliminary preparation of raw materials. For this purpose, it is suggested to use the energy of a part of reaction products through regenerative heating of initial raw materials to the temperature of spontaneous reaction beginning. For wood, this temperature is ~400°C. For the values $\alpha$ and $\varphi$, given in Table 14.1, calculations have been made to determine the available energy when cooling reaction products up to temperature of ~400°C (~670 K) and necessary to heat the raw materials to the same temperature.

The energy required for heating wood, air, and water to 100°C, its evaporation and heating of the formed steam was calculated. The results of the calculation are given in Table 14.2 as a value equal to the ratio of the required energy to the available one. These values show what part of the reaction products is sufficient to maintain it. One more characteristic of the described pyrolysis process can be the value of carbon residue. The

**Table 14.1** Equilibrium temperatures at various combinations of the degree of moisture of the initial raw material ($\varphi$) and the amount of supplied air ($\alpha$).

| φ | α | | | |
|---|---|---|---|---|
| | 0 | 0,1 | 0,2 | 0,3 |
| 0 | 1020 | 1095 | 1170 | 1310 |
| 0,1 | 1090 | 1140 | 1260 | 1500 |
| 0,2 | 1160 | 1340 | 1415 | |
| 0,3 | 1240 | 1380 | | |
| 0,4 | 1380 | 1420 | | |

**Table 14.2** Part of the products is sufficient to maintain the pyrolysis reaction.

| $\varphi$ | $\alpha$ | | | |
|---|---|---|---|---|
| | 0 | 0.1 | 0.2 | 0.3 |
| 0 | 0.67 | 0.61 | 0.55 | 0.44 |
| 0.1 | 0.82 | 0.73 | 0.58 | 0.41 |
| 0.2 | 0.88 | 0.61 | 0.53 | |
| 0.3 | 0.89 | 0.65 | | |
| 0.4 | 0.77 | 0.68 | | |

**Table 14.3** Carbon residue amount for different values $\varphi$ and $\alpha$.

| $\varphi$ | $\alpha$ | | | |
|---|---|---|---|---|
| | 0 | 0.1 | 0.2 | 0.3 |
| 0 | 0.59 | 0.28 | 0.11 | 0.02 |
| 0.1 | 0.41 | 0.17 | 0.04 | 0.002 |
| 0.2 | 0.26 | 0.04 | 0.005 | |
| 0.3 | 0.13 | 0.04 | | |
| 0.4 | 0.05 | 0.005 | | |

usage of various oxidizing agents for its decreasing significantly affects the energy value of reaction products. Air oxygen increases the temperature in the reaction zone and shifts reaction to the equilibrium zone. However, at the same time, the degree of the reaction products oxidation increases and their energy value decreases. The usage of the water vapors increases the content of synthesis gas components in reaction products and, accordingly, increases the reaction and energy value of reaction products. However, such a reaction requires large energy costs, as evidenced by the data in Table 14.2. At the same time, these data show the possibility of such a reaction without using oxygen in the air or with a small amount of it. Table 14.3 shows the results of calculation of the carbon residue amount for different values $\varphi$ and $\alpha$. The values are given in moles of carbon residue per one mole of the source substance. The given data show the possibility of complete oxidation of the

**Table 14.4** Carbon residue amount for different values $\varphi$ and $\alpha$.

| $\varphi$ | $\alpha$ | | | |
|---|---|---|---|---|
| | 0 | 0.1 | 0.2 | 0.3 |
| 0 | $C_1H_{1,44}O_{0,64}N_{0,009}$ | $C_{0,668}H_{0,962}O_{0,567}N_{0,531}$ | $C_{0,501}H_{0,722}O_{0,530}N_{0,792}$ | $C_{0,401}H_{0,578}O_{0,508}N_{0,950}$ |
| 0.1 | $C_{0,885}H_{1,504}O_{0,681}N_{0,008}$ | $C_{0,592}H_{1,006}O_{0,595}N_{0,528}$ | $C_{0,445}H_{0,756}O_{0,551}N_{0,790}$ | $C_{0,356}H_{0,605}O_{0,525}N_{0,948}$ |
| 0.2 | $C_{0,794}H_{1,556}O_{0,714}N_{0,007}$ | $C_{0,532}H_{1,042}O_{0,617}N_{0,527}$ | $C_{0,400}H_{0,783}O_{0,568}N_{0,788}$ | |
| 0.3 | $C_{0,714}H_{1,6}O_{0,743}N_{0,0064}$ | $C_{0,479}H_{1,073}O_{0,637}N_{0,525}$ | | |
| 0.4 | $C_{0,654}H_{1,634}O_{0,765}N_{0,0059}$ | $C_{0,439}H_{1,097}O_{0,652}N_{0,524}$ | | |

wet carbon residue of the source substance without oxygen in the air or at its minimum quantity.

The calculation results in Tables 14.1–14.3 were obtained for substances with gross formulas given in Table 14.4.

#### 14.2.3.4 Results of Pyrolysis Product Output Modeling

The generalized values given in Tables 14.1–14.3 are obtained based on detailed calculations with the determination of pyrolysis or gasification product composition depending on temperature. In the considered model case, unlike the real situation, the composition of the initial substance (pine wood) is known. Its moisture content is specified also. This makes it possible to identify the main features of the process. The following section shows how the suggested calculation methods can be applied to the case of the unknown composition of pyrolyzed organic matter.

As an example of detailed calculation results, let us consider the process of pyrolysis of pine wood (without air access) at two different humidity levels. Figure 14.3 shows a graph of the change in the composition of pyrolysis products ($\alpha = 0$) as a function of temperature for the case $\varphi = 0$ – absolutely dry wood does not exist. However, this case is interesting from the theoretical point of view of its extremity. Under equilibrium of the considered reaction, the enthalpy of initial raw materials (in the considered case, the known value) is equal to enthalpy of pyrolysis products. This allows determining the equilibrium temperature of the reaction. The results of the calculation of changes in the enthalpy (kJ/mole) of the reaction products are shown in Figure 14.4.

Similar results of pyrolysis ($\alpha = 0$) for wet wood for $\varphi = 0.4$ are shown in Figures 14.5 and 14.6.

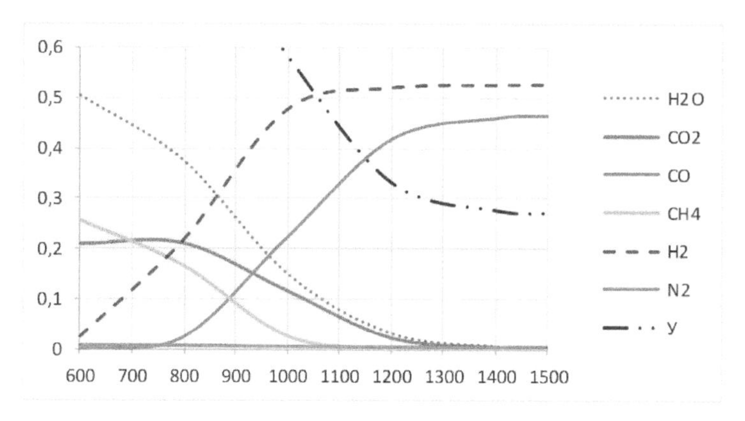


**Figure 14.3** Mole fractions of pyrolysis products depending on temperature (K) when $\varphi = 0$ and $\alpha = 0$.

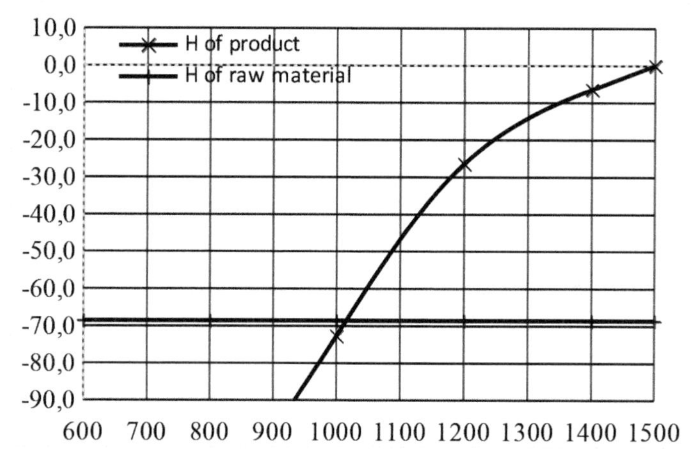


**Figure 14.4** Enthalpy of pyrolysis products (kJ/mole) depending on temperature (K) for $\varphi = 0$ and $\alpha = 0$.

Comparison of charts in Figures 14.4 and 14.6 shows that the equilibrium pyrolysis process for wet wood is carried out at higher temperatures (~1350 K) than for absolutely dry wood (~1000 K). The possibility of wet wood pyrolysis at such temperature without external heat sources is provided at the expense of the organization of the superadiabatic pyrolysis process according to the offered scheme (Figure 14.2). The possibility of high-temperature thermal reactions of damp organic substances without access to

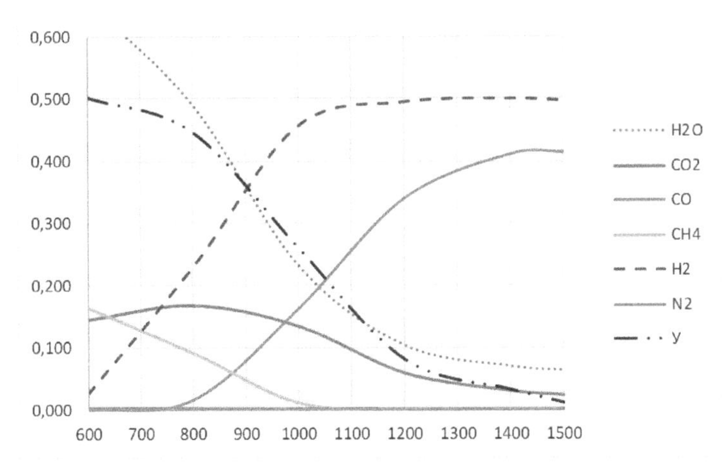


**Figure 14.5** Mole fractions of pyrolysis products depending on temperature (K) when $\varphi$ = 0.4 and $\alpha$ = 0.

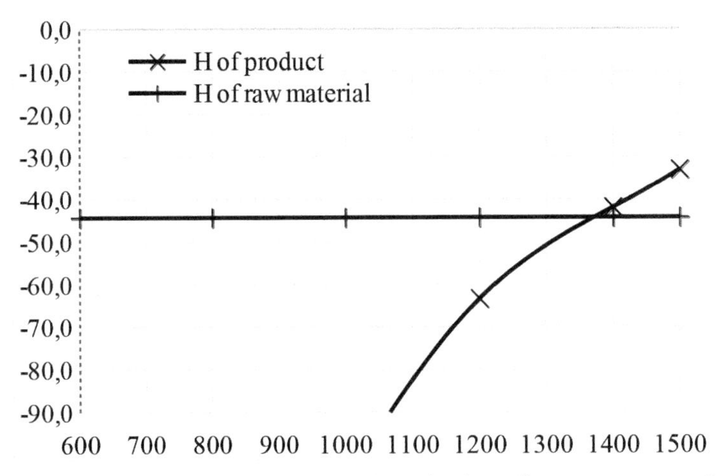


**Figure 14.6** Enthalpy of pyrolysis products (kJ/mole) depending on temperature (K) $\varphi$ = 0.4 and $\alpha$ = 0.

air is confirmed by the self-ignition of a bulk heap of coal, grain, wood waste, etc. It should be noted that the opposite situation is observed at the usual wood combustion: the combustion temperature of dry wood is higher than that of wet wood.

Comparison of calculation results displayed in Figures 14.3 and 14.5 shows approximately the same value of molar fractions of combustible gases:

$H_2$ and CO at the same temperatures. This is even though the carbon fraction in dry wood (Table 14.4) is higher than that in wet wood. In the pyrolysis of dry wood, the fraction of carbon residue in reaction products is high. Decrease of its share at the processing of damp raw materials occurs at the expense of oxidation of available water and oxygen from carbon dioxide by steams. Such a process causes the preservation of $H_2$ and CO share while reducing carbon share in the raw materials. It is especially noticeable when comparing the compositions of reaction products for corresponding temperatures of equilibrium processes in the first and second cases.

At the transition from pyrolysis to gasification also in the mode of superadiabatic filtration combustion even with a small amount of used air, the share of combustible gases $H_2$ and CO in reaction products sharply decreases. All other things being equal, the reaction temperature increases and the carbon residue gasification is almost complete. This reflects the data shown in Figures 14.7 and 14.8 for wood with humidity $\varphi = 0$ at air overflow coefficient $\alpha = 0.1$.

The results show that full gasification of the carbon residue during the thermal processing of even wet organic substances may require some air supply. To obtain the maximum yield of combustible gases, the amount of air must be kept to minimum. If the raw material composition is variable, the air supply process must be controlled. Moreover, this is possible only with a known composition of raw materials at any time moment.

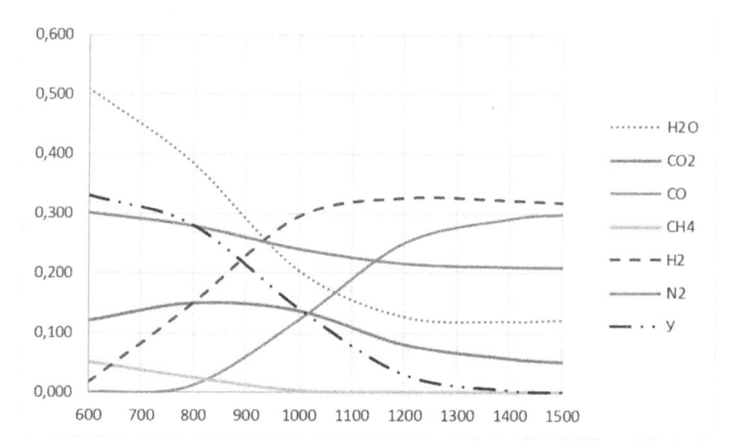


**Figure 14.7** Mole fractions of pyrolysis products depending on temperature (K) for $\varphi = 0.4$ and $\alpha = 0.1$.

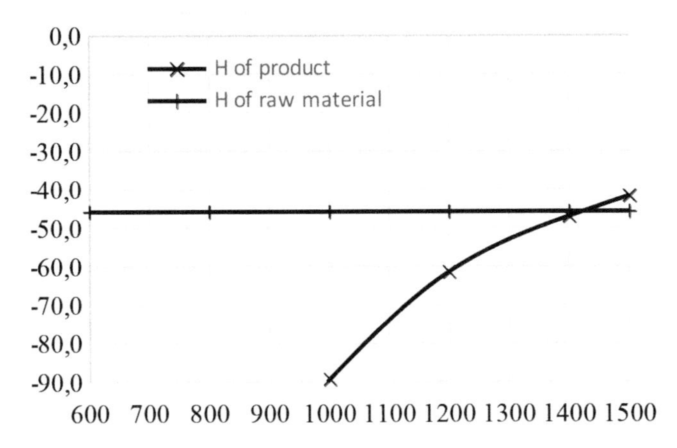


**Figure 14.8** Enthalpy of pyrolysis products (kJ/mole) depending on temperature (K) for $\varphi = 0.4$ and $\alpha = 0.1$.

## 14.3 Synthesis of the Plant Control System to Produce Product-Gas

### 14.3.1 The Control Method of Pyrolysis Technology in the Plant

As a rule, under real conditions, the composition of raw materials is unknown. Moreover, the composition of the substances being processed may change over time. For the possibility to process organic substances of any variable composition, a continuous pyrolysis unit can be used, the principal scheme of which is shown in Figure 14.9.

Construction of the main part of the plant – reactor III – corresponds to the scheme in Figure 14.2. The principle of operation and values of some parameters for stationary operation correspond to those described above. The feedstock is fed through collector I with a flow rate of $G_{ch}$ to the reactor input. Through collector II at the outlet of the reactor with the flow rate $Q_2$, hot product-gas is supplied to the general system, where it is divided into two flows with the flow rate: $Q_4$ – to the consumer; $Q_3$ – through the recirculation line to the reactor inlet V for drying and heating the raw materials. Through collector IV, the required amount of air with the flow rate $Q_1$ is supplied.

This facility allows the following:

1) at the stationary stage of work with the known composition and enthalpy of the processed substance completely gasify its organic component due to the calculation and regulation of the minimum required amount of air

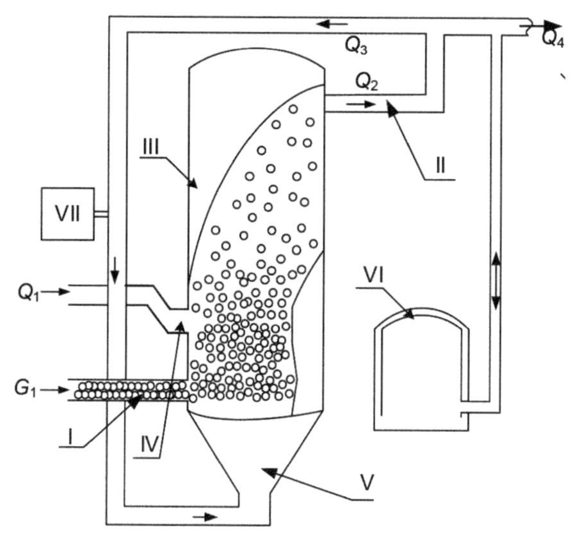


**Figure 14.9** Circuit diagram of the pyrolysis installation.

supply. In this case, the supply of a part (Table 14.2) of hot product-gas (II) from the output of the pyrolysis unit to its inlet (V) is calculated and regulated to ensure the superadiabatic mode of the process;

2) during operation, monitor the composition of the source substance with the help of a special device (VII) and switch to the mode of its determination when changing it;
3) in the transition to monitor with the help of special device (VII) the current composition of the initial substance and its enthalpy.

In the case of stationary operation and complete gasification of the organic component of the feedstock, the gross formula of the product-gas corresponds to the gross formula of the feedstock, for example, in the form given in Table 14.4. The product-gas composition and, consequently, its constancy are controlled through device VII. Its work is based on the method [13] for the determination of the gross formula of a mixture of combustible gases in the process of their combustion in a special device [16]. For this purpose, it is enough to measure the flow rates of the gas mixture, air, and temperature of combustion products in this device at different mixture flow. A gas mixture may also include non-combustible gases.

When fixing the change in product-gas composition, the unit is switched from gasification mode to combustion mode. This is expressed in the increase of air supply through collector IV to guarantee the transformation of raw materials with the changed unknown composition into gaseous products. Further in device VII (Figure 14.9), their gross formula and, accordingly, a new composition of the initial raw materials are determined. Further based on these data, the unit is transferred to the pyrolysis (gasification) mode by regulation of air supply following the new composition of the charge.

### 14.3.2 A Simulation Model of the Pyrolysis Plant Control System

Based on the above-described control method of the pyrolysis unit, a simulation mathematical model was developed, which describes the dynamic properties of the unit through the following channels:

1) airflow rate $Q_1$ – product-gas flow rate $Q_2$;
2) input $G_{ch}$ – product-gas flow rate $Q_2$;
3) recirculation gas flow $Q_3$ – product-gas flow $Q_2$;
4) airflow rate $Q_1$ – pyrolysis temperature $t$;
5) consumption of initial raw material $G_{ch}$ – pyrolysis temperature $t$;
6) recirculation gas consumption $Q_3$ – pyrolysis temperature $t$;
7) airflow rate $Q_1$ – product-gas composition $Q_n{}^w$;
8) a flow rate of initial raw material $G_{ch}$ – composition of product-gas $Q_n{}^w$.

The parameter diagram of the controlled device is shown in Figure 14.10.

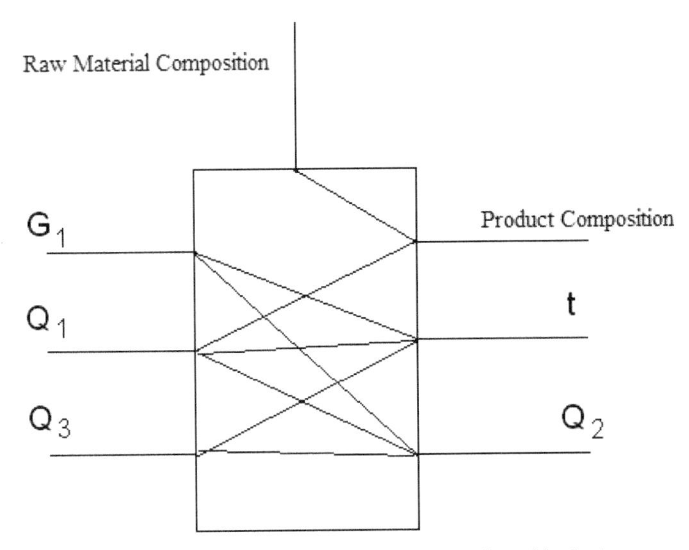


**Figure 14.10** Parametric diagram of the adjustable device.

The design of the pyrolysis plant is identical to that of a fluidized bed furnace. Fluidized bed furnaces have been used in power engineering since the 1960s; so their dynamic properties are well known. The works [20–24] present mathematical models of fluidized bed from fluidized bed temperature control and gas–air mixture volume at the layer outlet.

Traditionally, the dynamic properties of the control object are obtained by considering the equations of energy and mass conservation.

It is possible to write down for a pyrolysis unit by analogy with a fluidized bed furnace:

$$\frac{dT_L}{dt} S_L \rho_L c_L H_L = G_{\text{ch}} Q^w_{n\,r} + Q_1 h_1 + Q_3 Q^w_n - Q_2 Q^w_n - I_{\text{ash}} - I_R, \quad (14.6)$$

where $T_L$ – layer temperature and gas temperature;
$S_L$ – pyrolysis mirror area;
$\rho_L, \rho_g$ – bulk layer density and gas density;
$c_L, c_g$ – the material thermal capacity of the layer and gas thermal capacity;
$H_L$ – layer height;
$G_{ch} G_{\text{ch}}$ – input material consumption (charge);
$Q_1, Q_2, Q_3$ – air, product-gas, and recirculation gas consumption;
$Q^w_{n\,r}$ – raw material calorific value;
$Q^w_n$ – synthesis of gas and recirculation gas calorific value;
$I_{\text{ash}}, I_R$ – thermal losses with disposable ash and radiation.

After recording all the elements of Equation (14.6) in the form of increments, we will get the following equation:

$$\begin{aligned}\frac{dT_L}{dt} S_L \rho_L c_L H_L =& \Delta G_\phi \bar{Q}^{\eth}_{\dot{i}} + \bar{G}_\phi \Delta Q^{\eth}_{\dot{i}} + \Delta Q_1 \bar{c}_1 \bar{T}_1 + \Delta Q_3 \bar{c}_L \bar{T}_L \\ &+ \bar{Q}_3 \bar{c}_L \Delta T_L - \Delta Q_2 \bar{c}_L \bar{T} - \bar{Q}_2 \bar{c}_L \Delta T_L\end{aligned}$$

or

$$\begin{aligned}\frac{S_L \rho_L H_L}{\bar{Q}_2 - \bar{Q}_3} \frac{dT_L}{dt} + \Delta T_L =& \frac{\bar{c}_{\text{ch}} \left(T_L - \bar{T}_{\text{ch}}\right)}{\bar{c}_L \left(\bar{Q}_2 - \bar{Q}_3\right)} \Delta G_{\text{ch}} + \frac{\bar{G}_1}{\bar{c}_L \left(\bar{Q}_2 - \bar{Q}_3\right)} \Delta Q^w_n \\ &+ \frac{\bar{c}_1 \left(T_L - \bar{T}_1\right)}{\bar{c}_L \left(\bar{Q}_2 - \bar{Q}_3\right)} \Delta Q_1 + + \frac{\bar{T}_L}{\left(\bar{Q}_2 - \bar{Q}_3\right)} \Delta Q_3. \\ &- \frac{\bar{T}_L}{\left(\bar{Q}_2 - \bar{Q}_3\right)} \Delta Q_2\end{aligned} \quad (14.7)$$

It was stated above that the pyrolysis process should be carried out in such a way that the enthalpy of the raw materials and the gas synthesis are the

**Table 14.5** Table of transfer function presenting relationships between the input and output parameters.

| | Airflow rate $Q_1$, mole/s | input raw material flow $G_1$, mole/s | Recirculation gas flow $Q_3$, mole/s | Raw material calorific value $Q_{n\ \mathrm{ch}}^{w}$ |
|---|---|---|---|---|
| Product-gas flow $Q_2$, mole/s | $W(s) = \frac{1}{T_1 s+1}$ | $W(s) = \frac{1}{T_1 s+1}$ | $W(s) = \frac{1.83}{T_1 s+1}$ – | |
| Pyrolysis temperature $T_L$, °C | $W(s) = \frac{600e^{-\tau s}}{T_2 s+1}$ | $W(s) = \frac{300e^{-\tau s}}{T_2 s+1}$ | $W(s) = \frac{300e^{-\tau s}}{T_2 s+1}$ | $W(s) = \frac{200e^{-\tau s}}{T_2 s+1}$ |
| Product-gas calorific value $Q_n^w$ | $W(s) = \frac{k\left(\frac{Q_2}{G_1}\right)e^{-\tau s}}{T_1 s+1}$ | $W(s) = \frac{k\left(\frac{Q_2}{G_1}\right)e^{-\tau s}}{T_1 s+1}$ – | $W(s) = \frac{1 \cdot e^{-\tau s}}{T_1 s+1}$ | |

same. This is achieved by ensuring a superadiabatic regime for the pyrolysis process. Since the recirculation gas is part of the product-gas fed back into the unit, their calorific value is the same only concerning the pyrolysis temperature. Therefore, through the channels $\ll G_{ch} - T_L \gg$, $\ll Q_1 - T_L \gg$, $\ll Q_2 - T_L \gg$, and $\ll Q_3 - T_L \gg$, the regulated device is a first-order inertial link with a delay. The length of the inlet and outlet pipes determines the transport delay.

From the consideration of the law of mass conservation inside the pyrolysis unit

$$\frac{dm}{dt} = G_{\mathrm{ch}} + Q_1 + Q_3 - Q_2 \tag{14.8}$$

and the equations of chemical kinetics considered earlier, we can conclude that through the channels $\ll Q_1 - Q_2 \gg$, $\ll G_1 - Q_2 \gg$, and $\ll Q_3 - Q_2 \gg$, the controlled unit is an inertial link. The inertia of the object is determined by the speed of propagation of the pyrolytic decomposition front.

The relationships between the input and output parameters of the control device can be presented as a table of transfer functions. Using the model, which was described in 2.3.2, 2.3.3, and 2.3.4, the numerical values of the transfer coefficients of the transfer functions were determined.

Since the gross product-gas formula is the same as the feedstock and depends on the amount of air supplied, the product-gas composition is determined by the raw material composition and air consumption. As a parameter that characterizes the composition of product-gas, its calorific value has been chosen. The adjustable device through the channels "airflow – composition of product – gas" and "raw material flow – composition of product – gas" has non-linear properties. Based on the results of research [19], the quadratic dependence of the calorific value of product-gas on the ratio of consumption of raw materials and the air was accepted

$$Q_n^w = Q_{n0}^w \left(1 - k\left(Q_1 - G_r V_0\right)^2\right), \tag{14.9}$$

where $Q_n^w$ – a product-gas calorific value;

$Q_{n0}^w Q_{n0}^w -$ maximum possible calorific value, depending on the composition of the raw materials; at this stage, $Q_{n0}^w = 1$ is accepted;

$V_0 V_0 -$ air volume required for pyrolysis one mole of raw material. It depends on the raw material composition.

$k$ – the proportionality factor, determined by calculation or experiment for specific raw material composition. At this stage, $k = 0.2$ is accepted.

During the development of the mathematical model of pyrolysis process, the dependence of energy released during the raw materials oxidation by air on the consumption of raw materials and their calorific value was described by the following dependence:

$$Q = Q_{nch}^w \cdot \Delta G_1 + G_1 \cdot \Delta Q_{nch}^w + \Delta Q_n^w \cdot \Delta G_1. \tag{14.10}$$

Since time constants and time lag values are determined by the geometric dimensions of the unit and should be specified individually for each device, at this stage of the study, the next experimental data obtained in [25, 26, 27] are accepted: $T_1 = 400$ s; $T_2 = 500$ s; $\tau = 10$ s.

The structural diagram of the control object was shown in [28].

While studying the model of the control object, it was found out that through the channel $\ll Q_3 - Q_2 \gg$, such modes of operation are possible, when the installation goes from a stable state to an unstable one due to the positive feedback. Properties of the control object through the specified channel are determined by the ratio between the expenditure of raw materials $G_{ch}$ and the flow of recirculation gas $Q_3$.

In the production of product-gas after the pyrolysis unit "wet gasholder" is installed, it is marked as VI in Figure 14.9. It allows having a buffer tank for smoothing the flow of product-gas at sharp changes in its consumption. In addition, the level change in the gasometer can serve as another regulated parameter that shows the imbalance between the consumed and generated amount of gas product. The advantage of such a scheme is that there is no need to measure the flow of gas of variable composition.

Thus, in the object, the regulated parameters are:

1) liquid level in gasholder;
2) product-gas consumption;
3) pyrolysis temperature (product-gas);
4) composition of gas product.

The controlling effects are:

1) input raw material consumption;
2) airflow rate;
3) product-gas recirculation consumption.

External perturbations to the facility are:

1) specified product-gas consumption;
2) input raw material composition.

The main methods and control schemes for pyrolysis plants are known. The studied pyrolysis unit can be considered as part of those described in [29, 30, 31]. In contrast to the considered control systems, it is proposed to introduce a search circuit for the optimal air flow rate to maintain the pyrolysis temperature when the composition of the raw material changes.

The scheme of automatic control systems (ACS) of the pyrolysis plant was simulated in Matlab (license 1–4 AE K761327 ВД, № 308918). One imitation model is shown in Figure 14.11.

For synthesizing the extreme control system, the material was used from [32, 33].

The following designations are used in the figure:

1) *Subsystem* – the control object described above and whose structural diagram is shown in Figure 14.5;
2) *L_Control* –gasholder level controller;
3) *T2_Control* – pyrolysis temperature controller;
4) *Q2_Control* – product-gas flow controller;
5) *U1, Y, U* – chart output units blocks (*Scope*);
6) *QG, Q4* – system perturbation blocks;
7) *Lz, Tz, Q2z* – control assignors.

The lower group of seven blocks is an extreme regulation subsystem. This subsystem finds the maximum target function $Q_n^w = f(Q_1)$ by correcting the setpoint of the charge temperature controller *T*.

The algorithm of the ACS is as follows. The required amount of product-gas is determined by the level in the gas-holder (*L*) and is controlled by the

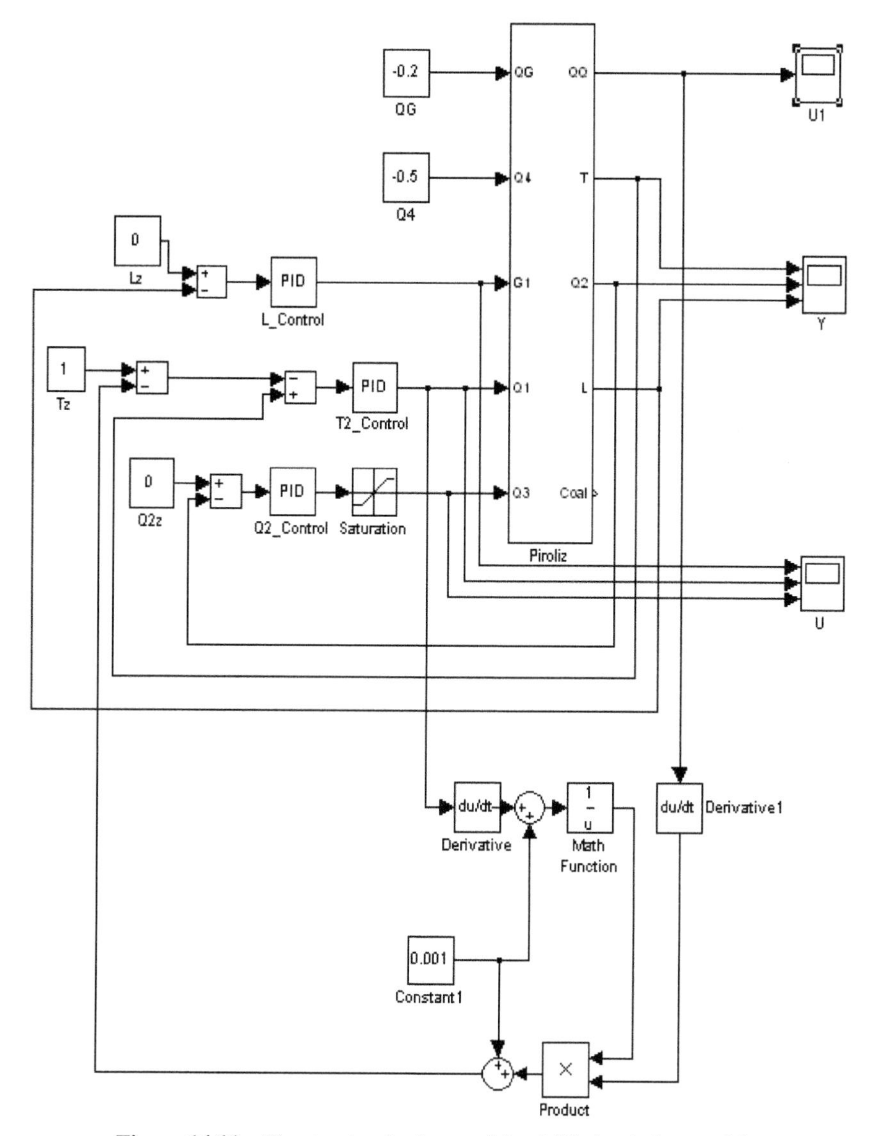


**Figure 14.11** The structural scheme of the ACS simulation model.

flow of raw materials ($G_1$). The control action is formed by the level controller *L_Control*. The product-gas flow rate at the plant outlet ($Q_2$) is stabilized by the recycle gas flow rate ($Q_3$). The control action is generated by the controller *Q2_Control*. The temperature of the pyrolysis process is stabilized by changing the air flow ($Q_1$). In these controllers, it is possible to use PI or P algorithm.

The control action is formed by the level controller *T2_Control*. The optimal amount of air depends on the composition of the raw material. To determine it, an extreme controller is used which maintains the maximum enthalpy of the product-gas (*QQ*). The specified regulator works by the method of direct determination of the derivative. The extreme derivative regulator, which changes its sign as it passes through the extremum, determines the direction of movement to it. When the derivative is equal to 0, the extremum is reached. Finding the derivative is carried out by dividing the block product derivative by the derivative (blocks derivative, derivative1).

### 14.3.3 Modeling Results of the Control Process by Pyrolysis Installation

Control processes at the application of perturbations by the flow of product-gas supplied to the consumer and the composition of the raw materials are shown in Figures 14.12–14.15.

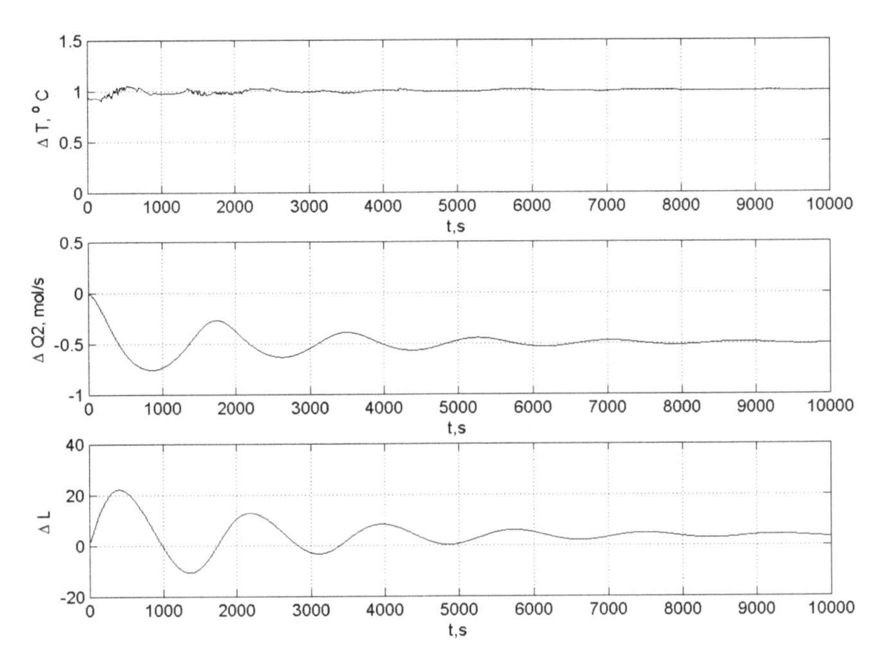


**Figure 14.12** Deviation of the adjustable values of the pyrolysis unit. Perturbation $Q_2 = -0.5$ mole/s, $\Delta Q^{w}_{nch} = -20\%$.

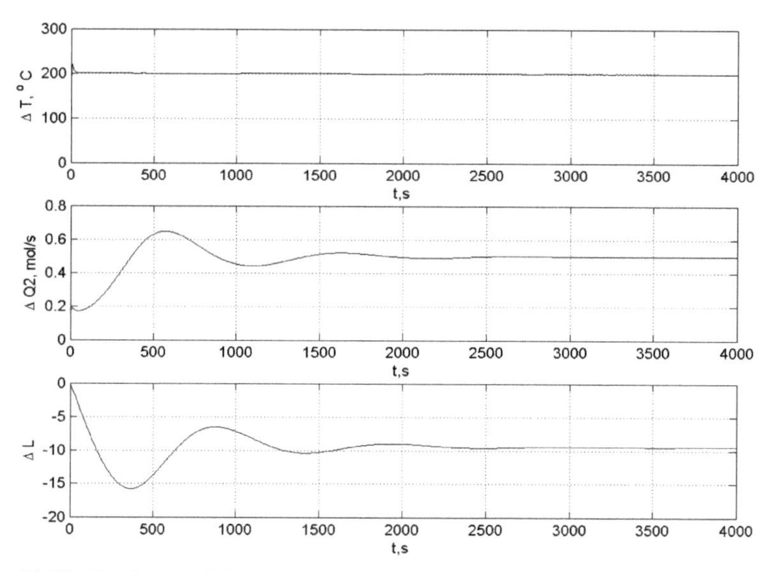


**Figure 14.13** Deviation of the adjustable values of the pyrolysis unit. Perturbation $Q_2$ = 0.5 mole/s, $\Delta Q^{w}_{nch} = -20\%$.

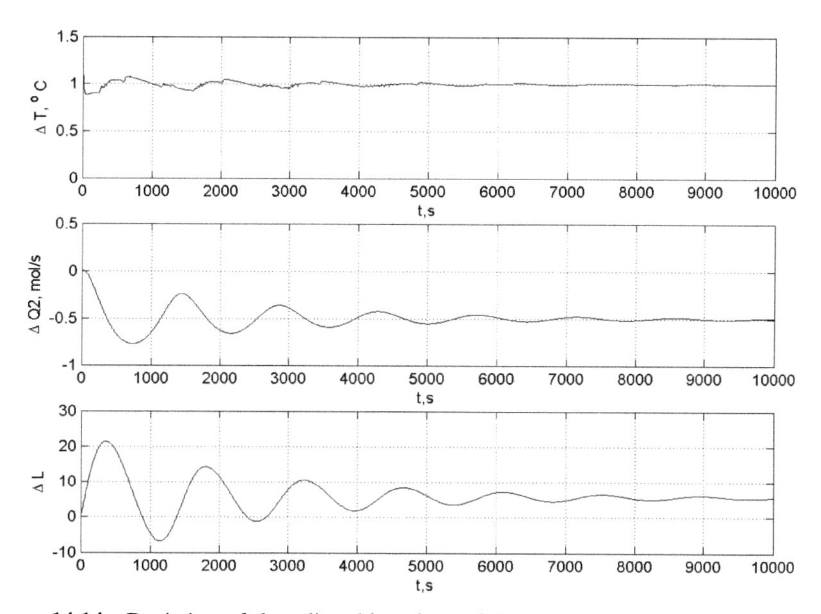


**Figure 14.14** Deviation of the adjustable values of the pyrolysis unit. Perturbation $Q_2$ = $-0.5$ mole/s, $\Delta Q^{w}_{nch} = 20\%$.

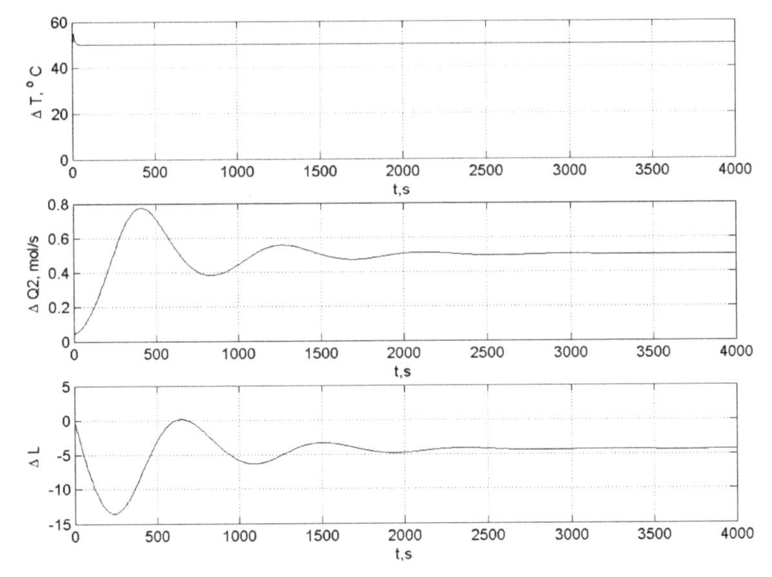


**Figure 14.15** Deviation of the adjustable values of the pyrolysis unit. Perturbation $Q_2$ = 0.5 mole/s, $\Delta Q^w_{n\mathrm{ch}} = 20\%$.

The following perturbations were applied to the system: change of product-gas consumption $Q_2 = \pm 0.5$ mole/s, change of initial raw materials composition, namely deviation of calorific value by 20% in the greater and less side: $Q^w_{n\ \mathrm{ch}} = \pm 0.2$. The perturbations were applied both separately

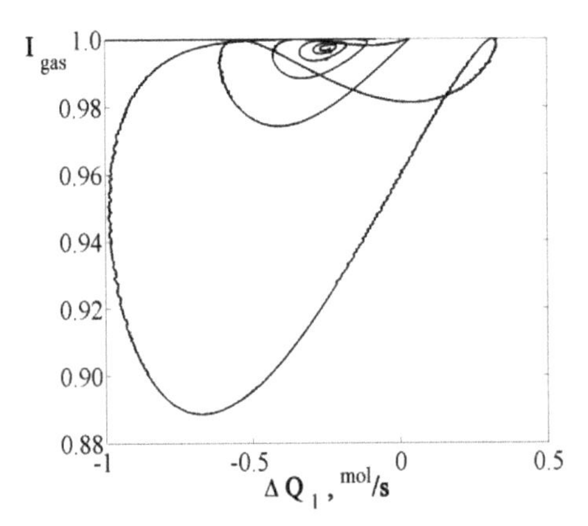


**Figure 14.16** The search for the maximum calorific value of the product-gas.Perturbation $Q_2$ = – 0.5 mole/s, $\Delta Q^w_{n\mathrm{ch}} = -20\%$.

and together. The presented graphs show the control transients only for joint perturbations.

The work of the optimization subsystem was also recorded during the ACS study. The search for the maximum calorific value of the product-gas is shown in Figures 14.16–14.19. The perturbations are the same as those mentioned above.

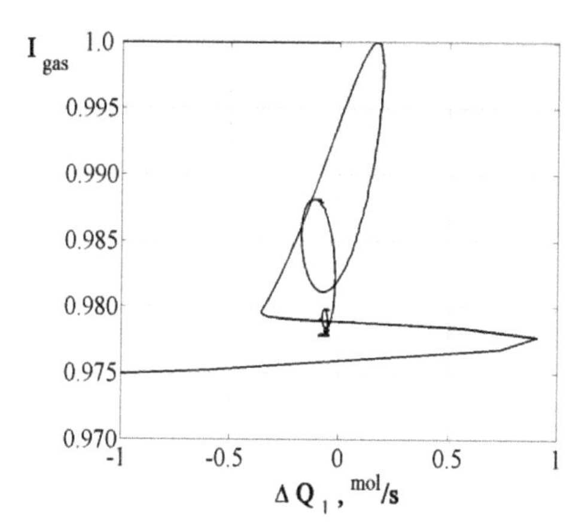


**Figure 14.17** The search for the maximum calorific value of the product-gas. Perturbation $Q_2$ = 0.5 mole/s, $\Delta Q^{w}_{nch} = -20\%$.

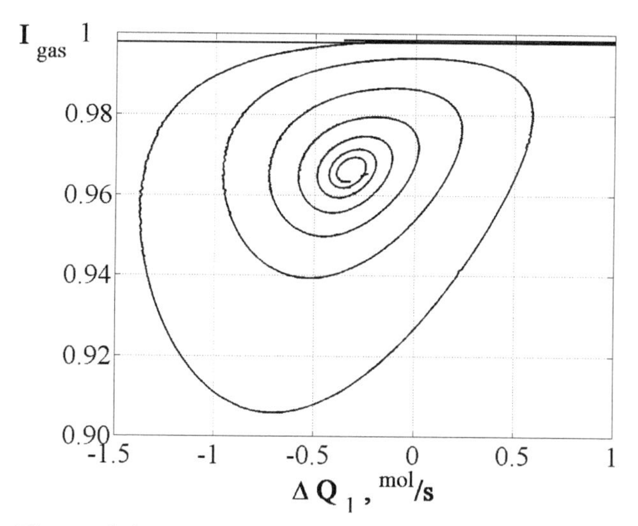


**Figure 14.18** The search for the maximum calorific value of the product-gas. Perturbation $Q_2$ = −0.5 mole/s, $\Delta Q^{w}_{nch} = 20\%$.

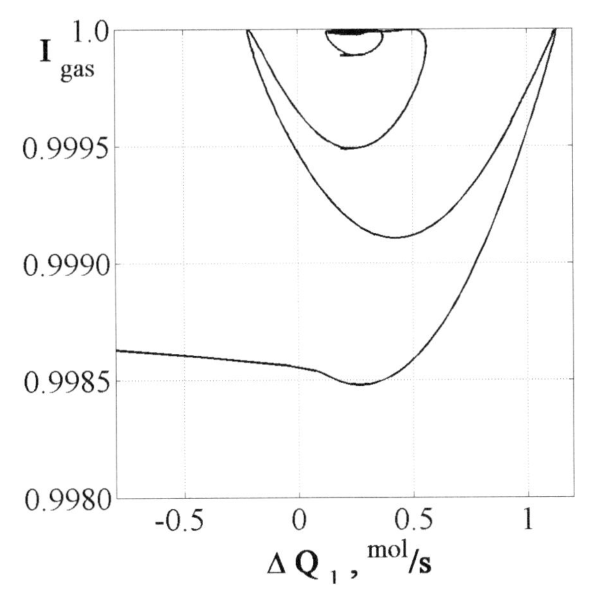


**Figure 14.19** The search for the maximum calorific value of the product-gas. Perturbation $Q_2$ = 0.5 mole/s, $\Delta Q^{w}_{nch} = 20\%$.

The figures show that the control system handles perturbations quite well. Furthermore, at multi-directional perturbations ($Q_2 = -0.5$ mole/s, $\Delta Q^{w}_{nch} = -20\%$and $Q_2$ = 0.5 mole/s, $\Delta Q^{w}_{nch} = 20\%$), the maximum value of the calorific value of product-gas is reached. At the same-directional perturbations ($Q_2 = -0.5$ mole/s, $\Delta Q^{w}_{nch} = 20\%$ and $Q_2$ = 0.5 mole/s, $\Delta Q^{w}_{nch} = -20\%$), the calorific value of the fuel is about 98% of the maximum possible value.

## 14.4 Results and Discussion

We live in the organic world, and, nowadays, we get most of our energy from the processing (burning) of certain types of organic substances. Their choice now is determined not by the type of substance but by the availability, significant amount, and constant fuel composition. Chemical and energy equipment is calculated on the use of raw materials of a certain composition and requires fine-tuning. A slight change in the composition of the same type of substance leads to failure or even a stoppage of technological processes. Thus, any of the refineries work only with a certain oil type. Boilers of power plants are calculated on the burning of certain types of coal and supply is tied to their specific fields. This is due to the usage of control systems designed for real-time steaming of changes in thermal and physical parameters and the inability to account for changes in the chemical composition of raw materials.

The proposed method of organic substances processing allows optimally transforming various organic substances into the gaseous state. Optimality is determined by the possibility of obtaining the maximum volume of the combustible gas mixture with maximum calorific value. Optimality is conditioned by the organization of the pyrolysis and gasification process in superadiabatic mode with the known composition of initial raw materials. The organization of such a thermal processing process allows using organic substances of high humidity. Such, in the most cases, are households and other wastes. Moreover, increasing humidity increases the share of combustible substances in the product-gas.

The composition of the raw materials in the recycling process can change in any way. Due to the introduction of a device for determining the composition of gaseous reaction products in real time in the control system (VII, Figure 14.9), the composition of raw materials is controlled. In the case of its change, the new composition is determined. Based on the received data, the unit is transferred to the optimal model with new parameters.

Based on the above description of the pyrolysis unit, an imitation mathematical model was developed, which describes the dynamic properties of the pyrolysis unit. Through the main channels, the adjustable unit is an inertial link of the first order with a delay. The geometric dimensions of the unit, which must be specified individually for each unit, determine the time constant and the value of the lag. However, it is possible to operate in such a way that the unit changes from a stable to an unstable state due to positive feedback. This is determined by the ratio between $G_{ch}$ raw material consumption and $Q_3$ recirculation gas consumption. The availability of regulators allows stabilizing the pyrolysis process at given values.

Modeling of ACS has shown that, at this stage, a set of single-circuit regulators is enough. Shutting off any of the regulators leads to the deterioration of regulation quality, but the system as a whole continues to function steadily.

In addition, in the process of modeling ACS, it was found out that the control system copes with perturbations quite well. Moreover, in case of multi-directional perturbations (reduction of product-gas consumption and decrease of calorific value or increase of product-gas consumption and increase of the calorific value of initial raw materials), the maximum value of product-gas is reached. At same-directional perturbations (decrease in consumption of a product-gas and increase in calorific power or increase in product-gas consumption and decrease in calorific power of initial raw materials), calorific power of fuel makes about 98% from the maximum possible value.

## 14.5 Conclusion

1) An imitation model of pyrolysis and gasification processes has been developed, which allows obtaining product-gas composition with maximum calorific value and with a minimum amount of carbon residue at the different components of raw materials. The minimum amount of carbon residue determines the possibility of obtaining the maximum possible volume of the combustible gas mixture. The maximum calorific value is determined by the organization of the processing process in a superadiabatic mode with the minimum possible use of air as an oxidizer.
2) The possibility of the existence of the model solutions for energy balance is determined. The calculated ratio of the energy required to maintain the pyrolysis process and the energy available in the reaction products indicates the possibility of organizing the process without additional air. Minimal quantity of additional oxidizer provides the maximum calorific value of formed product-gas.
3) The control method of the pyrolysis of organic raw materials is developed to provide the given flow rate of product-gas at the infliction of perturbations by the flow rate of synthesis gas and composition of initial raw materials.
4) The automated control system of the pyrolysis unit allowing to stabilize the technological process of product-gas production at the set technological values is synthesized.

## References

[1] Rousta K., Ekström K.M. (2013). "Assessing incorrect household waste sorting in a medium-sized Swedish city." Sustainability 2013, 5, 4349–4361.

[2] GNewspaper website The Guardian. "America's 'recycled' plastic waste is clogging landfills, survey finds." Retrieved from http://www.theguard

ian.com/us-news/2020/feb/18/americas-recycled-plastic-waste-is-clogging-landfills-survey-finds.

[3] Newspaper website The Washington Post "A giant wave of plastic garbage could flood the U.S., a study says" https://www.washingtonpost.com/news/energy-environment/wp/2018/06/20/a-giant-wave-of-plastic-garbage-could-flood-the-u-s-in-10-years-a-study-says/?noredirect=on

[4] Newspaper website DW. "Import or processing: how can Germany get rid of plastic waste?" (in Russian) https://www.dw.com/ru/импорт-или-переработка-как-германии-избавиться-от-пластиковых-отходов/a-47226164

[5] Newspaper website Spiegel "Nur 16 Prozent des Plastikmülls werden wiederverwendet" https://www.spiegel.de/wissenschaft/natur/plastikmuell-nur-16-prozent-werden-in-deutschland-wiederverwendet-a-1271125.html

[6] Perrot J.F., Subiantoro A. (2018). "Municipal Waste Management Strategy Review and Waste-to-Energy Potentials in New Zealand." Sustainability. 2018, 10(9), DOI: 10.3390/su10093114

[7] Seltenrich N. (2016). "Emerging Waste-to-Energy Technologies: Solid Waste Solution or Dead End¿' Environmental Health Perspectives, 2016, Vol. 124, No.6.

[8] Brunetkin O., Maksymov M.V., Maksymenko A., Maksymov M.M. (2019). Development of the unified model for identification of composition of products from incineration, gasification, and slow pyrolysis. EasternEuropean Journal of Enterprise Technologies, 2019, 4/6 (100), 25–31.

[9] Guizani C., Jeguirim M. , Valin S., Limousy L., Salvador S. (2017). "Biomass Chars: The Effects of Pyrolysis Conditions on Their Morphology, Structure, Chemical Properties and Reactivity." Energies 2017, 10(6), 796

[10] Akhmetov S.A., Serikov T.P., Kuzeev I.R., Bayazitov M.I. (2006). Technology and equipment of oil and gas refining processes. St Petersburg, Nedra. 2006, 868 pp. (in Russian)

[11] Nadia H.L., Leibbrandt N.H., Aboyade A.O., Knoetze J.H., Görgens J.F. (2013). Process efficiency of biofuel production via gasification and Fischer–Tropsch synthesis. Fuel, Volume 109, 2013, P. 484-492.

[12] Salganskiy Y.A., Fursov V.P., Glazov S.V., Salganskaya M.V., Manelis G.B. (2006). Model of vapor-air gasification of solid fuel in the filtration

mode . Physics of Combustion and Explosion, 2006, t. 42, No 1, s. 65-72. (in Russian)
[13] Brunetkin O., Davydov V., Butenko O., Lysiuk G., A. Bondarenko (2019). Determining the composition of burned gas using the method of constraints as a problem of model interpretation. Eastern-European Journal of Enterprise Technologics – 2019. – No3/6 (99). pp. 22–30. doi: 10.15587/1729–4061.2019.169219
[14] Polianchik Ye.V., Glazov S.V. (2015). 'Filtration combustion of carbon in the presence of endothermic oxidizing agents.' Physics of Combustion and Explosion – 2015. – t. 51, No5 – pp. 34–43. (in Russian)
[15] Matsevityi Yu.M., Alekhina S.V., Borukhov V.T., Zayats G.M., Kostikov A.O. (2017). Identification of the Thermal Conductivity Coefficient for Quasi-Stationary Two-Dimensional Heat Conduction Equations. Journal of Engineering Physics and Thermophysics. – 2017. – V. 90, No.6. – C. 1295-1310
[16] Maksimov, M.V., Brunetkin, O.I., Lysyuk, O.V., & Tarakhtiy, O.S. (2018). Installation for determining the composition of combustible gas in the process of combustion. Patent of Ukraine for invention. G01N 7/08 (2006.01), G01N 25/20 (2006.01). No. 120216; declared 22.12.2017; published 11.06.2018, No11.
[17] Lyamin V.J. Wood gasification . Moscow: Forestry industry, 1967. 264 p. (In Russian).
[18] Glushko, V.P., (Ed). "Thermodynamic and thermophysical properties of combustion products: a handbook" (Vols 1-6). Moscow: All-Union Institute of Scientific and Technical Information. 1973. vol. 3. 624 p. (in Russian)
[19] V.O. Davydov, A.V. Bondarenko. "The method of calculating the combustion temperature of any mixture of hydrocarbon fuel for any excess air." // Praci Odesskogo politechnicheskogo universitetu. Odessa, 2013. - No3 (42) — C. 98 — 102 (in Russian)
[20] Heat and mass-transfer modeling of an angled gas-jet LCVD system.Duty, C., Johnson, R. (et al.) // Applied Physics A, Springer, - 2003, -vol. 0, No 5, - P. 697-705
[21] Z. Fu, J. Wang and C. Yang, "Research on heat transfer function modeling of plastic waste pyrolysis gasification reaction kettle," 2017 Chinese Automation Congress (CAC), Jinan, 2017, pp. 2698-2701
[22] Wolfgang, Rodi, M. Mulas. Engineering Turbulence Modelling and Experiments 6 / Rodi, Wolfgang, M. Mulas. - Elsevier, 2005. – 1012 p.

[23] B.V. Gavrilenko, S.V. Neyezhmakov, "Mathematical modeling of the fluidized bed furnace of a mine autonomous air heater under non-stationary conditions" / B.V. Gavrilenko, S.V. Neezhmakov // Problems of operation of equipment of mine installations: a collection of scientific papers. - 2005. - p. 297 – 304]. (in Russian)

[24] B.V. Gavrilenko, "Synthesis of a mathematical model of a fluidized bed furnace of a mine air heater under non-stationary conditions for automatic control tasks" / B.V. Gavrilenko, S.V. Neyezhmakov // Modelling and information technology: coll. Science. pr./Nats. acad. Sciences of Ukraine, Inst. modelling in energy named after G.E. Pukhov. - Kyiv, 2010. - №. 57. - P. 164-173. (in Russian)

[25] Y. P. Kondratenko, O. V. Kozlov, "Mathematic Modeling of Reactor's Temperature Mode of Multiloop Pyrolysis Plant", in book: Lecture Notes in Business Information Processing: Modeling and Simulation in Engineering, Economics and Management, K. J. Engemann, A. M. Gil-Lafuente, J. M. Merigo, Eds., vol. 115. Berlin, Heidelberg: Springer-Verlag, pp. 178–187, 2012.

[26] Y. P. Kondratenko, O. V. Kozlov, G. V. Kondratenko, I. P. Atamanyuk, "Mathematical Model and Parametrical Identification of Ecopyrogenesis Plant Based on Soft Computing Techniques", in Complex Systems: Solutions and Challenges in Economics, Management and Engineering, Christian Berger-et.al. (Eds.), Book Series: Studies in Systems, Decision and Control, Vol. 125, Berlin, Heidelberg: Springer International Publishing, pp. 201-233, 2018.

[27] Y. P. Kondratenko, O. V. Kozlov, "Mathematical Model of Ecopyrogenesis Reactor with Fuzzy Parametrical Identification", in: Recent Developments and New Direction in Soft-Computing Foundations and Applications, Studies in Fuzziness and Soft Computing 342, Lotfi A. Zadeh et al. (Eds.). Berlin, Heidelberg: Springer-Verlag, pp. 439-451, 2016.

[28] M. Maksymov. The method of finding the most natural structure of a biotank power plant / M. Maksymov, K. Beglov, O. Maksymova, O. Maksymov // Proceedings of Odessa Polytechnic University, Issue 1(60), 2020 — p. 82-95.

[29] Y. P. Kondratenko, O. V. Kozlov, O. S. Gerasin, A. M. Topalov, O. V. Korobko, "Automation of Control Processes in Specialized Pyrolysis Complexes Based on Web SCADA Systems", in Proceedings of the 9th IEEE International Conference on Intelligent Data Acquisition

and Advanced Computing Systems: Technology and Applications (IDAACS), Vol. 1, Bucharest, Romania, pp. 107-112, 2017.

[30] A. Balestrino, F. Bassini and P. Pelacchi, "On the Control of a Pyrolysis Process" 2007 International Conference on Clean Electrical Power, Capri, 2007, pp. 409-414.

[31] M. Mircioiu, E. Cimpoeşu and C. Dimon, "Robust control and optimization for a petrochemical pyrolysis reactor," 18th Mediterranean Conference on Control and Automation, MED'10, Marrakech, 2010, pp. 1097-1102.

[32] Y.P. Kondratenko, A.A. Chikrii, V.F. Gubarev, J. Kacprzyk. (Eds). Advanced Control Techniques in Complex Engineering Systems: Theory and Applications. Dedicated to Professor Vsevolod M. Kuntsevich. Studies in Systems, Decision and Control, Vol. 203. Cham: Springer Nature Switzerland AG, 2019.

[33] V.M. Kuntsevich, V.M., V.F. Gubarev, V.F., Y.P. Kondratenko, Y.P., D.V. Lebedev, D.V., V.P. Lysenko, (Eds). Control Systems: Theory and Applications. Series in Automation, Control and Robotics, River Publishers, Gistrup, Delft, 2018.

# Index

# About the Editors

**Yuriy P. Kondratenko** received the Ph.D. degree in 1983 and the D.Sc. degree in 1994 in computer and control systems from Odessa National Polytechnic University.

He is the Doctor of Science (habil.), Professor, Honour Inventor of Ukraine (2008), Corr. Academician of Royal European Academy of Doctors – Barcelona 1914 (2000), and the Head of Intelligent Information Systems Department at Petro Mohyla Black Sea National University, Ukraine. He is the regional coordinator of Tempus (Cabriolet) and Erasmus+ (Aliot) projects, principal researcher of several international research projects with Spain, Germany, P.R. of China, and others. He is the author of more than 140 patents and 14 books (including edited monographs) published by Springer, River Publishers, World Scientific, Pergamon Press, Academic Verlag, etc. His research interests include intelligent decision support systems, automation, sensors and control systems, fuzzy logic, soft computing, modelling and simulation, robotics, and elements and devices of computing systems.

Dr. Kondratenko received several international grants and scholarships for conducting research at Institute of Automation of Chongqing University, P. R. China (1988–1989), Ruhr-University Bochum, Germany (2000 and 2010), Nazareth College, and Cleveland State University, USA (2003). In 2015, he received Fulbright grant for conducting research during nine months in USA (Cleveland State University, Department of Electrical Engineering and Computer Science). He is a member of the Scientific Committee of the National Council of Ukraine on Development of Science and Technology, National Committee of Ukrainian Association on Automatic Control, as well as GAMM, DAAAM, AMSE UAPL, and PBD-Honor Society of International Scholars, and a Visiting Lecturer at the universities in Rochester, Cleveland, Kassel, Vladivostok, and Warsaw. He is a member of Editorial Boards of journals such as *Journal of Automation and Information Sciences, International Journal of Computing, Eastern European Journal of Enterprise*

*Technologies*, *International Research and Review: Journal of Phi Beta Delta*, *Quantitative Methods in Economics*, and others.

**Vsevolod M. Kuntsevich** graduated from Kyiv Polytechnic Institute in 1952. He received the Ph.D. degree in 1959 and the D.Sc. degree in 1965.

He is a Professor (1967), Academician of the National Academy of Sciences of Ukraine (1992), and the Honorary Director of Space Research Institute of NASU, Kyiv, Ukraine. He worked with the Institute of Electrical Engineering from 1958 to 1963, with the Institute of Cybernetics from 1963 to 1996, and with the Space Research Institute from 1996. He is the founder of the National School in the field of discrete control systems, and he made a significant contribution to the development of modern theory of adaptive and robust control under uncertainty.

Dr. Kuntsevich is the Honored Figure of Science and Technology of Ukraine (1999), Laureate of the State Prize of the Ukrainian SSR (1978 and 1991) and Ukraine (2000) in the field of science and technology, and a recipient of the S. Lebedev Award (1987), V. Glushkov Award (1995), and V. Mikhalevich Award (2003). He serves as the editorial staff of several journals such as *Journal of Automation and Information Sciences, Cybernetics and Systems Analysis*. He is the author of 8 books and over 250 articles. He is the Chairman of the National Committee of Ukrainian Association on Automatic Control (NMO of IFAC).

**Arkadii A. Chikrii** received the Ph.D. degree in 1972 and the D.Sc. degree in 1979.

He is a Professor (1989), Academician of the National Academy of Sciences of Ukraine (2018), and Head of the department "Optimization of Controlled Processes" of the Institute of Cybernetics, NAS of Ukraine. After graduation from Ivan Franko Lviv University in 1968, has been working with the Institute of Cybernetics. He is a Professor of Taras Shevchenko Kyiv National University, Professor of Ihor Sykorski National Technical University, and Professor of Yurij Fedkovich Chernivtsi National University. He is a specialist in the field of applied nonlinear analysis, theory of extremal problems, mathematical theory of control, theory of dynamic games, theory of search for moving objects, and computer technologies for analysis of conflict situations. He is the disciple and follower of L.S. Pontryagin, N.N. Krasovskii, and B.N. Pshenichnyi. He is the author of 6 books and over 550 articles.

Dr. Chikrii is the Laureate of the State Prize of Ukraine in the field of science and technology (1999), a recipient of Glushkov Award (2003), and the Laureate of the State Prize of Ukraine in the field of education (2018). He is the Editor-in-Chief for the journal "*Problemy Upravleniya I Informatiki*" published in English in the USA under the title "*Journal of Automation and Information Sciences*" (since 2020).

**Vyacheslav F. Gubarev** has received the Ph.D. degree in 1971 and the D.Sc. degree in 1992 in system analysis and automatic control from Institute of Cybernetic of National Academy of Science, Ukraine.

He is the Doctor of Science (1992), Professor, Corresponding Member of National Academy of Science of Ukraine (2006), Professor of Mathematical Methods of System Analysis Department, Kiev National Technical University, Ukraine, and the Head of Control Department, Space Research Institute, National Academy of Science, Ukraine. He has taken part in several international grants with Russia Academy of Science, Moscow State University, and others. His research interests include mathematical modeling of complex systems, automatic control, estimation and identification, ill-posed mathematical problems, dynamic and control under uncertainty, and spacecraft control systems.

Dr. Gubarev serves as editorial staff of several journals such as *Journal of Automation and Information Sciences, Cybernetics and Systems Analysis*. He is Vice-Chairman of Ukrainian Association on Automatic Control which is NMO of IFAC.